Simulations in Nanobiotechnology

Simulations in Nanobiotechnology

Kilho Eom

CRC Press
Taylor & Francis Group
Boca Raton London New York

CRC Press is an imprint of the
Taylor & Francis Group, an **informa** business

CRC Press
Taylor & Francis Group
6000 Broken Sound Parkway NW, Suite 300
Boca Raton, FL 33487-2742

First issued in paperback 2018

ISBN-13: 978-1-4398-3504-3 (hbk)
ISBN-13: 978-1-138-37437-9 (pbk)

Library of Congress Cataloging-in-Publication Data

Simulations in nanobiotechnology / [edited by] Kilho Eom.
    p. ; cm.
    Includes bibliographical references and index.
    ISBN 978-1-4398-3504-3 (hardcover : alk. paper)
    I. Eom, Kilho.

    [DNLM: 1. Nanotechnology. 2. Computational Biology. 3. Computer Simulation. 4. Models, Theoretical. QT 36.5]
    610.28--dc23   2011036042

Visit the Taylor & Francis Web site at
http://www.taylorandfrancis.com

and the CRC Press Web site at
http://www.crcpress.com

… the truth shall make you free

John 8:32

Contents

SECTION I   Simulations in Biological Sciences

SECTION II   Simulations in Nanoscience and Nanotechnology

Preface

Computational simulations play a significant role in multiple research disciplines ranging from engineering, physics, and chemistry to biology. In particular, computational simulations have been recently highlighted in nanobiotechnology because of their ability to characterize the nanoscale behavior of various biological objects (e.g., deoxyribonucleic acid [DNA], ribonucleic acid [RNA], proteins, and cells) and nanomaterials (carbon nanotube [CNT], graphene, nanowire, etc.). Recent advances in various computational simulation techniques (e.g., atomistic simulation, coarse-grained [CG] simulation, multiscale simulation, and continuum simulation) complement the experimental studies in nanobiotechnology, since simulations are able to provide detailed insights, which are inaccessible with experimental techniques, into the nanoscale physics and mechanics of biomaterials or nanomaterials or both. Despite several monographs or books that describe the current state-of-arts in computational simulations, these monographs or books are still lacking the presentation of the broad exploitation of computational simulations for understanding the nanoscale phenomena of various objects ranging from biological systems to synthetic nanoscale materials. This led me to design and write a book that provides the state of the art in simulation-based nanoscale characterizations in nanoscience, nanotechnology, and biological science.

For writing this book, I remembered my graduate study when I was first exposed to molecular mechanics and CG modeling. At that time, it was not easy to find a book dedicated to the presentation of molecular mechanics and/or CG modeling techniques and their applications in nanoscale characterizations that are useful in understanding the fundamental principles of nanobiotechnology, albeit there were classical books that delineated the fundamentals of atomistic models and simulations such as molecular dynamics (MD). Although monographs or books that provide the principles of atomistic models and/or CG models and their applications in each specific discipline have appeared recently, it is still difficult to find a book that covers broad applications of such models and simulations in nanobiotechnology that require the use of principles from multiple academic disciplinary backgrounds. This experience resulted in my writing of this book. Furthermore, my experience in writing research articles in nanobiotechnology has enabled me to develop a book that is capable of comprehensively demonstrating nanoscale characterizations in disparate disciplines such as physics, mechanics, materials science, and biology; in order to achieve this objective, this book comprises contributions from researchers including myself, who write from their own expertise and research experiences in simulation-based nanoscale characterization in such disciplines.

I suggest that researchers and students who are unfamiliar with atomistic and/or CG models start with Chapter 1. This introductory chapter summarizes the efforts that have been made recently to conduct nanoscale characterizations in biological science, bioengineering, and nanotechnology, and establishes the fundamental principles of the various modeling schemes that can be employed for nanoscale characterizations, at multiple scales ranging from atomistic to continuum scales. The remainder of this book can be separated into two sections: (1) The first section details recent advances made in simulation-based nanoscale characterizations in biology, particularly the simulations of nanoscale dynamic behavior of biological molecules such as DNA, RNA, and proteins; and (2) The second section provides a discussion of recent attempts in the exploitation of computational simulations for nanoscale characterizations of inorganic nanoscale materials such as CNTs and graphenes as well as nano-/microscale devices such as microcantilevers.

Section I is aimed at presenting the simulation-based nanoscale characterizations in biological science. It consists of six chapters: Chapter 2, by Carr, Comer, and Aksimentiev, describes recent efforts in the MD simulation-based characterization of DNA/protein transport dynamics in the nanopore and nanochannel, the insights into which play a significant role in the development of a single-molecule experimental toolkit for future applications such as genome sequencing. Chapter 3, by Eom et al., delineates the state of the art in CG modeling techniques that are applicable to large protein complexes for understanding their dynamics, such as conformational changes related to the biological functions of proteins. Chapter 4, by Chen, presents recent advances made in continuum mechanics–based modeling of membrane proteins for understanding their deformation mechanisms in response to mechanical stimuli, which provide insights into the biological functions of membrane proteins and their related cellular functions. Chapter 5, by Hyeon, summarizes theoretical frameworks along with atomistic simulations, which gives rise to fundamental insights into the free energy landscape for biomolecules, especially protein folding. Chapter 6, by Liu, Ramachandra, and Guo, suggests the recent attempts to develop CG models for gaining insights into DNA transport dynamics inside nanopores. Chapter 7, by Eom, reviews the recent advances in computational modeling techniques that are applicable to protein structures at multiple spatial scales for understanding their mechanical properties relevant to the biological function of proteins.

Section II discusses recent advances made in modeling techniques and their applications in understanding the nanoscale behavior of inorganic nanoscale materials such as nanoparticles, CNTs, and graphene. Chapter 8, by Garcia, Sen, and Buehler, describes recent efforts in nature-inspired material design, particularly nacre-inspired mineralized structures, which provide insights into the fundamentals of future biomimetics-based material designs. Chapter 9, by Nahar, Pradhan, and Montenegro, summarizes recent advances made in atomistic simulation–based characterization of nanoparticles' optical properties and nanoparticle-based applications in therapeutics. Chapter 10, by Eom and Kwon, reviews the theoretical/computational frameworks along with experimental efforts in biochemical detection using resonant devices for future applications in early diagnostics. Chapter 11, by Duan, summarizes the theoretical/computational frameworks in microcantilever mechanics for studying surface characteristics as well as microcantilever-based sensing applications. Chapter 12, by Ke and Zheng, presents an overview of the recent advances made in experiment- and simulation-based characterizations of nanoscale adhesive properties. Chapter 13, by Ke and Wei, suggests the state of the art in theoretical frameworks with experimental efforts in the development of nanoresonators for future nanoscale device designs. Chapter 14, by Lu and Huang, delineates recent advances in theoretical and computational attempts in understanding the mechanical behavior of a graphene monolayer.

Despite my efforts to provide a comprehensive presentation of computational simulation–based characterizations relevant to the research area of nanobiotechnology, I apologize to readers for the possibility that some topics that may be of interest to them might not be properly covered in the book; but I am confident that this book gives a reader insights into not only the fundamentals of simulation-based characterizations in nanobiotechnology but also how to approach new and interesting problems in nanobiotechnology using the basic theoretical and computational frameworks demonstrated in this book. I hope that readers receive both personal and intellectual pleasure from reading this book and gaining knowledge of simulations in nanobiotechnology.

Kilho Eom

Acknowledgments

I express my deep gratitude to Lance Wobus, a senior editor with Taylor & Francis, without whom this book would not have been possible. I gratefully acknowledge the suggestions and guidance provided by Lance during the preparation of this book. Furthermore, I appreciate Kathryn Younce, the book's project coordinator, for her kindness in helping me prepare the book. I am also thankful to Patricia Roberson, who was a project coordinator prior to her recent retirement from CRC Press, for her assistance. Finally, I acknowledge the help and support provided by people working at CRC Press/Taylor & Francis who were involved in the various aspects of book production.

This book could not be published without the support of my colleagues, whose collaboration has helped to broaden my perspectives and experiences. It is impossible for me not to gratefully acknowledge Taeyun Kwon, who is always willing to share his insights into experimental research with me. Working with him is a pleasure in itself and has always led to wonderful achievements in my nanobiotechnology research. I must also acknowledge my other colleagues—Jaemoon Yang, Dae Sung Yoon, Harold S. Park, Sungsoo Na, Chang-Wan Kim, Changhong Ke, Sang Woo Lee, Markus J. Buehler, and Changbong Hyeon—for sharing their insights and experiences with me in various research areas. In particular, my collaboration with Jaemoon, who has enabled me to take a step into researching nanotechnology-based medical applications, is gratefully acknowledged. I appreciate Dae Sung for his suggestion of collaborations that will allow us to eventually pave the way for future nanotechnology-based biological applications. I am grateful to Harold for his willingness to share his insights into the simulations of nanoresonators, which resulted in writing with him of a review article on the subject. I gratefully acknowledge the contributions of Sungsoo, who has worked with me in protein modeling and was willing to adventure into the research topic of protein dynamics with me. It was a pleasure to collaborate with Chang-Wan, whom I have known since my graduate years at the University of Texas at Austin, Texas, and his expertise in finite element modeling has enriched my research topics of simulations in nanotechnology. It was a pleasure to have the chance of collaborating with Changhong in nanotechnology. I am happy to acknowledge Sang-Woo for his contributions regarding single-molecule experiments based on dielectrophoretic bioassay. I appreciate Markus for his invitation that allowed me to contribute to the special issue of a journal by writing a review article. I am thankful to Changbong for willingly sharing his insights into single-molecule pulling experiments that provide the fundamentals of protein folding. In addition, I acknowledge the National Research Foundation of Korea (NRF), which partially supports my research work and the preparation of this book under grant numbers NRF-2008-313-D00012, NRF-2009-0071246, and NRF-2010-0026223.

I further acknowledge the contributors of this book, whose kind contributions have enabled the realization of this book. I appreciate Aleksei Aksimentiev and his research group for their contributions in the simulations of DNA/protein transport dynamics through nanopores and nanochannels. It is a pleasure to acknowledge Xi Chen for his beautiful descriptions on the continuum modeling of membrane proteins. I happily acknowledge Changbong Hyeon for his excellent delineations of the theoretical frameworks useful in analyzing single-molecule pulling experiments resulting in a fundamental understanding of the free energy landscape relevant to protein folding. I am thankful to Yaling Liu and his research group for their contributions in the field of CG modeling of DNA transport dynamics in nanopores. I am happy to acknowledge Sultana N. Nahar and her colleagues for their contributions in MD simulation–based design of therapeutic nanoscale agents. I gratefully acknowledge Markus J. Buehler and his research group for their summary of diatom-inspired mineralized structures based on atomistic simulations. The contribution from Huiling Duan, who has done a nice job in describing the recent advances and theoretical frameworks in microcantilever-based applications in nanotechnology, is quite opportune. I am happy to acknowledge Changhong Ke and his research group, who have contributed two chapters in the book: one on experiments and simulations of nanoscale adhesive interactions, and the other on the state of the art in nanoscale resonators. Finally, I gratefully acknowledge Rui Huang and his research group, who excellently present recent advances in atomistic and continuum modeling of a graphene monolayer.

I am happy to acknowledge several graduate students who have researched with me and helped me prepare the book. Specifically, I appreciate two PhD students, Gwonchan Yoon and Jae-In Kim, for their help in the preparation of some figures in this book, and my experience in advising their research efforts (regarding molecular simulations) has been very useful and helpful in my research career and also in writing this book. I acknowledge Jinsung Park, Huihun Jung, Kuewhan Jang, Kihwan Nam, and Gyudo Lee for their assistance in the experimental research works supervised by Taeyun Kwon and myself, and discussions with them have increased my knowledge of nanobiotechnology. In addition, I acknowledge several former graduate students for their contributions to some parts of my research work.

My mentors in my graduate studies, who exposed me to the field of simulations in biological sciences and allowed me to delve into the research topics in nanobiotechnology, deserve special mention. In particular, I heartily appreciate Gregory J. Rodin who provided the guidance that helped me develop my research career. My graduate study under the supervision of Greg Rodin for five years was extremely worthwhile and it led me to pursue a research career in nanobiotechnology. Moreover, I gratefully acknowledge Dmitrii E. Makarov for his kind advice on molecular modeling and CG modeling for studying protein dynamics. My graduate study with Dima Makarov inspired me to study the atomistic simulations and CG modeling of biomolecules. My life was much enriched by the aforementioned two mentors as well as UT Austin alumni and friends who shared my life and experiences during my graduate study years. I specially acknowledge the following UT alumni and friends: Rajan Arora, Rahul Prashar, Mintae Kim, Jaeyoung Lim, Se-Hyuk Im, Serdal Kirmizialtin, Lei Huang, Pai-Chi Li, Jaewoo Kim, and numerous UT alumni.

Finally, I acknowledge my family for the love and support they have rendered in all my endeavors. First, it is my honor to acknowledge my wife for her benevolent support that enriches not only my research but also my life. Spending time with her is more pleasant than research; it has refreshed my thoughts and escalated my research capabilities. I gratefully acknowledge my parents for their endless support of me. Without their sacrifices and support, I would not have had a chance to develop my research career. I am very thankful to my parents-in-law for their support and interest in my research work and life. I am happy to acknowledge my little sister who has prayed for me. I appreciate my brother-in-law for spending time with me; these meetings are always pleasant. In addition, I acknowledge the pastors Soo-Il Yang, Jong-Wook Im, Sung-Sam Yeo, and Cephas Kye who have prayed for me and guided my Christian life.

Contributors

Abhijit Ramachandra
Department of Mechanical Engineering
  and Mechanics
Lehigh University
Bethlehem, Pennsylvania

Aleksei Aksimentiev
Department of Physics
University of Illinois at Urbana
  Champaign
Urbana, Illinois

and

Beckman Institute for Advanced
  Science and Technology
University of Illinois at Urbana
  Champaign
Urbana, Illinois

Markus J. Buehler
Laboratory for Atomistic and
  Molecular Mechanics
Department of Civil and
  Environmental Engineering
Massachusetts Institute of
  Technology
Cambridge, Massachusetts

Rogan Carr
Department of Physics
University of Illinois at
  Urbana Champaign
Urbana, Illinois

Xi Chen
Department of Earth and
  Environmental Engineering
Columbia University
New York

and

School of Aerospace
Xi'an Jiaotong University
Xi'an, China

Jeffrey Comer
Department of Physics
University of Illinois at Urbana
  Champaign
Urbana, Illinois

Huiling (H. L.) Duan
State Key Laboratory for Turbulence and
  Complex System, CAPT
Department of Mechanics and
  Aerospace Engineering
College of Engineering, Peking
  University
Beijing, China

Kilho Eom
Department of Mechanical Engineering
Korea University
Seoul, Republic of Korea

and

Institute for Molecular Sciences
Seoul, Republic of Korea

Andre P. Garcia
Laboratory for Atomistic and Molecular
    Mechanics
Department of Civil and Environmental
    Engineering
Massachusetts Institute of Technology
Cambridge, Massachusetts

Qingjiang Guo
Department of Mechanical Engineering
    and Mechanics
Lehigh University
Bethlehem, Pennsylvania

Rui Huang
Department of Aerospace
    Engineering and Engineering
    Mechanics
University of Texas at Austin
Austin, Texas

Changbong Hyeon
School of Computational Sciences
Korea Institute for Advanced Study
Seoul, Republic of Korea

Changhong Ke
Department of Mechanical
    Engineering
State University of New York at
    Binghamton
Binghamton, New York

Jae In Kim
Department of Mechanical
    Engineering
Korea University
Seoul, Republic of Korea

Taeyun Kwon
Department of Biomedical
    Engineering
Yonsei University
Wonju, Republic of Korea

and

Institute for Molecular Sciences
Seoul, Republic of Korea

Yaling Liu
Department of Mechanical Engineering
    and Mechanics
Lehigh University
Bethlehem, Pennsylvania

Qiang Lu
Department of Mechanical Engineering
University of Texas at Austin
Austin, Texas

Maximiliano Montenegro
College of Education (Facultad
    de Educaci)
The Pontifical Catholic University
    of Chile
Avda Vicua Mackenna
Santiago, Chile

Sungsoo Na
Department of Mechanical Engineering
Korea University
Seoul, Republic of Korea

Sultana N. Nahar
Department of Astronomy
The Ohio State University
Columbus, Ohio

Anil K. Pradhan
Department of Astronomy
The Ohio State University
Columbus, Ohio

Dipanjan Sen
Laboratory for Atomistic and Molecular
    Mechanics
Department of Civil Engineering
Massachusetts Institute of
    Technology
Cambridge, Massachusetts.

and

Department of Materials Science and
    Engineering
Massachusetts Institute of
    Technology
Cambridge, Massachusetts

Qing Wei
Department of Mechanical
    Engineering
State University of New York at
    Binghamton
Binghamton, New York

Meng Zheng
Department of Mechanical
    Engineering
State University of New York at
    Binghamton
Binghamton, New York

Gwonchan Yoon
Department of Mechanical Engineering
Korea University
Seoul, Republic of Korea

1 Introduction to Simulations in Nanobiotechnology

Kilho Eom

CONTENTS

1.1 THE NANOBIOTECHNOLOGY WORLD

It is not an exaggeration when one says that we live in the era of nanoscience/
nanotechnology, a branch of study that has a great impact on our lives indirectly.
For instance, people use smartphones that can perform multiple functions (as can the
personal computer); an ability is attributed to the emergence of miniaturized electronic
chips thanks to the microelectromechanical system (MEMS) technology and recently
nanofabrication technologies. In recent decades, genomic sequencing that can unveil

the origin of life has been made efficiently possible due to the development of nano-technology; For instance, nanopore-/nanochannel-based deoxyribonucleic acid (DNA) sequencing has endowed the cost-efficient genomic sequencing analysis. Furthermore, nanotechnology has enabled the unprecedented early diagnosis of diseases such as cancers, particularly the diagnosis of cancers at very early stages (see Section 1.1.2). Moreover, due to the discovery of nanoscale materials, it has recently been suggested that one is able to develop de novo materials that can exhibit unprecedented material properties (e.g., high stiffness and superelasticity), which cannot be achieved based on conventional engineering material design. It is very essential to gain fundamental insights into the nanoscale characteristics of various nanoscale objects ranging from biomolecules to nanomaterials for various applications in multiple disciplines.

In particular, over the last two decades significant advances have been made in technology, which enables one to gain unprecedented insights into physics, chemistry, and biology at the nanoscale. For instance, one of the technological advances in experimental apparatuses is the discovery of the "atomic force microscope" (AFM), which was suggested about two decades ago in order to probe surface morphology at small scales (e.g., nanometer length scales) [1]; later, it was widely utilized to probe the physical properties of various small-scale objects [2] such as biomolecules (e.g., protein, DNA, and ribonucleic acid [RNA]) [3–5] and novel nanomaterials [6,7] (e.g., nanowire, carbon nanotube [CNT], and graphene). For example, the AFM has allowed quantitative characterization of the mechanical properties of biomolecules, which has led to some fundamental insights into their functions. In particular, an AFM-based single-molecule pulling experiment [3–5,8,9] has recently received sig-nificant attention from the viewpoint that such an experiment is capable of providing not only insights into the response of proteins to a mechanical force but also the fun-damental principles of protein folding [10–12] related to the biological functions of proteins. Recently, AFM indentation [7] has been highlighted because of its ability to characterize the mechanical properties of nanomaterials. Specifically, AFM bend-ing experiments enable one to understand the size-dependent mechanical properties of nanowires [13–15] and to discover that some nanomaterials (e.g., graphene [16]) exhibit excellent mechanical properties (e.g., high elastic stiffness of values even up to ~1 TPa). These clearly demonstrate that technological advances (e.g., discovery of new experimental apparatuses such as the AFM) have paved the way to observing the physical phenomena at nanoscale, for example, protein unfolding, which opens a new avenue in the fundamental understanding of the physics of various objects such as biological materials and nanomaterials.

The subsections of this section briefly overview and present attempts that have recently been made in order to characterize the nanoscale behavior of small-scale systems such as biological systems (e.g., biomolecule, subcellular system, and cell) and nanomaterials.

1.1.1 SINGLE-MOLECULE EXPERIMENTS: CHARACTERIZATION OF BIOMOLECULES

The last decade witnessed the appearance of single-molecule experiments, which is attributed to the development of novel experimental apparatuses such as AFM

[3–5], the optical tweezer [17], and the dielectrophoretic tweezer [18]. The key idea of single-molecule experiments is to apply a mechanical force to a biomolecule in such a way that one end of the biomolecule is chemically attached to a substrate while the other end is chemically conjugated to a probe (e.g., AFM tip, or nanobead of tweezer) that can be displaced (Figure 1.1). When the probe attached to the end of the biomolecule is displaced, the resulting force exerted on the biomolecule is measured using Hooke's law, $f(t) = k(vt - e)$, where $f(t)$ is the force exerted on a biomolecule at time t, e is the extension of a biomolecule, k is the force constant of the probe, and v is the pulling rate.

One of the pioneering single-molecule experiments is the mechanical stretching of a DNA molecule [17,19,20]. Single-molecule pulling experiments conducted on a DNA molecule provide the force–displacement relation (similar to stress–strain relation) for the DNA molecule. It is shown that a DNA molecule exhibits nonlinear response to a mechanical force [19]. This unique nonlinear mechanical response for DNA molecules sheds light on the polymer theory, the wormlike chain (WLC) model [21,22]. Specifically, the WLC model is able to capture the mechanical behavior of DNA molecules characterized by

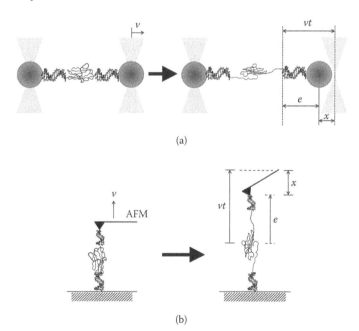

(a)

(b)

FIGURE 1.1 Schematic illustration of single-molecule experiments: (a) Optical tweezer–based single-molecule pulling experiments: When a laser-trapped nanoparticle is moved with a constant velocity v, the biomolecule conjugated to nanoparticles undergoes mechanical extension of the amount e (Note that x is the distance between a nanoparticle and the laser focus). (b) AFM-based single-molecule pulling experiments: When AFM cantilever tip is displaced with a constant velocity v, the cantilever tip is bent due to the force acting on the biomolecule; the cantilever bending deflection is denoted as x and then the mechanical extension of the biomolecule is given by $e = |x - vt|$.

single-molecule pulling experiments (Figure 1.2a). The WLC model provides the following force–displacement relation:

$$f = \frac{k_B T}{l_p} \left[\frac{1}{4} \left(1 - \frac{e}{l_0} \right)^{-2} - \frac{1}{4} + \frac{e}{l_0} \right] \tag{1.1}$$

where f is the force exerted on a DNA molecule, e is the extension of the DNA molecule, k_B is Boltzmann's constant, T is the absolute temperature, l_0 is the contour length of the molecule, and l_p is the persistence length of a DNA molecule. Here, the persistence length l_p is related to bending rigidity D such as $D = l_p k_B T$. It is implied that single-molecule pulling experiments are able to validate the physical model (polymer theory) that is suitable to describe the mechanical behavior of biomolecules such as DNA. Recently, Wiggins et al. [23] considered AFM images of short DNA chains in order to check whether the WLC model is able to depict the bending behavior of short DNA chains or not. They showed that the WLC model does not suffice for dictating the mechanical behavior of short DNA chains and developed the novel polymer chain theory referred to as the "subelastic chain" (SEC) model [24], which is able to describe the bending response of short DNA chains that is observed in AFM imaging. In recent years, the ability of the SEC model to describe the mechanics of short DNA chains has been validated by computational simulations such as molecular dynamics (MD) simulations [25]. It is implied that AFM-based single-molecule imaging and/or computational simulations such as MD simulations can be regarded as a toolkit that validates the previous physical theories/models (e.g., polymer theory) and suggests novel physical models of molecules.

Over the last decade, single-molecule pulling experiments have been highlighted for the mechanical characterization of protein unfolding phenomena. In particular, unlike the mechanical stretching of DNA chains the mechanical response of a protein to a force is quite unique in that a certain amount of a force can unfold the three-dimensional folding structure of proteins [3,21]. Specifically, when a force exerted on a protein reaches a critical value, unfolding of the folded protein domain is initiated and this unfolding is reflected in the force–extension curve such that the unfolding of a domain leads to a force drop in the force–extension curve [3,26] (Figure 1.2b). The force–extension curve for a protein shows that before mechanical unfolding occurs the mechanical behavior of a protein is well described by the WLC model. This has led researchers [27,28] to develop the physical model of a protein structure in such a way that a protein folded domain is dictated by the WLC model, whereas the probability of unfolding the domain is depicted by Bell's theory [29], which originally demonstrated the kinetics of rupture of chemical bonds. In other words, protein unfolding is ascribed to the rupture of hydrogen bonds that sustain the three-dimensional structure of a protein [30]. Bell's theory suggests that the kinetics of bond rupture is represented in the following form [31]:

$$k(f) = \Omega \exp\left[-\frac{U_0 - fx}{k_B T} \right] \tag{1.2}$$

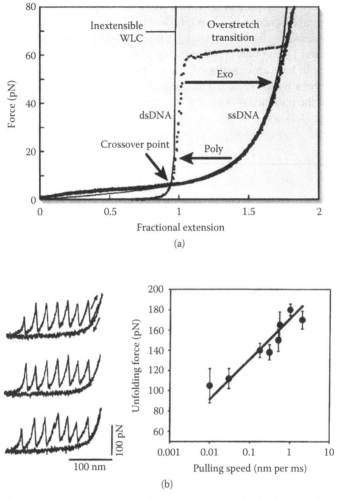

FIGURE 1.2 Force–extension curves for biomolecules obtained from single-molecule experiments: (a) Force–extension curves for a deoxyribonucleic acid (DNA) molecule that is stretched by an optical tweezer. When double-stranded DNA (dsDNA) is mechanically extended, there is an overstretch transition state (at a force around ~67 pN) found in the force–extension curve. For both dsDNA and single-stranded DNA (ssDNA), the force–extension curve is well fitted to the wormlike chain (WLC) model. This figure is adopted with permission from Ref. [17], Bustamante, C., Z. Bryant, and S. B. Smith. 2003. *Nature* 421:423. © Nature Publishing Group, Macmillan Publisher Ltd. (b) Force–extension curves for a protein repeat that is pulled by an atomic force microscope: The sawtooth pattern observed in the force–extension curve is attributed to a rupture of folded domains. The force that induced the rupture of a folded domain, referred to as "unfolding force," is linearly proportional to the logarithm of pulling speed, which is consistent with the theoretical predictions of Bell's model given by Equation 1.3. (Adapted from Bustamante, C., Z. Bryant, and S. B. Smith. 2003. *Nature* 421:423. © Nature Publishing Group, Macmillan Publisher Ltd.; Oberhauser, A. F., P. E. Marszalek, H. P. Erickson, and J. M. Fernandez. 1998. *Nature* 393:181. © Nature Publishing Group, Macmillan Publisher Ltd. With permission.)

where $k(f)$ is the kinetic rate of bond rupture in the presence of a mechanical force f, Ω is the natural frequency of a bond, U_0 is the internal energy of a protein, x is the mechanical extension (displacement), k_B is Boltzmann's constant, and T is the absolute temperature. The probability of breaking a bond is balanced with the decrease in the probability to have an intact bond, i.e., $p_b(f) = k(f)S(f) = -(df/dt)(dS/df)$, where $p_b(f)$ is a probability to break a bond, and $S(f)$ is the probability of exhibiting an intact bond in the presence of a force f. The most probable force to break a bond can be computed as follows:

$$\langle f \rangle = \int_0^\infty f p_b(f)\, df = \frac{k_B T}{x_b} \ln\left[\left(\frac{df}{dt}\right)\left(\frac{x_b}{k_0 k_B T}\right)\right] \tag{1.3}$$

where x_b is the difference between reaction coordinates corresponding to denatured and bonded states, respectively, and k_0 is reaction rate at zero force given as $k_0 = \Omega$ $\exp(-U_0/k_B T)$. Equation 1.3 demonstrates that the mean force to unfold the domain (or chemical bond), which can be measured from single-molecule pulling experiments, can provide quantitative insights into the free energy landscape relevant to protein folding (or chemical bond), i.e., x_b and U_0 (equivalent to k_0) [32] that plays a critical role in biological processes such as disease expression due to protein misfolding [10].

In the last decade, single-molecule pulling experiments have been considered to validate the nonequilibrium dynamics theories. One of the recently suggested physics theories dealing with nonequilibrium processes is "Jarzynski's equality" [33], which relates the work done during a nonequilibrium process to equilibrium free energy differences between two states:

$$\exp\left(-\frac{\Delta G}{k_B T}\right) = \left\langle \exp\left(-\frac{W}{k_B T}\right)\right\rangle \tag{1.4}$$

where ΔG and W represent the free energy difference between two equilibrium states and the work done during the nonequilibrium process from one equilibrium to another, respectively, and an angle bracket $< >$ indicates the ensemble average. Recently, Bustamante and coworkers [34] considered the single-molecule pulling experiments of RNA hairpins in order to verify Jarzynski's equality (Figure 1.3). In particular, an optical tweezer was used to stretch an RNA hairpin with a very slow pulling rate for measuring the free energy difference between folded and unfolded states, since the stretching process with a very slow pulling rate is nearly an equilibrium process so that the free energy difference is almost identical to the work done during the stretching process [35]. This measured free energy difference between folded and unfolded states is compared with that evaluated from Jarzynski's equality based on the stretching of an RNA hairpin with a fast pulling rate. It is shown that Jarzynski's equality is valid and robust to extract the free energy landscape for protein folding based on force–extension curves (which measure work done during the mechanical extension) obtained from single-molecule pulling experiments [36–38]. In a similar spirit, the single-molecule pulling experiments of RNA hairpins [39] have been reconsidered to verify Crooks' theorem [40] (Figure 1.4), which provides

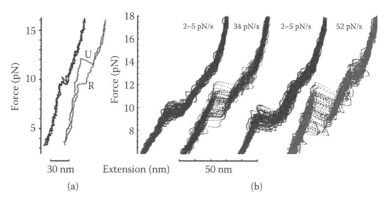

(a)   (b)

FIGURE 1.3 Optical tweezer–based stretching of a ribonucleic acid (RNA) hairpin: (a) Typical force–extension curves for unfolding and refolding of an RNA hairpin with a loading rate of 2 pN/s (left) or 5 pN/s (right). (b) unfolding and refolding force–extension curves of an RNA hairpin with various loading rates ranging from 2 to 52 pN/s. (Adapted from Liphardt, J., S. Dumont, S. B. Smith, I. Tinoco, and C. Bustamante. 2002. *Science* 296:1832. © The American Association for the Advancement of Science. With permission.)

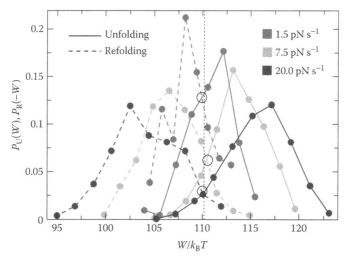

FIGURE 1.4 Single-molecule pulling experiment–based verification of Crooks' theorem: Optical tweezer–based single-molecule pulling experiments on a ribonucleic acid (RNA) hairpin provides force–extension curves for both unfolding and refolding processes and the related work done by unfolding and refolding processes. Probability distributions of the work done by unfolding and that done by refolding are obtained from repeated single-molecule pulling experiments. Circles show the points where the probability distribution of work done during the unfolding process becomes equal to that done during the refolding process. The values of work at these points correspond to the free energy differences between folded and unfolded states for the RNA hairpin. (Adapted from Collin, D., F. Ritort, C. Jarzynski, S. B. Smith, I. Tinoco, and C. Bustamante. 2005. *Nature* 437:231. © Nature Publishing Group, Macmillan Publisher Ltd. With permission.)

the free energy difference between two equilibrium states (i.e., folded and unfolded states) from the following relation:

$$\frac{P_U(W)}{P_R(-W)} = \exp\left(\frac{W - \Delta G}{k_B T}\right) \qquad (1.5)$$

Here, $P_U(W)$ is the probability distribution of the work done during the mechanical unfolding process, $P_R(-W)$ is the probability distribution of the work done during the refolding process, and ΔG is the free energy difference between folded and unfolded states.

Single-molecule pulling experiments have recently been highlighted for studying protein folding processes, which have not been clearly elucidated until now. Fernandez and coworkers [41] employed AFM single-molecule pulling experiments in order to observe the refolding of denatured protein (Figure 1.5). Specifically, a folded protein domain is mechanically stretched by an AFM and then a mechanical force is removed so as to induce the refolding of a denatured protein domain. The AFM single-molecule

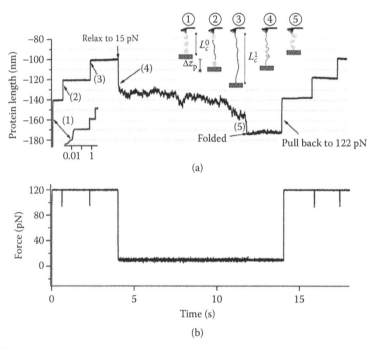

FIGURE 1.5 Force clamp–based single-molecule pulling experiments on the refolding dynamics of denatured protein: (a) A constant force of 122 pN is applied to three repeated ubiquitin domains. The stepwise force–extension curve is ascribed to the unfolding of folded ubiquitin domains such that the force plateau corresponds to the unfolding force. When force applied to the fully denatured ubiquitin domain is abruptly decreased from 122 to 15 pN, the end-to-end distance is decreased with fluctuation behaviors and it finally reaches the end-to-end distance corresponding to the fully folded ubiquitin domain. (b) Force acting on the repeated ubiquitin domains during the unfolding and refolding is shown. (Adapted from Fernandez, J. M., and H. Li. 2004. *Science* 303:1674. © The American Association for the Advancement of Science. With permission.)

experiments remarkably show that the protein folding trajectory can be quantitatively described such that folding time is dependent on the mechanical force applied to a protein. In a similar spirit, Thirumalai and coworkers [42] computationally studied the refolding process for a denatured protein in the presence of a small force (e.g., 1 pN to <100 pN) based on coarse-grained (CG) MD simulations of an RNA hairpin. The study showed that folding time is exponentially decreasing with respect to the mechanical force, which implies that the kinetics of protein folding can be captured by Bell's model depicted in Equation 1.2. This suggests that the refolding of a denatured protein can provide direct insight into the free energy landscape for the protein folding. Recently, Bustamante and coworkers [43] have utilized optical tweezer–based single-molecule experiments in order to understand the protein folding process. In particular, they have found that there might be several folding routes based on their observation of the refolding process of denatured protein. In addition, a recent study [44] reports that folding routes can be quantitatively described using "fluorescence resonance energy transfer" (FRET) bioassay. Remarkably, FRET bioassay is able to describe the dynamics of small molecule–mediated protein folding. It is implied that single-molecule techniques are quite essential for getting a fundamental understanding of protein folding that is highly correlated with the biological functions of proteins.

1.1.2 Characterization of Subcellular Materials: Implications for Medical Diagnosis

The characterization of protein materials at various length scales ranging from subcellular to cellular levels has come to play a significant role in medical diagnosis. In the past decades, diagnosis of diseases has been dependent on analyses of images of cells in order to sort out possible malfunctioning cells (i.e., tumors), whose shape is quite different from that of normal cells. However, the image-based separation of cancerous cells from normal cells is not possible at the early stage of disease expression. This is because at the early stage of disease expression, the shape of abnormal cells (i.e., cancerous cells) is indistinguishable from that of normal cells. In recent years, AFM nanoindentation has been employed to characterize the mechanical properties of both normal and abnormal cells, and it is realized that cancerous cells are notably more flexible than normal cells although the shape of the latter is similar to that of cancerous cells [45,46]. This indicates AFM nanoindentation is able to distinguish cancerous cells from normal cells even at the early stage of disease expression. This implies mechanical characterization of cells at nano- or microscales can lead to early diagnosis, which cannot be achieved with current medical diagnosis techniques of specific diseases such as cancers. In a recent study, Gimzewski and coworkers [47] took into account AFM nanoindentation in order to evaluate mechanical stiffness of various cancerous cells in various ranges depending on cancer types, patients' age, and patients' sex (Table 1.1). It has been remarkably shown that all cancerous cells are more flexible by about four times than normal cells, which suggests that mechanical characterization can be a novel route for the early diagnosis of cancers. Moreover, Sokolov and coworkers [48] have characterized the mechanical properties of cell surfaces using AFM nanoindentation. Their hypothesis in differentiating

TABLE 1.1

Patient Characteristics and Cytological Diagnosis versus Nanomechanical Diagnosis Based on Mechanical Measurements

Case Number	Age (years)/Sex	Clinical History	Cytological Diagnosis of Pleural Fluid	Stiffness (kPa): Tumor	Stiffness (kPa): Normal
1	52/Female	Non–small cell carcinoma of the lung	Positive for metastatic malignant cells	0.56 ± 0.09	2.10 ± 0.79
2	60/Female	Non–small cell carcinoma of the lung	Positive for metastatic malignant cells	0.52 ± 0.12	2.05 ± 0.87
3	49/Female	Breast ductal adenocarcinoma	Positive for metastatic malignant cells	0.50 ± 0.08	1.93 ± 0.50
4	85/Male	Pancreatic adenocarcinoma	Positive for metastatic malignant cells	0.54 ± 0.08	0.54 ± 0.12
5	40/Male	Liver cirrhosis	Negative for malignant cells		1.86 ± 0.50
6	47/Male	Fever and hepatic failure	Negative for malignant cells		1.75 ± 0.61
7	92/Female	Anasarca peripheral edema	Negative for malignant cells		2.09 ± 0.98

Source: Cross, S. E., Y.-S. Jin, J. Rao, and J. K. Gimzewski. 2007. *Nat Nanotechnol* 2:780. With permission.

cancerous cells from normal cells using measurements of the mechanical properties of cell surfaces is attributed to the fact that specific molecular chains (e.g., receptor molecules) are expressed on the surface of cancerous cells, whereas they do not appear on the surface of normal cells. It is remarkably shown that, based on polymer theory, mechanical characterization of a cell surface allows the separation of cancerous cells from normal cells (Figure 1.6a).

Moreover, a recent study [49,50] reports that mechanical characterization of cartilage enables early detection of pathological changes in osteoarthritis, which is related to a degenerative joint disease among older people (Figure 1.6b). The biological function of cartilage is critically dependent on the interplay of molecules that are constituents of the extracellular matrix. In particular, collagen fibrils and proteoglycans are two principal components that play a critical role in determining the physiological properties (e.g., mechanical stiffness) of cartilage of the extracellular matrix. The onset of osteoarthritis is ascribed to an imbalance between catabolic and anabolic processes, which causes degradation of the extracellular matrix. This degradation process is initiated by the depletion of proteoglycans followed by the degradation of collagen fibrils. It is hypothesized that this degradation process significantly affects mechanical properties of the extracellular matrix. This hypothesis has been experimentally validated by AFM nanoindentation, which shows that at the early stage of the degradation process the extracellular matrix becomes more rigid due to entanglement and disarrangement of fibrils. On the other hand, at a certain stage the extracellular matrix becomes more flexible owing to the disruption of the mesh network of collagen fibrils. This indicates mechanical characterization of the extracellular matrix can be considered a nanomechanical toolkit that enables early diagnosis of the onset of osteoarthritis.

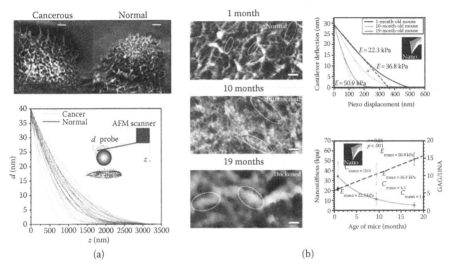

(a)   (b)

FIGURE 1.6 Atomic force microscope (AFM)-based nanoscale characterization of subcellular materials: (a) Scanning electron microscope (SEM) images of cellular surfaces for cancerous cell (left upper panel) and normal cell (right upper panel): The SEM image shows that cancerous cells exhibit different cellular surface from that of normal cells, which is attributed to the expression of specific proteins (marker proteins) on the cancerous cells surface. The AFM nanoindentation experiments provide the force–extension (F–z) curves of cellular surfaces, where the force (F) is related to the cantilever bending deflection (d) such as $F = kd$ with k being the AFM cantilever stiffness. (b) The AFM images of articular cartilages of mice at various ages: 1 month (left upper panel), 10 months (left middle panel), and 19 months (left lower panel). The AFM images show that on aging the articular cartilage is significantly degraded. Right upper panel shows the AFM nanoindentation of articular cartilages at different stages. The degraded articular cartilage on aging exhibits increased stiffness, whereas the concentration of glycosaminoglycan is decreased on aging (right lower panel). (c) The AFM images of the nuclear envelope (NE) of the cell nucleus for both normal cell (left panels) and cell experiencing apoptosis (right panels). The AFM indentation map (middle and lower panels) clearly elucidates that cellular apoptosis significantly reduces the stiffness of NE of the cell nucleus. (d) Proposed hierarchical structures of β-lactoglobulin fibrils (upper panel that shows the schematics) and their corresponding AFM images of fibrils (middle panel). The AFM imaging–based analysis provides the relationship between fibril height (thickness) and persistence length (equivalent to the bending stiffness) of the fibril or the helical pitch of the fibril (lower panel). (Adapted from Iyer, S., R. M. Gaikwad, V. Subba Rao, C. D. Woodworth, and I. Sokolov. 2009. *Nat Nanotechnol* 4:389. © Nature Publishing Group, Macmillan Publisher Ltd.; Stolz, M., R. Gottardi, R. Raiteri, S. Miot, I. Martin, R. Imer, U. Staufer et al. 2009. *Nat Nanotechnol* 4:186. © Nature Publishing Group, Macmillan Publisher Ltd.; Kramer, A., I. Liashkovich, H. Oberleithner, S. Ludwig, I. Mazur, and V. Shahin. 2008. *Proc Natl Acad Sci U S A* 105:11236. © National Academy of Sciences; Adamcik, J., J.-M. Jung, J. Flakowski, P. De Los Rios, G. Dietler, and R. Mezzenga. 2010. *Nat Nanotechnol* 5:423. © Nature Publishing Group, Macmillan Publisher Ltd. With permissions.)

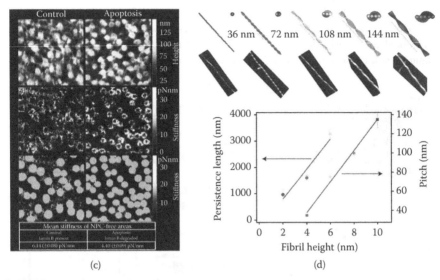

(c)   (d)

FIGURE 1.6   *(Continued)*

Today, mechanical characterization of cells not only distinguishes cancerous cells from normal cells as mentioned earlier but also provides fundamental insights into a cell's functional cycles such as apoptosis. For instance, a recent study [51] reports that apoptosis of a cell has been well characterized based on the measurement of mechanical properties of the nuclear envelope (NE) of a cell nucleus (Figure 1.6c). In particular, when a cell undergoes apoptosis, this is followed by the sequential destruction of key cellular elements: the NE including nuclear pore complex (NPC) proteins and the nuclear lamina filaments that are believed to be associated with the mechanical stability of the NE. The change in the mechanical stability of the NE was recently characterized based on AFM nanoindentation. Specifically, it was shown that the injection of cytochrome c into the cytosol of a cell, which results in cellular apoptosis, leads to a change in the mechanical stiffness of the NE, which is accompanied by morphological changes in the NE, particularly changes in the structure of NPCs.

Moreover, amyloid fibrils that are strongly related to disease expression, such as the expression of Alzheimer's disease, have been considered recently to characterize their nanomechanical behavior with respect to the pattern of self-assembly or self-aggregation or both [52]. Remarkably, the self-aggregation of mechanically weak protein domains results in an increase in the mechanical stiffness of amyloid fibrils, which may be attributed to the geometrical confinement of hydrogen bonds between aggregated protein domains. For instance, a recent study [53] reports that AFM-based imaging of amyloid fibrils has enabled the measurement of the mechanical properties of amyloid fibrils as a function of their hierarchical structures (Figure 1.6d). In particular, the persistence length l_p (related to the bending rigidity D) of amyloid fibrils is strongly dependent on aggregation pattern, as shown in Figure 1.6d. It is shown that the number of aggregated filaments dominantly governs the hierarchical structure of amyloid fibrils and consequently the nanomechanical properties (i.e., persistence length). This implies the nanomechanical characterization of amyloid fibrils provides fundamental insights into how self-aggregation and/or self-assembly of weak proteins significantly enhances the mechanical properties of protein filaments, which highlights that the expression of undesired self-aggregated protein filaments leads to disease expression (e.g., type 2 diabetes) through the replacement of specific cells (e.g., β cell) with stiff protein filaments at specific sites (e.g., pancreas). This sheds light on why nanomechanical characterization is essential for a fundamental understanding of protein materials that are critically involved in disease expression.

1.1.3 CHARACTERIZATION OF NANOSCALE MATERIALS: UNIQUE MATERIAL PROPERTIES

Since the discovery of CNT in 1991 [54], there have been a lot of attempts to develop or discover novel nanoscale materials that can be employed for various applications as nanoscale devices (e.g., sensors) [55–60], therapeutic agents [61–64], and so forth. For nanomaterials to be exploited for such applications, it is essential that the physical properties (e.g., mechanical property, electric property, and optical property) of various nanomaterials are characterized. Unlike properties of a bulk material at macroscopic scales, the physical properties of nanomaterials are very sensitive to the atomic structure of the nanomaterial. For instance, minuscule structure defects of a CNT may lead to significant changes in physical properties [65] such as mechanical stiffness [66] and electric conductance of materials. Attempts to characterize the material properties of nanoscale materials are aimed at gaining fundamental insights into the relationship between atomic structure and material properties, highlighting the need to construct a theoretical framework that provides insight into the structure–property relationship for nanoscale materials as has been done for bulk materials [67,68] at macroscopic scales.

The pioneering attempt to characterize the mechanical properties of nanostructures was implemented using the CNT. Specifically, an individual CNT was prepared in such a way that one end of the CNT was clamped onto the substrate and the other free end was displaced by an AFM tip in order to induce a bending deformation of the CNT for measuring its bending rigidity [69]. The force–extension curve (equivalent to the stress–strain curve) for CNT, which undergoes the bending deformation, was well depicted by the classical elasticity theory—Euler–Bernoulli beam theory [70]. This indicates that classical continuum mechanics enables the interpretation of experimentally observed mechanics of nanostructures. More remarkably, AFM indentation of cantilevered CNT based on the Euler–Bernoulli beam theory provides that the elastic modulus (for bending deformation) of CNT is ~1 TPa [69,71,72], which is much larger than the corresponding value for any other conventional engineering materials. This implies CNT can be regarded as a novel material that can exhibit excellent mechanical properties, which may result in the replacement of conventional engineering materials with nanomaterials when high elastic stiffness is required for the development of engineering systems. For instance, CNTs may be utilized for the construction of an aircraft wing or body, which requires both high elastic stiffness and low mass density for the material used for construction. Moreover, because of low mass density and high stiffness, CNTs are very useful in the development of nanoscale high-frequency devices [73].

The last decade witnessed the development of a novel carbon-based nanomaterial, that is, the atomic layer of carbons renowned as graphene [74–76]. It was found that graphene

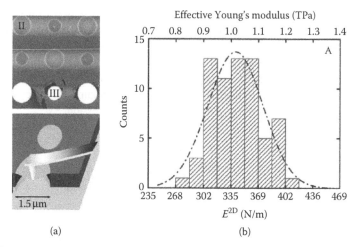

(a)   (b)

FIGURE 1.7 Atomic force microscope (AFM) indentation–based nanoscale characterization of a graphene monolayer: (a) Scanning electron microscope (SEM) image of graphene monolayer suspended on a substrate that exhibits circular holes (upper panel); schematic illustration of the AFM indentation of the graphene monolayer suspended on a circular hole (lower panel). (b) Measuring elastic modulus of the graphene monolayer. (Adapted from Lee, C., X. Wei, J. W. Kysar, and J. Hone. 2008. *Science* 321:385. © The American Association for the Advancement of Science. With permission.)

exhibits outstanding electronic properties that are much more beneficial than those of other bulk materials [74]. Recently, for furthering the applications of graphenes, an attempt was made to characterize the mechanical properties of graphenes using AFM nanoindentation. In particular, Hone and coworkers [16] indented a graphene monolayer that was deposited onto a substrate that has a circular hole (Figure 1.7a). The force–extension curve (stress–strain curve), which was obtained by indentation of graphene at the center of the hole, fitted well with the theoretical predictions based on classical elasticity theory—thin film mechanics [77]. It was shown that graphene obeys nonlinear elastic behavior such that the stress–strain relation for a graphene monolayer is given by $\sigma = E\varepsilon + D\varepsilon^2$, where σ and ε represent stress and strain, respectively; and E and D indicate linear elastic stiffness and nonlinear elastic stiffness, respectively. The nonlinear elastic properties for graphene were measured based on force–extension curve (for the estimation of E) and fracture toughness (for the evaluation of D). This unique nonlinear mechanical property of graphene, which was measured from AFM indentation based on classical elasticity theory, was recently taken into account theoretically using an "atomistic model" in order to validate classical elasticity theory that was used to extract the mechanical properties of graphene [78]. Specifically, the atomistic model was utilized to derive the elastic constants (E and D) such that the elastic constants were computed from the second derivative of the strain energy computed from the model, which suggests that classical elasticity theory in the subset of continuum mechanics theory is sufficient to depict the mechanical behavior of graphene. Interestingly, it was found that the linear elastic stiffness, E, of a graphene monolayer is ~1 TPa (Figure 1.7b), which is comparable to that of CNT.

Since carbon-based nanomaterials such as CNTs and graphenes exhibit excellent mechanical properties as described earlier in this section, such nanomaterials are regarded as building blocks for constructing large-scale materials. Specifically, there have been recent attempts to build large-scale structures by using chemical synthesis that allows the chemical binding of individual CNTs. For instance, a recent study [79] reports that the cross-linking of individual CNTs leads to the development of a thin film whose thickness is ~1 nm, implying that individual CNTs can be utilized for the fabrication of nanocomposites. It was shown that the mechanical properties of a CNT-based framework (i.e., thin film) can be improved by introducing covalent bonds that connect individual CNTs (Figure 1.8a). This indicates that the mechanical properties of a CNT-based novel material are

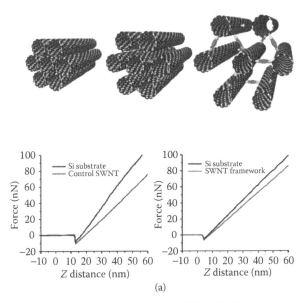

FIGURE 1.8 Mechanical characterization of nanobuilding block–based nanoscale hierarchical structures: (a) Schematic illustrations of single-walled carbon nanotube (SWNT)-based design of membrane structures using different cross-links (upper panel). The force-displacement curves (obtained from atomic force microscope [AFM] nanoindentation) for SWNT buckypaper fabricated from an assembly of SWNTs using van der Waals interactions (left lower panel) and SWNT-based membrane synthesized from the assembly of SWNTs using covalent bond cross-linking (right lower panel) are presented. This clearly elucidates that the interaction between SWNTs is a key parameter for determining the mechanical properties of SWNT-based membranes. (b) Mechanical characterization of irradiation-induced cross-linked multiwalled carbon nanotube (MWNT). Schematic illustration of the atomistic structure of irradiation driven cross-linked MWNT is shown in the upper panel. Load fraction on the inner shell of an MWNT versus strain for (5,5)/(10,10) double-walled nanotubes containing six Frenkel pair cross-links, six interstitial cross-links, and six divacancy cross-links is shown (left lower panel). The load fraction on the inner shell versus strain for (5,5)/(10,10) nanotubes containing different numbers of Frenkel pair cross-links is presented (middle and right lower panels). (Adapted from Song, C., T. Kwon, J.-H. Han, M. Shandell, and M. S. Strano. 2009. *Nano Lett* 9:4279. © American Chemical Society; Peng, B., M. Locascio, P. Zapol, S. Li, S. L. Mielke, G. C. Schatz, and H. D. Espinosa. 2008. *Nat Nanotechnol* 3:626. © Nature Publishing Group, Macmillan Publisher Ltd. With permissions.)

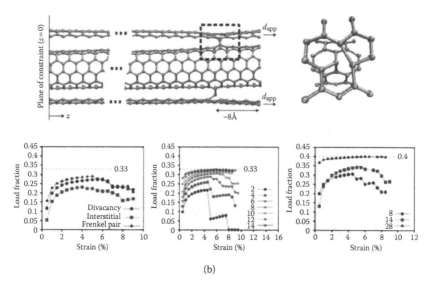

FIGURE 1.8 (*Continued*)

dependent on not only the mechanical behavior of CNTs (as a building block) but also the property of chemical bonds that cross-link the CNTs. Moreover, Espinosa and coworkers [80] report the mechanical properties of multiwalled CNTs (MWCNTs) and improvement of their mechanical properties by the introduction of irradiation that induces cross-links between individual CNTs. Moreover, they showed that the mechanical properties of MWCNTs can be tailored by irradiation, which also induces structural defects on CNTs (Figure 1.8b). It is implied that the mechanical properties of MWCNTs can be manipulated by controlling the chemical bonds between individual CNTs as well as the structural defects. Recently, researchers [81] proved that the mechanical properties of CNT bundles (i.e., fiber) can be tuned by using DNA, which binds individual CNTs. It was shown that the mechanical properties of the fiber can be tailored by hydration and/or ion concentration, which can affect DNA-mediated interactions between individual CNTs. These experimental efforts to develop a novel composite material using nanostructures such as CNTs suggest that it is possible to fabricate the material as well as to tune the material properties by controlling the interactions between individual nano-building blocks such as CNTs. This sheds light on the materiomics that paves the way for the development of de novo materials whose mechanical properties can be tailored by controlling the chemical bonds that govern the interaction between individual building blocks.

Solid nanowires (e.g., silicon and copper nanowires) have been developed as a nanoscale material whose cross-sectional dimension is on the order of nanometer scales [82]. Unlike the bulk material, a solid nanowire has a structural feature such that the nanowire possesses a large surface-to-volume ratio, which may result in anomalous material properties. For instance, the large surface-to-volume ratio for nanomaterials imply that the mechanical deformation of nanomaterials is controlled by not only strain energy but also surface energy, which is defined as an energetic cost to create a new surface through scaling down [83,84]. In other words, the deformation of nanomaterials is governed by bulk stress (derived from strain energy) as well as surface stress. Here, surface stress is defined as $\tau = \partial^2 U_S/\partial \varepsilon^2$ [85], where U_S is surface energy and ε is mechanical strain. It is implied that surface stress plays a critical role in the mechanical deformation of solid nanowires. A previous study by Diao et al. [86] reports that a metal nanowire that exhibits a very large surface-to-volume ratio undergoes the phase (structural) transformation attributed to surface stress. In recent years, it was found that the mechanical properties of solid nanowires are dependent on their cross-sectional dimension such that when the cross-sectional dimension of a solid nanowire approaches sub-50 nm, surface stress significantly changes the elastic properties and causes them to deviate from the bulk properties [13–15,87–91] (for details, refer to the review by Park et al. [92] that discusses the size dependence of the elastic properties of solid nanowires). This implies that in order to understand the unique mechanical properties of nanomaterials, it is a priori requisite to develop a theoretical framework that is able to explain the origin of anomalous mechanical behavior of nanomaterials, which is mainly ascribed to surface effects.

1.1.4 DESIGN OF DE NOVO FUNCTIONAL MATERIALS INSPIRED FROM BIOLOGY: BIOMIMETICS AND MATERIOMICS

The last decade witnessed studies focusing on the unique material properties of biological materials such as spider silk protein. Specifically, spider silk exhibits remarkable, unique mechanical properties in that it possesses both superelasticity and high fracture toughness simultaneously, and this behavior is quite different from that of conventional engineering materials [93,94] (Figure 1.9). Spider silk can be extended by about ~100% strain depending on hydration, and it also has a fracture toughness that is much higher than the value for any other human-made composite material such as Kevlar. This mechanical property is highly correlated with the biological function of spider silk; the ability of spider silk to capture a flying prey is attributed to its high extensibility that allows the transformation of kinetic energy of the flying prey into heat dissipation. Over a decade, there have been attempts to understand how spider silk can exhibit its unique mechanical properties—superelasticity as well as high fracture toughness—based on the establishment of a theoretical framework that enables the garnering of fundamental insights into the structure–property–function relation. For instance, Termonia [95] attributed the remarkable mechanical properties of spider silk to its molecular structure; spider silk consists of β-sheet crystals, which are responsible for its high fracture toughness, and amorphous molecular chains, which plays a role in its superelasticity (Figure 1.10a). Although Termonia's model [95] is able to capture the features observed in experiments involving the mechanical extension of spider silk, the assumption of the model has been contradicted by a recent experimental work [96] showing that spider silk protein

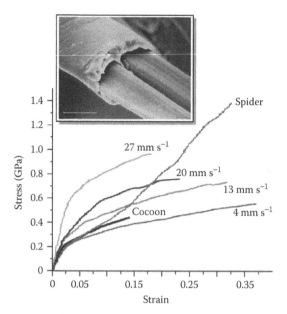

FIGURE 1.9 Mechanical characterization of various silks: Inset shows unwashed native silkworm silk. Stress–strain curves for washed and degummed single-filament silks of *Bombyx mori* silkworm under different stretching speeds ranging from 4 to 27 mm/s are shown. (Adapted from Shao, Z., and F. Vollrath. 2002. *Nature* 418:741. © Nature Publishing Group, Macmillan Publisher Ltd. With permission.)

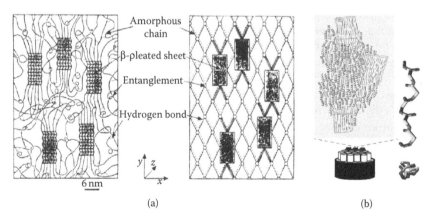

FIGURE 1.10 Molecular architecture of spider silk: (a) Schematic illustrations of the molecular structure of spider silk (left panel) and its corresponding physical model (right panel) suggested by the researcher Termonia. (b) Schematic diagram of the atomistic structure of spider silk. (Adapted from Termonia, Y. 1994. *Macromolecules* 27:7378. © American Chemical Society; van Beek, J. D., S. Hess, F. Vollrath, and B. H. Meier. 2002. *Proc Natl Acad Sci U S A* 99:10266. © National Academy of Sciences. With permission.)

comprises β-sheet crystals and well-ordered α helices (Figure 1.10b). Recently, researchers [97,98] have showed that the high stiffness and toughness of β-sheet crystal is attributed to the hydrogen bonds between β sheets and their geometric confinement of such hydrogen bonds. Researchers at the Massachusetts Institute of Technology (MIT) [99] have also developed the CG molecular model of spider silk in order to study the role of molecular structures and their hierarchies on mechanical properties of spider silk. The mechanical behavior of spider silk proteins in the different hierarchies has been experimentally studied, and it is observed that the mechanical responses of spider silks in different hierarchies can be captured by exponential law for the relationship between force and extension obtained from AFM single-molecule pulling experiments [100]. A recent study [101] has reconsidered the experimental result [100] in order to construct a theoretical model for spider silk. The researchers of the study have conjectured the hierarchical structure is a key engineering parameter that determines the mechanical response of spider silk. As described here, fundamental insights into the structure–property–function relationship can provide an idea of not only how some biological materials exhibit excellent mechanical properties but also how mechanical properties are manipulated by controlling molecular architecture (e.g., geometry of hydrogen bonds).

This basic concept of controlling the mechanical properties of materials by controlling their molecular architectures was recently realized. A recent pioneering study [102] has illuminated that proteins can be used to develop de novo materials (i.e., thin film) that possess high mechanical stiffness comparable to that of CNT buckypaper. In particular, amyloid proteins were considered for the development of thin films, because the aggregation of amyloid proteins can lead to increased mechanical stiffness (Figure 1.11). Specifically, when weak protein domains are aggregated, the geometric confinement of hydrogen bonds results in a remarkable enhancement of mechanical toughness of the aggregated proteins [98]. Based on this observation, a thin film was fabricated by depositing amyloid proteins (with aggregation) on a substrate. In a similar spirit, researchers [103] have recently developed a novel material, an artificial elastomeric protein, by the combination of specific protein domains such as GB1 and resilin in such a way that these two domains are connected by photochemically generated cross-links. It was shown that such an artificial elastomeric protein exhibits mechanical properties comparable to those exhibited by muscle fibers. Particularly, this artificial elastomeric protein has excellent viscoelastic properties, which can be controlled by denaturing the protein folded domain GB1. This indicates one can develop novel materials using proteins by controlling the folding states (i.e., geometric confinement of hydrogen bonds).

In summary, as presented here, a fundamental understanding of the relationship between molecular structure and mechanical properties of materials may enable the design of novel materials that can perform excellent functions. This sheds light on the structure–property–function relationship of biological materials for the design of de novo materials at multiple length scales ranging from nano- to millimeter scales; this design is recently referred to as "materiomics."

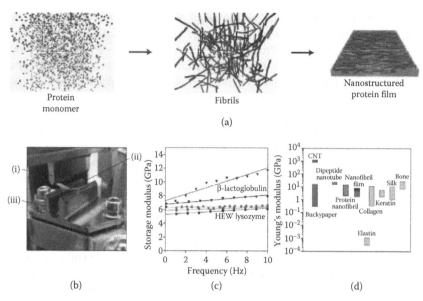

FIGURE 1.11 Mechanical characterization of protein amyloid-based biomimetic materials: (a) Schematic illustrations of the synthesis of biomimetic materials using protein amyloid fibrils. (b) microindentation experimental setup. (c) storage modulus versus frequency for two different films synthesized from β-lactoglobulin and hen egg white lysozyme. (d) mechanical properties of various materials including biological materials (protein fibrils, collagen, etc.), nanomaterials (carbon nanotube [CNT], CNT buckypaper, etc.), and biomimetic materials. (Adapted from Knowles, T. P. J., T. W. Oppenheim, A. K. Buell, D. Y. Chirgadze, and M. E. Welland. 2010. *Nat Nanotechnol* 5:204. © Nature Publishing Group, Macmillan Publisher Ltd. With permission.)

1.2 HOW TO UNDERSTAND NANOSCALE PHENOMENA

As described in Section 1.1, a lot of efforts have been made to develop novel nanoscale materials such as nanowires, biologically inspired materials, graphene, and CNTs, as well as to experimentally characterize the material properties of such nanoscale materials and of biological materials such as proteins. Even though these experimental efforts allow the characterization of nanoscale objects such as biological materials and nanomaterials, they lack fundamental insights into why such objects exhibit anomalous material properties that are quite different from those of bulk materials. For instance, although the AFM-based nanomechanical experiment has quantitatively described the size-dependent mechanical properties of nanowires, the physical origin of such size-dependent mechanical behavior cannot be well understood in such an experiment; this can be elucidated from computational simulations because computational simulations provide insights into the detailed mechanisms of the mechanical deformation behavior of nanomaterials, particularly the mechanics of atomic structures in response to mechanical deformation, which cannot be dictated by experiments. This indicates that computational simulations based on physical models at multiple length scales complement the experimental efforts to characterize nanoscale material properties. This section is purposed to present the basic concepts on how to model nanoscale objects for further applications in computational simulation–based virtual experiments.

Since the spatial scales of nanomaterials and biological materials span nanometer to micrometer length scales, appropriate physics models and their related simulations at characteristic spatial and temporal scales must be employed (Figure 1.12). In particular, if one utilizes a very refined model such as a quantum mechanics–based model, such as density functional theory (DFT) [104], it is computationally prohibitive to simulate the molecular system. This indicates that DFT may be restricted to the simulation of atomic systems whose spatial scales are limited to Angstrom scales or to subnanometer scales. In order to overcome this computational restriction for characterizing a molecular system, atomistic simulations [105] have been developed that are able to cover the simulation of molecular systems whose spatial scales range from nanometer to sub-100 nm scales. For structures whose spatial length scales are larger than micrometer scales (e.g., subcellular or cellular levels), the continuum elasticity–based model is able to delineate physical behavior. For instance, continuum elasticity model–based simulations, which are usually referred to as "finite

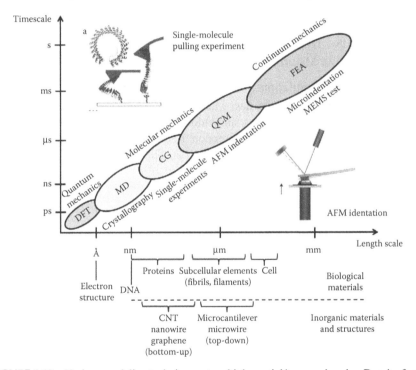

FIGURE 1.12 Various modeling techniques at multiple spatial/temporal scales: Density function theory (DFT) has been widely used for simulation at sub-1 nm length scales and sub-1 ps timescales. Molecular dynamics simulations have been extensively employed for studying the dynamics of nanoscale objects (at length scales ranging from 1 nm to <100 nm and a timescale of ≤1 μs). Coarse-grained (CG) modeling has been recently developed to simulate the dynamics of objects at length scales of ~1 μm and timescales of ~10 μs. The multiscale model, referred to as the quasicontinuum method, has been considered for analyzing the mechanical and dynamical behaviors of objects at length scales of ~1 μm to <1 mm and timescales of <1 ms. Finite element analysis based on the continuum mechanics theory has been broadly employed over more than three decades to study the mechanical and dynamical behaviors of macroscopic objects.

element analysis" (FEA) [106,107], are successful in describing mechanical response of subcellular materials such as viral capsids [108,109]. Even though the physical properties of structures at specific length scales are well characterized by specific model-based simulations (e.g., DFT, atomistic simulation, and FEA simulation), a conventional computational model that can capture the physical phenomena at multiple length scales ranging from nanometer to micrometer length scales is not available. There have been many recent efforts to develop de novo computational models that are computationally efficient in capturing not only microscopic behavior but also macroscopic response. These computational models are denoted as "multiscale models" [110] because they can capture the physical phenomena at multiple length scales. Sections 1.2.1 through 1.2.4 are dedicated to describing computational models at different length scales ranging from atomic scale to macroscopic scale. In these sections, review of the quantum mechanics–based model (e.g., DFT) is skipped as this topic is out of the scope of this book. For an overview of the quantum mechanics–based model, the readers can refer to the work by Sholl and Steckel [104].

1.2.1 MODELING TECHNIQUES AT ATOMIC SCALES: MOLECULAR MODELS AND SIMULATIONS

1.2.1.1 Atomistic Simulations: Molecular Dynamics Simulations

The physical model at atomic scales, which is usually referred to as the atomistic model, and atomistic model–based simulations known as atomistic simulations or MD simulations are discussed in this section. Let us briefly consider the motion of a diatomic molecule as presented in Figure 1.13a. The motion of such a system can

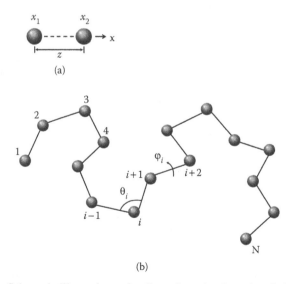

FIGURE 1.13 Schematic illustrations of a diatomic molecule and a chain molecule: The distance between two atoms is denoted as z. (a) Schematic illustration of a diatomic molecule. (b) schematic illustration of a chain molecule (e.g., protein molecule). The black solid line represents covalent bonds.

be described using internal coordinates, that is, interatomic distance denoted as z. The potential for an interaction between two atoms is denoted as $V(z)$. The equilibrium interatomic distance for the diatomic molecule can be computed from the principle that the potential energy reaches a minimum at the equilibrium state, which is depicted by the relation $\partial V(z)/\partial z = 0$. The fluctuation behavior of the diatomic molecule can be described by the equation of motion given by

$$m\frac{d^2 z(t)}{dt^2} + \gamma\frac{dz(t)}{dt} = -\frac{\partial V}{\partial z} + f_{th}(t) \tag{1.6}$$

where m is the effective mass of the diatomic molecule, γ is the friction coefficient that captures the atomic friction between the diatomic molecule and the environment (e.g., solvent), and $f_{th}(t)$ is the random force (Gaussian white noise) that results from thermal fluctuations. Here, Gaussian white noise satisfies the following condition:

$$f_{th}(t)f_{th}(\tau) = 2\pi k_B T\gamma^{-1}\delta(t-\tau) \tag{1.7}$$

Here, $\delta(t)$ is the Dirac delta function. The thermal fluctuation behavior of the diatomic molecule can be straightforwardly described by solving the differential equation of motion given in Equation 1.6. The details of how to numerically solve the equation of motion are presented as follows:

The principle of atomistic simulation is well depicted by the basic concept for describing the motion of the diatomic molecule; but atomistic simulations are purposed to describe the equation of motion for N atoms, whereas the motion of the diatomic molecule can be depicted as a one-dimensional (1D) equation of motion as given in Equation 1.6. Before implementing MD simulation, the equilibrium position has to be determined for further simulations such as the mechanical stretching of molecules. As is the case for the diatomic molecule, the equilibrium configuration can be found by minimizing the potential energy V prescribed to N atoms. In general, the potential energy V consists of strain energies for covalent bond stretching, bending of bond angle, and twist of dihedral angle, and the nonbonded interactions such as electrostatic repulsion and/or van der Waals (vdW) interactions.

$$V = \sum_{j=1}^{N-1}\frac{k_b}{2}\left(a_j - a_j^0\right)^2 + \sum_{j=1}^{n-2}\frac{k_\theta}{2}\left(\theta_j - \theta_j^0\right)^2 + \sum_{j=1}^{N-3}\frac{k_\phi}{2}\left(\phi_j - \phi_j^0\right)^2$$
$$+ \sum_{\substack{i<j}}^{N}\frac{Dq_iq_j}{r_{ij}} + \sum_{\substack{i<j}}^{N}4\varepsilon_0\left[\left(\frac{\sigma}{r_{ij}}\right)^6 - \left(\frac{\sigma}{r_{ij}}\right)^{12}\right] \tag{1.8}$$

where k_b, k_θ, and k_ϕ indicate the force constants corresponding to the stretching of a covalent bond, the bending of bond angle, and the twist of dihedral angle, respectively; a_j is the bond length defined as $a_j \equiv |\mathbf{a}_j| = |\mathbf{r}_j - \mathbf{r}_{j-1}|$ with $\mathbf{r}_i$ being the Cartesian coordinate of the jth atom; θ_j is the bond angle defined as $\cos\theta_j = \mathbf{a}_i \cdot \mathbf{a}_{i+1}/a_i a_{i+1}$; ϕ_j is the dihedral angle given as $\cos\phi_j = (\mathbf{a}_i \times \mathbf{a}_{i+1}) \cdot \mathbf{a}_{i+2}/|(\mathbf{a}_i \times \mathbf{a}_{i+1}) \cdot \mathbf{a}_{i+2}|$ (Figure 1.13b); r_{ij} is distance between the ith and jth atoms; q_i is the electrostatic charge for the *i*th atom; ε_0 is the energy for vdW interactions; σ is the length scale for vdW interactions; and superscript 0 represents the equilibrium configuration. In general, the parameters for potential energy, for example, force constants, can be determined in comparison with the atomic motions predicted from first principles such as DFT calculations. Once a potential energy V is prescribed to a molecular system, the equation of motion that describes the dynamics of the molecular system is given as follows:

$$M_i \frac{d^2 \mathbf{r}_i(t)}{dt^2} + D_{ij} \frac{d\mathbf{r}_i(t)}{dt} = -\nabla_i V(\mathbf{R}) + \mathbf{f}_i(t) \tag{1.9}$$

where M_i is molecular weight of the *i*th atom; D_{ij} indicates the hydrodynamic interactions between the *i*th and *j*th atoms; $\mathbf{R}$ represents the atomic coordinates for a molecular system, denoted as $\mathbf{R} = [\mathbf{r}_1, ..., \mathbf{r}_N]$; and $\mathbf{f}_i(t)$ is the Gaussian white noise acting on the *i*th atom. In order to solve the equation of motion depicted in Equation 1.9 for characterizing the dynamics of the molecular system, numerical integration schemes have been developed to numerically integrate the equation of motion. The Verlet algorithm [111], which is widely used for numerical integration, is briefly reviewed here. By denoting the atomic coordinates and atomic velocities of a molecular system at time t as $\mathbf{R}(t)$ and $\mathbf{V}(t)$, respectively, the atomic coordinates and atomic velocities at a time $t + \Delta t$ (where Δt is a time interval that is usually on the order of femtosecond regime) are computed as follows:

$$\mathbf{R}(t + \Delta t) = \mathbf{R}(t) + \Delta t \mathbf{V}(t) + \frac{(\Delta t)^2}{2} \mathbf{A}(t) \tag{1.10a}$$

$$\mathbf{V}(t + \Delta t) = \mathbf{V}(t) + \frac{\Delta t}{2} [\mathbf{A}(t) + \mathbf{A}(t + \Delta t)] \tag{1.10b}$$

where $\mathbf{A}(t)$ denotes atomic accelerations, that is, $\mathbf{A}(t) = d^2 \mathbf{R}(t)/dt^2$, which can be measured from Equation 1.9. As depicted by Equations 1.10a and b, the numerical integration scheme coupled with the equation of motion given by Equation 1.9 provides the trajectories of all atoms of a molecular system as a function of time. These trajectories allow the quantitative description of the dynamic behavior of a molecular system.

For the last two decades, MD simulations have been extensively exploited for understanding not only the dynamic behavior of nanoscale objects, particularly biological materials such as proteins [112–114], but also the interpretation of single-molecule pulling experiments [115]. Specifically, despite the ability of single-molecule pulling experiments to provide an insight into the mechanical behavior of biomolecules, these experiments are unable to suggest the fundamental principles of how a mechanical force induces the structural deformation of a protein such as the unfolding mechanism that is closely related to the biological function of proteins (Figure 1.14). This lack in the fundamental understanding of the detailed mechanisms of structural deformation (e.g., unfolding) of proteins can be resolved by exploiting MD simulations. In particular, MD simulations can provide detailed trajectories of all atoms of a protein as a function of time when a mechanical force is

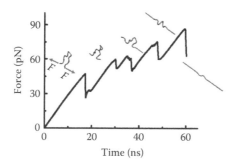

FIGURE 1.14 Molecular dynamics simulations of the mechanical stretching of a β-hairpin: The force–extension curve for the β-hairpin shows the bond rupture mechanisms that correspond with a force drop in the force–extension curve. The force-induced conformational changes for the β-hairpin are also shown at each bond rupture event.

applied to the protein, and thus MD simulations are a versatile toolkit that is able to depict the time-dependent atomic structure changes (deformations), the insights into which cannot be gained from single-molecule pulling experiments. For simulating the mechanical stretching of a biomolecule with a loading apparatus such as an AFM and/or an optical tweezer, the effective potential field V_{eff} must include the effect of harmonic strain energy due to loading apparatus: $V_{eff} = V + (K/2)(e - vt)^2$, where V is the potential energy of a molecule given by Equation 1.8, K is the stiffness of a loading apparatus, e is the end-to-end distance of a biomolecule defined as $e(t) = |\mathbf{r}_N(t) - \mathbf{r}_1(t)|$, and v is the pulling rate. This effective potential energy is used for constructing the equation of motion depicted in Equation 1.9. Numerical integration provides the relationship between the end-to-end distance $e(t)$ and the force $F(t)$ defined as $F(t) = K[e(t) - vt]$, and this force–extension relation, that is, $F(t)$ versus $e(t)$, computed from MD simulations can be compared with that obtained from single-molecule pulling experiments.

1.2.1.2 Normal Mode Analysis: Statistical Mechanics–Based Model

As discussed in Section 1.2.1.1, the computational expense of MD simulation is dependent on the timescale relevant to the dynamics of a molecular system, because the process that is the most time consuming is numerical integration of the equation of motion as well as computation of the atomic force $-\nabla_i V(\mathbf{R})$, which is the key component of the equation of motion as depicted in Equation 1.9. This indicates that when a long-timescale simulation is required for understanding the dynamics of a molecule, MD simulation may be computationally unfavorable.

Over the last two or three decades, normal mode analysis (NMA) [116–119] has received much attention due to development of the theoretical framework of statistical mechanics [120–122] that is able to explain the underlying mechanism of molecular behaviors such as fluctuation behaviors. In order to elucidate the fundamentals of statistical mechanics, we first take into account a diatomic molecule because the motion of a diatomic molecule can be described by a 1D equation of motion. For the small deformation of a diatomic molecule, the interaction energy V between two atoms can be approximated as a harmonic energy, that is, $V(z) = (k/2)z^2$, where z is the interatomic distance and k is the force constant for harmonic interaction energy. The equation of motion for each atom is represented in the following form:

$$m_1 \ddot{x}_1 = k(x_1 - x_2) \tag{1.11a}$$

$$m_1 \ddot{x}_2 = k(x_2 - x_1) \tag{1.11b}$$

where a dot represents the time derivative and x_j is the 1D coordinate of the jth atom (here, $j = 1$ or 2). It should be noted that the interatomic distance z can be expressed in terms of the atomic coordinates x_j in the relation $z = x_2 - x_1$ when $x_2 > x_1$. Therefore, the equation of motion can be described in terms of interatomic distance as follows:

$$\ddot{z} \equiv \ddot{x}_2 - \ddot{x}_1 = \frac{k}{m_{\text{eff}}} z \tag{1.12}$$

Here, m_{eff} is the effective mass defined as $m_{\text{eff}} = m_1 m_2/(m_1 + m_2)$. The equation of motion given by Equation 1.2 implies that the effective potential energy V and kinetic energy P are given by $V = (k/2)z^2$ and $P = (m_{\text{eff}}/2)v^2$, where $v = dz/dt$. The statistical mechanics theory [120–122] describes that the probability of finding a specific position and velocity (z, v) on the phase space is

$$p(z, v) = \frac{1}{Z} \exp\left[-\frac{H}{k_B T}\right] = \frac{1}{Z} \exp\left[-\frac{P + V}{k_B T}\right] \tag{1.13}$$

where $p(z, v)$ is the probability of finding the specific position and velocity (z, v) on the phase space, H the Hamiltonian of a diatomic molecule such as $H = P + V$, k_B Boltzmann's constant, T the absolute temperature, and Z the partition function. It must be noted that the probability of finding the specific phase space (z, v) obeys Boltzmann's distribution, which implies that the probability is governed by a diffusion-like process on the phase space, that is, the probability function $p(z, v)$ satisfies Smoluchowski's equation, which describes the diffusion process on the phase space [123]. The fluctuation behavior of a diatomic molecule can be depicted by the quantities $<z(t)>$ and $<z^2(t)>$, where an angle bracket $<\ >$ indicates the time average (or equivalently, ensemble average) of a variable. These ensemble averages $<z>$ and $<z^2>$ can be computed as follows:

$$\langle z \rangle = \int_0^\infty \int_0^\infty z p(z, v) \, dz \, dv = 0 \tag{1.14a}$$

$$\langle z^2 \rangle = \int_0^\infty \int_0^\infty z^2 p(z, v) \, dz \, dv = \frac{k_B T}{k} \tag{1.14b}$$

It is interesting to note that from statistical mechanics theory the mean-square interatomic distance is determined by the thermal energy $k_B T$ and the stiffness of a diatomic molecule.

Now, let us consider the general case, where a molecule that is composed of N atoms is considered. The equation of motion for such a molecule is represented as follows:

$$\mathbf{M\ddot{q}} = \mathbf{Kq} \tag{1.15}$$

where $\mathbf{\ddot{q}}$ denotes the atomic coordinates of a molecule, $\mathbf{M}$ is the mass matrix whose components are the molecular weight of the atoms in the molecule, and $\mathbf{K}$ is the stiffness (Hessian) matrix for the molecule. Here, it must be noted that the potential energy V of a molecule is approximated as a harmonic potential such that $V = (1/2) \mathbf{q}^t \mathbf{Kq}$, where superscript t indicates the transpose of a vector. Using spectral decomposition, that is, $\mathbf{K} = \mathbf{R}^t \Lambda \mathbf{R}$, where $\mathbf{R}$ is a modal matrix and Λ is a diagonal matrix whose component is an eigenvalue of $\mathbf{K}$, the potential energy V can be expressed as follows:

$$V = \sum_{j=1}^{3N-6} \frac{1}{2} \lambda_j y_j^2 \tag{1.16}$$

where λ_j is the jth eigenvalue of $\mathbf{K}$ and y_j is the component of a vector $\mathbf{y}$ defined as $\mathbf{y} = \mathbf{Rq}$. Since the contributions of each normal mode to the potential energy is independent of one another, the probability distribution to find the specific state $\mathbf{y}$ obeys Boltzmann's distribution, that is, the equipartition theorem [120]:

$$p(\mathbf{y}) = \frac{1}{Z} \exp\left[-\frac{\sum_{k=1}^{3N-6} \lambda_j y_j^2}{2 k_B T} \right] \tag{1.17}$$

where $p(\mathbf{y})$ is the probability distribution to find the state $\mathbf{y}$. Using Equation 1.14 and the equipartition theorem, the ensemble averages $<y_j>$ and $<y_j^2>$ can be estimated as follows:

$$\langle y_j \rangle = 0 \tag{1.18a}$$

$$\langle y_j^2 \rangle = k_B T / \lambda_j \tag{1.18b}$$

Using the transformation $\mathbf{y} = \mathbf{Rq}$ (or equivalently, $\mathbf{q} = \mathbf{R}^t \mathbf{y}$), the ensemble averages $<\mathbf{q}>$ and $<\mathbf{q} \otimes \mathbf{q}>$ are given as follows:

$$\langle \mathbf{q} \rangle = 0 \tag{1.19a}$$

$$\langle \mathbf{q} \otimes \mathbf{q} \rangle = \sum_{j=1}^{3N-6} \frac{k_B T}{\lambda_j} \mathbf{W}_j \otimes \mathbf{W}_j \tag{1.19b}$$

where $\mathbf{w}_j$ is the jth column of the modal matrix $\mathbf{R}$. As elucidated in Equation 1.19, the fluctuation behavior such as mean-square fluctuation (i.e., $<\mathbf{q} \otimes \mathbf{q}>$) can be easily computed once the eigenvalues and their corresponding normal modes are estimated based on the stiffness matrix **K**, which is the second derivative of potential energy V with respect to atomic coordinates. This indicates NMA allows quantitative descriptions of the fluctuation behavior of a molecular system.

1.2.2 MODELING AT MESOSCALES: COARSE-GRAINED MODELING

As problem size (or equivalently, the size of a molecule that is of interest to simulations) increases, the atomistic model that can be employed for MD simulations and/or NMA-based simulations becomes computationally intractable due to the need for computationally expensive processes (e.g., computing force and/or stiffness matrix by differentiation of potential energy with respect to atomic coordinates, numerical integrations, etc.). In order to make large molecules computationally feasible, efforts have been made [124] to develop novel models that enable one to perform a computationally efficient analysis on the dynamics of such molecules; these models must also be able to capture the dynamic characteristics that are comparable to the conventional atomistic model. Specifically, efforts that have been made recently in this direction involve the development of a physically relevant model that takes into account the reduced degrees of freedom rather than considering all the atoms of a molecule. In particular, a group of atoms that undergo rigid body–like motion is modeled as a single bead (Figure 1.15), and then the empirical potential field that can be prescribed to beads is appropriately developed. This modeling scheme is usually referred to as "coarse graining (CG)" or as the CG model, since a CG model generally coarsens the details of atomic structures.

The procedure to construct a CG model from the atomistic model can be described as follows: As briefly demonstrated in the previous paragraph, a group of atoms is replaced with a single bead, as long as such atoms experience similar motion.

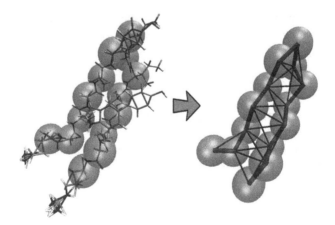

FIGURE 1.15   Schematic illustrations of the coarse-graining strategy: The left panel shows the atomistic structure of a β-hairpin. During the coarse-graining process, a group of some atoms that undergo collective dynamics is replaced by a single pseudo atom represented as a bead in the right panel. In this case (i.e., β-hairpin), atoms belonging to a single amino acid sequence are grouped as a pseudo atom.

In particular, a group of atoms is selected in such a way that the atoms of the group satisfy the following conditions:

$$\chi^2 = \frac{1}{T}\int_0^T \sum_{i\in\Omega} \sum_{j\ge i\in\Omega} |\mathbf{r}_i(t) - \mathbf{r}_j(t)|^2 \, dt < \delta \tag{1.20}$$

where Ω is the group of atoms, $\mathbf{r}_i(t)$ is the Cartesian coordinates of the ith atom at time t, T is the simulation timescale (e.g., ~1 ns), and δ is the cutoff value. Here, the trajectories of atoms (i.e., $\mathbf{r}_j$) are computed from a short-timescale MD simulation. Moreover, the position of a bead is usually located at the mass center of a set of atoms. Once a set of atoms are selected to define a CG bead, the effective potential field prescribed to CG beads must be constructed. Inspired from the refined atomistic model, the effective potential field, which describes the interactions between CG beads, consists of bonded and nonbonded interactions. In particular, the effective potential field V is represented in the following form:

$$V = \sum_{i=1}^{n-1} \frac{k_D}{2}\left(D_i - D_i^0\right)^2 + \sum_{i=1}^{n-2} \frac{k_\Omega}{2}\left(\Omega_i - \Omega_1^0\right)^2 + \sum_{i=1}^{n-3} \frac{k_\Phi}{2}\left(\Phi_i - \Phi_i^0\right)^2$$
$$+ \sum_{i=1}^{n-1}\sum_{j=i+1}^{n} \frac{AQ_iQ_j}{R_{ij}} + \sum_{i=1}^{n-1}\sum_{j=i+1}^{n} 4E_0\left[\left(\frac{\Xi}{R_{ij}}\right)^6 - \left(\frac{\Xi}{R_{ij}}\right)^{12}\right] \tag{1.21}$$

where k_D, k_Ω, and k_Φ represent the force constants for strain energies corresponding to the stretching of pseudo–covalent bond, bending of pseudo–bond angle, and twist of pseudo–dihedral angle, respectively; n is the total number of CG beads; D_i is the length of a pseudobond given by $D_i = |\mathbf{D}_i| = |\mathbf{R}_{i+1} - \mathbf{R}_i|$ with $\mathbf{R}_i$ being the position vector of the ith CG bead; Ω_i is the pseudo–bond angle defined as $\cos\Omega_i = \mathbf{D}_i \cdot \mathbf{D}_{i+1}/D_iD_{i+1}$; Φ_i is the pseudo–dihedral angle expressed as $\cos\Phi_i = (\mathbf{D}_i \times \mathbf{D}_{i+1}) \cdot \mathbf{D}_{i+2}/|(\mathbf{D}_i \times \mathbf{D}_{i+1}) \cdot \mathbf{D}_{i+2}|$; Q_i is the effective charge acting on the ith CG bead; R_{ij} is the distance between the ith and jth CG beads; E_0 is the energy scale for effective vdW interactions between CG beads; and Ξ is the length scale for effective vdW interactions. One of the important processes in CG modeling is to determine parameters (e.g., k_D, k_Ω, k_Φ, Q_i, E_0, and Ξ) for the effective potential field. In general, such parameters are empirically determined in such a way that the trajectories of CG beads, which describe their motion, are compared with those predicted from refined atomistic simulations such as MD simulations. Moreover, the effective potential can be simplified such that some insignificant energy contributing to the motion of the CG model is neglected. For instance, mechanical deformation of carbon-based nanomaterials such as CNT and graphene are attributed to the deformation of a covalent bond [125–127], implying that bonded interactions play a dominant role in the mechanical deformation of carbon-based nanomaterials, whereas nonbonded interactions (e.g., electrostatic repulsion and vdW interactions) do not.

Once CG models are established as aforementioned, the dynamic behavior of CG models can be depicted from two major simulation techniques: (1) the MD-like numerical integration of the equation of motion, which provides the trajectories of

CG beads; and (2) application of NMA to the CG model, which can suggest an insight into the fluctuation behavior that is described as mean-square displacement.

1.2.3   MODELING STRATEGY AT MACROSCOPIC SCALES: CONTINUUM MECHANICS–BASED MODELING TECHNIQUES

As the size of an object of interest reaches micrometer length scales, it becomes impossible to computationally simulate the dynamics of the object using atomistic model–based simulations (such as MD simulations and/or NMA) and even CG model–based simulations. The computational infeasibility of atomistic simulation and CG models in analyzing the dynamics of objects at macroscopic length scales (e.g., >1 μm length scale) is ascribed to the fact that the timescale accessible with atomistic simulations and CG models is several orders of magnitude smaller than the values relevant to the dynamic behavior of macroscopic objects. Furthermore, atomistic simulation encounters computational prohibition in the estimation of atomic force and/or atomic stiffness matrix.

Classically, simulations of the dynamics (and/or mechanics) of macroscopic objects have been widely conducted based on the theoretical framework of continuum mechanics, which results in continuum mechanics–based simulations. Before performing an overview of continuum mechanics–based simulations, the classical continuum mechanics theory, particularly linear elasticity [128], is briefly reviewed here. For convenience, an elastic object (solid) that can undergo structural deformation due to a mechanical force (Figure 1.16a) is considered. Structural deformation of an object is described by the displacement field denoted as $u_j(\mathbf{r})$, which is a function of the coordinate $\mathbf{r}$ of an arbitrary point belonging to the object. Here, an index j indicates the jth Cartesian directions, that is, $j = x$, y, or z. The dimensionless measure to quantify structural deformation is strain e_{ij}, which is defined as $e_{ij} = (1/2)(\partial u_i/\partial r_j + \partial u_j/\partial r_i)$, where r_j is the jth Cartesian component of the coordinate $\mathbf{r}$. Here, it is noted that repeated indices indicate summation (Einstein's summation rule). Further, it is assumed that structural deformation of an elastic solid is small enough for the solid to undergo geometric linear elastic deformation. Assuming that the solid obeys Hooke's law (i.e., linear relationship between deformation and the applied force field), the stress τ_{ij}, which is the dimensionless quantity defined as the force acting on a unit area of the cross section, can be obtained as $\tau_{ij} = C_{ijkl}e_{kl}$, where C_{ijkl} indicates the elastic modulus tensor. For an isotropic material that is presumed for theoretical convenience, the elastic modulus tensor C_{ijkl} is given by $C_{ijkl} = \lambda\delta_{ij}\delta_{kl} + \mu(\delta_{ik}\delta_{jl} + \delta_{il}\delta_{jk})$, where λ and μ are Lamé constants, and δ_{ij} is the Kronecker delta defined as $\delta_{ij} = 1$ if $i = j$ and otherwise $\delta_{ij} = 0$. It must be noted that the stress field τ_{ij} should satisfy the equation of motion given by $\partial\tau_{ij}/\partial r_j + \rho p_i = \rho(\partial^2 u_i/\partial t^2)$, where ρ is the mass density and p_i is the ith Cartesian component of the body force vector $\mathbf{p}$. Therefore, from the strain–displacement relation and Hooke's law, the equation of motion can be described in terms of the displacement field $u_j(\mathbf{r})$:

$$(\lambda+\mu)\frac{\partial^2 u_k}{\partial r_i \partial r_k}+\mu\frac{\partial^2 u_i}{\partial r_k \partial r_k}+\rho p_i = \rho\frac{\partial^2 u_i}{\partial t^2} \tag{1.22}$$

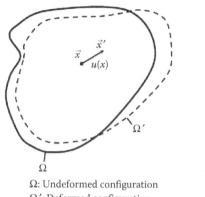

Ω: Undeformed configuration
Ω': Deformed configuration
$u(x)$: Displacement vector
$u(\vec{x}) = \vec{x}' - \vec{x}$

(a)

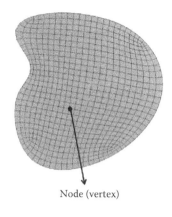

Node (vertex)

(b)

FIGURE 1.16 Schematic illustrations of a continuous medium and finite element–based discretization of a continuous medium: (a) Schematic illustration of a continuous medium: Black solid line shows the undeformed configuration, whereas the black dotted line represents the deformed configuration. Here, $\vec{x}$ and $\vec{x}'$ denote the position vectors at point x in the undeformed and deformed configurations, respectively. $u(x)$ is the displacement vector that represents the deformation of a continuous medium. (b) Schematic illustration of the finite element-based discretization of a continuous medium.

This equation of motion is well known as "Navier's equation." The deformation of an elastic solid can be described by solving Navier's equation with appropriate boundary conditions.

In order to quantitatively understand the mechanical deformation of an elastic solid at macroscopic scales (at least larger than micrometer length scales), Navier's equation, which is analytically intractable, must be solved. One of the routes to solve Navier's equation is to exploit FEA, whose basic principle is to numerically solve the equation based on the discretization of a continuous medium. As shown in Figure 1.16b, an elastic solid is discretized such that it consists of discretized elements. The displacement field $u_j(\mathbf{r})$ is approximated as follows [129]:

$$u_j(\mathbf{r}) = \sum_{\alpha=1}^{N} U_\alpha N_\alpha^j(\mathbf{r}) \equiv U_\alpha N_\alpha^j \qquad (1.23)$$

where U_α is displacement at node (vertex) α of an element, N is the total number of nodes for an element, and $N_\alpha^j(\mathbf{r})$ is the shape (basis) function that is a priori assumed. Here, an index j indicates Cartesian directions (i.e., $j = x$, y, or z), whereas a Greek index represents the node index. Accordingly, the strain field $e_{ij}(\mathbf{r})$ can be represented in the following form:

$$e_{ij}(\mathbf{r}) = \frac{1}{2} \sum_{\alpha=1}^{N} U_\alpha \left[\frac{\partial N_\alpha^i(\mathbf{r})}{\partial r_j} + \frac{\partial N_\alpha^j(\mathbf{r})}{\partial r_i} \right] \equiv U_\alpha B_{ij}^\alpha \qquad (1.24)$$

where B_{ij}^α represents the matrix that provides the strain–displacement relationship. From Hooke's law, the stress field can be expressed as follows:

$$\tau_{ij} = (\lambda \delta_{ij} B_{kk}^\alpha + 2\mu B_{ij}^\alpha) U_\alpha \equiv G_{ik} B_{kj}^\alpha U_\alpha \tag{1.25}$$

Here, as aforementioned, an isotropic material property is presumed for an elastic solid. In general, FEA accounts for a weak form of the differential equation rather than directly considering the differential equation (i.e., Navier's equation) described in Equation 1.22. Internal virtual work is given as follows:

$$\delta\Pi_{int} = \int_V \tau_{ij} \delta e_{ij} \, dV = \sum_{\alpha=1}^{N} \left[\sum_{\beta=1}^{N} B_{ij}^\alpha G_{ik} B_{jk}^\beta \, dV \right] U_\beta \delta U_\alpha \tag{1.26}$$

where V is volume of the elastic solid; $\delta\Pi_{int}$ indicates internal virtual work, that is, variation of internal energy due to structural deformation; δe_{ij} is virtual strain (i.e., variation of a strain due to an infinitesimal deformation of a solid); and δU_α is virtual nodal displacement. The external virtual work due to body force is

$$\delta\Pi_{ext} = \int_V p_i \delta u_i \, dV = \left[\sum_{\alpha=1}^{N} \int_V p_i N_\alpha^i \, dV \right] \delta U_\alpha \tag{1.27}$$

where δu_j is virtual displacement (i.e., infinitesimal variation of a displacement field). By the principle of virtual work, that is, $\delta\Pi_{int} = \delta\Pi_{ext}$, the equation of motion (i.e., Navier's equation) can be transformed into a numerical linear equation represented in the following form:

$$K_{\alpha\beta} U_\beta = F_\beta \tag{1.28a}$$

where $K_{\alpha\beta}$ and F_β represent the stiffness matrix and the force vector, respectively, which are described as follows:

$$K_{\alpha\beta} = \int_V B_{ij}^\alpha(\mathbf{r}) G_{ik} B_{jk}^\beta(\mathbf{r}) \, dV \tag{1.28b}$$

$$F_\alpha = \int_V p_i N_\alpha^i(\mathbf{r}) \, dV \tag{1.28c}$$

It is implied that FEA transforms the partial differential equation into a numerical linear equation that can be computationally feasible. It is noted that in this section only the FEA strategy that is applied to an elastic solid whose deformation is governed by linear elasticity is reviewed. In general, FEA enables numerical analyses on various physical problems such as problems in electromagnetics, solid mechanics, fluid mechanics, and heat transfer as long as such physical problems can be delineated by partial differential equations (e.g., Maxwell's equation for electromagnetics, Navier's equation for solid mechanics or fluid mechanics, Laplace's equation for heat transfer).

1.2.4 MODELING TECHNIQUES AT DISPARATE, MULTIPLE SCALES: MULTISCALE MODELS

As presented in Section 1.2.1 and 1.2.2, atomistic models and CG models are only suitable for simulations of objects whose length (spatial) scales are sub-100 nm, whereas continuum mechanics–based simulations such as FEA are usually applicable to continuous media whose spatial scales are a few orders of magnitude larger than nanometer length scales. This indicates that it is required to develop a theoretical model that can be employed for the computational simulation of an object whose spatial scale ranges between sub-100 nm and micrometer length scales. An effort made by researchers over approximately 15 years aimed at coupling the features of continuum mechanics and atomistic models. This modeling strategy is known as the "quasicontinuum method" (QCM) [130]; in this method, a specific region of an object that cannot be analyzed by a classical continuum mechanics–based simulation is modeled as an atomistic model, whereas the remaining region of the object is computationally treated by a classical continuum mechanics–based simulation. Consider the following case for instance: A classical continuum mechanics–based simulation has failed in gaining insight into the dynamics of crack propagation in an elastic solid, particularly into the dynamics of a crack tip. To resolve the infeasibility of continuum mechanics theory to analyze the dynamics of this crack tip, the region near the crack tip is modeled as a refined atomistic model (e.g., DFT and/or MD), whereas the continuum model is employed to model the remaining region of the solid. Another example is to simulate the dynamics of protein-DNA interactions that is inaccessible with atomistic model. To resolve computational inefficacy for atomistic model, small proteins are modeled as atomistic model while a continuum model is exploited to depict a long DNA chain (Figure 1.17).

Here, a review of the basic principle of multiscale modeling by accounting for the mechanical deformation of a solid is feasible. Consider the case where a solid is deformed by a mechanical force and a specific region (e.g., crack tip) undergoes anomalous dynamics or deformation (e.g., crack propagation) that cannot be depicted by a conventional continuum mechanics theory or its related models. It is presumed deformation of the remaining region can be dictated by a continuum mechanics–based simulation (i.e., FEA), the details of which are presented in Section 1.2.3. The key to multiscale modeling strategy is to combine the atomistic model of a specific region Ω_A, whose deformation cannot be simulated by FEA, and the continuum mechanics model (i.e., FEA) of the remaining region. In particular, the coupling of the atomistic model to the continuum model can be straightforward if the atomistic model that describes a region Ω_A can be transformed into an FEA-like linear equation, for example, Equation 1.28. In general, the key to an atomistic model is utilization of the empirical potential field that is described by Equation 1.8. Here, empirical potential is denoted as as $V_{\mathrm{eff}}(\mathbf{r}_j)$, where $\mathbf{r}_j$ is the position vector of an atom belonging to a region Ω_A that is modeled as an atomistic model. Now, the strain energy density ψ that is defined as follows can be introduced:

$$\sum_{j=1}^{n} V_{\mathrm{eff}}(\mathbf{r}_j) = \int_{V_0} \Psi(\mathrm{W})\, \mathrm{d}V \tag{1.29}$$

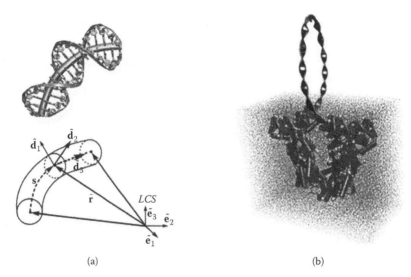

(a)   (b)

FIGURE 1.17 Analysis of DNA-protein interactions using the quasicontinuum model: (a) Continuum modeling descriptions of a DNA molecule is presented. In particular, the atomistic model of DNA is shown in the upper panel, while a continuum-model-based description (i.e., elastic rod model) of DNA is depicted in the lower panel. (b) A twisted DNA loop described as an elastic rod model was held by a lac repressor depicted as an atomistic model. (Adapted from Phillips, R., M. Dittrich, and K. Schulten. 2002. *Annu Rev Mater Res* 32:219. With permission.)

where n is the total number of atoms belonging to the region Ω_A; V_0 is volume of the region Ω_A; and $\mathbf{W}$ is the stretch tensor defined as $W_{ij} = F_{ki}F_{kj}$, where F_{ij} is the deformation gradient tensor defined as $F_{ij} = \partial u_i(\mathbf{r})/\partial \mathbf{r}_j$ with $\mathbf{r}$ being the position vector of a point belonging to the region Ω_A, u_i being the displacement field, and an index i (or j) indicates the Cartesian direction (i.e., $i = x, y$, or z). The Piola–Kirchhoff stress tensor T_{ij} is defined as $T_{ij} = \partial \psi(\mathbf{W})/\partial W_{ij}$, and the tangent modulus tensor $\mathbf{C}$ is represented in the form $C_{ijkl} = \partial T_{ij}/\partial W_{kl} = \partial^2 \psi(\mathbf{W})/\partial W_{ij}\partial W_{kl}$. Similar to FEA, the discretization provides the displacement field $u_i(\mathbf{r})$ in the form $u_i(\mathbf{r}) = U_\alpha N_\alpha^i(\mathbf{r})$, where U_α is nodal displacement at node α, N_α^i is the a priori defined shape (basis) function, Greek index α indicates the index of a node, and repeated indices represent Einstein's summation rule. Based on the definition of strain energy density ψ, the internal virtual work due to the infinitesimal deformation of a region Ω_A is

$$\delta \Pi_{int} = \left[\int_V L_{ki}^\alpha(\mathbf{r}) T_{ij} L_{kj}^\beta(\mathbf{r}) \, dV \right] U_\beta \delta U_\alpha \equiv K_{\alpha\beta} U_\beta \delta U_\alpha \qquad (1.30)$$

where L_{ij}^α is the tensor that provides the relationship between deformation gradient F_{ij} and nodal displacement U_α such that $L_{ij}^\alpha = \partial N_\alpha^i(\mathbf{r})/\partial r_j$, δU_α is virtual nodal displacement (i.e., infinitesimal variation of nodal displacement), and $K_{\alpha\beta}$ is the stiffness matrix for a region Ω_A. This implies that mechanical deformation of a region Ω_A, which is described by an atomistic model (i.e., atomic potential field V_{eff}), can be depicted by an FEA-like linear equation given by Equation 1.30, which allows the coupling of a region Ω_A to the remaining region of an elastic solid resulting in a set of linear equations that are computationally efficient and tractable.

REFERENCES

1. Binnig, G., C. F. Quate, and C. Gerber. 1986. Atomic force microscope. *Phys Rev Lett* 56:930.
2. Colton, R. J., D. R. Baselt, Y. F. Dufrene, J. B. D. Green, and G. U. Lee. 1997. Scanning probe microscopy. *Curr Opin Chem Biol* 1:370.
3. Carrion-Vazquez, M., A. F. Oberhauser, T. E. Fisher, P. E. Marszalek, H. Li, and J. M. Fernandez. 2000. Mechanical design of proteins studied by single-molecule force spectroscopy and protein engineering. *Prog Biophys Mol Biol* 74:63.
4. Neuman, K. C., T. Lionnet, and J. F. Allemand. 2007. Single-molecule micromanipulation techniques. *Annu Rev Mater Res* 37:33.
5. Muller, D. J., M. Krieg, D. Alsteens, and Y. F. Dufrene. 2009. New frontiers in atomic force microscopy: analyzing interactions from single-molecules to cells. *Curr Opin Biotechnol* 20:4.
6. Dickinson, M. E., and J. P. Schirer. 2009. Probing more than the surface. *Mater Today* 12:46.
7. Huey, B. D. 2007. AFM and acoustics: Fast, quantitative nanomechanical mapping. *Annu Rev Mater Res* 37:351.
8. Muller, D. J., and Y. F. Dufrene. 2008. Atomic force microscopy as a multifunctional molecular toolbox in nanobiotechnology. *Nat Nanotechnol* 3:261.
9. Hinterdorfer, P., and Y. F. Dufrene. 2006. Detection and localization of single molecular recognition events using atomic force microscopy. *Nat Meth* 3:347.
10. Borgia, A., P. M. Williams, and J. Clarke. 2008. Single-molecule studies of protein folding. *Annu Rev Biochem* 77:101.
11. Li, P. T. X., J. Vieregg, and I. Tinoco. 2008. How RNA unfolds and refolds. *Annu Rev Biochem* 77:77.
12. Forman, J. R., and J. Clarke. 2007. Mechanical unfolding of proteins: insights into biology, structure and folding. *Curr Opin Struct Biol* 17:58.
13. Cuenot, S., C. Fretigny, S. Demoustier-Champagne, and B. Nysten. 2004. Surface tension effect on the mechanical properties of nanomaterials measured by atomic force microscopy. *Phys Rev B* 69:165410.
14. Agrawal, R., B. Peng, E. E. Gdoutos, and H. D. Espinosa. 2008. Elasticity size effects in ZnO nanowires: A combined experimental computational approach. *Nano Lett* 8:3668.
15. Agrawal, R., B. Peng, and H. D. Espinosa. 2009. Experimental-computational investigation of ZnO nanowires strength and fracture. *Nano Lett.*
16. Lee, C., X. Wei, J. W. Kysar, and J. Hone. 2008. Measurement of the elastic properties and intrinsic strength of monolayer graphene. *Science* 321:385.
17. Bustamante, C., Z. Bryant, and S. B. Smith. 2003. Ten years of tension: Single-molecule DNA mechanics. *Nature* 421:423.
18. Baek, S. H., W. J. Chang, J. Y. Baek, D. S. Yoon, R. Bashir, and S. W. Lee. 2009. Dielectrophoretic technique for measurement of chemical and biological interactions. *Anal Chem* 81:7737.
19. Bustamante, C., J. F. Marko, E. D. Siggia, and S. Smith. 1994. Entropic elasticity of lambda-phage DNA. *Science* 265:1599.
20. Baumann, C., S. Smith, V. Bloomfield, and C. Bustamante. 1997. Ionic effects on the elasticity of single DNA molecules. *Proc Natl Acad Sci U S A* 94:6185.

21. Strick, T. R., M. N. Dessinges, G. Charvin, N. H. Dekker, J. F. Allemand, D. Bensimon, and V. Croquette. 2003. Stretching of macromolecules and proteins. *Rep Prog Phys* 66:1.

22. Marko, J. F., and E. D. Siggia. 1995. Stretching DNA. *Macromolecules* 28:8759.

23. Wiggins, P. A., T. van der Heijden, F. Moreno-Herrero, A. Spakowitz, R. Phillips, J. Widom, C. Dekker, and P. C. Nelson. 2006. High flexibility of DNA on short length scales probed by atomic force microscopy. *Nat Nanotechnol* 1:137.

24. Wiggins, P. A., and P. C. Nelson. 2006. Generalized theory of semiflexible polymers. *Phys Rev E* 73:031906.

25. Mazur, A. K. 2007. Wormlike chain theory and bending of short DNA. *Phys Rev Lett* 98:218102.

26. Oberhauser, A. F., P. E. Marszalek, H. P. Erickson, and J. M. Fernandez. 1998. The molecular elasticity of the extracellular matrix protein tenascin. *Nature* 393:181.

27. Rief, M., J. M. Fernandez, and H. E. Gaub. 1998. Elastically coupled two-level systems as a model for biopolymer extensibility. *Phys Rev Lett* 81:4764.

28. Staple, D. B., S. H. Payne, A. L. C. Reddin, and H. J. Kreuzer. 2008. Model for stretching and unfolding the giant multidomain muscle protein using single-molecule force spectroscopy. *Phys Rev Lett* 101:248301.

29. Bell, G. I. 1978. Models for the specific adhesion of cells to cell. *Science* 200:618.

30. Lu, H., B. Isralewitz, A. Krammer, V. Vogel, and K. Schulten. 1998. Unfolding of titin immunoglobulin domains by steered molecular dynamics simulation. *Biophys J* 75:662.

31. Frauenfelder, H., S. G. Sligar, and P. G. Wolynes. 1991. The energy landscapes and motions of proteins. *Science* 254:1598.

32. Ackbarow, T., X. Chen, S. Keten, and M. J. Buehler. 2007. Hierarchies, multiple energy barriers, and robustness govern the fracture mechanics of {alpha}-helical and beta-sheet protein domains. *Proc Natl Acad Sci U S A* 104:16410.

33. Jarzynski, C. 1997. Nonequilibrium equality for free energy differences. *Phys Rev Lett* 78:2690.

34. Liphardt, J., S. Dumont, S. B. Smith, I. Tinoco, and C. Bustamante. 2002. Equilibrium information from nonequilibrium measurements in an experimental test of Jarzynski's equality. *Science* 296:1832.

35. Gebhardt, J. C. M., T. Bornschlogl, and M. Rief. 2010. Full distance-resolved folding energy landscape of one single protein molecule. *Proc Natl Acad Sci U S A* 107:2013.

36. Hummer, G., and A. Szabo. 2005. Free energy surfaces from single-molecule force spectroscopy. *Acc Chem Res* 38:504.

37. Hummer, G., and A. Szabo. 2001. Free energy reconstruction from nonequilibrium single-molecule pulling experiments. *Proc Natl Acad Sci U S A* 98:3658.

38. Harris, N. C., Y. Song, and C.-H. Kiang. 2007. Experimental free energy surface reconstruction from single-molecule force spectroscopy using Jarzynski's equality. *Phys Rev Lett* 99:068101.

39. Collin, D., F. Ritort, C. Jarzynski, S. B. Smith, I. Tinoco, and C. Bustamante. 2005. Verification of the Crooks fluctuation theorem and recovery of RNA folding free energies. *Nature* 437:231.

40. Crooks, G. E. 1999. Entropy production fluctuation theorem and the nonequilibrium work relation for free energy differences. *Phys Rev E* 60:2721.

41. Fernandez, J. M., and H. Li. 2004. Force-clamp spectroscopy monitors the folding trajectory of a single protein. *Science* 303:1674.

42. Hyeon, C., G. Morrison, D. L. Pincus, and D. Thirumalai. 2009. Refolding dynamics of stretched biopolymers upon force quench. *Proc Natl Acad Sci U S A* 106:20288.

43. Cecconi, C., E. A. Shank, C. Bustamante, and S. Marqusee. 2005. Direct observation of the three-state folding of a single protein molecule. *Science* 309:2057.

44. Karunatilaka, K. S., A. Solem, A. M. Pyle, and D. Rueda. 2010. Single-molecule analysis of Mss116-mediated group II intron folding. *Nature* 467:935.

45. Suresh, S. 2007. Biomechanics and biophysics of cancer cells. *Acta Mater* 55:3989.
46. Suresh, S. 2007. Nanomedicine: Elastic clues in cancer detection. *Nat Nanotechnol* 2:748.
47. Cross, S. E., Y.-S. Jin, J. Rao, and J. K. Gimzewski. 2007. Nanomechanical analysis of cells from cancer patients. *Nat Nanotechnol* 2:780.
48. Iyer, S., R. M. Gaikwad, V. Subba Rao, C. D. Woodworth, and I. Sokolov. 2009. Atomic force microscopy detects differences in the surface brush of normal and cancerous cells. *Nat Nanotechnol* 4:389.
49. Stolz, M., R. Raiteri, A. U. Daniels, M. R. VanLandingham, W. Baschong, and U. Aebi. 2004. Dynamic elastic modulus of porcine articular cartilage determined at two different levels of tissue organized by indentation-type atomic force microscopy. *Biophys J* 86:3269.
50. Stolz, M., R. Gottardi, R. Raiteri, S. Miot, I. Martin, R. Imer, U. Staufer et al. 2009. Early detection of aging cartilage and osteoarthritis in mice and patient samples using atomic force microscopy. *Nat Nanotechnol* 4:186.
51. Kramer, A., I. Liashkovich, H. Oberleithner, S. Ludwig, I. Mazur, and V. Shahin. 2008. Apoptosis leads to a degradation of vital components of active nuclear transport and a dissociation of the nuclear lamina. *Proc Natl Acad Sci U S A* 105:11236.
52. Buehler, M. J., and Y. C. Yung. 2009. Deformation and failure of protein materials in physiologically extreme conditions and disease. *Nat Mater* 8:175.
53. Adamcik, J., J.-M. Jung, J. Flakowski, P. De Los Rios, G. Dietler, and R. Mezzenga. 2010. Understanding amyloid aggregation by statistical analysis of atomic force microscopy images. *Nat Nanotechnol* 5:423.
54. Iijima, S. 1991. Helical Microtubules of Graphtic Carbon. *Nature* 354:56.
55. Yang, W., K. Ratinac, S. Ringer, P. Thordarson, J. Gooding, and F. Braet. 2010. Carbon nanomaterials in biosensors: Should you use nanotubes or graphene? *Angew Chem Int Ed* 49:2114.
56. Patolsky, F., G. Zheng, and C. M. Lieber. 2006. Nanowire-based biosensors. *Anal Chem* 78:4260.
57. Waggoner, P. S., and H. G. Craighead. 2007. Micro- and nanomechanical sensors for environmental, chemical, and biological detection. *Lab Chip* 7:1238.
58. Carrascosa, L. G., M. Moreno, M. Alvarez, and L. M. Lechuga. 2006. Nanomechanical biosensors: a new sensing tool. *TrAC Trends Anal Chem* 25:196.
59. Craighead, H. G. 2000. Nanoelectromechanical systems. *Science* 290:1532.
60. Mirkin, C. A., and C. M. Niemeyer. 2007. *Nanobiotechnology II: More Concepts and Applications*. Weinheim, Germany: Wiley-VCH Verlag GmbH & Co.
61. Rosi, N. L., and C. A. Mirkin. 2005. Nanostructures in biodiagnostics. *Chem Rev* 105:1547.
62. Liong, M., J. Lu, M. Kovochich, T. Xia, S. G. Ruehm, A. E. Nel, F. Tamanoi, and J. I. Zink. 2008. Multifunctional inorganic nanoparticles for imaging, targeting, and drug delivery. *ACS Nano* 2:889.
63. Peer, D., J. M. Karp, S. Hong, O. C. Farokhzad, R. Margalit, and R. Langer. 2007. Nanocarriers as an emerging platform for cancer therapy. *Nat Nanotechnol* 2:751.
64. Yang, J., C.-H. Lee, H.-J. Ko, J.-S. Suh, H.-G. Yoon, K. Lee, Y.-M. Huh, and S. Haam. 2007. Multifunctional magneto-polymeric nanohybrids for targeted detection and synergistic therapeutic effects on breast cancer. *Angew Chem Int Ed* 46:8836.
65. Schatz, G. C. 2007. Using theory and computation to model nanoscale properties. *Proc Natl Acad Sci U S A* 104:6885.
66. Mielke, S. L., T. Belytschko, and G. C. Schatz. 2007. Nanoscale fracture mechanics. *Annu Rev Phys Chem* 58:185.
67. Gibson, L. J., and M. F. Ashby. 1999. *Cellular Solids: Structure and Properties*. Cambridge, UK: Cambridge University Press.

68. Phillips, R. 2001. *Crystals, Defects, and Microstructures: Modeling Across Scales.* Cambridge, UK: Cambridge University Press.
69. Wong, E. W., P. E. Sheehan, and C. M. Lieber. 1997. Nanobeam mechanics: Elasticity, strength, and toughness of nanorods and nanotubes. *Science* 277:1971.
70. Gere, J. M. 2003. *Mechanics of Materials.* Thomson Learning: Belmont, CA.
71. Lu, J. P. 1997. Elastic properties of carbon nanotubes and nanoropes. *Phys Rev Lett* 79:1297.
72. Poncharal, P., Z. L. Wang, D. Ugarte, and W. A. de Heer. 1999. Electrostatic deflections and electromechanical resonances of carbon nanotubes. *Science* 283:1513.
73. Peng, H. B., C. W. Chang, S. Aloni, T. D. Yuzvinsky, and A. Zettl. 2006. Ultrahigh frequency nanotube resonators. *Phys Rev Lett* 97:087203.
74. Novoselov, K. S., A. K. Geim, S. V. Morozov, D. Jiang, Y. Zhang, S. V. Dubonos, I. V. Grigorieva, and A. A. Firsov. 2004. Electric field effect in atomically thin carbon films. *Science* 306:666.
75. Geim, A. K., and K. S. Novoselov. 2007. The rise of graphene. *Nat Mater* 6:183.
76. Geim, A. K. 2009. Graphene: Status and prospects. *Science* 324:1530.
77. Komaragiri, U., M. R. Begley, and J. G. Simmonds. 2005. The mechanical response of freestanding Circular elastic films under point and pressure loads. *J Appl Mech* 72:203.
78. Cadelano, E., P. L. Palla, S. Giordano, and L. Colombo. 2009. Nonlinear elasticity of monolayer graphene. *Phys Rev Lett* 102:235502.
79. Song, C., T. Kwon, J.-H. Han, M. Shandell, and M. S. Strano. 2009. Controllable synthesis of single-walled Carbon Nanotube Framework Membranes and Capsules. *Nano Lett* 9:4279.
80. Peng, B., M. Locascio, P. Zapol, S. Li, S. L. Mielke, G. C. Schatz, and H. D. Espinosa. 2008. Measurements of near-ultimate strength for multiwalled carbon nanotubes and irradiation-induced crosslinking improvements. *Nat Nanotechnol* 3:626.
81. Lee, C. K., S. R. Shin, J. Y. Mun, S.-S. Han, I. So, J.-H. Jeon, T. M. Kang et al. 2009. Tough supersoft sponge fibers with tunable stiffness from a DNA self-assembly technique. *Angew Chem Int Ed* 48:5116.
82. Appell, D. 2002. Nanotechnology: Wired for success. *Nature* 419:553.
83. Haiss, W. 2001. Surface stress of clean and adsorbate-covered solids. *Rep Prog Phys* 64:591.
84. Ibach, H. 1997. The role of surface stress in reconstruction, epitaxial growth, and stabilization of mesoscopic structures. *Surf Sci Rep* 29:193.
85. Freund, L. B., and S. Suresh. 2003. *Thin Film Materials.* Cambridge, UK: Cambridge University Press.
86. Diao, J., K. Gall, and M. L. Dunn. 2003. Surface-stress-induced phase transformation in metal nanowires. *Nat Mater* 2:656.
87. Miller, R. E., and V. B. Shenoy. 2000. Size-dependent elastic properties of nanosized structural elements. *Nanotechnology* 11:139.
88. Yun, G., and H. S. Park. 2009. Surface stress effects on the bending properties of fcc metal nanowires. *Phys Rev B* 79:195421.
89. Wang, G., and X. Li. 2007. Size dependency of the elastic modulus of ZnO nanowires: Surface stress effect. *Appl Phys Lett* 91:231912.
90. He, J., and C. M. Lilley. 2008. Surface effect on the elastic behavior of static bending nanowires. *Nano Lett* 8:1798.
91. Chen, C. Q., Y. Shi, Y. S. Zhang, J. Zhu, and Y. J. Yan. 2006. Size dependence of Young's modulus in ZnO nanowires. *Phys Rev Lett* 96:075505.
92. Park, H. S., W. Cai, H. D. Espinosa, and H. Huang. 2009. Mechanics of crystalline nanowires. *MRS Bull* 34:178.
93. Gosline, J., P. Guerette, C. Ortlepp, and K. Savage. 1999. The mechanical design of spider silks: from fibroin sequence to mechanical function. *J Exp Biol* 202:3295.

94. Shao, Z., and F. Vollrath. 2002. Surprising strength of of silkworm silk. *Nature* 418:741.
95. Termonia, Y. 1994. Molecular modeling of spider silk elasticity. *Macromolecules* 27:7378.
96. van Beek, J. D., S. Hess, F. Vollrath, and B. H. Meier. 2002. The molecular structure of spider dragline silk: Folding and orientation of the protein backbone. *Proc Natl Acad Sci U S A* 99:10266.
97. Keten, S., Z. Xu, B. Ihle, and M. J. Buehler. 2010. Nanoconfinement controls stiffness, strength and mechanical toughness of [beta]-sheet crystals in silk. *Nat Mater* 9:359.
98. Knowles, T. P., A. W. Fitzpatrick, S. Meehan, H. R. Mott, M. Vendruscolo, C. M. Dobson, and M. E. Welland. 2007. Role of intermolecular forces in defining material properties of protein nanofibrils. *Science* 318:1900.
99. Nova, A., S. Keten, N. M. Pugno, A. Redaelli, and M. J. Buehler. 2010. Molecular and nanostructural mechanisms of deformation, strength and toughness of spider silk fibrils. *Nano Lett* 10:2626.
100. Becker, N., E. Oroudjev, S. Mutz, J. P. Cleveland, P. K. Hansma, C. Y. Hayashi, D. E. Makarov, and H. G. Hansma. 2003. Molecular nanosprings in spider capture-silk threads. *Nat Mater* 2:278.
101. Zhou, H., and Y. Zhang. 2005. Hierarchical chain model of spider capture silk elasticity. *Phys Rev Lett* 94:028104.
102. Knowles, T. P. J., T. W. Oppenheim, A. K. Buell, D. Y. Chirgadze, and M. E. Welland. 2010. Nanostructured films from hierarchical self-assembly of amyloidogenic proteins. *Nat Nanotechnol* 5:204.
103. Lv, S., D. M. Dudek, Y. Cao, M. M. Balamurali, J. Gosline, and H. Li. 2010. Designed biomaterials to mimic the mechanical properties of muscles. *Nature* 465:69.
104. Sholl, D., and J. A. Steckel. 2009. *Density Functional Theory: A Practical Introduction.* Boboken, NJ: Wiley Interscience.
105. Allen, M. P., and D. J. Tildesley. 1987. *Computer Simulation of Liquids.* Oxford: Oxford University Press.
106. Zienkiewicz, O. C., and R. L. Taylor. 1989. *The Finite Element Method: Basic Formulation and Linear Problems.* London: McGraw-Hill.
107. Bathe, K.-J. 1996. *Finite Element Procedures.* Eaglewood Cliffs, NJ: Prentice Hall.
108. Michel, J. P., I. L. Ivanovska, M. M. Gibbons, W. S. Klug, C. M. Knobler, G. J. L. Wuite, and C. F. Schmidt. 2006. Nanoindentation studies of full and empty viral capsids and the effects of capsid protein mutations on elasticity and strength. *Proc Natl Acad Sci U S A* 103:6184.
109. Roos, W. H., R. Bruinsma, and G. J. L. Wuite. 2010. Physical virology. *Nat Phys* 6:733.
110. Phillips, R., M. Dittrich, and K. Schulten. 2002. Quasicontinuum representation of atomic-scale mechanics: From proteins to dislocations. *Annu Rev Mater Res* 32:219.
111. Verlet, L. 1967. Computer "experiments" on classical fluids. I. thermodynamical properties of Lennard-Jones molecules. *Phys Rev* 159:98.
112. McCammon, J. A., and S. Harvey. 1987. *Dynamics of Proteins and Nucleic Acids.* Cambridge: Cambridge University Press.
113. Karplus, M., and G. A. Petsko. 1990. Molecular dynamics simulations in biology. *Nature* 347:631.
114. Karplus, M., and J. A. McCammon. 2002. Molecular dynamics simulations of biomolecules. *Nat Struct Mol Biol* 9:646.
115. Sotomayor, M., and K. Schulten. 2007. Single-molecule experiments in vitro and in silico. *Science* 316:1144.
116. Brooks, B., and M. Karplus. 1983. Harmonic dynamics of proteins: Normal modes and fluctuations in bovine pancreatic trypsin inhibitor. *Proc Natl Acad Sci U S A* 80:6571.
117. Ma, J. P. 2005. Usefulness and limitations of normal mode analysis in modeling dynamics of biomolecular complexes. *Structure* 13:373.

118. Bahar, I., T. R. Lezon, A. Bakan, and I. H. Shrivastava. 2010. Normal mode analysis of biomolecular structures: Functional mechanisms of membrane proteins. *Chem Rev* 110:1463.
119. Hayward, S., and N. Go. 1995. Collective variable description of native protein dynamics. *Annu Rev Phys Chem* 46:223.
120. Weiner, J. H. 1983. *Statistical Mechanics of Elasticity*. Mineola, NY: Dover publication.
121. Chandler, D. 1987. *Introduction to Modern Statistical Mechanics*. Oxford, UK: Oxford University Press.
122. Reichl, L. E. 2009. *A Modern Course in Statistical Physics*. Weinheim, Germany: Wiley-VCH.
123. Coffey, W. T., Y. P. Kalmykov, and J. T. Waldron. 1996. *Langevin Equation with Applications in Physics, Chemistry, and Electrical Engineering*. World Scientific Publishing Co: Singapore.
124. Voth, G. A. 2009. *Coarse-Graining of Condensed Phase and Biomolecular Systems*. Boca Raton, FL: CRC Press.
125. Li, C., and T.-W. Chou. 2003. "Single-walled carbon nanotubes as ultrahigh frequency nanomechanical resonators" *Phys Rev B* 68:073405.
126. Li, C., and T.-W. Chou. 2004. "Vibrational behaviors of multiwalled-carbon-nanotube-based nanomechanical resonators" *Appl Phys Lett* 84:121.
127. Liu, B., H. Jiang, Y. Huang, S. Qu, M. F. Yu, and K. C. Hwang. 2005. "Atomic-scale finite element method in multiscale computation with applications to carbon nanotubes" *Phys Rev B* 72:035435.
128. Gould, P. L. 1983. *Introduction to Linear Elasticity*. New York, NY: Springer-Verlag.
129. Oden, J. T., E. B. Becker, and G. F. Carey. 1981. *Finite Elements: An Introduction (Volume 1)*. Eaglewood Cliffs, New Jersey: Prentice Hall.
130. Tadmor, E. B., and R. E. Miller. 2009. Quasicontinuum. http://www.qcmethod.com (accessed June 8, 2011).

Section I

Simulations in Biological Sciences

2 Modeling the Interface between Biological and Synthetic Components in Hybrid Nanosystems

Rogan Carr, Jeffrey Comer, and Aleksei Aksimentiev

CONTENTS

2.1 INTRODUCTION

Hybrid structures that combine biological and inorganic materials are ubiquitous in the biosphere, from the carbonated hydroxyapatite that makes up the inorganic structure of bones to the calcium carbonate created by coral polyps and coralline algae that form the structure of coral reefs. The ability to design materials that are biocompatible, or able to be incorporated into organisms without provoking an immune response, is important for medicine and therapeutics, for example, in prosthetics and implanted medical devices such as pacemakers. However, designing synthetic materials to interact favorably with biological components is made difficult by the fact that many biological processes, including immune response, are dictated by interactions between biomolecules and their environment on the nanometer scale. Therefore, our ability to fabricate materials with nanometer precision would provide hope for a future in which man-made devices can easily interface with the biological world.

The technology driving the miniaturization of materials manufacturing has advanced to the point where it is now possible to manufacture synthetic device structures on the nanometer scale. Researchers can create surfaces patterned with 2- to 5-nm diamond grains [1], nanopores smaller than the width of duplex DNA [2], macroscopic-length channels with cross-sections smaller than 10 nm [3], and hollow polyoxometalate macromolecular spheres measuring 4 nm across [4]. Although manufacturing structures with feature sizes of several nanometers is now possible, how these structures interact with biomolecules remains mostly uncharted territory. The watery environments necessary for biology and the function of biomolecules are incompatible with the imaging methods usually applied to the nanoscale, such as electron microscopy; thus, we must turn to computational models to visualize and quantify the nanoscale details of the interactions between biomolecules and synthetic materials. With accurate models of this sort in hand, it will be possible to optimize next-generation devices that incorporate biological and synthetic nano-structured components and manufacture them on a commercial scale.

In this chapter, we draw upon our past research experience to detail how molecu-lar dynamics (MD) can be used as a tool to provide insight into the design and function of nanodevices that are embedded in biological systems, biological systems that are embedded in synthetic nanostructures, and hybrid devices that incorporate both biological and synthetic components. To this end, we describe novel, contem-porary efforts to model the interface between biological and synthetic materials in (1) immunosurfaces used for detection of live bacteria; (2) nanofluidic systems for protein transport; (3) solid-state nanopores for sequencing DNA; and (4) synthetic analogs of biological ion channels.

2.2 IMMUNOSURFACES FOR BACTERIAL DETECTION

The immunosurface is an example of a hybrid material that combines biological components from an organism's immune system, that is, antibodies, with inorganic nanostructures to perform sensing and detection of foreign substances. A schematic illustration of an immunosurface is shown in Figure 2.1a. Antibodies are covalently tethered to an inorganic substrate so that the antibodies remain exposed in solution to preserve their activity and stay bound to a particular region of the substrate. This also prevents aggregation of the antibodies, as occurs in solution.

Proteins, such as antibodies, tend to bind to inorganic surfaces strongly and non-specifically [5]. To prevent adhesion of the antibodies to substrates, the inorganic substrates are functionalized, or covalently coated, with a layer of protective mole-cules known as the functionalization layer. This functionalization layer also provides the binding site for the molecules that tether the antibodies to the surface.

Antibodies are a powerful addition to a sensor because they bind their targets strongly and specifically in a way that is currently not possible to achieve with a purely synthetic device. Antibodies, or immunoglobulins, are proteins typically composed of two heavy chains and two light chains, which are cross-linked with disulfide bonds to form a Y-shape. The tips of the "Y" are known as variable regions because the amino acid sequence varies greatly between antibodies in the same organism. This high variability gives an antibody its high specificity for the ligand, or antigen, to which it binds. However, if an antibody binds to an inorganic surface, it will rapidly lose activity as the protein becomes unstructured or oriented in a way that the antigen cannot diffuse into the variable region.

In collaboration with the research groups of Bashir, Hamers, Carlisle, and King, we have been working to create an immunosurface capable of detecting different

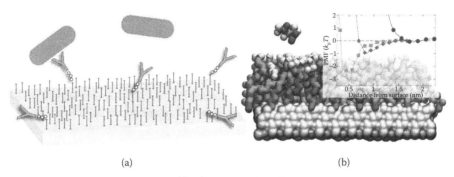

(a)   (b)

FIGURE 2.1 Immunosurfaces for bacterial detection. (a) A schematic showing an immunosurface that consists of a solid-state substrate, such as glass or diamond (shown as a gray slab), functionalized with organic coating (lines) to which linker molecules (circles) anchor antibodies (Ys) against pathogens like bacteria (ovals). The individual components are not drawn to scale. (b) A close-up view of an all-atom model of an immunosurface used for molecular dynamics simulations. The inorganic substrate (ultra nanocrystalline diamond [UNCD]) is shown as a gray molecular surface, the aminodecane functionalization layer is shown as gray chains with nitrogen in dark gray and hydrogen in white, and an asparagine peptide is shown as tubes with carbon in gray, nitrogen in dark gray, oxygen in black, and hydrogen in white. The water and ions that fill the system are not shown. The inset plots the potential of mean force (PMF) for bringing an asparagine molecule from the bulk solution to the functionalized UNCD (circles), functionalized silica (diamonds), or degraded functionalized silica substrate (squares). (Adapted from Radadia, A. D. et al. 2011. *Adv Funct Mater* 21:1040–1050.)

strains of bacteria for monitoring the purity of water and food supplies [1]. To develop stable bacteria-capturing immunosurfaces, we built and modeled immunosurfaces created by attaching antibodies to functionalized ultra nanocrystalline diamond (UNCD) and glass substrates. UNCD substrates are novel films composed of 2- to 5-nm-diameter diamond grains on which alkene functionalization layers show excellent stability [6]. For comparison to a standard substrate, immunosurfaces were also made on Corning GAPS II amine-terminated glass (GAPSG) slides [1]. To test the long-term stability of the immunosurfaces, each immunosurface was submerged in phosphate buffered saline and monitored over a period of 2 weeks. The half-life of the antibody activity was found to be around 4 days for the GAPSG substrate as compared to about 11 days for the UNCD substrate. While there was about 25% loss of antibodies from the GAPSG substrate in the first week and 13% loss from the UNCD substrate in the second week, these losses alone are not enough to explain the magnitude of degradation in immunosurface performance [1].

One explanation for this loss in performance is that the antibodies could be binding to the substrates, somehow bypassing the functionalization layer meant to prevent binding. To determine if protein–substrate interactions could be driving the degradation of the immunosurfaces, we created all-atom models of the functionalized UNCD and silica substrates submerged in a buffer solution. Modeling antibody proteins with all-atom MD simulations is challenging because antibodies are large and complex proteins, which require long-timescale simulations even to study diffusive motion. To simplify the problem, we have studied protein–substrate interactions using individual amino acids in the place of an entire antibody protein. By studying the substrate interactions with one amino acid from each of the basic types—positively charged, negatively charged, polar, nonpolar, and aromatic—we can approximate the interaction of a given protein with the substrate by looking at the amino-acid composition of the protein surface [5]. A typical simulation system is shown in Figure 2.1b. The all-atom model consists of an amino acid in solution above a functionalized substrate.

To quantify the affinity of amino acids for the two substrates, we computed the potential of mean force (PMF) for each representative type of amino acid as a function of the distance from the substrate [1]. The PMF calculated in such a way can be thought of as the change in free energy for bringing the amino acids from bulk solution to the respective substrates. If the PMF decreases as the amino acid nears the surface, then the interaction is attractive, and the amino acid will bind to the surface if the interaction is greater than about $1 k_BT$ (the thermal energy); if the PMF increases as the amino acid approaches the surface, then the interaction is repulsive on average, and the amino acid will tend not to bind. As an example, the inset in Figure 2.1b, we have plotted the PMF for asparagine near a functionalized UNCD substrate, a GAPSG substrate, and a degraded GAPSG substrate, where 25% of the functionalization layer has been removed. The interaction of asparagine with the functionalized UNCD substrate can be seen to be repulsive, as the PMF increases as the amino acid approaches the surface. The interaction of asparagine with the GAPSG substrate is slightly attractive, as the PMF decreases to $-1 k_BT$. However, the PMF between asparagine and the degraded GAPSG substrate decreases to less than $-3 k_BT$, signifying that the amino acid will bind much more strongly on the degraded substrate.

By performing PMF calculations for bringing one amino acid of each type from bulk solution to each substrate, we can quantitatively characterize the interactions of proteins with the substrates. Furthermore, the data we get through our MD simulations shows us not only the binding affinities but also the origin of the differences between the amino acid–substrate interactions. By simulating the functionalized substrates submerged in a buffer solution without the presence of protein, we can characterize the dynamics of the functionalized substrates for different functionalization densities. In the case of the functionalized UNCD and GAPSG substrates, we find that the UNCD substrate remains dry, but water pockets form in the functionalization layer of GAPSG and increase in size for lower functionalization layer densities, providing a route for protein–substrate binding [1].

In this work, we performed MD simulations of model systems to quantitatively characterize the energetics and dynamics of antibodies on inorganic substrates. This method of simulating a model system to identify dynamical features and following up with free-energy calculations to quantify specific interactions is a robust way to characterize the properties of the interface between solutes in solution and inorganic surfaces. Here, we characterized the interface between functionalized substrates and biological molecules in solution, but such methods could readily be adapted to study the interface between any solutes and any type of surface, inorganic or biological.

2.3 MODELING TRANSPORT THROUGH NANOCHANNELS

Riding the wave of miniaturization innovations developed for the electronics indus-try, researchers are seeking to miniaturize the wet laboratory, which would allow processes like chemical synthesis, sample purification, and protein crystallization to be performed on single microchips and perhaps even enable processes hereto-fore impossible. Although the technology required to fabricate nanoscale wires for transporting electrons is well developed, transporting water and water-borne sol-utes through nanoscale pipes in nanofluidic devices remains a significant challenge. Nanofluidic devices are composed of reservoirs connected by a network of "pipes," called nanochannels, which have at least one dimension smaller than 100 nm. To force fluid to flow through these channels, pressure gradients, chemical gradients, and electric fields can be used. As the size of nanochannels becomes comparable to the size of transported solutes, the surface of the channels will play an active role in transport of solutes; see Figure 2.2. The interactions of the solutes with the surface can drive device degradation through nonspecific adsorption, surface accumulation, and even clogging of the nanochannel [7]. To understand such behavior and enable rapid progress in device design, it will be necessary to not only be able to create devices on such a scale but to accurately model and predict the device behavior.

As force fields for inorganic compounds that are compatible with biomolecular force fields have become available, such as those for silica [8,9], silicon nitride [10,11], and titanium dioxide [12], we can now model the solute–surface interface of nanofluidic devices in all-atom detail. While complete nanofluidic devices are too large to be simulated by all-atom MD, it is possible to simulate representative parts of nanofluidic devices to model the transport of chemicals and biomolecules and their interactions with the surfaces of the device. Both electroosmotic [13] and pressure-driven [5] transport of chemicals through the channels are amenable to the MD method.

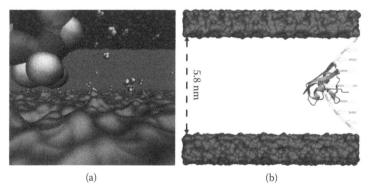

(a) (b)

FIGURE 2.2 Transport of solutes through a nanochannel. (a) A molecule's-eye view of a nanochannel. The small constriction of the nanochannel means that the channel surfaces play an important role in solute transport. (b) Molecular dynamics simulation of pressure-driven transport of proteins through a nanochannel. The small protein takes up a significant part of the nanochannel's constriction. A typical water velocity profile is shown schematically as dashed lines and arrows. (Adapted from Carr, R. et al. 2011. *IEEE Trans Nanotechnol* 10:75–82.)

To characterize the trajectories that a biomolecule takes when transported through a silica nanochannel by a pressure gradient, we have simulated this process using the all-atom MD method [5]. Figure 2.2b illustrates the setup of a typical simulation. As the protein is transported through the channel in the direction of the flow, it diffuses laterally across the channel and interacts with the silica surface. Once in contact with the surface, the protein rolls along the surface and eventually binds the surface, becoming immobile. From this state, the protein can be forced to unbind by gradually increasing the pressure gradient in the channel. The unbinding pressure gradient gives a qualitative measure of the strength of the protein–silica surface interaction. Repeating such simulations multiple times we found that the protein orients differently with respect to the surface each time it binds the surface, but always binds strongly, showing that the protein–surface binding is nonspecific. The binding of the protein to the inorganic substrate was not driven by any particular amino acid. Rather, the strength of the binding varied depending on the atomic-scale features of the inorganic surface at points where binding occurred as well as the identity of the amino acids of the protein participating in the binding [5]. In order to properly model the transport of proteins and other solutes through nanofluidic devices, it will be necessary to characterize the atomic-scale features of the inorganic surfaces, such as surface roughness, charge distribution, and hydrophobicity, and understand how each contributes to strength of the binding.

To fully characterize the interactions between an inorganic surface and a protein, we would ideally determine the two-dimensional (2D) landscape of binding free energy for each amino acid over a representative fragment of the surface. Knowing such 2D landscapes and the amino acid content of the protein surface, we could characterize the binding free energy of the entire protein to the inorganic surface in statistical terms. Currently, such calculations are prohibitively expensive for heterogeneous surfaces, including the one we used to construct our nanochannel. To demonstrate that such an approach is in principle possible, we characterized the interactions of individual amino acids with a featureless surface (referred to as a "phantom" surface [14]), which models a frictionless hydrophobic material.

To determine the binding free energy for each of the 20 amino acids, we have computed the PMF of individual amino acids as a function of the distance between the amino acid and the phantom surface. A schematic overview of these calculations is shown in Figure 2.3. Typical conformations adopted near the phantom surface by two amino acids, phenylalanine (Phe) and isoleucine (Ile), are shown in Figure 2.3a. In these conformations, the nonpolar groups of the amino acids are located near the surface, whereas their polar backbones are submerged in the solution.

The PMF as a function of the distance from the surface for the two amino acids, Phe and Ile, is shown in Figure 2.3b. This PMF reveals the change in free energy for bringing an amino acid from bulk solution to the phantom surface. Far from the surface, the PMF is flat, as expected for a molecule surrounded by water. As the molecule approaches the surface, the PMF first decreases considerably and then increases rapidly where the steric clashes between the inorganic surface and the protein dominate. The change in the PMF from the flat region to the minimum value

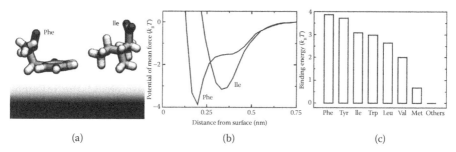

(a)   (b)   (c)

FIGURE 2.3 Calculations of amino-acid binding to a featureless, frictionless "phantom" surface. (a) Typical configurations adopted by phenylalanine (Phe) and isoleucine (Ile) near the phantom surface. Carbon atoms are shown in gray, nitrogen in dark gray, oxygen in black, and hydrogen in white. The phantom surface is shown in black; the interface region is shown as a fade from black to white. (b) The PMF of Phe and Ile as a function of the distance of their α-carbon atoms to the phantom surface. (c) The maximum depth of the PMF curve for all amino acids near the phantom surface (see text).

near the surface is considered to be the binding energy for the particular amino acid. We have calculated such PMF curves for all 20 amino acids; the values of their PMF minima are shown in Figure 2.3c.

The PMF curves of 13 amino acids (denoted as "others" in Figure 2.3c) did not exhibit a minimum near the surface, indicating that binding to the phantom surface was not energetically favorable. The PMFs of Phe, Tyr, Ile, Tre, Leu, and Val displayed relatively deep minima (>2 $k_B T$), whereas the binding energy of methionine (Met) was rather weak, 0.67 $k_B T$. A common feature of the amino acids that displayed strong affinity for the phantom surface was either a long nonpolar side chain or an aromatic ring. Thus, no charged or polar amino acids were found to bind to the phantom surface except tyrosine (Tyr), which contains a hydrophobic phenyl ring. Furthermore, amino acids with relatively short nonpolar side chains, such as alanine (Ala), also did not bind to the surface. Such calculations, if repeated with a heterogeneous, all-atom substrate in place of a phantom surface, could reveal the effects of surface heterogeneity and atomic-scale roughness on protein binding.

Calculating the PMF for bringing amino acids from a solution onto a surface, as we have done for the phantom surface here, and in Section 2.2, not only characterizes the strength of amino acid binding to the surface but gives us a free-energy profile for the interaction as a function of distance from the surface. This free-energy profile takes into account the effects of the solution and features of the surface, and therefore could be used to create coarse-grain models for solute transport through nanochannels that implicitly take all-atom details into account. Indeed, models have already been built for processes like the transport of ions through ion channel proteins using the electrostatic profile inside the protein pore [15]. The extension of such models to the transport of solutes through nanofluidic systems using intersolute PMFs and the solute–surface PMFs, as we have calculated here, would be of great use for modeling nanofluidic transport systems on the same temporal and spatial scales as in experiment.

2.4 TRANSLOCATION OF DNA THROUGH SYNTHETIC NANOPORES

DNA is the data storage medium of all living things, containing information for constructing the cell's machinery, proteins, and ribozymes, as well as information pertaining to the organization and transcription of the DNA itself. All this information is encoded in the sequence of nucleotides (A, C, G, or T), which make up DNA. Because each nucleotide is about 1 nm along its longest dimension and differs from the others by only a few atoms, determining the sequence of single DNA molecules falls within the scope of nanobiotechnology.

Inexpensive DNA sequencing promises to make the information in DNA available for use in personal medicine, with numerous potential benefits to the efficacy and efficiency of health care [16–18]. Of the methods being developed to permit the use of DNA sequencing in personal medicine [17,19], nanopore methods, which involve threading DNA through nanoscale pores [20–26], arguably have the greatest potential to revolutionize sequencing in terms of speed, flexibility, and cost. In 2009, the National Human Genome Research Institute (National Institutes of Health) awarded a total of 19.353 million dollars to 10 groups for the development of advanced sequencing technology [27]. More than half of the resources—10.753 million dollars—were awarded to four projects involving the development of nanopore technology for DNA sequencing.

Figure 2.4 shows a diagram of a so-called DNA translocation experiment. Electrodes are immersed in electrolytic solution on each side of the membrane, allowing a transmembrane bias voltage to be imposed. When DNA molecules, which are negatively charged, are added to the solution on the side of the membrane containing the negative electrode, some molecules are forced through the pore by the electric field and enter the compartment containing the positive electrode. As single molecules of nucleic acids pass through the pore, step-like transients in the ion current through the pore are measured between the electrodes. The duration and magnitude of these transients can be used to determine the translocating molecule's length [28], orientation [29,30], and nucleotide sequence details [21,24,31–39]. Although detection of the sequence through the ion current was the first method to be proposed for reading sequence information in nanopores [31], other methods have also been suggested, including the use of fluorescent markers [24], measurement of the electrical signal induced on electrodes embedded in the pore walls [26,40], and transverse tunneling current measurements [22,23].

In the development of nanopore sequencing technology, computation has permitted the visualization of the process of nanopore translocation [10,11,14,41–43] and the prediction of signals that are to be used for sequencing DNA [11,14,22,40,44,45], such as ion currents. However, simulations of DNA translocation experiments are complicated by the fact that we must treat phenomena that occur on a range of temporal and spatial scales.

The capture of DNA molecules by the electric field near the pore is a slow, diffusion-limited process [46]. However, once within the capture radius [46], DNA molecules are actively driven toward the pore, typically on a millisecond timescale. Once a DNA molecule enters the pore opening, the high electric field near the pore constriction

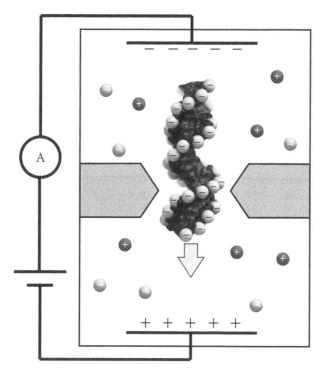

FIGURE 2.4 Schematic of a typical DNA–nanopore experiment. A thin insulating membrane separates a chamber into two electrolyte-filled compartments. An external electric potential is applied transverse to the membrane, causing migration of negatively charged DNA from the compartment containing the negative electrode to the compartment containing the positive electrode via a small pore in the membrane. This external potential also causes a current of ions to flow through the pore, which can be measured by sensitive electronics. The ion current is sensitive to atomic-scale features of the DNA near the constriction of the pore; thus, the current measurements can be used to detect the DNA and to give information about its conformation and sequence.

($\sim 10^5$ V/cm) causes the DNA to begin threading through the pore in ~ 1–100 ns [11,14]. As for spatial scales, the dimensions of the pore constriction can be near the diameter of DNA (or smaller); therefore, the atomic details of the nanopore surface can affect the DNA motion [10,11]. Any method for determining the sequence-dependent ion current clearly must also have atomic-scale spatial resolution.

The minimum timescale for identifying a single DNA base depends on the method used to read the identity of the base. To discriminate a single DNA base using the ion current or the signal induced by the DNA at electrodes embedded in a nanopore, the measurement time must be long enough to average over noise due to fluctuations of the DNA conformation, rearrangements of ions and water molecules near the DNA, and other electronic or thermal noise associated with the measurement devices. It seems unlikely that accurate base discrimination using any methods proposed thus far can be performed in less than 100 ns.

No one computational method can currently cover the wide range of spatial and temporal scales relevant to capture, threading, and readout of the DNA sequence. One way to deal with the range of spatial and temporal scales pertinent to DNA translocation experiments and nanopore sequencing is to separate the problem into three parts, for which different computational approaches can be used. First, using models much coarser than all-atom MD, we determine how the DNA molecules approach the pore and what conformations they are likely to have when they reach the opening [11]. Next, given the initial conditions generated by coarser methods, we can perform all-atom MD simulations to determine how translocation through the pore begins and how the DNA threads through the pore [10,11,14,43]. We can then derive the distribution of conformations of the DNA near the constriction from these simulations. Finally, using each of the probable conformations, we estimate the signals in ion current and electric potential that might be used to sequence the DNA using methods more efficient than all-atom MD due to neglect of the explicit dynamics of the solvent and DNA [40].

Outside the capture radius, the electric field is sufficiently low that it does not significantly affect the DNA structure. Thus, we can expect the conformation of a DNA molecule that is much longer than its persistence length (~0.8 nm for single-stranded DNA [47,48] and 30–70 nm for double-helical B-form DNA, depending on the ionic strength of the solution [46,49]) to be similar to a random coil. A random coil may be somewhat of an idealization for single-stranded DNA, as MD simulations suggest that interactions between nucleotides can lead to a disordered structure that is nevertheless more compact than a random coil [11]. Furthermore, as the DNA molecules approach the nanopore, it is possible that the inhomogeneous electric field in the vicinity of the pore distorts the equilibrium structure [46]. However, the conformation of the DNA undergoes dramatic changes once it enters the pore, so approximating the equilibrium structure of the molecule as a random coil may be sufficient to determine initial conditions for subsequent all-atom MD simulations.

In this spirit, we have used an extensible freely jointed chain model [47] to determine plausible initial conditions for a hairpin DNA comprised of a short (10 base pair) double-helical portion and a longer unpaired portion of varying lengths [11]. Principally, we wished to determine whether there was a significant probability of the DNA to be loaded into the pore with the double-helical portion leading rather than the unpaired portion. Modeling the double-helical portion as a hard cylinder with a diameter of 2.2 nm and length of 3.4 nm and the unpaired portion as a self-avoiding extensible freely jointed chain with a diameter of 0.8 nm, we used Metropolis Monte Carlo [50] and the moves described by Zhang et al. [47] to generate thousands of independent conformations. Figure 2.5 shows a number of conformations for unpaired regions of different lengths. To estimate the probability of translocation of the DNA with the double-helical portion leading, we calculated the fraction of conformations for which the double-helical portion was closest to a distant nanopore. We found that for coils of 25, 50, and 100 nucleotides, the probabilities that the double-helical portion was closest to the nanopore were 36%, 28%, and 20%, respectively, with errors of <1%.

In the example described above, the extensible freely jointed chain model was used merely to establish the plausibility of certain initial conditions. However, it is possible to generate initial conditions explicitly using such a model. The model could

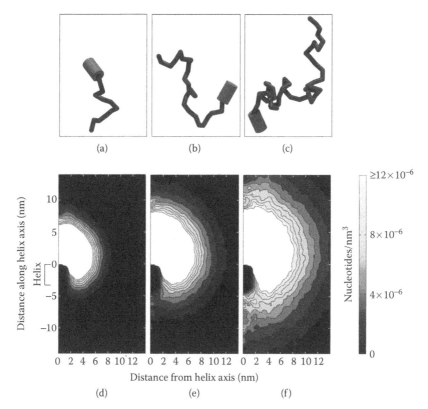

FIGURE 2.5 Generation of DNA structures using a Monte Carlo method. (a–c) Typical conformations generated by a Monte Carlo procedure for hairpin DNA molecules having unpaired regions with lengths of 25 (a), 50 (b), or 100 (c) nucleotides attached to a dou-ble-helical portion of 10 base pairs. (d–f) The density of nucleotides averaged more than 90,000 three-dimensional conformations of hairpin DNA with the three different coil lengths shown as a function of the distance from the axis of the double helix and the distance along the axis of the double helix. The position of the double-helical portion is indicated at the left. (Reprinted from Comer, J., V. Dimitrov, Q. Zhao, G. Timp, and A. Aksimentiev. 2009. *Biophys J* 96(2):593–608. With permission.)

be further improved by describing the interactions between the nucleotides more realistically (such as including electrostatic effects [47]) and simulating the applica-tion of an inhomogeneous electric field to the equilibrium structure.

Given realistic initial conformations, all-atom MD permits us to "see" how DNA threads through the pore with atomic resolution. MD simulations can therefore be essential in interpreting DNA translocation experiments. For example, we found that changing the minimum diameter of the pore constriction from 1.3 to 1.6 nm (just one or two atomic layers) resulted in the translocation of the hairpin DNA in very differ-ent manners [43]. As shown in Figure 2.6, we observed unzipping of the bases of the double helix for a 1.3-nm-diameter pore, wherein the DNA passed through the con-striction as a single strand. However, for a 1.6-nm-diameter pore, we found that the double-helical portion of the hairpin DNA was able to pass through the pore without

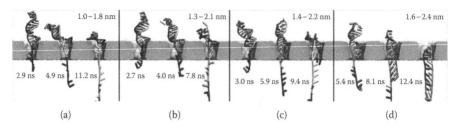

FIGURE 2.6 Dependence of hairpin DNA translocation mode on pore size. (a–d) Steered-MD simulations of hairpin DNA translocation through nanopores of different sizes. Hairpin DNA is shown in the cartoon representation, the solid-state membrane is shown as a grey molecular surface, and water and ions are not shown. Each panel contains three snapshots of the same system taken from an MD trajectory. For each pore, its bottom and top cross-section diameters are listed in the corresponding panel. For the three smaller pores (a–c), the translocation occurs by unzipping of the base pairs one by one. For the largest pore (d), the base pairs do not unzip, but the two strands pass through the constriction simultaneously with large distortions of the B-form double helix. (Reprinted from Zhao, Q., J. Comer, V. Dimitrov, A. Aksimentiev, and G. Timp. 2008. *Nucleic Acids Res* 36(5):1532–41. With permission.)

complete dissociation of all base pairs. Further simulations led to predictions about the likely modes of translocation as a function of the transmembrane voltage and minimum pore diameter [11].

Having an atomically-detailed description of DNA passage through nanopores of various sizes, it is possible to estimate what signals may be measured in experiments. MD simulations have been used to associate different levels of ion currents with different DNA–nanopore configurations [11,43]. In these examples, differences in the values of ion current were quite large, on the order of 1 nA. However, discriminating DNA sequences may require differentiating currents that differ by ~1–100 pA [21,31]. Calculating ion current essentially involves counting ion permeation events; therefore, to obtain precise estimates of ion current, we must observe many ion permeation events and, hence, run long simulations.

Coarser representations, in which the dynamics of water or ions are not explicitly modeled, can permit efficient simulation of large numbers of systems on spatial and temporal scales inaccessible to all-atom MD. For example, using DNA conformations extracted from all-atom MD, numerical solutions to the Poisson–Boltzmann equation provided estimates of the signals induced by A, T, G, and C bases on electrodes embedded in a nanopore [40]. Brownian dynamics, in which ions are explicitly modeled, whereas water molecules are not, has been used to efficiently simulate the passage of ions through narrow channels [15,51–53] and could also be used to estimate sequence-dependent ion current through nanopores containing DNA.

2.5 SYNTHETIC ION CHANNELS EMBEDDED IN BIOLOGICAL MEMBRANES

While up to now we have discussed hybrid systems that combine biological molecules and synthetic membranes, there has been a great scientific and medical interest in designing synthetic systems that can be incorporated into biological organisms.

Phospholipid bilayers and the proteins that reside in them make up the structures by which cells separate themselves from the surrounding environment and create internal compartments such as vesicles or organelles. Model phospholipid bilayer membranes that mimic biological membranes can be created in a laboratory and used as parts of hybrid devices [54]. Engineering synthetic systems that can operate in biological membranes would enable researchers and engineers to directly probe fundamental cellular processes and modify such processes for use in bionanotechnology.

A lipid bilayer membrane is virtually impervious to ions, water, soluble nutrients, and proteins. Therefore, a special class of proteins residing in the cell's membranes, known as ion channels, are required to transport ions in and out of the cell and cellular organelles. The use of synthetic devices to mimic biological ion channels would permit a step-by-step study of the atomic-scale properties that allow the channels to perform their biological function, for example, selective transport of specific ions. Furthermore, such devices would assist in the development of synthetic systems for biotechnology by exploiting the same mechanisms found in proteins. One promising candidate for such a device is the porous polyoxomolybdate (POM) nanocapsule [4].

Porous POM nanocapsules [4] are novel spherical macromolecules similar in size to the width of a biological membrane, with nanoscale features that resemble those of ion channel proteins. POM nanocapsules, shown in Figure 2.7a, are made of 12 pentagonal units connected to create a hollow spherical molecule with 20 circular pores on the surface of the capsule. These complex molecules can operate in striking similarity to biological ion channels: inorganic ions can flow in and out of the pores, which can be modified to tune the ion selectivity and reversibly blocked to control

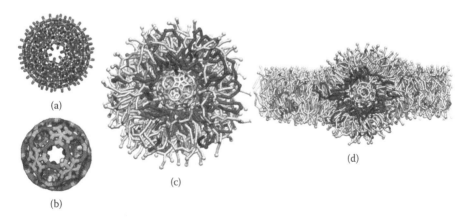

(a)

(b)

(c)

(d)

FIGURE 2.7 Embedding a porous polyoxomolybdate (POM) nanocapsule—a synthetic ion channel—in a lipid bilayer membrane. (a) All-atom structure of the porous POM nanocapsule, oriented to show a pore in the center. Molybdenum atoms are shown in white, oxygen in dark gray, and sulfur in light gray. (b) Coarse-grain model of the POM nanocapsule. Different types of coarse-grain beads are shown in different colors. (c) Structure of the capsule liposome, cutaway to show the nanocapsule (white) surrounded by a shell of positively charged surfactant (dark gray) and a secondary shell of phospholipid (white). (d) Fusion of a POM-liposome with a lipid bilayer, shown cutaway to reveal the nanocapsule in the center of the bilayer. (Adapted from Carr, R. et al. 2008. *Nano Lett* 8:3916–21.)

the flow of ions [4,55]. Assuming they could be inserted into biological membranes, these properties would make porous POM nanocapsules promising candidates for synthetic models of ion channel proteins. However, the POM nanocapsules are highly charged, which prevents them from readily incorporating themselves into the hydrophobic interior of the lipid bilayer membrane. Working in collaboration with the research groups of Weinstock, Sivaprasadarao, and Müller, we used MD to develop a method for inserting POM nanocapsules into phospholipid bilayer membranes [56].

For this research project, we used the coarse-grain MD technique, which enabled the simulation of a large lipid bilayer on microsecond timescales, a feat that is currently prohibitively expensive with all-atom MD. Coarse-grain MD is similar to all-atom MD, except that molecules are represented by clusters of particles, each of which corresponding to multiple atoms (usually four or five heavy atoms) [57,58] rather than to single atoms as in the case of all-atom MD. For example, a butane molecule (C_4H_{10}) would be represented by one bead and an octane molecule (C_8H_{18}) by two beads; a cluster of several water molecules would be condensed into a single bead having dynamical properties similar to water. Figure 2.7a and b show an all-atom and a coarse-grain representation of the POM nanocapsule. The main features of the macromolecule are captured by both models, but the coarse-grain model is much simpler, as groups of atoms have been averaged into single beads. The coarse-grain model permits more efficient simulation because (1) fewer particles are used to represent the system, and (2) the removal of short-timescale degrees of freedom (such as those pertaining to hydrogen atoms) and use of smoother potentials allow single simulation steps to span larger times (10–20 fs as opposed to 1–2 fs).

The protocol we developed for inserting the POM nanocapsules into lipid bilayer membranes relied on the observation that these macromolecules can be encapsulated with a charged surfactant to form a hydrophobic ball-like structure [59]. Using our coarse-grained model, we were able to reproduce such behavior [56]. Furthermore, we found that a mixture of a POM capsule, surfactant, and phospholipid would self-assemble into a multilayer liposome-like structure, as shown in Figure 2.7c [56]. The encapsulation of the porous POM nanocapsule in layers of surfactant and phospholipid not only makes the highly charged macromolecule soluble and nonpolar but also provides a delivery mechanism for its insertion into lipid bilayer membranes. Indeed, we have shown when the liposomal structure fuses with a lipid bilayer membrane, the fusion process can deliver the capsule to the center of the membrane [56], as shown in Figure 2.7d. However, the nanocapsule remains surrounded by surfactant and phospholipid, preventing it from accessing the solution on both sides of the membrane and thus acting as a transmembane pore. Continuing work involves decorating the capsule with covalently linked moieties similar to the surfactant, which would allow its insertion into the bilayer without blocking the pores of the capsule.

In this section, we have reviewed our methods for developing a protocol for inserting synthetic ion channels into lipid bilayer membranes. We used coarse-grained MD to quickly develop and evaluate experimental methods and protocols. Such an approach can work well for hybrid biological–inorganic systems in which an all-atom force field is not always available and the processes of interest are out of reach of conventional all-atom MD approaches. Once the methods and protocols have been thoroughly studied through coarse-grain MD, the system can then be refined to all-atom detail to model specific processes that rely on atomic-scale features.

REFERENCES

1. Radadia, A. D., C. J. Stavis, R. Carr, H. Zeng, W. P. King, J. A. Carlisle, A. Aksimentiev, R. J. Hamers, and R. Bashir. 2011. Control of nanoscale environment to improve stability of immobilized proteins on diamond surfaces. *Adv Funct Mater* 21:1040–1050.
2. Ho, C., R. Qiao, A. Chatterjee, R. J. Timp, N. R. Aluru, and G. Timp. 2005. Electrolytic transport through a synthetic nanometer-diameter pore. *Proc Natl Acad Sci U S A* 102:10445–50.
3. Haneveld, J., N. R. Tas, N. Brunets, H. V. Jansen, and M. Elwenspoek. 2008. Capillary filling of sub 10 nm nanochannels. *J Appl Phys* 104:014309.
4. Mueller, A., S. K. Das, S. Talismanov, S. Roy, E. Beckmann, H. Boegge, M. Schmidtmann et al. 2003. Trapping cations in specific positions in tuneable artificial cell channels: New nanochemistry perspectives. *Angew Chem Int Ed* 42:5039–44.
5. Carr, R., J. Comer, M. D. Ginsberg, and A. Aksimentiev. 2011. Modeling pressure-driven transport of proteins through a nanochannel. *IEEE Trans Nanotechnol* 10:75–82.
6. Stavis, C., T. L. Clare, J. E. Butler, A. D. Radadia, R. Carr, H. Zeng, W. P. King et al. 2011. Surface chemistry special feature: Surface functionalization of thin-film diamond for highly stable and selective biological interfaces. *Proc Natl Acad Sci U S A* 108:983–8.
7. Tegenfeldt, J. O., C. Prinz, H. Cao, R. L. Huang, R. H. Austin, S. Y. Chou, E. C. Cox, and J. C. Sturm. 2004. Micro- and nanofluidics for DNA analysis. *Anal Bioanal Chem* 378:1678–92.
8. Cruz-Chu, E. R., A. Aksimentiev, and K. Schulten. 2006. Water-silica force field for simulating nanodevices. *J Phys Chem B* 110:21497–508.
9. Hassanali, A. A., and S. J. Singer. 2007. Model for the water–amorphous silica interface: The undissociated surface. *J Phys Chem B* 111:11181–93.
10. Aksimentiev, A., J. B. Heng, G. Timp, and K. Schulten. 2004. Microscopic kinetics of DNA translocation through synthetic nanopores. *Biophys J* 87:2086–97.
11. Comer, J., V. Dimitrov, Q. Zhao, G. Timp, and A. Aksimentiev. 2009. Microscopic mechanics of hairpin DNA translocation through synthetic nanopores. *Biophys J* 96:593–608.
12. Park, J. H., and N. R. Aluru. 2009. Temperature-dependent wettability on a titanium dioxide surface. *Mol Simul* 35:31–7.
13. Lorenz, C. D., P. S. Crozier, J. A. Anderson, and A. Travesset. 2008. Molecular dynamics of ionic transport and electrokinetic effects in realistic silica channels. *J Phys Chem C* 112:10222–32.
14. Heng, J. B., A. Aksimentiev, C. Ho, P. Marks, Y. V. Grinkova, S. Sligar, K. Schulten, and G. Timp. 2006. The electromechanics of DNA in a synthetic nanopore. *Biophys J* 90:1098–106.
15. Im, W., S. Seefeld, and B. Roux. 2000. A grand canonical Monte Carlo–Brownian dynamics algorithm for simulating ion channels. *Biophys J* 79:788–801.
16. Guttmacher, A. E., and F. S. Collins. 2002. Genomic medicine—A primer. *N Engl J Med* 347:1512–20.
17. Shendure, J., R. D. Mitra, C. Varma, and G. M. Church. 2004. Advanced sequencing technologies: Methods and goals. *Nat Rev Genet* 5:335–44.
18. Metzker, M. L. 2005. Emerging technologies in DNA sequencing. *Genome Res* 15:1767–76.

19. Shendure, J., and H. Ji. 2008. Next-generation DNA sequencing. *Nat Biotechnol* 26:1135–45.
20. Branton, D., D. W. Deamer, A. Marziali, H. Bayley, S. A. Benner, T. Butler, M. Di Ventra et al. 2008. The potential and challenges of nanopore sequencing. *Nat Biotechnol* 26:1146–53.
21. Clarke, J., H. C. Wu, L. Jayasinghe, A. Patel, S. Reid, and H. Bayley. 2009. Continuous base identification for single-molecule nanopore DNA sequencing. *Nat Nanotechnol* 4:265–70.
22. Zwolak, M., and M. Di Ventra. 2008. Colloquium: Physical approaches to DNA sequencing and detection. *Rev Mod Phys* 80:141–65.
23. Lagerqvist, J., M. Zwolak, and M. Di Ventra. 2006. Fast DNA sequencing via transverse electronic transport. *Nano Lett* 6:779–82.
24. Soni, G. V., and A. Meller. 2007. Progress toward ultrafast DNA sequencing using solid-state nanopores. *Clin Chem* 53:1996–2001.
25. Zhao, Q., G. Sigalov, V. Dimitrov, B. Dorvel, U. Mirsaidov, S. Sligar, A. Aksimentiev, and G. Timp. 2007. Detecting SNPs using a synthetic nanopore. *Nano Lett* 7:1680–5.
26. Sigalov, G., J. Comer, G. Timp, and A. Aksimentiev. 2008. Detection of DNA sequence using an alternating electric field in a nanopore capacitor. *Nano Lett* 8:56–63.
27. Advanced Sequencing Technology Awards, 2009. http://www.genome.gov/27534236.
28. Kasianowicz, J. J., E. Brandin, D. Branton, and D. W. Deamer. 1996. Characterization of individual polynucleotide molecules using a membrane channel. *Proc Natl Acad Sci U S A* 93:13770–3.
29. Mathé, J., A. Aksimentiev, D. R. Nelson, K. Schulten, and A. Meller. 2005. Orientation discrimination of single stranded DNA inside the α-hemolysin membrane channel. *Proc Natl Acad Sci U S A* 102:12377–82.
30. Butler, T. Z., J. H. Gundlach, and M. A. Trol. 2006. Determination of RNA orientation during translocation through a biological nanopore. *Biophys J* 90:190199.
31. Akeson, M., D. Branton, J. J. Kasianowicz, E. Brandin, and D. W. Deamer. 1999. Microsecond time-scale discrimination among polycytidylic acid, polyadenylic acid, and polyuridylic acid as homopolymers or as segments within singe RNA molecules. *Biophys J* 77:3227–33.
32. Meller, A., L. Nivon, E. Brandin, J. Golovchenko, and D. Branton. 2000. Rapid nanopore discrimination between single polynucleotide molecules. *Proc Natl Acad Sci U S A* 97:1079–84.
33. Vercoutere, W., S. Winters-Hilt, H. Olsen, D. Deamer, D. Haussler, and M. Akeson. 2001. Rapid discrimination among individual DNA hairpin molecules at single-nucleotide resolution using an ion channel. *Nat Biotechnol* 19:248–52.
34. Vercoutere, W. A., S. Winters-Hilt, V. S. DeGuzman, D. Deamer, S. E. Ridino, J. T. Rodgers, H. E. Olsen, A. Marziali, and M. Akeson. 2003. Discrimination among individual Watson-Crick base pairs at the termini of single DNA hairpin molecules. *Nucleic Acids Res* 31:1311–8.
35. Nakane, J., M. Wiggin, and A. Marziali. 2004. A nanosensor for transmembrane capture and identification of single nucleic acid molecules. *Biophys J* 87:615–21.
36. Ashkenasy, N., J. Sánchez-Quesada, H. Bayley, and M. R. Ghadiri. 2005. Recognizing a single base in an individual DNA strand: A step toward DNA sequencing in nanopores. *Angew Chem Int Ed* 44:1401–4.
37. Cockroft, S. L., J. Chu, M. Amorin, and M. R. Ghadiri. 2008. A single-molecule nanopore device detects DNA polymerase activity with single-nucleotide resolution. *J Am Chem Soc* 130:818–20.
38. Timp, W., U. M. Mirsaidov, D. Wang, J. Comer, A. Aksimentiev, and G. Timp. 2010. Nanopore sequencing: Electrical measurements of the code of life. *IEEE Trans Nanotechnol* 9:281–94.

39. Derrington, I. M., T. Z. Butler, M. D. Collins, E. Manrao, M. Pavlenok, M. Niederweis, and J. H. Gundlach. 2010. Nanopore DNA sequencing with MspA. *Proc Natl Acad Sci U S A* 107:16060.

40. Gracheva, M. E., A. Aksimentiev, and J.-P. Leburton. 2006. Electrical signatures of single-stranded DNA with single base mutations in a nanopore capacitor. *Nanotechnology* 17:3160–5.

41. Wells, D. B., V. Abramkina, and A. Aksimentiev. 2007. Exploring transmembrane transport through α-hemolysin with grid-steered molecular dynamics. *J Chem Phys* 127:125101.

42. Muthukumar, M. 2007. Mechanism of DNA transport through pores. *Annu Rev Biophys Biomol Struct* 36:435–50.

43. Zhao, Q., J. Comer, V. Dimitrov, A. Aksimentiev, and G. Timp. 2008. Stretching and unzipping nucleic acid hairpins using a synthetic nanopore. *Nucleic Acids Res* 36:1532–41.

44. Aksimentiev, A., I. A. Balabin, R. H. Fillingame, and K. Schulten. 2004. Insights into the molecular mechanism of rotation in the Fo sector of ATP synthase. *Biophys J* 86:1332–44.

45. Aksimentiev, A. 2010. Deciphering ionic current signatures of DNA transport through a nanopore. *Nanoscale* 2:468–83.

46. Chen, P., J. Gu, E. Brandin, Y.-R. Kim, Q. Wang, and D. Branton. 2004. Probing single DNA molecule transport using fabricated nanopores. *Nano Lett* 4:2293–8.

47. Zhang, Y., H. Zhou, and Z.-C. Ou-Yang. 2001. Stretching single-stranded DNA: Interplay of electrostatic, base-pairing, and base-pair stacking interactions. *Biophys J* 81:1133–43.

48. Tinland, B., A. Pluen, J. Sturm, and G. Weill. 1997. Persistence length of single-stranded DNA. *Macromolecules* 30:5763–5.

49. Baumann, S. G., S. B. Smith, V. A. Bloomfield, and C. Bustamante. 1997. Ionic effects on the elasticity of single DNA molecules. *Proc Natl Acad Sci U S A* 94:6185–90.

50. Metropolis, N., M. Rosenbluth, A. Rosenbluth, A. Teller, and E. Teller. 1953. Equation of state calculations by fast computing machines. *J Chem Phys* 21:1087–92.

51. Im, W., and B. Roux. 2002. Ion permeation and selectivity of OmpF porin: A theoretical study based on molecular dynamics, Brownian dynamics, and continuum electrodiffusion theory. *J Mol Biol* 322:851–69.

52. Noskov, S. Y., S. Berneche, and B. Roux. 2004. Control of ion selectivity in potassium channels by electrostatic and dynamic properties of carbonyl ligands. *Nature* 431:830–4.

53. Egwolf, B., Y. Luo, D. E. Walters, and B. Roux. 2010. Ion selectivity of α-hemolysin with β-cyclodextrin adapter. II. Multi-ion effects studied with grand canonical Monte Carlo/Brownian dynamics simulations. *J Phys Chem B* 114:2901–9.

54. Venkatesan, B. M., J. Polans, J. Comer, S. Sridhar, D. Wendell, A. Aksimentiev and R. Bashir. 2011. Lipid bilayer coated Al_2O_3 nanopore sensors: Towards a hybrid biological solid-state nanopore. *Biomedical Microdevices*. DOI:10.1007/s10544-011-9537-3.

55. Merca, A., E. T. Haupt, T. Mitra, H. Bögge, D. Rehder, and A. Müller. 2007. Mimicking biological cation-transport based on sphere-surface supramolecular chemistry: Simultaneous interaction of porous capsules with molecular plugs and passing cations. *Chemistry* 13:7650–8.

56. Carr, R., I. A. Weinstock, A. Sivaprasadarao, A. Müller, and A. Aksimentiev. 2008. Synthetic ion channels via self-assembly: A route for embedding porous polyoxometalate nanocapsules in lipid bilayer membranes. *Nano Lett* 8:3916–21.

57. Shih, A. Y., A. Arkhipov, P. L. Freddolino, and K. Schulten. 2006. A coarse grained protein-lipid model with application to lipoprotein particles. *J Phys Chem B* 110:3674–84.

58. Marrink, S. J., H. J. Risselada, S. Yefimov, D. P. Tieleman, and A. H. de Vries. 2007. The MARTINI forcefield: Coarse grained model for biomolecular simulations. *J Phys Chem B* 111:7812–24.

59. Volkmer, D., A. Du Chesne, D. G. Kurth, H. Schnablegger, P. Lehmann, M. J. Koop, and A. Muller. 2000. Toward nanodevices: Synthesis and characterization of the nanoporous surfactant-encapsulated Keplerate $(DODA)_{40}(NH_4)_2[(H_2O)Mo_{132}O_{372}(CH_3COO)_{30}(H_2O)_{72}]$. *J Am Chem Soc* 122:1995–8.

3 Coarse-Grained Modeling of Large Protein Complexes for Understanding Their Conformational Dynamics

Kilho Eom, Gwonchan Yoon,*
Jae In Kim,* and Sungsoo Na

CONTENTS

* These authors (G. Y. and J. I. K.) made equal contribution.

3.1 INTRODUCTION

Characterization of protein structures and their dynamic behavior, which are related to biological functions of proteins [1–6], are receiving much attention. For instance, an insight into the dynamic regulation of enzymes relevant to drug design was recently gained from computational models such as atomistic simulations and/or coarse-grained models [7–9]. Specifically, it is essential to characterize the conformational transitions of enzymes upon ligand binding for de novo design of inhibitors, that is, drug molecules. Until recently, the quantitative description of conformational changes of proteins and/or the conformational transition pathway was still a challenging problem.

Conformational dynamics of proteins has been well tackled by experimental approaches using X-ray crystallography [10], nuclear magnetic resonance (NMR) [11], and/or cryo-electron microscopy (cryo-EM) [12,13]. Such experimental techniques (except cryo-EM) have enabled the depiction of protein structures at atomistic scale, and such protein structures are deposited in the Protein Data Bank (PDB; http://www.pdb.org). However, such experimental approaches do not provide a detailed mechanism of conformational dynamics, albeit they can sometimes provide intermediate conformation during the conformational transitions. In recent years, a single-molecule imaging technique such as fluorescence energy resonance transfer (FRET) [14,15] bioassay has been reported for its ability to provide an insight into conformational dynamics of proteins by tagging the fluorescent dyes to residues belonging to two different protein domains that undergo significant conformational changes. Nevertheless, FRET bioassay is still insufficient to provide insight into conformational dynamics of large protein complexes that consist of multiple protein domains.

Although the aforementioned experimental approaches are sometimes able to capture a feature of conformational changes of proteins, the conformational transition pathway of proteins is still not accessible with such experimental approaches because the timescale available for experimental equipment is relatively longer compared to that of conformational fluctuations of proteins. In recent decades, in order to reveal the conformational transition pathway, computational approaches have extensively been used due to the increase of computing power of the CPU, which doubles every 1.5 years (Moore's law). This implies that computational simulation can describe protein dynamics such as conformational transitions, which is not well understood by experimental approaches.

Molecular dynamics (MD) simulation is one of the important simulation toolkits that can describe protein dynamics [16,17]. MD simulation historically dates back to the 1950s, in which Alder and Wainwright considered the Langevin dynamics of fluid molecules to characterize the dynamics of simple liquid [18]. Since then, MD simulation has been a popular simulation toolkit that enables the quantitative description of a motion of atomistic and/or molecular structures. Remarkably, Karplus et al. [19] first showed the potential of MD simulation to provide insight into the collective dynamics of a small protein. As a consequence, MD simulation has been widely used for studying the dynamic behavior of biological molecules such as DNA, RNA, and/or proteins. The principle of MD simulation is to numerically solve the equation of motion for all atoms with a presumed empirical potential field. Here, the parameters for the empirical potential field are usually determined by comparison with quantum mechanical calculations such as the density functional theory calculations. Numerical integration of the equation of motion provides the trajectories of all atoms, and consequently, ensemble-averaged quantities such as mean-square fluctuations relevant to the description of collective dynamics. Although MD simulation suggests a detailed mechanism of protein dynamics, it exhibits the computational restriction that the timescale of MD simulation is currently in the order of ~1 nanosecond (=10^{-9} seconds), which is much shorter than the timescale of conformational changes [20]. This indicates that MD simulation is inappropriate for studying conformational changes. Moreover, the short timescale of MD simulation prevents us from gaining detailed insight into the force-induced conformational change, particularly protein unfolding mechanics, in comparison with experiments, since computational restrictions on MD simulation can be resolved by hiring unrealistic loading rate much larger than that used in experiment [21,22]. This implies that MD simulation may provide a different insight into force-driven conformational transition mechanisms when compared with experimental results.

Over the last decade, normal mode analysis (NMA) [1–3,23–25] has been introduced as an alternative computational toolkit. Compared to MD simulation, NMA can simulate the protein dynamics at a longer timescale. The fundamental aspect of NMA is to solve the eigenvalue problem $\mathbf{Ku} = \lambda\mathbf{Mu}$, where $\mathbf{K}$ is the Hessian (stiffness) matrix that is the second derivative of the empirical potential field with respect to atomic coordinates, $\mathbf{M}$ is the mass matrix, $\mathbf{u}$ is the eigenvector (normal mode), and λ is the eigenvalue (related to natural frequency of a system). NMA presumes that protein dynamics can be approximated as harmonic dynamics around the equilibrium conformation, which has to be numerically found from energy minimization of empirical potential field. Finding the equilibrium conformation and the Hessian matrix is the most time-consuming process. Although NMA is computationally more efficient than MD simulation, NMA encounters a computational challenge to simulate the dynamics of biological macromolecules such as the viral capsid.

To overcome the computational restrictions encountered in MD simulation and/or NMA, coarse-graining strategy has been considered and has received much attention. The key idea of coarse graining is to reduce the degrees of freedom of molecular structures such that a coarse-grained bead as a pseudo atom represents a group of several atoms [26–28]. In coarse-grained models, the empirical potential field prescribed to pseudo atoms has to be well defined. Moreover, another coarse-graining strategy is to simplify the empirical potential field. Case and Teeter [29] have shown that low-frequency motions relevant to biological functions of proteins are insensitive to details of the empirical potential field. This implies that protein dynamics are governed by native topology rather than details of empirical potential [30]. It is attributed to the fact that most proteins are globular proteins, whose vibration motion is similar to that of lattice material with a different scaling exponent [31–33]. The role of native topology in protein dynamics has been also validated by Lu and Ma [30] such that perturbation of the Hessian matrix does not affect low-frequency motion of proteins, as long as the native topology of proteins is not significantly perturbed. This suggests that coarse graining can be acceptable as long as the native topology is well treated.

The topology-based coarse-grained model dates back to the late 1970s, in which Gō et al. [34–36] suggested a simple model, renowned as the Gō model, that considers α-carbon atoms of a backbone chain with a simplified empirical potential field comprising the covalent bond stretch and nonbonded interaction for native contact. The Gō model was first used for studying the protein folding mechanism [37]. This indicates that the topology-based coarse-grained model unveils the protein folding mechanism. Similarly, Eaton et al. [38–40] developed the topology-based coarse-grained model, referred to as the Munoz–Eaton model (ME model), where pseudo atoms representing the residues of a backbone chain are prescribed by simplified Ising-like empirical potential to describe the protein topology. Eaton et al. [38,39] have shown that their topology-based coarse-grained model has successfully described the folding mechanism for the RNA hairpin and its relevant free-energy landscape. Moreover, the ME model has been reconsidered for studying the mechanical unfolding mechanism of proteins [41]. In recent years, Thirumalai et al. [42,43] have reported the self-organized polymer (SOP) model to describe protein structures for studying the dynamic behavior of proteins. The SOP model is similar to the Gō model, while the SOP model adopts a more detailed empirical potential field than the Gō model. Specifically, the SOP model uses the polymer chain model with appropriate treatment of native contacts such that nonbonded interaction (i.e., Lennard–Jones potential) and electrostatic repulsion are prescribed to native contacts. The SOP model has been recently reported for gaining insight into the protein unfolding mechanism [42–45] that is quantitatively comparable to that obtained from atomic force microscope (AFM) single-molecule pulling experiments.

In recent decades, Tirion [46] had introduced the more simplified, topology-based coarse-grained model known as the elastic network model (ENM) [26,46–50], inspired from the Gō model. Specifically, the ENM has been suggested in such a way that residues within a neighborhood are connected by harmonic springs with an identical force constant. In other words, the ENM presumes that the empirical potential field is the harmonic potential prescribed to native contacts. It is remarkably shown that the thermal fluctuation behavior of protein predicted from the ENM are quantitatively comparable to that obtained from experiments such as X-ray crystallography and/or NMR spectroscopy. This indicates that the ENM is versatile to represent the protein topology that plays a critical role in protein dynamics. Remarkably, the ENM provides low-frequency normal modes that are highly correlated with conformational transitions obtained from experiments [51–53]. This sheds light on the ENM for studying the conformational changes of proteins using low-frequency normal modes of the ENM.

This chapter addresses the current state of the art in computational models for protein structures and their related conformational transitions. The remainder of this chapter is organized as follows: Section 3.2 briefly summarizes the generic molecular modeling strategies such as MD simulation and NMA; Section 3.3 overviews the topology-based models such as the ME model, the Gō model, and SOP model; Section 3.4 describes the key idea of the ENM as well as its applications; Section 3.5 provides the currently suggested de novo ENMs such as the atomistic model–based ENM, coarse-grained ENM, and multiscale ENM. How to construct a novel ENM that can reflect the more realistic physics at atomic scale is of theoretical interest; Section 3.6 suggests a computational strategy to computationally improve NMA. Herein, we provide the review of our attempt to computationally enhance NMA for proteins using component mode synthesis (CMS); and the concluding Section 3.7 summarizes the current state of the art in coarse-graining strategies applicable to large protein complexes for understanding their dynamic behavior related to biological functions.

3.2 MOLECULAR SIMULATIONS

3.2.1 MOLECULAR DYNAMICS SIMULATIONS

The principle of MD simulation is to solve the equation of motion for all atoms described as follows [54]:

$$m_i \frac{d^2}{dt^2} \mathbf{r}_i = -\frac{\partial}{\partial \mathbf{r}_i} V(\mathbf{r}_1, \ldots, \mathbf{r}_N) \tag{3.1}$$

where m_i, $\mathbf{r}_i$, and V represent the molecular weight of the ith atom, the position vector of the ith atom, and the empirical potential field as a function of atomic coordinates, respectively, and N is the total number of atoms. In case of constant temperature simulation (NVT ensemble), the equation of motion given by Equation 3.1 is updated by including the Nosé-Hoover thermostat [18] as follows:

$$m_i \frac{d^2 \mathbf{r}_i(t)}{dt^2} = -\frac{\partial}{\partial \mathbf{r}_i} V(\mathbf{r}_1, \ldots, \mathbf{r}_N) + m_i \kappa \left[\frac{T_0}{T(t)} - 1 \right] \frac{d\mathbf{r}_i(t)}{dt} \tag{3.2}$$

Here, κ is an arbitrary frictional drag parameter (inverse time constant) chosen as the coupling parameter that determines the timescale relevant to temperature fluctuation, T_0 is the mean temperature (prescribed temperature in the simulation), and $T(t)$ is the temperature at time t computed as

$$T(t) = \frac{1}{(3N - n)k_B} \sum_{i=1}^{N} m_i \mathbf{v}_i(t) \cdot \mathbf{v}_i(t) \tag{3.3}$$

Herein, $(3N - n)$ is the degrees of freedom of unrestrained atoms, k_B is Boltzmann's constant, and $\mathbf{v}_i(t)$ is the velocity of ith atom, that is, $\mathbf{v}_i(t) = d\mathbf{r}_i(t)/dt$.

The atomic motion of protein structures depends on the details of the empirical potential field. Several empirical potential fields have been suggested such as CHARMM and AMBER. The empirical potential field for protein structures consists of energies, which correspond to the covalent bond stretch (E_s), bending of bond angle (E_b), twist of dihedral angle (E_t), nonbonded interaction such as Lennard–Jones interaction (E_{LJ}), and electrostatic repulsion (E_{el}) [54,55].

$$\begin{aligned} V &= E_s + E_b + E_t + E_{LJ} + E_{el} \\ &= \sum_{i=1}^{N-1} \frac{k_c}{2} \left(d_{i,i+1} - d_{i,i+1}^0 \right)^2 + \sum_{i=1}^{N-2} \frac{k_b}{2} \left(\theta_i - \theta_i^0 \right)^2 + \sum_{i=1}^{N-3} k_\varphi \left[\cos(n\varphi_i + \delta) \right] \\ &\quad + \sum_{i=1}^{N-1} \sum_{j=i+1}^{N} 4\varepsilon \left[\left(\frac{\sigma}{d_{ij}} \right)^6 - \left(\frac{\sigma}{d_{ij}} \right)^{12} \right] + \sum_{i=1}^{N-1} \sum_{j=i+1}^{N} D \frac{q_i q_j}{d_{ij}} \end{aligned} \tag{3.4}$$

where d_{ij} is the distance between the ith atom and jth atom, k_c is the force constant of the covalent bond stretch, θ_i is the bending angle formed by two adjacent bond vectors, k_b is the force constant of bending energy for the bending angle, φ_i is the dihedral angle, that is, the twist angle of a bond vector with respect to the plane formed by other two adjacent bond vectors, k_φ is the force constant for torsion energy of the dihedral angle, ε is the energy parameter for nonbonded interaction (for native contact), σ is the length scale for nonbonded interaction, q_i is the electrical charge for the ith atom, and superscript 0 indicates the equilibrium state.

The atomic details on protein dynamics can be depicted through numerical integration of the equation of motion given by Equation 3.1 or 3.2. Here, we skip the review on numerical integration schemes, which have been well explained in references [18,54]. Numerical solution to the equation of motion provides the trajectories of all atoms with respect to time t. Based on such trajectories, the mean-square fluctuation for the ith atom is given by $<|\mathbf{r}_i(t) - \mathbf{r}_i^0|^2>$, where $< >$ indicates the ensemble average (or time average), and $\mathbf{r}_i^0 = <\mathbf{r}_i(t)>$. Such mean-square fluctuation can be compared with the Debye-Waller factor available from experiments with a relation of $B_i = (8\pi^2/3)<|\mathbf{r}_i(t) - \mathbf{r}_i^0|^2>$. For a description of the correlated motion, we can consider fluctuation matrix $\mathbf{C}$ comprising 3×3 block matrix $\mathbf{C}_{ij}$ defined as $\mathbf{C}_{ij} = \left\langle \left[r_i(t) - r_i^0 \right] \otimes \left[r_j(t) - r_j^0 \right] \right\rangle$. Here, the symbol $\otimes$ represents the tensor product between two vectors. A diagonalization of fluctuation matrix $\mathbf{C}$ provides a fluctuation behavior to that spanned by normal mode space, which allows us to gain insight into the role of low-frequency modes on the fluctuation dynamics. This procedure is referred to as *essential dynamics* [56–58], which enables the prediction of significant fluctuation motion along the low-frequency normal mode.

In the case of mechanical stretching of biomolecules, the work done by mechanical stretching has to be included in the effective potential field. For stretching of biomolecules with constant pulling velocity, the effective potential field is given as follows [59–61]:

$$V^* = V + \frac{k_L}{2}(R - vt)^2 \tag{3.5}$$

where k_L is the force constant for the loading apparatus such as an AFM or optical laser tweezer, R is the distance between two atoms at termini, that is, $R(t) = |\mathbf{r}_1(t) - \mathbf{r}_N(t)|$, and v is the pulling speed. In the case of mechanical stretching of biomolecules with constant force, the effective potential field is represented in the following form:

$$V^* = V - \mathbf{F} \cdot \mathbf{R}(t) \tag{3.6}$$

Here, $\mathbf{F}$ is the constant force vector applied at termini, and $\mathbf{R}(t)$ is the distance vector connecting two atoms at which a force $\mathbf{F}$ is applied. Numerical integration suggests the force-extension curve, that is, the relation between end-to-end distance $R(t)$ and mechanical force $F(t)$, which can be compared with that obtained from single-molecule pulling experiments.

3.2.2 Normal Mode Analysis

NMA [1–3,23–25] presumes that protein dynamics can be approximated as harmonic dynamics, referred to as *quasiharmonic* approximation [24]. The equation of motion for quasiharmonic approximation becomes

$$m_i \frac{d^2 \mathbf{r}_i(t)}{dt^2} + \sum_{j=1}^{N} \mathbf{K}_{ij}\mathbf{r}_j(t) = 0 \tag{3.7}$$

where $\mathbf{K}_{ij}$ is the 3×3 Hessian (stiffness) matrix defined as $\mathbf{K}_{ij} = \partial^2 V(\mathbf{r}_1, \mathbf{r}_2, \ldots, \mathbf{r}_N)/\partial\mathbf{r}_i\partial\mathbf{r}_j$. With the introduction of normal mode space, the position vector $\mathbf{r}_i(t)$ can be represented in the form of $\mathbf{r}_i(t) = \mathbf{q}_i\cos(\omega t + \delta)$, where ω and $\mathbf{q}_i$ indicate the natural frequency and its corresponding normal mode, respectively. As a consequence, the equation of motion is transformed into the eigenvalue problem given as [62]

$$\sum_{j=1}^{N} \mathbf{K}_{ij}\mathbf{q}_j = -\omega^2 m_i \mathbf{q}_i \equiv \lambda\mathbf{q}_i \tag{3.8a}$$

Here, λ is an eigenvalue with a relation of $\lambda = -\omega^2 m_i$. Equation 3.8a can be represented in the matrix form:

$$\mathbf{K}\mathbf{q} = \lambda\mathbf{q} \tag{3.8b}$$

where $\mathbf{K}$ is the $3N \times 3N$ Hessian matrix, and $\mathbf{q}$ is $3N \times 1$ normal mode vector.

Equilibrium statistical mechanics theory provides that the probability distribution of normal mode obeys Boltzmann's distribution [63–65].

$$p(\mathbf{q}^1,\ldots,\mathbf{q}^{3N-6}) = \frac{1}{Z}\exp\left[-\frac{1}{2k_B T}\sum_{j=1}^{3N-6}\lambda_j\mathbf{q}^j\cdot\mathbf{q}^j\right] \tag{3.9a}$$

Here, $\mathbf{q}^k$ is the kth normal mode, k_B is Boltzmann's constant, T is the absolute temperature, Z is the partition function, and enumeration of mode index excludes six zero normal modes corresponding to rigid body motions. Herein, the partition function Z is defined as

$$Z = \int d\mathbf{q}^1 \ldots d\mathbf{q}^{3N-6}\exp\left[-\frac{1}{2k_B T}\sum_{j=1}^{3N-6}\lambda_j\mathbf{q}^j\cdot\mathbf{q}^j\right] \tag{3.9b}$$

Using the probability distribution given by Equation 3.9, we have the following ensemble-averaged quantities:

$$\langle\mathbf{q}^k\rangle = \frac{1}{Z}\int d\mathbf{q}^1 \ldots d\mathbf{q}^{3N-6}\left[\mathbf{q}^k p(\mathbf{q}^1,\ldots,\mathbf{q}^{3N-6})\right] = 0 \tag{3.10a}$$

$$\langle\mathbf{q}^i\mathbf{q}^j\rangle = \frac{1}{Z}\int d\mathbf{q}^1 \ldots d\mathbf{q}^{3N-6}[\mathbf{q}^i\mathbf{q}^j p(\mathbf{q}^1,\ldots,\mathbf{q}^{3N-6})] = k_B T\lambda_i^{-1}\delta_{ij} \tag{3.10b}$$

Herein, we have used the Gaussian integral and the orthogonality conditions for normal modes. The fluctuation matrix $\mathbf{C}$ can be easily computed from the transformation of $<\mathbf{q}^i\mathbf{q}^j>$ spanned by normal mode space into the Cartesian coordinate space.

$$\mathbf{C} \equiv \left\langle \left(\mathbf{r} - \langle\mathbf{r}\rangle\right) \otimes \left(\mathbf{r} - \langle\mathbf{r}\rangle\right) \right\rangle = k_B T \mathbf{Q}^\dagger \Lambda^{-1} \mathbf{Q} \equiv \sum_{j=1}^{3N-6} \frac{k_B T}{\lambda} \mathbf{q}^j \otimes \mathbf{q}^j \tag{3.11}$$

where $\mathbf{r}$ is the atomic coordinates such as $\mathbf{r} = [\mathbf{r}_1 \ \mathbf{r}_2 \ldots \mathbf{r}_N]$.

3.3 TOPOLOGY-BASED COARSE-GRAINED MODELS AND THEIR APPLICATIONS

The principle of coarse graining is to describe molecular structures using pseudo atoms that represent the group of several atoms. One of the popular coarse-grained models to describe protein structures is the Gō-like model, in which protein structures are represented using the α-carbon atoms of the protein backbone. To depict the protein dynamics, the effective potential field prescribed to the coarse-grained model has to be well defined such that this effective potential field is able to capture the significant dynamic characteristics. For the Gō-like model, the effective potential field prescribed to α-carbon atoms consists of the covalent bond stretch and nonbonded interaction for native contacts (see Figure 3.1a).

The robustness of Gō-like coarse-grained models has been validated by considering the protein folding problem [37,44,66], protein fluctuation dynamics [34–36,67], and protein unfolding mechanics [68]. The ability of the coarse-grained model to describe protein mechanics is attributed to the finding that protein mechanics are dominated by protein native topology. The role of the protein native topology on the dynamics is suggested by Case and Teeter [29], who reported that the low-frequency motion of a protein is insensitive to the details of the empirical potential field. Similarly, Lu and Ma [30] have suggested that perturbation of the Hessian matrix does not significantly affect the low-frequency motion as long as the perturbation of the Hessian matrix does not distort the protein native topology. This supports that the topology-based coarse-grained model enables the computationally efficient analysis of protein dynamics. Moreover, the relationship between protein topology and protein dynamics may be ascribed to the densely packed system (for globular protein) [31–33]. ben-Avraham [31] showed that density of state resembles that of Debye solids except for the scaling law, implying that packing density related to native topology is responsible for protein dynamics. This also confirms that topology-based coarse-grained models are acceptable for studying protein dynamics.

3.3.1 THE GŌ MODEL

The Gō model dates to the late 1970s, when Gō et al. [34–37] studied protein dynamics such as protein folding dynamics and/or fluctuation dynamics. An empirical potential prescribed to α-carbon atoms is given by [68]

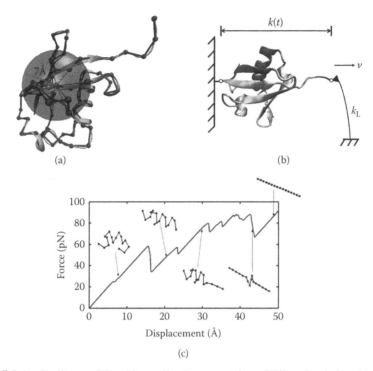

(a)

(b)

(c)

FIGURE 3.1 Gō-like model and its application to protein unfolding simulation. (a) A protein structure (shown as ribbon type) is represented by α-carbon atoms (dictated as beads) of a backbone chain. Two consecutive α-carbon atoms are connected by a covalent bond represented as a solid line. α-carbon atoms within the neighborhood (characterized by cutoff distance of 7 Å) interact with one another via Lennard–Jones potential. (b) Schematic illustration of single-molecule pulling experiment of a protein domain is presented. Here, a protein structure can be represented by a Gō-like model while a atomic force microscopy cantilever is modeled as a harmonic spring connected to a protein domain. (c) Force-extension relation for a single protein domain α helix is presented when it is stretched with constant pulling speed. Here, we have used a Gō model to describe an α helix.

$$V = \sum_{i=1}^{N-1} \left[\frac{k_1}{2} \left(r_{i,i+1} - r_{i,i+1}^0 \right)^2 + \frac{k_2}{4} \left(r_{i,i+1} - r_{i,i+1}^0 \right)^4 \right] + \sum_{i=1}^{N-1} \sum_{j=i+1}^{N} 4\varepsilon \left[\left(\frac{\sigma}{r_{ij}} \right)^6 - \left(\frac{\sigma}{r_{ij}} \right)^{12} \right] \quad (3.12)$$

where r_{ij} is the distance between the ith α-carbon atom and jth α-carbon atom.

The Gō model has been recently revisited for studying the protein unfolding mechanics (see Figure 3.1b). For instance, Cieplak et al. [69–72] have intensively employed the Gō model for simulating a mechanical unfolding of a protein. In their work [69–72], the mechanical unfolding mechanism of small proteins has been well characterized by the Gō model. Until recently, they have simulated the mechanical unfolding mechanism of ~10^4 small proteins to find the optimal molecular structure that possesses high resilience to mechanical loading [73]. Recently, Buehler et al. [74] have used the Gō-like model for studying the mechanical response of intermediate filaments—tetramer and dimer. Their work [74] surprisingly showed that mechanical response of tetramer is quite different from that of dimer, which is attributed to intermolecular interactions between dimers in tetramer. This indicates that the hierarchy in protein material plays a key role in mechanical response.

Thirumalai et al. [42,43] have developed the SOP model, inspired from the Gō model. Unlike the Gō model, the SOP model considers a more refined empirical potential field prescribed to α-carbon atoms. Specifically, the strain energy for the covalent bond stretch is depicted by finite extensible nonlinear elastic potential, whereas nonbonded interaction for native contact is dictated by Lennard-Jones potential and electrostatic repulsion. Thirumalai et al. [42] have studied the mechanical unfolding pathway of green fluorescence protein (GFP) in order to gain an insight into the folding pathway of GFP. It is interestingly suggested that there may be two major possible unfolding pathways for GFP, implying the ruggedness of the free-energy landscape for protein folding [42]. Dima et al. [45] have taken into account the mechanical unfolding mechanism of tubulin monomer and/or tubulin dimer using the SOP model. Their study [45] suggests that the mechanical unfolding pathway of tubulin dimer is quite different from that of tubulin monomer. This indicates that the mechanical behavior of tubulin dimer is governed by not only the folding topology of monomer but also the intermolecular interactions between tubulin monomers. Moreover, Thirumalai et al. [44,66] have studied the force-quenched refolding mechanism of proteins. Their work [44,66] and other experimental works [75,76] suggest that an insight into protein folding dynamics can be gained from single-molecule pulling experiments [44,66,75,76].

3.3.2 Munoz–Eaton Model

Munoz and Eaton [38,39] provided a simple, statistical model that is able to describe the folding mechanism of a β hairpin (see Figure 3.2a). Their model is similar to the Gō-like model in that native contacts are significantly treated. The ME model is inspired from the Ising model [64,65], which uses the binary number for configuration. Specifically, in the ME model, the binary value s_i has been introduced in such a way that values of s_i ($s_i = 0$ or 1) indicate the native bond ($s_i = 1$) or unfolded bond ($s_i = 0$). The Hamiltonian of the ME model is represented in the form of

$$H(\{s_k\}) = \sum_{i=1}^{N-1} \sum_{j=i+1}^{N} \varepsilon_{ij} \Delta_{ij} \prod_{k=i}^{j} s_k - T \sum_{i=1}^{N} s_i \Delta q_i \qquad (3.13)$$

where ε_{ij} is the binding energy of a contact between bonds i and j, Δ_{ij} is the quantity to represent the state of contact, that is, $\Delta_{ij} = 1$ if two bonds i and j form a native contact; otherwise, $\Delta_{ij} = 0$, Δq_i is the entropic cost for the native bond, and T is the absolute temperature. Herein, the Hamiltonian of the ME model consists of binding energy to form a native contact and entropic cost to maintain the native bond.

Using single-sequence approximation, the partition function Z for the ME model is given by

$$Z = 1 + \sum_{j=1}^{N} \sum_{i=1}^{N-j+1} w_{j,i} \qquad (3.14)$$

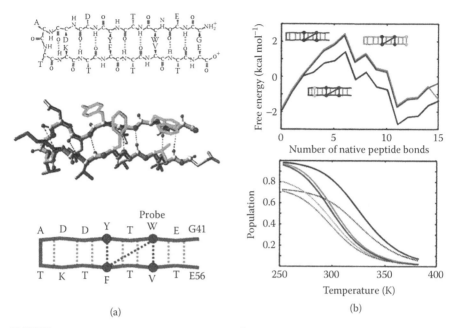

(a)   (b)

FIGURE 3.2 Munoz–Eaton (ME) model for a β hairpin. (a) Schematic representation of a β hairpin: (top) the chemical structure of β hairpin, (middle) molecular structure, and (bottom) ME model are presented. Thick solid line in the ME model shows the backbone chain, while the black-dotted line indicates the native bonds, and the gray-dotted line represents the native contacts. (b) (Top) Free-energy landscape for a β hairpin is obtained from the ME model. (Bottom) Population of folded domain with respect to denatured domain is computed from the ME model. Herein, population equal to 1 indicates the fully folded structure, whereas population equal to 0 shows the completely denatured structure. (Adapted from Munoz, V., E. R. Henry, J. Hofrichter, and W. A. Eaton. 1998. *Proc Natl Acad Sci U S A* 95:5872–9. With permission.)

Here, the weight $w_{j,i}$ is defined as

$$w_{j,i} = \exp\left[-\frac{1}{k_{\mathrm{B}}T} \left(\sum_{k=i}^{j-1} \sum_{l=k+1}^{j} \varepsilon_{kl} \Delta_{kl} - T \sum_{k=i}^{j} \Delta q_i \right) \right] \qquad (3.15)$$

The equilibrium probability for all-coil conformation is $P_{0,0} = 1/Z$, and the equilibrium probability for other conformations is $P_{j,i} = w_{j,i}/Z$. Further, the free energy E_{f} for a given conformation is obtained from a relation of $E_{\mathrm{f}} = -k_{\mathrm{B}}T \ln Z$ (see Figure 3.2b).

3.4 ELASTIC NETWORK MODEL AND ITS APPLICATIONS

3.4.1 THEORY AND MODEL

The topology-based coarse-grained models such as the Gō-like model have motivated the development of the more simplified coarse-grained model. Specifically, because protein dynamics is dominated by native topology rather than details of the potential field [29–31], Tirion [46] had simplified the Gō-like model in such a way

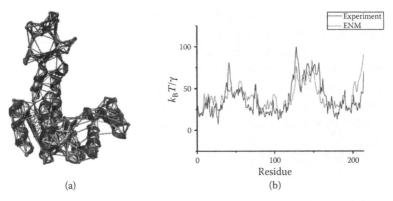

(a)   (b)

FIGURE 3.3 Elastic network model (ENM) of adenylate kinase. (a) Schematic illustration of an ENM to represent the protein structure. Here, the ENM structure (represented by lines that indicate the harmonic interaction between α-carbon atoms) is embedded on a molecular structure shown as a ribbon that is generated from visual molecular dynamics. (b) Debye–Waller B-factors for adenylate kinase computed from the ENM is compared to that obtained from experiment. It indicates that the ENM is sufficient to represent the protein's native topology that plays a critical role on protein dynamics.

that native contacts are prescribed by harmonic interaction potential. In other words, as shown in Figure 3.3a, the ENM regards protein structures as a harmonic spring network such that the residues within a neighborhood (dictated by cutoff distance) are connected by harmonic springs with an identical force constant. Accordingly, the potential field for an ENM is represented in the form of [26,46–49,77]

$$V = \frac{\gamma}{2} \sum_{i=1}^{N-1} \sum_{j=i+1}^{N} \left(\left| \mathbf{r}_i - \mathbf{r}_j \right| - \left| \mathbf{r}_i^0 - \mathbf{r}_j^0 \right| \right)^2 \cdot U\left(r_c - \left| \mathbf{r}_i^0 - \mathbf{r}_j^0 \right| \right) \qquad (3.16)$$

where γ is a force constant, N is the total number of residues, $\mathbf{r}_i$ is the atomic coordinates for the ith residue, r_c is the cutoff distance, $U(x)$ is the Heaviside unit step function defined as $U(x) = 0$ if $x < 0$; otherwise, $U(x) = 1$, and superscript 0 indicates the equilibrium (native) conformation. The potential energy V can be in the form of

$$V = \frac{1}{2} \sum_{i=1}^{N-1} \sum_{j=i+1}^{N} \left(\mathbf{r}_i - \mathbf{r}_i^0 \right)^t \mathbf{H}_{ij} \left(\mathbf{r}_j - \mathbf{r}_j^0 \right) \equiv \frac{1}{2} \mathbf{u}^t \mathbf{H} \mathbf{u} \qquad (3.17)$$

Here, $\mathbf{H}_{ij}$, $\mathbf{H}$, and $\mathbf{u}$ represent the 3×3 block Hessian matrix, the $3N \times 3N$ Hessian matrix, and the atomic displacement vector, respectively, and the symbol t indicates a transpose of a vector. Herein, the block Hessian matrix $\mathbf{H}_{ij}$ (for $i \neq j$) is given as [47,78]

$$\mathbf{H}_{ij} = -\gamma U\left(r_c - \left| \mathbf{r}_i^0 - \mathbf{r}_j^0 \right| \right) \frac{\left(\mathbf{r}_i^0 - \mathbf{r}_j^0 \right) \otimes \left(\mathbf{r}_i^0 - \mathbf{r}_j^0 \right)}{\left| \mathbf{r}_i^0 - \mathbf{r}_j^0 \right|^2} \qquad (3.18)$$

The Hessian matrix $\mathbf{H}$ comprises block Hessian matrices $\mathbf{H}_{ij}$ such that an off-diagonal block Hessian is $\mathbf{H}_{ij}$ (see Equation 3.18) and the diagonal block Hessian is given as $\mathbf{H}_{ii} = -\sum_{j \neq i}^{N} \mathbf{H}_{ij}$. Consequently, using statistical mechanics theory in NMA (see Section 3.2.2), the fluctuation matrix can be straightforwardly computed as follows [63–65]:

$$\mathbf{C}_{ij} \equiv \left\langle \left(\mathbf{r}_i(t) - \left\langle \mathbf{r}_i(t) \right\rangle \right) \otimes \left(\mathbf{r}_j(t) - \left\langle \mathbf{r}_j(t) \right\rangle \right) \right\rangle = \sum_{n=7}^{3N} \frac{k_B T}{\lambda_n} (\mathbf{v}_n \otimes \mathbf{v}_n)_{ij} \tag{3.19}$$

where the angle bracket < > indicates the ensemble (time) average, k_B is Boltzmann's constant, T is the absolute temperature, and λ_n and $\mathbf{v}_n$ correspond to the eigenvalue and its corresponding normal mode for Hessian matrix $\mathbf{H}$, respectively. Herein, the summation excludes six zero normal modes corresponding to rigid body motions. The B-factor for residue i can be estimated from the ENM (see Figure 3.3b) as follows:

$$B_i = \frac{8\pi^2}{3} (tr\mathbf{D}_{ii}) \tag{3.20}$$

Here, tr indicates a trace of a matrix. Further, the correlated motion could be predicted from the ENM such that the correlated motion between two residues i and j is defined as

$$c_{ij} = \frac{\left\langle \left(\mathbf{r}_i(t) - \left\langle \mathbf{r}_i(t) \right\rangle \right) \left(\mathbf{r}_j(t) - \left\langle \mathbf{r}_j(t) \right\rangle \right) \right\rangle}{\sqrt{\left(\mathbf{r}_i(t) - \left\langle \mathbf{r}_i(t) \right\rangle \right)^2 \left(\mathbf{r}_j(t) - \left\langle \mathbf{r}_j(t) \right\rangle \right)^2}} = \frac{tr\mathbf{D}_{ij}}{\sqrt{(tr\mathbf{D}_{ii})(tr\mathbf{D}_{jj})}} \tag{3.21}$$

A value of c_{ij} close to 1 indicates the correlated motion between two residues i and j, whereas the value of c_{ij} close to -1 shows the anticorrelated motion between two such residues, and a value of c_{ij} approaching 0 implies the uncorrelated motion between two such residues.

In a similar manner, Bahar et al. [48,77] introduced the Gaussian network model (GNM), which may be regarded as a one-dimensional version of the ENM, since isotropic fluctuation is assumed. The potential energy of the GNM is represented in the form of

$$V = \frac{1}{2} \sum_{i=1}^{N-1} \sum_{j=i+1}^{N} u_i \Gamma_{ij} u_j \tag{3.22}$$

where u_i is the magnitude of fluctuation for residue i, that is, $u_i = |\mathbf{r}_i - \mathbf{r}_i^0|$, and Γ_{ij} is the $N \times N$ Hessian matrix given as [79]

$$\Gamma_{ij} = -\gamma(1 - \delta_{ij}) U\left(r_c - \left| \mathbf{r}_i^0 - \mathbf{r}_j^0 \right| \right) - \delta_{ij} \sum_{k \neq i}^{N} \Gamma_{ik} \tag{3.23}$$

Here, δ_{ij} is the Kronecker delta defined as $\delta_{ij} = 1$ if $i = j$; otherwise, $\delta_{ij} = 0$. The fluctuation matrix C_{ij} is given by

$$C_{ij} = \left\langle \left(\mathbf{r}_i(t) - \langle \mathbf{r}_i(t) \rangle \right) \left(\mathbf{r}_j(t) - \langle \mathbf{r}_j(t) \rangle \right) \right\rangle = \sum_{n=2}^{N} \frac{\kappa_n}{k_B T} (\mathbf{z}_n \otimes \mathbf{z}_n)_{ij} \qquad (3.24)$$

where κ_n and $\mathbf{z}_n$ are the eigenvalue and its corresponding normal mode for Hessian matrix Γ, respectively. Herein, there is a single zero normal mode corresponding to the rigid body motion. The B-factor for residue i is thus represented in the form of $B_i = (8\pi^2/3)C_{ii}$. Further, the cross-correlation, which indicates the correlation between motions of two residues, is given by $c_{ij} = C_{ij}/(C_{ii}C_{jj})^{1/2}$.

Zheng [80] has recently introduced the generalized anisotropic network model (G-ANM), which unifies the GNM and ENM, to provide a more realistic fluctuation behavior close to that obtained from experiment. The potential energy of the G-ANM is obtained from combining potential energies of the GNM and ENM as follows [80]:

$$V = \frac{1}{2} \sum_{i=1}^{N-1} \sum_{j=i+1}^{N} \mathbf{u}_i{}' \left(\alpha \Gamma_{ij}^{\mathrm{GNM}} \mathbf{I}_3 + (1-\alpha) \mathbf{H}_{ij}^{\mathrm{ANM}} \right) \mathbf{u}_j \qquad (3.25)$$

where α is the weight parameter to couple GNM and ENM, $\Gamma_{ij}^{\mathrm{GNM}}$ is the Kirchhoff matrix (stiffness matrix) for GNM, $\mathbf{I}_3$ is the 3×3 identity matrix, and $\mathbf{H}_{ij}^{\mathrm{ANM}}$ is the Hessian matrix for ENM. G-ANM improves the predictions on thermal fluctuation as well as anisotropic fluctuation related to normal modes involved in the conformational transitions [80].

3.4.2 APPLICATIONS OF ELASTIC NETWORK MODELS: UNDERSTANDING OF CONFORMATIONAL TRANSITIONS

The ENM has been recently highlighted for studying conformational transitions. Tama and Sanejouand [51] have shown that low-frequency normal modes of the ENM are highly correlated with conformational changes. This sheds light on the ENM for its potential in describing conformational transitions. For instance, Bahar et al. [53,81] have employed low-frequency normal modes of the ENM to describe the conformational transitions (Figure 3.4). It has been remarkably shown that the perturbation of an unbound structure along its low-frequency normal mode results in a conformation close to that at ligand-bound state. This supports the "population shift model" [82,83], which presumes that the intrinsic dynamics of an unbound structure possess the pre-existing equilibrium states corresponding to a ligand-bound structure. This underlies the role of protein dynamics in a quantitative understanding of ligand-binding and/ or protein–protein interactions. Similarly, Karplus et al. [84] have predicted intermediate conformations of myosin V using projection of a displacement vector corresponding to conformational transition to the normal mode space at an unbound state. Findings show that low-frequency normal modes are sufficient to depict the conformational transitions for motor protein [84]. Further, Wolynes et al. [85] have studied the conformational transitions of adenylate kinase using low-frequency normal mode along which protein structures are perturbed. It is shown that the breakage of bonds

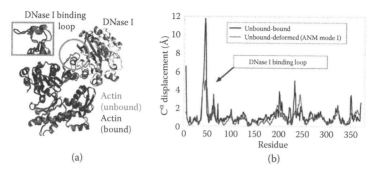

(a)　　　　　　　　　　　　　　(b)

FIGURE 3.4 Conformational transitions of actin upon DNase I binding analyzed from the elastic network model (ENM). (a) Molecular structure of actin in bound state (shown as black ribbon) and/or bound state (indicated as gray ribbon). (b) Displacement of α-carbon atoms due to the conformational transition is well fitted to that predicted from the ENM. Here, a prediction of conformational transition using the ENM is implemented such that the protein structure is perturbed along the low-frequency normal modes of the ENM. This indicates that low-frequency motions of the unbound structure are highly correlated with conformational transition. (Adapted from Tobi, D., and I. Bahar. 2005. *Proc Natl Acad Sci U S A* 102:18908–13. With permission.)

is accompanied during the conformational transition of adenylate kinase, which is referred to as "cracking" in the conformational changes [85,86].

Karlpus et al. [87] have adopted the quantum mechanical approach using the mixture of two distinct potential fields at two equilibrium conformations to understand the conformational transitions of proteins. Specifically, the potential energy at intermediate conformation is assumed to be a mixture of two harmonic potentials (i.e., ENM potentials) corresponding to unbound and bound states, respectively. In a similar manner, Takagi et al. [88] have developed the dual Gō model (DGM) by combining two Gō-like potentials at two distinct conformations. Furthermore, Hummer et al. [89] have reported the mixed elastic network model (MENM) by mixing two ENM potentials at the unbound and the bound states. Herein, the mixing rule adopted by Hummer et al. [89] is different from that used by Karplus et al. [87].

Jernigan et al. [90] established the elastic network interpolation (ENI) to predict conformational transition pathways. In their model, interresidue distances at intermediate state, which were obtained from interpolation between two equilibrium conformations, are used to find a conformational transition pathway along which the strain energy (harmonic potential) due to such transition is minimized. The ENI is capable of predicting conformational transition pathways for the GroEL complex, which is comparable to those predicted by molecular simulations. Kidera et al. [91,92] have recently employed the linear response theory with the ENM potential to understand the conformational changes and their relation to low-frequency motions. Low-frequency motion for proteins is found to play a key role in the conformational changes. In a similar manner, Demirel and Lesk [93] have used the statistical mechanics theory with the ENM to gain insight into molecular binding forces. It is provided that molecular forces obtained from their model allow the identification of binding sites. Further, Erman et al. [94,95] have used the relative fluctuation behavior computed from the ENM for identifying functional sites such as binding sites and/or active sites. Moreover, Bahar et al. [96] have used the low-frequency fluctuation behavior obtained from the ENM to find the possible binding sites and/or active sites. This implies that allosteric signaling driven by ligand binding may be governed by low-frequency motion.

There have been recent attempts to couple low-frequency normal modes computed from the ENM to atomistic models such as molecular dynamics for predicting the conformational transitions. For example, Bahar et al. [97] have used low-frequency normal modes obtained from the ENM to induce the conformational changes, whereas MD has been used to find the intermediate, equilibrium conformations. Similarly, Doruker et al. [98] have considered low-frequency normal modes of the ENM to drive the conformational changes, whereas they have used Monte Carlo simulation to find the equilibrium, intermediate conformations.

The ENM has recently been taken into account in studying the protein unfolding mechanics. Rief and Dietz [99] have recently developed an elastic bond network model (EBNM) by coupling the ENM and Bell's model [22,100–103] to depict the rupture of native contacts. Specifically, the ENM is stretched with the application of a mechanical loading, and then internal forces acting on the native contacts are calculated to estimate the probability to find a ruptured native contact. Here, Bell's model describes that the kinetic rate for mechanical unfolding is given by $k(f) = k_0\exp(fx_u/k_BT)$ [22,100–103], where k_0 is the unfolding rate at zero force, f is the mechanical force acting on a native contact, x_u is the width (reaction coordinate) of the energy barrier for a contact, k_B is Boltzmann's constant, and T is the absolute temperature. It is shown that the probability distribution for an unfolding force computed from the EBNM is quantitatively similar to that obtained from AFM single-molecule experiments (Figure 3.5). Recently,

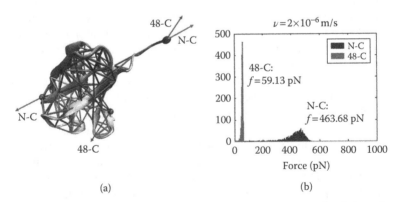

(a)   (b)

FIGURE 3.5 Application of the elastic network model (ENM) to a simulation of protein unfolding mechanics. (a) Schematic illustration of simulating protein unfolding using the ENM. Here, the protein structure represented by the ENM is stretched with constant pulling speed such that two specific residues (e.g., termini) are pulled. An arrow tagged with N-C shows the mechanical force acting on two termini residues. An arrow tagged with 48-C arrow indicates the mechanical force acting on two residues such as residue 48 and termini C. (b) Histogram of unfolding forces with respect to two different pulling geometries. Here, pulling speed is prescribed as 2×10^{-6} m/s. It is shown that ubiquitin exhibits excellent mechanical resistance when a mechanical force is exerted on two termini. On the other hand, ubiquitin has different mechanical resistance when a mechanical force is prescribed on residue 48 and termini C. This indicates the anisotropic mechanical properties of ubiquitin.

Sacquin-Mora and Lavery [104] have used Brownian dynamics simulation with the ENM potential to study the mechanical response of proteins. Moreover, Eom et al. [105,106] have scrutinized the role of topology of cross-links (equivalent to folding topology) in the mechanical unfolding mechanism of the cross-linked polymer chain model using the Gaussian chain model coupled with Bell's model to describe the bond-rupture mechanism. It was found that a specific cross-link topology resembling the parallel strands enhances the mechanical stability. These examples imply that the ENM may be applicable for studying a protein unfolding mechanism such as the unfolding pathway. However, it should be recognized that Bell's model is theoretically limited such that the bond rupture by large force, which instantaneously removes the energy barrier, should be accounted for based on more realistic bond-rupture models [107–109].

3.5 DE NOVO ELASTIC NETWORK MODELS

3.5.1 ELASTIC NETWORK MODELS ESTABLISHED FROM MOLECULAR SIMULATIONS

The ENM [26,46–50] is a phenomenological model such that it is constructed using two empirical parameters such as a force constant and cutoff distance. Although the ENM does not distinguish covalent bond from native contact, an interaction network describing the short-range interaction between neighborhood residues is sufficient to provide the fluctuation behavior. This is attributed to the fact that protein dynamics are dominated by native topology but insensitive to the details of the potential field [29]. Furthermore, a protein structure is a densely packed system (i.e., globular proteins), and its density of state resembles that of lattice solids [31–33], indicating that protein structures could be modeled as a lattice structure dictated by short-range interactions. This implies that the ENM properly captures a feature of the protein topology that plays a role in protein dynamics [3,50].

Recently, to develop a more realistic ENM, there have been attempts to establish an ENM based on atomistic simulations such as MD simulations. Ming and Wall [110] have reported that the unimodal density of state predicted from the ENM is different from the bimodal density of state evaluated from MD simulations. They developed the backbone-enhanced network model [110] in such a way that force constants for covalent bond and native bond are determined from Kullback–Leibler (KL) divergence, which compares an elastic network structure with an atomistic structure. Moreover, it has been shown that KL divergence enables the identification of possible functional sites such as binding sites [111].

Smith and Moritsugu [112,113] have attempted to develop the atomistic model–based ENM referred to as Realistic Extended Algorithm via Covariance Hessian (REACH). Specifically, instead of using empirical parameters such as cutoff distance, REACH allows the establishment of an ENM extracted from MD simulations in such a way that the Hessian matrix for the elastic network is obtained from the pseudo inverse of the fluctuation matrix, that is, the covariance Hessian. Here, the fluctuation behavior predicted from MD simulation is analyzed using principal component analysis [56,58], identical to NMA, such that an eigenvalue problem is applied to the fluctuation matrix to find low-frequency normal modes. It should be recognized that REACH provides the unphysical force constant for some elastic springs due to anharmonic effect. This indicates that the Hessian matrix computed from REACH includes both harmonic and anharmonic effects. It is still challenging to decouple the harmonic dynamics from anharmonic motions in the REACH model.

Recently, Lu and Ma [114] have reported the minimalist network model (mini-NM) extracted from molecular NMA. The mini-NM is constructed from block normal mode analysis (BNMA) [115] with an anharmonic potential field such as CHARMM, and then symmetry conditions and positive definiteness are applied to the Hessian of BNMA to establish the Hessian matrix, which excludes the anharmonic effect. It is shown that the mini-NM provides low-frequency normal modes similar to those computed from atomistic models. Moreover, the mini-NM allows not only the development of the residue-level ENM but also the coarse-grained ENM comprising a few nodal points to describe the group of several residues.

Voth et al. [116] have developed the heterogeneous elastic network model (hetero-ENM) by using a fluctuation matrix computed from MD simulation. Specifically, at the beginning, in the hetero-ENM, it is assumed that all residues are connected by harmonic springs with an identical spring constant. Subsequently, force constants for entropic springs are updated by comparison between interresidue fluctuations obtained from MD simulation and hetero-ENM. This process is repeated until hetero-ENM provides the B-factor quantitatively comparable to that computed from MD simulation. It is remarkably shown that hetero-ENM established from short timescale MD simulation provides low-frequency normal modes comparable to those obtained from long timescale MD simulation.

3.5.2 COARSE-GRAINED ELASTIC NETWORK MODELS

Protein dynamics has unique features such as collective dynamics [36,67,117] and correlated motion [118]. For instance, the motion of a protein dimer is well described by a hinge-bending motion such that some residues connecting two monomers act as a hinge responsible for a motion. Further, residues belonging to a protein domain move in similar directions, indicating the collective motion of residues belonging to a domain [119,120]. This leads to the modeling concept that protein dynamics may be described using few degrees of freedom rather than total number of residues.

Doruker et al. [121,122] have suggested the coarse-grained ENM in such a way that less nodal points rather than all residues were used to describe protein structures. Further, an elastic network was constructed by connecting neighborhood nodal points. Herein, the cutoff distance to describe the network topology was empirically determined using polymer theory, whereas the force constant was obtained from fitting the B-factor obtained from the coarse-grained ENM to that provided from experiment (usually available in the Protein Data Bank). Similarly, Jernigan et al. [123] have reported the multiscale elastic network in such a way that a region near the functional site is described by an atomic-scale elastic network, whereas the remaining region is depicted by a residue-level elastic network.

In this section, we review our current attempts [49,79,124] on systematic coarse-graining strategies that enable construction of elastic networks described by few degrees of freedom (Figure 3.6). We have sorted the residues into two

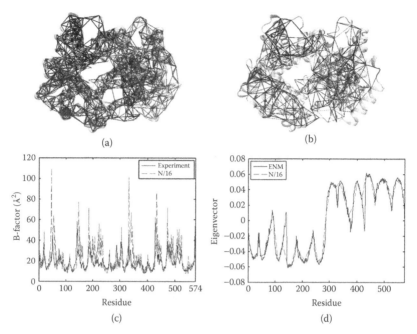

FIGURE 3.6 Elastic network model and coarse-grained elastic network model (ENM) structures. (a) A hemoglobin structure (shown as a ribbon) is represented by an ENM (indicated as lines to represent the harmonic interactions between α-carbon atoms). (b) A coarse-grained ENM is presented to describe hemoglobin. Here, a coarse-grained ENM is established such that $N/4$ nodal points are connected by harmonic springs, where N is the total number of residues. (c) Debye-Waller B-factors computed from the ENM and coarse-grained ENM using $N/16$ nodal points are presented. It is shown that a coarse-grained ENM provides the thermal fluctuation behavior of hemoglobin, quantitatively comparable to that predicted from an ENM. (d) Lowest frequency normal mode predicted from a coarse-grained ENM is quantitatively comparable to that computed from an ENM. It is shown that coarse graining of an ENM allows computationally efficient analysis on protein dynamics with computational accuracy.

groups: (1) *master residues*—residues that are maintained during the coarse graining such that the coarse-grained elastic network is described by a harmonic spring network comprising such residues; and (2) *slave residues*—residues that are eliminated during the coarse graining. For instance, for construction of a coarse-grained elastic network consisting of $N/2$ nodal points, where N is the total number of residues, a set of residue indices $[1, 3, …, N/2 − 1]$ is regarded as *master residues*, whereas the remaining residues are regarded as *slave residues*, or vice versa. As a consequence, the potential energy can be represented in the following form:

$$V = \frac{1}{2}\mathbf{u}^t\mathbf{H}\mathbf{u} = \frac{1}{2}\begin{bmatrix} \mathbf{u}_M^t & \mathbf{u}_S^t \end{bmatrix}\begin{bmatrix} \mathbf{H}_{MM} & \mathbf{H}_{MS} \\ \mathbf{H}_{SM} & \mathbf{H}_{SS} \end{bmatrix}\begin{bmatrix} \mathbf{u}_M \\ \mathbf{u}_S \end{bmatrix} \tag{3.26}$$

where $\mathbf{u}_M$ and $\mathbf{u}_S$ indicate the displacement fields for master residues and slave residues, respectively. Herein, $\mathbf{H}_{MM}$ dictates the harmonic interactions between master residues, while $\mathbf{H}_{MS}$ (or $\mathbf{H}_{SM}$) depicts the harmonic interaction between master residue and slave residue, and $\mathbf{H}_{SS}$ describes the harmonic interactions between slave residues. Our coarse-graining method is inspired from skeletonization [125], that is, low-rank approximation. It is assumed that the fluctuation behavior of slave residues is ignored, so that we have the following relation:

$$\frac{\partial V}{\partial \mathbf{u}_S} = \mathbf{H}_{SM}\mathbf{u}_M + \mathbf{H}_{SS}\mathbf{u}_S = 0 \qquad (3.27)$$

Accordingly, the effective potential field V^* for a coarse-grained ENM can be obtained by removing $\mathbf{u}_S$ using Equation 3.27.

$$V^* = \frac{1}{2}\mathbf{u}_M^t \bar{\mathbf{H}}\mathbf{u}_M = \frac{1}{2}\mathbf{u}_M^t \left[\mathbf{H}_{MM} - \mathbf{H}_{MS}\mathbf{H}_{SS}^{-1}\mathbf{H}_{SM} \right] \mathbf{u}_M \qquad (3.28)$$

Here, $\bar{\mathbf{H}}$ is the effective Hessian matrix for a coarse-grained ENM. It is shown that B-factors obtained from a coarse-grained ENM are quantitatively comparable to those obtained from experiment and/or a residue-level ENM (Figure 3.6c). Moreover, a coarse-grained ENM suggests low-frequency normal modes quantitatively comparable to those obtained from an ENM (Figure 3.6d). In addition, a correlated motion is well described by a coarse-grained ENM such that collective motion of a domain as well as correlated motion between domains predicted from a coarse-grained ENM is similar to those computed from the ENM (not shown). These indicate that the coarse-grained ENM is capable of the computationally efficient analysis on protein dynamics such as low-frequency motions relevant to functional motions and/or conformational changes.

3.5.3 Multiscale Elastic Network Model

Though a coarse-grained ENM [49,79,124] is able to describe large protein dynamics with computational efficiency, the dynamic behavior of functional sites such as active and/or binding sites may not be captured in such a coarse-grained ENM due to the coarse-graining scheme, which coarsens the interactions near the functional sites. In this section, the multiscale modeling concept applied to protein structures is overviewed.

Here, we have developed the ENM-based multiscale model, referred to as the multiscale network model (multi-NM) [126], in such a way that a region near functional sites is dictated by refined network models (e.g., ENM), while the remaining region is depicted by a coarse-grained ENM (see Figure 3.7a). Specifically, a high-resolution structure is established in such a way that residues within a neighbor of specific residues of interest are selected for a refined structure (i.e., ENM), while the other residues far from specific residues of interest are taken into account for low-resolution structures. Further, we have introduced the interface structure that bridges a high-resolution structure to a low-resolution structure. Herein, the modeling parameters such as force constant and cutoff distance of high-resolution structures are identical to those usually used in the ENM. The key of the multi-NM is how to determine the modeling parameters of low-resolution structure and interface.

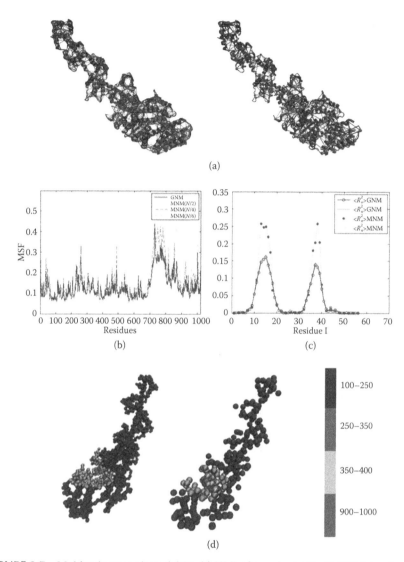

FIGURE 3.7 Multiscale network model (multi-NM) of myosin. (a) (Left) ENM structure of myosin and (right) multi-NM of myosin are presented. Here, multi-NM is constructed such that a region near binding sites is depicted by a refined network model (e.g., ENM), while the remaining region is dictated by a coarse-grained network model. (b) Mean-square fluctuation of myosin obtained from an ENM and multi-NMs using different coarse-graining schemes. It is shown that fluctuation behavior is well predicted by the multi-NM. (c) Relative fluctuation of specific residue is computed from an ENM and/or multi-NM. It is shown that relative fluctuation obtained from the multi-NM is quantitatively comparable to that computed from the ENM. This indicates that the multi-NM is able to characterize the dynamic behavior of binding sites, quantitatively comparable to that described by the ENM. (d) Kullback-Leibler divergence is presented using an (left) ENM and (right) multi-NM. It suggests that the multi-NM is capable of identifying the dynamic behavior of binding sites. (Adapted from Jang, H., S. Na, and K. Eom. 2009. *J Chem Phys* 131:245106. With permission.)

For interface, modeling parameters are obtained from linear interpolation between high-resolution and low-resolution structures. For instance, the cutoff distance for interface, R_c^{int}, is determined in such a way that a range of interaction, which is dictated by the volume of a sphere whose radius is cutoff distance, for an interface is computed from linear interpolation between ranges of interaction for high-resolution structures and low-resolution structures [121].

$$R_c^{int} = \left[\frac{\left(R_c^H \right)^3 + \left(R_c^L \right)^3}{2} \right]^{1/3} \tag{3.29}$$

where R_c^H and R_c^L represent the cutoff distances for high-resolution and low-resolution structures, respectively. Herein, the cutoff distance of a high-resolution structure, R_c^H, is identical to that of an ENM (e.g., $R_c^H = \sim 10$ Å), whereas the cutoff distance of a low-resolution structure, R_c^L, is determined from the polymer theory (see below). Similarly, the force constant of the interface, γ_{int}, is obtained from the following relation:

$$\gamma_{int} = \gamma_H + \frac{\left(\gamma_H - \gamma_L \right)}{R_c^H - R_c^L} \left(R_c^{int} - R_c^H \right) \tag{3.30}$$

Here, γ_H and γ_L are force constants of high-resolution structures and low-resolution structures, respectively. Here, a force constant of a high-resolution structure is equal to that used in an ENM, whereas a force constant of low-resolution structure is determined using KL divergence (for a detailed description, see next paragraph).

The low-resolution structure is depicted using N_s nodal points far fewer than the total number of residues belonging to such structures. For a nodal point to represent N_p residues, the radius of gyration is given as $R_g = ap^b$, where $b = \sim 0.6$ from polymer theory [121,127,128] with excluded volume effect. As a consequence, the cutoff distance for the low-resolution structure is in the form of $R_c^L = Ap^b$. The force constant of the low-resolution structure is determined by minimizing a KL divergence, which is able to compare the two structures, MNM and ENM. Here, KL divergence is defined as [110,111]

$$D_{KL} = \sum_{i=1}^{3N_s} \left[\log \left(\frac{\omega_i^{multi-NM}}{\overline{\omega}_i^{ENM}} \right) + \frac{1}{2k_B T} \left(\overline{\omega}_i^{ENM} \right)^2 \left| \overline{\mathbf{u}}_i^{ENM} \cdot \Delta \mathbf{X} \right| \right.$$
$$\left. + \frac{1}{2} \sum_{j=1}^{3N_s} \left(\frac{\overline{\omega}_i^{ENM}}{\omega_i^{multi-NM}} \left| \mathbf{u}_i^{multi-NM} \cdot \overline{\mathbf{u}}_i^{ENM} \right| \right)^2 - \frac{1}{2} \right] \tag{3.31}$$

where $\omega_i^{multi-NM}$ and $\mathbf{u}_i^{multi-NM}$ represent the natural frequency and its corresponding normal mode, respectively, for the MNM at the ith mode. $\overline{\omega}_i^{ENM}$ and $\overline{\mathbf{u}}_i^{ENM}$ indicate the natural frequency and its corresponding normal mode, respectively, for a coarse-grained ENM extracted from an ENM (see Section 3.5.2). $\Delta \mathbf{X}$ is the difference between position vectors for nodal points between the multi-NM and ENM.

For validation of the multi-NM, two model proteins such as proteinase inhibitor (pdb: 1aap) and myosin (pdb: 1dfk for unbound structures and 1dfl for ligand-bound structures) are taken into account. It is shown that the B-factor representing the fluctuation behavior is well predicted by the multi-NM, highly correlated with that obtained from the ENM or experiment (Figure 3.7b). Moreover, it is clearly seen that low-frequency normal modes of the multi-NM are quantitatively comparable to those computed from ENM. Since the multi-NM allows for characterization of specific functional sites of interest, we have considered the relative fluctuation defined as $<(\Delta R_{ij})^2>^{1/2} = [<(\Delta r_i)>^2 + <(\Delta r_j)>^2 - 2 <\Delta \mathbf{r}_i \cdot \Delta \mathbf{r}_j>]^{1/2}$, where ΔR_{ij} is the fluctuation of interresidue distance between residues i and j, $<(\Delta r_i)>^2$ is the mean-square fluctuation of the ith nodal point (or residue or α-carbon atom) representing the mobility, and $<\Delta \mathbf{r}_i \cdot \Delta \mathbf{r}_j>$ is the correlation between motions of the ith nodal point and jth nodal point. A large value of $<(\Delta R_{ij})^2>^{1/2}$ indicates a highly anticorrelated motion between the ith and jth nodal points. A recent study [94] states that a binding site exhibits the maximum of relative fluctuation with respect to termini, that is, $<(\Delta R_{iN})^2>^{1/2}$. It is interestingly found that, in both the multi-NM and ENM, a relative fluctuation with respect to termini reaches the maximum at a binding site (Figure 3.7c). This implies that the dynamic characteristics of functional sites are well conserved in the multi-NM. Moreover, we have also considered the KL divergence that allows the identification of functional sites such that KL divergence becomes a maximum at a binding site. It is shown that KL divergence for a binding site computed from the multi-NM is quantitatively comparable to that estimated from the ENM (Figure 3.7d). This indicates that the dynamic behavior of functional sites can be well depicted by multi-NM.

3.6 FAST NORMAL MODE ANALYSIS: COMPONENT MODE SYNTHESIS

Over the last decade, NMA has received much attention because of its ability to quantitatively provide insight into the conformational transitions of proteins [1,2,23,129,130]. Furthermore, normal mode-based description on the conformational transitions supports a theoretical hypothesis on the binding mechanism [7], known as the *population shift model* [82]. For more than 50 years, Koshland's perspective on the binding mechanism, referred to as the *induced fit model*, has been accepted and adopted in textbooks for the explanation of the binding mechanism [82]. Recent computational studies [7–9] have shown that low-frequency normal modes are highly correlated with conformational changes upon ligand binding. This highlights the population shift model demonstrating that intrinsic dynamics of unbound structures capture the multiple preexisting equilibrium conformations, one of which corresponds to ligand-bound structures. It is implied that NMA plays a pivotal role in understanding conformational transitions as well as in predicting ligand-bound structures.

As the size of proteins is increasing, there is a computational issue in NMA that is key to reducing the computational burden encountered in NMA. For decades, there have been attempts to develop a computationally favorable algorithm applicable to NMA. For instance, Component Mode Synthesis (CMS) [131,132] and the substrate synthesis method (SSM) [62] have been used for computational analysis of vibration behavior of large engineering structures such as aircrafts and buildings. Attempts such as CMS and SSM could be applicable to describe protein dynamics. Recently, Ma et al. [133] have used SSM for large protein structures to describe its low-frequency motions. Moreover, we have recently reported that CMS also enables the computationally efficient analysis of low-frequency function motions of large protein structures [78,134]. In this section, we have restricted our discussion to only the application of CMS to large protein complexes.

The fundamental of CMS is to decompose protein structures into several substructures, at which NMA is applied rather than implementation of NMA to a whole protein structure. For clear demonstration, as shown in Figure 3.8a, we consider the decomposition of protein structures into two substructures. Here, it should be recognized that a motion of substructure A is constrained by substructure B, and vice versa, in such a way that the residues belonging to the interface between two substructures should exhibit a continuous displacement field. This geometric constraint

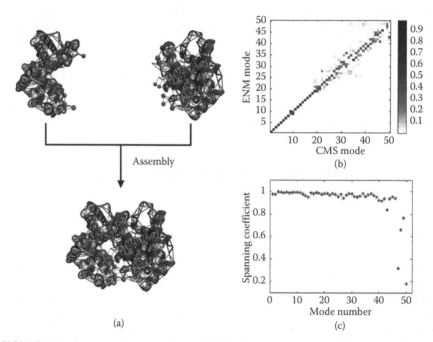

FIGURE 3.8 Component mode synthesis (CMS) of a protein structure. (a) Schematic demonstration of CMS applied to a protein structure. Here, a protein structure is decomposed to two substructures, for each of which the normal mode analysis is implemented. Once natural frequencies and their corresponding normal modes are obtained for each substructure, such frequencies and normal modes are assembled using geometric constraints to α-carbon atoms (shown as beads) at the interface between two substructures. (b) Correlation between normal modes computed from the ENM and CMS. It is shown that CMS provides the functional low-frequency normal modes, quantitatively comparable to those computed from the ENM. (c) Spanning coefficient to represent how many normal mode spaces of CMS are described by the 50 low-frequency normal modes of the ENM. Low-frequency normal modes of CMS are similar to those of the ENM. This implies that CMS can provide functional low-frequency motions with computational efficiency.

for the interface is a generic computational scheme in structural dynamics rather than prescribing the domain–domain interactions directly. Now, for convenience, we relax the geometric constraint at this moment, and then the potential energy for a whole protein structure is given by

$$V' = \frac{1}{2}\left(\mathbf{u}_A^t \mathbf{K}_A \mathbf{u}_A + \mathbf{u}_B^t \mathbf{K}_B \mathbf{u}_B\right) \tag{3.32}$$

where the prime indicates that geometric constraints are not imposed at this moment, $\mathbf{u}_i$ and $\mathbf{K}_i$ represent the displacement field and the stiffness matrix, respectively, for substructure i ($i = A$ or B), and superscript t denotes the transpose of a column vector. Similarly, the kinetic energy is represented in the form of

$$T' = \frac{1}{2}\left(\dot{\mathbf{u}}_A^t \mathbf{M}_A \dot{\mathbf{u}}_A + \dot{\mathbf{u}}_B^t \mathbf{M}_B \mathbf{u}_B\right) \tag{3.33}$$

Here, the dot shows the time derivative, and $\mathbf{M}_i$ is the mass matrix for substructure i. Subsequently, we introduce the linear transformation such that displacement field $\mathbf{u}_i$ is assumed in the form of $\mathbf{u}_i(\mathbf{x}, t) = \Psi_i(\mathbf{x}) \cdot \mathbf{v}_i(t)$, where $\Psi_i(\mathbf{x})$ is the matrix whose column vector is the eigenvector of $\mathbf{K}_i$. That is, $\Psi_i(\mathbf{x})$ satisfies the eigenvalue problem such as $\mathbf{K}_i \Psi_i(\mathbf{x}) = \Psi_i(\mathbf{x})\Lambda_i$, where Λ_i is the diagonal matrix whose component is the eigenvalue of $\mathbf{K}_i$. With such a transformation, the potential energy and the kinetic energy without implementation of geometric constraint are given as follows:

$$V' = \frac{1}{2}\begin{bmatrix} \mathbf{v}_A^t & \mathbf{v}_B^t \end{bmatrix}\begin{bmatrix} \Lambda_A & \mathbf{0} \\ \mathbf{0} & \Lambda_B \end{bmatrix}\begin{bmatrix} \mathbf{v}_A \\ \mathbf{v}_B \end{bmatrix} \equiv \frac{1}{2}\mathbf{v}^t \Lambda \mathbf{v} \tag{3.34a}$$

$$T' = \frac{1}{2}\begin{bmatrix} \dot{\mathbf{v}}_A^t & \dot{\mathbf{v}}_B^t \end{bmatrix}\begin{bmatrix} \psi_A^t \mathbf{M}_A \psi_A & \mathbf{0} \\ \mathbf{0} & \psi_B^t \mathbf{M}_B \psi_B \end{bmatrix}\begin{bmatrix} \dot{\mathbf{v}}_A \\ \dot{\mathbf{v}}_B \end{bmatrix} \equiv \frac{1}{2}\dot{\mathbf{v}}^t \mathbf{Q}\dot{\mathbf{v}} \tag{3.34b}$$

Here, the displacement field vector spanned in normal mode space, that is, $\mathbf{v}(t)$, has more degrees of freedom than $3N$, where N is the total number of residues, since a geometric constraint is not imposed at this moment.

Let us now consider the geometric constraints in order to assemble protein structures for describing the motion of a whole protein structure. As mentioned earlier, the geometric constraint is that the displacement field for the interface should be continuous. Such constraint can be denoted as $\mathbf{Pv} = \mathbf{0}$. Because $\mathbf{v}$ has the redundancy due to redundant enumeration of residues belonging to the interface, a vector $\mathbf{v}$ can be separated into independent variable $\mathbf{w}(t)$ and dependent variable $\mathbf{z}(t)$. Then, the constraint equation becomes $\mathbf{Pv} = \mathbf{P}_1\mathbf{w}(t) + \mathbf{P}_2\mathbf{z}(t) = 0$, which leads to a relation of

$$\mathbf{v} = \begin{bmatrix} \mathbf{w}(t) \\ \mathbf{z}(t) \end{bmatrix} = \begin{bmatrix} \mathbf{w}(t) \\ -\mathbf{P}_2^{-1}\mathbf{P}_1\mathbf{w}(t) \end{bmatrix} = \begin{bmatrix} \mathbf{I} \\ -\mathbf{P}_2^{-1}\mathbf{P}_1 \end{bmatrix}\mathbf{w}(t) \equiv \mathbf{Bw}(t) \tag{3.35}$$

where $\mathbf{B}$ is a constraint matrix. Consequently, with application of constraints, the potential energy and the kinetic energy, respectively, become

$$V = \frac{1}{2}\mathbf{w}^t(\mathbf{B}^t\Lambda\mathbf{B})\mathbf{w} \equiv \frac{1}{2}\mathbf{w}^t\mathbf{D}\mathbf{w} \tag{3.36a}$$

$$T = \frac{1}{2}\dot{\mathbf{w}}^t\mathbf{B}^t\mathbf{Q}\mathbf{B}\dot{\mathbf{w}} \equiv \frac{1}{2}\dot{\mathbf{w}}^t\mathbf{S}\dot{\mathbf{w}} \tag{3.36b}$$

Here, $\mathbf{D}$ and $\mathbf{S}$ represent the stiffness matrix and the mass matrix, respectively, represented in the space spanned by the normal modes of the substructures. NMA provides an eigenvalue problem such as $\mathbf{D}\mathbf{U} = \mathbf{S}\mathbf{U}\Omega$, where $\mathbf{U}$ is the modal matrix and Ω is the diagonal matrix whose component is an eigenvalue of a given protein structure. For the description of low-frequency motions, the modal matrix $\mathbf{U}$ has to be transformed into the matrix $\mathbf{V}$, whose column vector represents the normal modes, such as $\mathbf{V} = \Psi\mathbf{B}\mathbf{U}$, where $\Psi^t = [\Psi_A^t \ \Psi_B^t]$.

For robustness of CMS-based description on protein dynamics, we have considered hemoglobin as a model protein comprising four protein domains (575 residues). It is shown that the Debye–Waller factor for hemoglobin is well described by CMS, quantitatively comparable to original NMA (not shown). Here, we have used an ENM, for hemoglobin structures, with a spring constant of 0.886 kcal/mol · Å^2 and cutoff distance of 7 Å. Moreover, the lowest frequency normal mode computed from CMS is identical to that obtained from original NMA. Further, for quantitative description of the correlation between normal modes computed from CMS and original NMA, we have taken into account an overlap defined as $\chi_{ij} = \mathbf{v}_i^{NMA} \cdot \mathbf{v}_j^{CMS}$, where $\mathbf{v}_i^{NMA}$ and $\mathbf{v}_j^{CMS}$ represent the ith normal mode estimated from the original NMA and the jth normal mode evaluated from CMS, respectively. A value of χ_{ij} close to 1 indicates a high correlation between the ith normal mode computed from NMA and jth normal mode estimated from CMS. Figure 3.8b depicts that a map χ_{ij} is almost a diagonal matrix whose component is close to 1, indicating that normal modes computed from CMS are highly correlated with those calculated from NMA. For a quantitative comparison between CMS and NMA, we have also employed the parameter referred to as the *spanning coefficient* defined as $\mu_j = \sum_{i=1}^{M}\chi_{ij}$, where we set $M = 50$. The spanning coefficient μ_i indicates how much the ith normal mode of CMS can be described by the 50 normal modes of NMA. It is provided that the low-frequency normal modes of CMS can be almost depicted by the space spanned by the 50 normal modes of NMA (Figure 3.8c). This implies that an insight into a low-frequency motion relevant to conformational transitions can be efficiently gained from CMS.

3.7   OUTLOOK AND PERSPECTIVES

We have demonstrated the current state of the art in coarse-grained modeling schemes applicable to protein structures. Coarse graining is based on the principle that protein dynamics is governed by native topology rather than details of the atomic potential field. Over recent decades, coarse-grained models have been developed such that native topology has been well treated. For instance, topology-based models such as the ME model and Gō-like models have been widely used for studying protein folding and/or mechanical unfolding. Such topology-based models have led to the development of the ENM (first suggested by Tirion and later by many research groups), which allows computationally efficient description on conformational dynamics. It is remarkably shown that low-frequency normal modes of the ENM are highly correlated with conformational transitions, which supports the recently suggested binding mechanisms such as the *population shift model*. We have described the applications of the ENM as well as the novel elastic networks based on MD simulations and/or model reductions.

Ma et al. [12] and Brooks et al. [3] reported that the ENM can be constructed from a cryo-EM image, which provides a protein's shape rather than detailed information of atomic coordinates. This indicates that a coarse-grained model may be developed without using atomic coordinates as long as low-resolution structures are available from cryo-EM. Moreover, it has been recently provided that a finite element framework can be applicable to the molecular structure of proteins [135–137]. It is implied that an atomic coordinate-based coarse-grained model may be coupled to a finite element framework for computationally efficient modeling of protein structures. This sheds light on the applicability of multiscale modeling concepts such as the quasicontinuum model (QCM) [138–140] to computational simulation of large protein dynamics and mechanics. Here, the QCM has been broadly employed for studying the mechanical deformation of nanoscale structures such as metallic lattices [139,140]. To the best of our knowledge, multiscale modeling concepts such as the QCM have not been considered for modeling large protein structures for simulating their dynamic behavior.

Moreover, various types of elastic networks ranging from Tirion's model to coarse-grained elastic networks could be applicable for studying the conformational transitions of proteins in such a way that protein structures may be perturbed along the functional, low-frequency normal modes of such elastic networks. Recently, it has been suggested that the perturbation method of using low-frequency normal modes of the elastic network can be coupled to an all-atom MD simulation for predictions on conformational transition pathways. Specifically, a structural perturbation at relatively long timescale is implemented using low-frequency normal modes, while short timescale all-atom MD simulation is conducted to find equilibrium, intermediate conformations. This indicates that short timescale MD simulation could be coupled to ENM normal modes for gaining insight into conformational transitions.

ACKNOWLEDGMENTS

K. E. gratefully acknowledges the support by National Research Foundation of Korea (NRF) under Grant Nos. NRF-2009-0071246 and NRF-2010-0026223. S. N. appreciates the financial support by the NRF under Grant No. NRF-2008-314-D00012.

REFERENCES

1. Bahar, I., and A. J. Rader. 2005. Coarse-grained normal mode analysis in structural biology. *Curr Opin Struct Biol* 15:586–92.
2. Bahar, I., T. R. Lezon, A. Bakan, and I. H. Shrivastava. 2010. Normal mode analysis of biomolecular structures: Functional mechanisms of membrane proteins. *Chem Rev* 110:1463–97.

3. Tama, F., and C. L. Brooks. 2006. Symmetry, form, and shape: Guiding principles for robustness in macromolecular machines. *Annu Rev Biophys Biomol Struct* 35:115–33.

4. Smith, J. C. 1991. Protein dynamics: Comparison of simulations with inelastic neutron scattering experiments. *Q Rev Biophys* 24:227.

5. Engel, A., and H. E. Gaub. 2008. Structure and mechanics of membrane proteins. *Annu Rev Biochem* 77:127–48.

6. Hammes-Schiffer, S., and S. J. Benkovic. 2006. Relating protein motion to catalysis. *Annu Rev Biochem* 75:519–41.

7. Bahar, I., C. Chennubhotla, and D. Tobi. 2007. Intrinsic dynamics of enzymes in the unbound state and relation to allosteric regulation. *Curr Opin Struct Biol* 17:633–40.

8. Bakan, A., and I. Bahar. 2009. The intrinsic dynamics of enzymes plays a dominant role in determining the structural changes induced upon inhibitor binding. *Proc Natl Acad Sci U S A* 106:14349–54.

9. Balabin, I. A., W. Yang, and D. N. Beratan. 2009. Coarse-grained modeling of allosteric regulation in protein receptors. *Proc Natl Acad Sci U S A* 106:14253–8.

10. Halle, B. 2002. Flexibility and packing in proteins. *Proc Natl Acad Sci U S A* 99:1274–9.

11. Rose, G. D., P. J. Fleming, J. R. Banavar, and A. Maritan. 2006. A backbone-based theory of protein folding. *Proc Natl Acad Sci U S A* 103:16623–33.

12. Ming, D., Y. F. Kong, M. A. Lambert, Z. Huang, and J. P. Ma. 2002. How to describe protein motion without amino acid sequence and atomic coordinates. *Proc Natl Acad Sci U S A* 99:8620–5.

13. Ranson, N. A., D. K. Clare, G. W. Farr, D. Houldershaw, A. L. Horwich, and H. R. Saibil. 2006. Allosteric signaling of ATP hydrolysis in GroEL-GroES complexes. *Nat Struct Mol Biol* 13:147–52.

14. Joo, C., H. Balci, Y. Ishitsuka, C. Buranachai, and T. Ha. 2008. Advances in single-molecule fluorescence methods for molecular biology. *Annu Rev Biochem* 77:51–76.

15. Roy, R., S. Hohng, and T. Ha. 2008. A practical guide to single-molecule FRET. *Nat Methods* 5:507–16.

16. Karplus, M., and G. A. Petsko. 1990. Molecular dynamics simulations in biology. *Nature* 347:631.

17. Karplus, M., and J. A. McCammon. 2002. Molecular dynamics simulations of biomolecules. *Nat Struct Mol Biol* 9:646–52.

18. Allen, M. P., and D. J. Tildesley. 1987. *Computer Simulation of Liquids*. Oxford: Oxford University Press.

19. McCammon, J. A., B. R. Gelin, and M. Karplus. 1977. Dynamics of folded proteins. *Nature* 267:585–90.

20. Elber, R. 2005. Long-timescale simulation methods. *Curr Opin Struct Biol* 15:151–6.

21. Eom, K., J. Yang, J. Park, G. Yoon, Y. Sohn, S. Park, D. Yoon, S. Na, and T. Kwon. 2009. Experimental and computational characterization of biological liquid crystals: A review of single-molecule bioassays. *Int J Mol Sci* 10:4009–32.

22. Evans, E., and K. Ritchie. 1997. Dynamic strength of molecular adhesion bonds. *Biophys J* 72:1541–55.

23. Cui, Q., and I. Bahar. 2005. *Normal Mode Analysis: Theory and Applications to Biological and Chemical Systems*. Boca Raton, FL: CRC Press.

24. Brooks, B., and M. Karplus. 1983. Harmonic dynamics of proteins: Normal modes and fluctuations in bovine pancreatic trypsin inhibitor. *Proc Natl Acad Sci U S A* 80:6571–5.

25. Tama, F. 2003. Normal mode analysis with simplified models to investigate the global dynamics of biological systems. *Protein Pept Lett* 10:119–32.

26. Eom, K., G. Yoon, J.-I. Kim, and S. Na. 2010. Coarse-grained elastic models of protein structures for understanding their mechanics and dynamics. *J Comput Theor Nanosci* 7:1210–26.

27. Ayton, G. S., W. G. Noid, and G. A. Voth. 2007. Multiscale modeling of biomolecular systems: In serial and in parallel. *Curr Opin Struct Biol* 17:192–8.
28. Ayton, G. S., E. Lyman, and G. A. Voth. 2009. Hierarchical coarse-graining strategy for protein-membrane systems to access mesoscopic scales. *Faraday Discuss* 144:347–57.
29. Teeter, M. M., and D. A. Case. 1990. Harmonic and quasiharmonic descriptions of crambin. *J Phys Chem* 94:8091–7.
30. Lu, M. Y., and J. P. Ma. 2005. The role of shape in determining molecular motions. *Biophys J* 89:2395–401.
31. ben-Avraham, D. 1993. Vibrational normal-mode spectrum of globular proteins. *Phys Rev B* 47:14559–60.
32. Potestio, R., F. Caccioli, and P. Vivo. 2009. Random matrix approach to collective behavior and bulk universality in protein dynamics. *Phys Rev Lett* 103:268101.
33. Ciliberti, S., P. De Los Rios, and F. Piazza. 2006. Glasslike structure of globular proteins and the boson peak. *Phys Rev Lett* 96:198103.
34. Go, M., and N. Go. 1976. Fluctuations of an alpha-helix. *Biopolymers* 15:1119–27.
35. Ueda, Y., and N. Go. 1976. Theory of large-amplitude conformational fluctuations in native globular proteins. Independent fluctuating site model. *Int J Pept Protein Res* 8:551–8.
36. Noguti, T., and N. Go. 1982. Collective variable description of small-amplitude conformational fluctuations in a globular protein. *Nature* 296:776–8.
37. Ueda, Y., H. Taketomi, and N. Go. 1978. Studies on protein folding, unfolding, and fluctuations by computer simulation. II. A. Three-dimensional lattice model of lysozyme. *Biopolymers* 17:1531–48.
38. Munoz, V., P. A. Thompson, J. Hofrichter, and W. A. Eaton. 1997. Folding dynamics and mechanism of [beta]-hairpin formation. *Nature* 390:196–9.
39. Munoz, V., E. R. Henry, J. Hofrichter, and W. A. Eaton. 1998. A statistical mechanical model for beta-hairpin kinetics. *Proc Natl Acad Sci U S A* 95:5872–9.
40. Bruscolini, P., and A. Pelizzola. 2002. Exact solution of the Munoz-Eaton model for protein folding. *Phys Rev Lett* 88:258101.
41. Imparato, A., A. Pelizzola, and M. Zamparo. 2007. Ising-like model for protein mechanical unfolding. *Phys Rev Lett* 98:148102–4.
42. Mickler, M., R. I. Dima, H. Dietz, C. Hyeon, D. Thirumalai, and M. Rief. 2007. Revealing the bifurcation in the unfolding pathways of GFP by using single-molecule experiments and simulations. *Proc Natl Acad Sci U S A* 104:20268–73.
43. Hyeon, C., G. Morrison, and D. Thirumalai. 2008. Force-dependent hopping rates of RNA hairpins can be estimated from accurate measurement of the folding landscapes. *Proc Natl Acad Sci U S A* 105:9604–9.
44. Hyeon, C., G. Morrison, D. L. Pincus, and D. Thirumalai. 2009. Refolding dynamics of stretched biopolymers upon force quench. *Proc Natl Acad Sci U S A* 106:20288–93.
45. Dima, R. I., and H. Joshi. 2008. Probing the origin of tubulin rigidity with molecular simulations. *Proc Natl Acad Sci U S A* 105:15743–8.
46. Tirion, M. M. 1996. Large amplitude elastic motions in proteins from a single-parameter, atomic analysis. *Phys Rev Lett* 77:1905–8.
47. Atilgan, A. R., S. R. Durell, R. L. Jernigan, M. C. Demirel, O. Keskin, and I. Bahar. 2001. Anisotropy of fluctuation dynamics of proteins with an elastic network model. *Biophys J* 80:505–15.
48. Haliloglu, T., I. Bahar, and B. Erman. 1997. Gaussian dynamics of folded proteins. *Phys Rev Lett* 79:3090–3.
49. Eom, K., and S. Na. 2009. Coarse-grained structural model of protein molecules. In *Computational Biology: New Research*, ed. A. S. Russe, 193–213. New York: Nova Science Publisher.
50. Bahar, I., T. R. Lezon, L.-W. Yang, and E. Eyal. 2010. Global dynamics of proteins: Bridging between structure and function. *Annu Rev Biophys* 39:23–42.

51. Tama, F., and Y. H. Sanejouand. 2001. Conformational change of proteins arising from normal mode calculations. *Protein Eng* 14:1–6.
52. Valadie, H., J. J. Lacapcre, Y. H. Sanejouand, and C. Etchebest. 2003. Dynamical properties of the MscL of Escherichia coli: A normal mode analysis. *J Mol Biol* 332:657–74.
53. Tobi, D., and I. Bahar. 2005. Structural changes involved in protein binding correlate with intrinsic motions of proteins in the unbound state. *Proc Natl Acad Sci U S A* 102:18908–13.
54. McCammon, J. A., and S. Harvey. 1987. *Dynamics of Proteins and Nucleic Acids.* Cambridge: Cambridge University Press.
55. Brooks, C. L., M. Karplus, and B. M. Pettit. 1988. Proteins: A theoretical perspective of dynamics, structure, and thermodynamics. *Adv Chem Phys* 71:1–249.
56. Amadei, A., A. B. M. Linssen, and H. J. C. Berendsen. 1993. Essential dynamics of proteins. *Proteins Struct Funct Genet* 17:412–25.
57. Rueda, M., P. Chacon, and M. Orozco. 2007. Thorough validation of protein normal mode analysis: A comparative study with essential dynamics. *Structure* 15:565–75.
58. Yang, L., G. Song, A. Carriquiry, and R. L. Jernigan. 2008. Close correspondence between the motions from principal component analysis of multiple HIV-1 protease structures and elastic network modes. *Structure* 16:321–30.
59. Sotomayor, M., and K. Schulten. 2007. Single-molecule experiments in vitro and in silico. *Science* 316:1144–8.
60. Lu, H., and K. Schulten. 1999. Steered molecular dynamics simulations of force-induced protein domain unfolding. *Proteins* 35:453.
61. Lu, H., B. Isralewitz, A. Krammer, V. Vogel, and K. Schulten. 1998. Unfolding of titin immunoglobulin domains by steered molecular dynamics simulation. *Biophys J* 75:662.
62. Meirovitch, L. 1980. *Computational Methods in Structural Dynamics.* Rockville, MD: Sijthoff & Noordhoff.
63. Weiner, J. H. 1983. *Statistical Mechanics of Elasticity.* Mineola, NY: Dover publication.
64. Chandler, D. 1987. *Introduction to Modern Statistical Mechanics.* Oxford, UK: Oxford University Press.
65. Reichl, L. E. 2009. *A Modern Course in Statistical Physics.* Weinheim, Germany: Wiley-VCH.
66. Hyeon, C., and D. Thirumalai. 2008. Multiple probes are required to explore and control the rugged energy landscape of RNA hairpins. *J Am Chem Soc* 130:1538–9.
67. Hayward, S., and N. Go. 1995. Collective variable description of native protein dynamics. *Annu Rev Phys Chem* 46:223–50.
68. Sulkowska, J. I., and M. Cieplak. 2008. Selection of optimal variants of Go-like models of proteins through studies of stretching. *Biophys J* 95:3174–91.
69. Cieplak, M., T. X. Hoang, and M. O. Robbins. 2002. Thermal folding and mechanical unfolding pathways of protein secondary structures. *Proteins Struct Funct Genet* 49:104–13.
70. Cieplak, M., T. X. Hoang, and M. O. Robbins. 2002. Folding and stretching in a Go-like model of titin. *Proteins Struct Funct Genet* 49:114–24.
71. Cieplak, M., A. Pastore, and T. X. Hoang. 2005. Mechanical properties of the domains of titin in a Go-like model. *J Chem Phys* 122:054906.
72. Cieplak, M., T. X. Hoang, and M. O. Robbins. 2004. Thermal effects in stretching of Go-like models of titin and secondary structures. *Proteins Struct Funct Bioinfo* 56:285–97.
73. Sikora, M., J. I. Sułkowska, and M. Cieplak. 2009. Mechanical strength of 17,134 model proteins and cysteine slipknots. *PLoS Comput Biol* 5:e1000547.
74. Qin, Z., L. Kreplak, and M. J. Buehler. 2009. Hierarchical structure controls nanomechanical properties of vimentin intermediate filaments. *PLoS One* 4:e7294.
75. Fernandez, J. M., and H. Li. 2004. Force-clamp spectroscopy monitors the folding trajectory of a single protein. *Science* 303:1674.

76. Schlierf, M., and M. Rief. 2009. Surprising simplicity in the single-molecule folding mechanics of proteins. *Angew Chem Int Ed* 48:820–2.

77. Bahar, I., A. R. Atilgan, M. C. Demirel, and B. Erman. 1998. Vibrational dynamics of folded proteins: Significance of slow and fast motions in relation to function and stability. *Phys Rev Lett* 80:2733–6.

78. Kim, J.-I., S. Na, and K. Eom. 2009. Large protein dynamics described by hierarchical component mode synthesis. *J Chem Theor Comput* 5:1931–9.

79. Eom, K., S.-C. Baek, J.-H. Ahn, and S. Na. 2007. Coarse-graining of protein structures for normal mode studies. *J Comput Chem* 28:1400–10.

80. Zheng, W. 2008. A unification of the elastic network model and the Gaussian network model for optimal description of protein conformational motions and fluctuations. *Biophys J* 94:3853–7.

81. Xu, C. Y., D. Tobi, and I. Bahar. 2003. Allosteric changes in protein structure computed by a simple mechanical model: Hemoglobin T <-> R2 transition. *J Mol Biol* 333:153–68.

82. Boehr, D. D., R. Nussinov, and P. E. Wright. 2009. The role of dynamic conformational ensembles in biomolecular recognition. *Nat Chem Biol* 5:789–96.

83. del Sol, A., C.-J. Tsai, B. Ma, and R. Nussinov. 2009. The origin of allosteric functional modulation: Multiple pre-existing pathways. *Structure* 17:1042–50.

84. Cecchini, M., A. Houdusse, and M. Karplus. 2008. Allosteric communication in myosin V: From small conformational changes to large directed movements. *PLoS Comput Biol* 4:e1000129.

85. Miyashita, O., J. N. Onuchic, and P. G. Wolynes. 2003. Nonlinear elasticity, proteinquakes, and the energy landscapes of functional transitions in proteins. *Proc Natl Acad Sci U S A* 100:12570–5.

86. Whitford, P. C., O. Miyashita, Y. Levy, and J. N. Onuchic. 2007. Conformational transitions of adenylate kinase: Switching by cracking. *J Mol Biol* 366:1661–71.

87. Maragakis, P., and M. Karplus. 2005. Large amplitude conformational change in proteins explored with a plastic network model: Adenylate kinase. *J Mol Biol* 352:807–22.

88. Takagi, F., and M. Kikuchi. 2007. Structural change and nucleotide dissociation of myosin motor domain: Dual Go model simulation. *Biophys J* 93:3820–7.

89. Zheng, W., B. R. Brooks, and G. Hummer. 2007. Protein conformational transitions explored by mixed elastic network models. *Protein Struct Funct Bioinf* 69:43–57.

90. Kim, M. K., R. L. Jernigan, and G. S. Chirikjian. 2005. Rigid-cluster models of conformational transitions in macromolecular machines and assemblies. *Biophys J* 89:43–55.

91. Ikeguchi, M., J. Ueno, M. Sato, and A. Kidera. 2005. Protein structural change upon ligand binding: Linear response theory. *Phys Rev Lett* 94:078102.

92. Omori, S., S. Fuchigami, M. Ikeguchi, and A. Kidera. 2009. Linear response theory in dihedral angle space for protein structural change upon ligand binding. *J Comput Chem* 30:2602–8.

93. Demirel, M. C., and A. M. Lesk. 2005. Molecular forces in antibody maturation. *Phys Rev Lett* 95:208106.

94. Haliloglu, T., E. Seyrek, and B. Erman. 2008. Prediction of binding sites in receptorligand complexes with the Gaussian network model. *Phys Rev Lett* 100:228102.

95. Haliloglu, T., and B. Erman. 2009. Analysis of correlations between energy and residue fluctuations in native proteins and determination of specific sites for binding. *Phys Rev Lett* 102:088103.

96. Yang, L.-W., and I. Bahar. 2005. Coupling between catalytic site and collective dynamics: A requirement for mechanochemical activity of enzymes. *Structure* 13:893–904.

97. Isin, B., K. Schulten, E. Tajkhorshid, and I. Bahar. 2008. Mechanism of signal propagation upon retinal isomerization: Insights from molecular dynamics simulations of rhodopsin restrained by normal modes. *Biophys J* 95:789–803.

98. Kantarci-Carsibasi, N., T. Haliloglu, and P. Doruker. 2008. Conformational transition pathways explored by Monte-Carlo simulation integrated with collective modes. *Biophys J* 95:5862–73.

99. Dietz, H., and M. Rief. 2008. Elastic bond network model for protein unfolding mechanics. *Phys Rev Lett* 100:098101.

100. Bell, G. I. 1978. Models for the specific adhesion of cells to cell. *Science* 200:618–27.

101. Prezhdo, O. V., and Y. V. Pereverzev. 2009. Theoretical aspects of the biological catch bond. *Acc Chem Res* 42:693–703.

102. Evans, E. 2001. Probing the relation between force-life time and chemistry in single molecular bonds. *Annu Rev Biophys Biomol Struct* 30:105.

103. Evans, E. A., and D. A. Calderwood. 2007. Forces and bond dynamics in cell adhesion. *Science* 316:1148–53.

104. Sacquin-Mora, S., and R. Lavery. 2009. Modeling the mechanical response of proteins to anisotropic deformation. *Chemphyschem* 10:115–8.

105. Eom, K., P. C. Li, D. E. Makarov, and G. J. Rodin. 2003. Relationship between the mechanical properties and topology of cross-linked polymer molecules: Parallel strands maximize the strength of model polymers and protein domains. *J Phys Chem B* 107:8730–3.

106. Eom, K., D. E. Makarov, and G. J. Rodin. 2005. Theoretical studies of the kinetics of mechanical unfolding of cross-linked polymer chains and their implications for single-molecule pulling experiments. *Phys Rev E* 71:021904.

107. Dudko, O. K., A. E. Filippov, J. Klafter, and M. Urbakh. 2003. Beyond the conventional description of dynamic force spectroscopy of adhesion bonds. *Proc Natl Acad Sci U S A* 100:11378–81.

108. Dudko, O. K., G. Hummer, and A. Szabo. 2006. Intrinsic rates and activation free energies from single-molecule pulling experiments. *Phys Rev Lett* 96:108101.

109. Dudko, O. K. 2009. Single-molecule mechanics: New insights from the escape-over-a-barrier problem. *Proc Natl Acad Sci U S A* 106:8795–6.

110. Ming, D., and M. E. Wall. 2005. Allostery in a coarse-grained model of protein dynamics. *Phys Rev Lett* 95:198103.

111. Ming, D., and M. E. Wall. 2005. Quantifying allosteric effects in proteins. *Protein Struct Funct Bioinf* 59:697–707.

112. Moritsugu, K., and J. C. Smith. 2007. Coarse-grained biomolecular simulation with REACH: Realistic extension algorithm via covariance Hessian. *Biophys J* 93:3460–9.

113. Moritsugu, K., and J. C. Smith. 2008. REACH coarse-grained biomolecular simulation: Transferability between different protein structural classes. *Biophys J* 95:1639–48.

114. Lu, M., and J. Ma. 2008. A minimalist network model for coarse-grained normal mode analysis and its application to biomolecular x-ray crystallography. *Proc Natl Acad Sci U S A* 105:15358–63.

115. Tama, F., F. X. Gadea, O. Marques, and Y. H. Sanejouand. 2000. Building-block approach for determining low-frequency normal modes of macromolecules. *Proteins Struct Funct Genet* 41:1–7.

116. Lyman, E., J. Pfaendtner, and G. A. Voth. 2008. Systematic multiscale parameterization of heterogeneous elastic network models of proteins. *Biophys J* 95:4183–92.

117. Leherte, L., and D. P. Vercauteren. 2008. Collective motions in protein structures: Applications of elastic network models built from electron density distributions. *Comput Phys Commun* 179:171–80.

118. Van Wynsberghe, A. W., and Q. Cui. 2006. Interpreting correlated motions using normal mode analysis. *Structure* 14:1647–53.

119. Song, G., and R. L. Jernigan. 2006. An enhanced elastic network model to represent the motions of domain-swapped proteins. *Protein Struct Funct Bioinf* 63:197–209.

120. Navizet, I., R. Lavery, and R. L. Jernigan. 2004. Myosin flexibility: Structural domains and collective vibrations. *Proteins Struct Funct Genet* 54:384–93.
121. Kurkcuoglu, O., R. L. Jernigan, and P. Doruker. 2004. Mixed levels of coarse-graining of large proteins using elastic network model succeeds in extracting the slowest motions. *Polymer* 45:649–57.
122. Kurkcuoglu, O., R. L. Jernigan, and P. Doruker. 2005. Collective dynamics of large proteins from mixed coarse-grained elastic network model. *QSAR Comb Sci* 24:443–8.
123. Kurkcuoglu, O., O. T. Turgut, S. Cansu, R. L. Jernigan, and P. Doruker. 2009. Focused functional dynamics of supramolecules by use of a mixed-resolution elastic network model. *Biophys J* 97:1178–87.
124. Eom, K., J. H. Ahn, S. C. Baek, J. I. Kim, and S. Na. 2007. Robust reduction method for biomolecules modeling. *CMC Comput Mater Continua* 6:35–42.
125. Cheng, H., Z. Gimbutas, P. G. Martinsson, and V. Rokhlin. 2005. On the compression of low rank matrices. *SIAM J Sci Comput* 26:1389–404.
126. Jang, H., S. Na, and K. Eom. 2009. Multiscale network model for large protein dynamics. *J Chem Phys* 131:245106.
127. Doi, M., and S. F. Edwards. 1986. *The Theory of Polymer Dynamics*. New York: Oxford University Press.
128. Flory, P. J. 1953. *Principles of Polymer Chemistry*. New York: Cornell University Press.
129. Tozzini, V. 2005. Coarse-grained models for proteins. *Curr Opin Struct Biol* 15:144–50.
130. Ma, J. P. 2005. Usefulness and limitations of normal mode analysis in modeling dynamics of biomolecular complexes. *Structure* 13:373–80.
131. Bhat, R. B. 1985. Component mode synthesis in modal testing of structures. *J Sound Vibr* 101:271–2.
132. Sung, S. H., and D. J. Nefske. 1986. Component mode synthesis of a vehicle structural-acoustic system model. *AIAA J* 24:1021–6.
133. Ming, D., Y. Kong, Y. Wu, and J. Ma. 2003. Substructure synthesis method for simulating large molecular complexes. *Proc Natl Acad Sci U S A* 100:104–9.
134. Kim, J.-I., K. Eom, M.-K. Kwak, and S. Na. 2008. Application of component mode synthesis to protein structure for dynamic analysis. *CMC Comput Mater Continua* 8:67–73.
135. Tang, Y., G. Cao, X. Chen, J. Yoo, A. Yethiraj, and Q. Cui. 2006. A finite element framework for studying the mechanical response of macromolecules: Application to the gating of the mechanosensitive channel MscL. *Biophys J* 91:1248–63.
136. Chen, X., Q. Cui, Y. Tang, J. Yoo, and A. Yethiraj. 2008. Gating mechanisms of mechanosensitive channels of large conductance, I: A continuum mechanics-based hierarchical framework. *Biophys J* 95:563–80.
137. Bathe, M. 2008. A finite element framework for computation of protein normal modes and mechanical response. *Protein Struct Funct Bioinf* 70:1595–609.
138. Phillips, R., M. Dittrich, and K. Schulten. 2002. Quasicontinuum representation of atomic-scale mechanics: From proteins to dislocations. *Annu Rev Mater Res* 32:219–33.
139. Hayes, R. L., G. Ho, M. Ortiz, and E. A. Carter. 2006. Prediction of dislocation nucleation during nanoindentation of Al3 Mg by the orbital-free density functional theory local quasicontinuum method. *Philos Mag* 86:2343–58.
140. Fago, M., R. L. Hayes, E. A. Carter, and M. Ortiz. 2004. Density-functional-theory-based local quasicontinuum method: Prediction of dislocation nucleation. *Phys Rev B* 70:100102.

4 Continuum Modeling and Simulation of Membrane Proteins

Xi Chen

CONTENTS

4.1 INTRODUCTION

The biological and biomedical importance of membrane proteins can hardly be overstated. More than 30% of all genes encode membrane proteins and 60% of approved drug targets are membrane proteins. With recent breakthroughs in protein purification and crystallization, the number of medium- to high-resolution structures of membrane proteins is increasing at a remarkable pace. These structures provide the basis for understanding many fundamentally important biological processes at the molecular level, leading to a golden age in membrane protein research.

An emerging theme in the study of membrane proteins is that the membrane environment, which was once considered only a passive cellular component, can play a highly active role in dictating these proteins' structure and/or activity [1–4]. This is particularly compelling for mechanosensitive (MS) channels [2–7] (Figure 4.1), which are membrane proteins that open a pore in response to mechanical stress in the membrane, allowing the transport of cellular matter due to chemical potential gradients. The detailed mechanisms by which MS channels sense and convert mechanical deformation into biological signals provide a fascinating research challenge. The pioneering thermodynamics studies by Phillips et al. [8–10] using continuum mechanical models argue that the membrane deformation penalty can be the major energy component behind the gating of MS channels; a concept that was supported by the experimental observation that the gating threshold of MS channels can be modulated using membranes consisting of lipids with different lengths of tails. Using the same line of reasoning, Phillips et al. [10] made the intriguing proposal that MS channels (and any membrane proteins that undergo large-scale

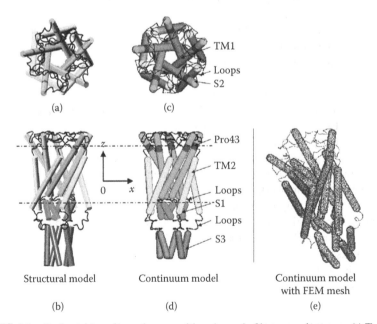

FIGURE 4.1 *Escherichia coli*-mechanosensitive channel of large conductance: (a) Top view and (b) side view of the closed structure of the homology model; (c) top view and (d) side view of the molecular dynamics-decorated finite element method model; and (e) mesh of the protein. The dashed line indicates the initial location of the lipid membrane. The protein components include the transmembrane TM1 bundle and TM2 subunits, and cytoplasmic S1 helices and S3 helices; they are connected by periplasmic and cytoplasmic loops.

structural transitions) exhibit a significant degree of cooperativity when they are within ~10 nm, which can be a typical value in cellular membranes. Experimental verification of cooperativity, however, remains scarce.

The involvement of the membrane environment in dictating the structure and/or activity of membrane proteins and presence of critical issues such as cooperativity call for an efficient multiscale computational approach. On the one hand, although remarkable progress has been made in atomistic simulation methodologies, it is still difficult to study the structural responses of membrane proteins using atomistic models alone since the accessible length scales and timescales are typically limited to 100 nm and 100 ns, respectively, whereas the cellular dimension is well beyond 1 μm and most functional structural transitions occur at the millisecond timescale or longer. On the other hand, although phenomenological models and thermodynamics approaches treating proteins and membranes as continuum mechanical objects with simple geometries can be highly instructive [1,5,8–12], a more quantitative analysis of membrane protein functions requires the development of refined models with sufficient details that treat both the protein and the membrane environment on an equal footing. Therefore, the fundamental challenge is to bridge the aforementioned gaps by establishing a computational framework that is sufficiently coarse grained to treat large length scales and timescales [2,7,13,14] at the same time including sufficient detail to faithfully capture the characteristics of the specific system.

An efficient multiscale framework that can both capture sufficient molecular details and deal with complex loadings over multiple scales is undoubtedly a valuable supplement to the standard experimental and modeling methods. In recent studies [15–17], my research group and colleagues have developed a top-down approach that effectively models biomolecules and their assemblies as integrated structures in an attempt to bridge the aforementioned gaps, allowing the deduction of physical mechanistic insights into mechanotransduction.

Using a representative system of *Escherichia coli*–MS channel of large conductance (MscL) as an example, current understandings of mechanotransduction based on experimental analysis, theoretical modeling, and numerical simulations are reviewed in this chapter. Since a set of up-to-date and extensive reviews was recently published on the experimental studies of MS channels in volumes 58 and 59 of the journal *Current Topics in Membrane*, this chapter mainly focuses on the computational aspects, especially insights into MscL using a continuum-based framework [15–17]. I believe many of these insights are highly relevant to the analysis of a host of other biomolecular systems, such as the MS channels in higher organisms. Further improvements of this multiscale model are also discussed in this chapter, and it is envisioned that such a refined framework will find great value in the study of mechanobiology.

4.2 OVERVIEW OF MECHANOSENSITIVE CHANNELS

Although the function of many transmembrane proteins has been shown to be sensitive to perturbations in the membrane [18], MS channels exhibit a most striking response to mechanical stress in the membrane and, therefore, they form a focal point in the proposed research. In bacteria, MS channels (MscLs) respond to load perturbation applied to the cell membrane or other membrane-associated components and act as "safety valves" by allowing the permeation of small ions and water molecules at high osmotic pressures [19]. So far, one of the most studied MS channels is *E. coli*–MscL, which has been chosen as a model system thanks to its ubiquity and simple structure. In Figure 4.1a and b (top and side views) the structure of *E. coli*–MscL in the closed state is shown, which was developed by homology modeling [20–22] based on the X-ray crystal structure of MscL in the bacteria *Mycobacterium tuberculosis* (Tb) [23] and available experimental constraints [24]. The structure of *E. coli*–MscL has fivefold symmetry, and residues on the top of transmembrane helices are connected by periplasmic loops, whereas those at the bottom of the transmembrane helices are linked to cytoplasmic helices via cytoplasmic loops. Among the transmembrane helices that directly interact with the membrane, the TM1 bundle consists of five longer helices that form an inner gate and the five TM2 helices form the outer bundle. The dashed line in Figure 4.1b indicates the approximate location of lipid membrane. In *E. coli*–MscL, TM1 and TM2 helices correspond to the residues Asn 15–Gly 50 and Val 77–Glu 107, respectively. There is a break in TM1 due to Pro43 (Figure 4.1) near the top of the TM1 helix, and in the literature the segment above Pro43 is sometimes referred to as S2 helices [20]. The cytoplasmic domain comprises bundles formed by S1 helices and S3 helices, which correspond to the residues Ile 3–Met 12 and Lys 117–Arg 135, respectively. It must be noted that the configurations of S3 helices are different in the homology models shown by Sukharev, Durell, and Guy [22] and Sukharev and Anishkin [21], where the S3 helices are longer in the newer model; in this chapter, the structure described by Sukharev and Anishkin [21] is used as the closed configuration. Among the three inner helical assemblies formed by TM1, S1, and S3 helix bundles, the transmembrane pore enclosed by TM1 helices is the most important and determines the ion flux that passes through; the size of the pore can be estimated by measuring the electric current [24]. An effective radius of the channel can be calculated from the area of the pentagon projection on the membrane plane formed by the principal axes of the TM1 helix bundle, which is ~6.5 Å for *E. coli*–MscL in its closed state.

The gating transitions of MS channels are triggered by external stimuli through their lipid environments. Membrane-activated gating behaviors were first observed by Kung's group when carrying out patch-clamp experiments on lipid vesicles [25]; since cytoskeleton and other membrane proteins were removed a priori, the mechanical deformation of lipid was demonstrated to be important during MS channel gating. The crucial role of membrane was also emphasized based on a thermodynamics analysis [26], where the free energy of lipid bilayer deformation was estimated as having the same order of magnitude as the energy barrier required for gating. Other experiments [27–29] and numerical studies [30,31] showed that MS channels are sensitive to lipid composition. Therefore, a proper analysis of the gating of MS channels requires explicit consideration of the lipid membrane.

A lipid bilayer membrane comprises phospholipids, and it forms a natural barrier between the inside and the outside of a cell to control materials exchange [32]. The exposed head groups of phospholipids are hydrophilic, whereas the tails are hydrophobic toward the center of the lipid bilayer. An example of dilauroylphosphatidylethanolamine lipid is shown in Figure 4.2a, into which an *E. coli*–MscL is inserted.

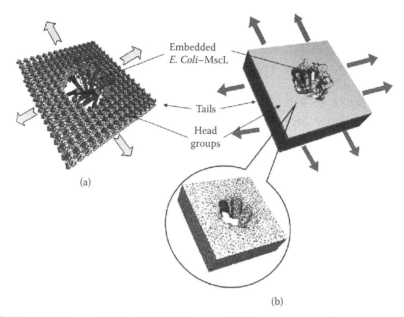

(a)

(b)

FIGURE 4.2 The assembled protein–lipid system: (a) Cartoon representation of *Escherichia coli*–mechanosensitive channel of large conductance (*E. coli*–MscL) and all-atom representation of a lipid, with a schematic of equibiaxial loading applied at the lipid boundary. (b) Continuum mechanics–based model of the *E. coli*–MscL lipid system. The finite element mesh of the trilayer lipid is shown in the zoomed-in view.

Under physiological conditions, this lipid is usually found in the fluid phase where it is incapable of bearing shear stress and of sustaining large strains; area expansion may rupture the membrane [33]. Nevertheless, within an elastic or a viscoelastic framework the effective mechanical properties of lipids (on tension, bending, etc.) can be estimated using relevant loads and deformation perturbations, and these can be used for modeling purposes [26]. For instance, the area expansion coefficient of a lipid bilayer is membrane tension per relative change in unit area, which is related to an effective Young's modulus of the membrane. Lipid membrane is nearly incompressible [34], which implies an effective Poisson's ratio close to 0.5. Besides in-plane tension, bending of lipid bilayer affects its curvature, and this contribution from an effective bending stiffness has been suggested to be important in the gating of MS channels [12]. This lipid property is inhomogeneous across the thickness and distinct peaks of lateral pressure have been characterized in the head group and tail regions as the area of the lipid is changed [30,35]; modification of the pressure profile can lead to different channel gating characteristics [9,36–38].

A living cell is subjected to diverse external stimuli [33] including, without being limited to, steady state contacts, high-frequency vibrations, and fluid shear stresses. These external stimuli may be superimposed with those stimuli generated internally, such as cytoskeletal polymerization and osmotic and hemodynamic pressure. It is interesting to investigate which stimulus is more relevant to mechanotransduction pathways. Since the forces acting on an MS channel are transferred through the lipid, the effect of various membrane deformation modes (e.g., in-plane tension and bending) on gating transition must be explored. The schematic in Figure 4.2a shows an example of equibiaxial tension, which is the most studied membrane deformation mode that can be triggered by variations in osmotic pressure.

4.3 LITERATURE REVIEWS OF MECHANOSENSITIVE CHANNELS OF LARGE CONDUCTANCE

This Section summarizes the efforts that have recently been made to characterize the mechanotransduction of MscL. In particular, this section presents not only the experimental attempts but also the computational approaches in gaining insights into the functions of MscL.

4.3.1 OVERVIEW OF EXPERIMENTAL INVESTIGATIONS

Patch-clamp experiments are a major tool for exploring the mechanism of cellular mechanotransduction. Mechanosensitive channels in the bacterial spheroplasts of *E. coli* were discovered by Martinac et al. [39] using this technique, which revealed pressure-activated, voltage-dependent, and ion-selective properties. By electrophysiological characterization, the gating behaviors of *E. coli*–MscL were observed also in reconstituted lipid vesicles [25]. The relationship between channel opening probabilities and membrane tension was reported for MscL by Sukharev et al. [40] and a five-subconductance-state model was established, which primarily attributed tension-dependent conformational transition to the pore area variation that occurred between the closed state and the first subconductance state [24].

Successful cloning and purification of Tb-MscL led to useful protein crystals and a solution of the X-ray structure at moderate resolutions (~3 Å) for the closed state. Availability of this structure was instrumental to the development of the homology model of *E. coli*–MscL, and this structure also guided many mutagenesis studies; the model further made it possible to carry out atomistic simulations to explore the mechanism of MscL. Nevertheless, it is worth emphasizing that due to the difference between crystallization and physiological conditions X-ray structures of membrane proteins may not correspond to their functional states (also see Section 4.2), not to mention the technical challenges associated with the refinement of X-ray structures at moderate resolutions. In the originally refined MscL structure (protein data bank [PDB] code is 1MSL), for example, the C-terminal S3 helix bundle has the unusual feature that hydrophobic groups face outward while several negatively charged residues cluster together. Anishkin et al. [41] modified this part of the structure based on the structure of pentameric cartilage protein, and the resulting conformation was shown to be consistent with results of disulfide trapping experiments. A revision of the X-ray structure (PDB code is 2OAR) [7,42] also supports the new conformation of S3. Recently, the modified model was confirmed by Maurer et al. [43] using combined computational and mutation studies. In a separate mutation study, Kloda, Ghazi, and Martinac [44] showed that it is likely the orientation of the C-terminal S3 helix bundle is pH dependent and an RKKEE cluster in that region in fact might function as a pH sensor.

The importance of various residues has been probed with numerous mutation studies. For example, the potential functional regions of MscL were revealed to involve Glu 56 and Lys 31 [45]; Gly 14 may serve as a pivot for the swinglike motion of the N-terminus [46]. Hydrophobicity of the residue Gly 22, which is located at the narrowest part of *E. coli*–MscL, was shown to play an important role in pressure sensitivity [47]. Loss-of-function mutations [48] found that the most severe mutations involve hydrophobic-to-hydrophilic substitutions at the rim of the funnel in MscL, which confirmed the idea that MscL receives forces from the membrane at the rim. Analysis of several gain-of-function mutants (e.g., V23D) [49] illustrated that transition from a dehydrated state to the water-filled state of a pore is likely to be the rate-limiting step in the gating of wild-type (WT) MscL, emphasizing that protein solvation could contribute substantially to the energetics of the gating process. Cutting or mutation of periplasmic loops, especially the residues that face the membrane, facilitates channel opening [50,51]. Taken together, these results paint a mechanistic picture in which the gating energetics is due to a combination of protein hydration, elastic loops rearrangements, and surrounding lipid membrane deformation.

Based on experimental constraints and known structural features of transmembrane proteins, structural models for the gating transition of Tb-MscL and *E. coli*–MscL upon equibiaxial tension have been proposed [22]. These models include 13 conformational states ranging from the closed state (with an effective pore radius, a, of about 6.5 Å) to an open conformation (when the maximum conductance can be measured experimentally) with $a = 19$ Å; these structural models are consistent with results from cysteine cross-linking experiments [20]. Subsequently, structural rearrangements in large prokaryotic MscL were determined by Perozo and coworkers [28,52] using electron paramagnetic resonance (EPR) spectroscopy and site-directed spin labeling (SDSL). The diameter change of the channel pentamer between its closed and open states was also characterized by a fluorescence resonance energy transfer (FRET) spectroscopic analysis [53]. Although tilts and rotations of transmembrane helices are features in the models of both Sukharev, Durell, and Guy [22] and Perozo and Rees [7], the magnitude of TM1 rotation is very different; the degree of TM1 rotation is counterclockwise (viewing from the cytoplasmic side) and slight in the study by Sukharev, Durell, and Guy [22], whereas the rotation is 110° clockwise in the study by Perozo and Rees [7]. Results of more recent in vivo substituted-cysteine accessibility method [54] and metal binding assays seem to be more consistent with the model based on EPR studies [7].

4.3.2 REVIEWS OF ANALYTICAL MODELING EFFORTS

Theoretical studies are useful for postulating principles behind the gating transition of MscL. A gating-by-tilting model was proposed based on thermodynamics as an alternative mechanism for dilatational gating [12]. By considering possible deformation mechanisms (e.g., membrane tension and torque), a thermodynamics formulation was presented by Wiggins and Philips [26] who established a pioneering lipid-centric model in which the system energy is dominated by hydrophobic mismatch and lipid tension while ignoring the contribution from protein deformation. This model was improved by incorporating other triggers, such as the change of membrane curvature and midplane deformation between closed and open states [9]; more recently, the model was further extended to study gating cooperativity [55].

Although these thermodynamics-based models provide useful insights into the common features of MS channels, the lack of sufficient structural details makes it difficult to evaluate their validity for a specific system (which is particularly important in biological systems where certain atomistic features are crucial to structure and function). For example, proteins were treated as either rigid or overly simplified objects [9,26,55] and thus their deformation energy contributions were largely ignored. Moreover, key parameters in these models were usually not derived from detailed simulations or experiments.

The elastic network model (ENM) [56] is a useful and efficient analytical alternative that incorporates the most salient structural details of proteins. In the ENM, the atomic structure of a macromolecule is simplified as a network of elastic springs, and a harmonic spring is applied to any pair of atoms (often calcium only) within a specified cutoff distance. Numerous studies have shown that the ENM can reliably reproduce low-frequency modes of proteins, which are often highly correlated with functional transitions. For MscL in particular, Valadie et al. [57] showed that the conformational transitions among the first few structures in the model of Sukharev, Durell, and Guy [22] can be described with three low-frequency modes only. The character of these modes clearly indicates an irislike movement involving both tilt and twist rotation of the transmembrane helices. Low-frequency modes can also be used in a linear response theory [58] to predict structural changes upon application of a force originating from the membrane. For *E. coli*–MscL, the structural changes upon equibiaxial tension predicted by the ENM [17] agreed qualitatively with structural models [21]. This protocol, however, requires the application of explicit force or displacement boundary conditions on certain atoms, whose selection is not straightforward [16,17].

4.3.3 PREVIOUS NUMERICAL APPROACHES

Numerical simulation is a powerful approach for exploring the fundamental principles of mechanobiology. Compared with analytical modeling, numerical simulations can incorporate sufficient structural details to propose or verify various mechanistic hypotheses as well as to improve the theoretical model such as theormodynamic formulation. In addition, numerical experiments can be manipulated in a well-defined way to help in both interpreting existing experimental data and designing (or stimulating the use of) new experiments.

All-atom simulations based on molecular dynamics (MD) are in principle capable of depicting gating transitions at the finest scale [59]. However, they are prohibitively expensive especially when the protein, lipid membrane, and the surrounding solvent are explicitly considered [60]. An MD simulation was first applied to study the gating of Tb-MscL [61], while this simulation is computationally limited due to its accessible timescale of 3 ns; in such timescale (i.e., 3 ns) the pore size barely changed during the simulation. Another 20 ns of simulation resulted in a similar behavior where the pore remained closed [62]. To accelerate structural transitions, protein deformation was decoupled from the membrane, and external forces were applied directly to the protein so as to assist gating. Steered MD (SMD) was used to study *E. coli*–MscL upon equibiaxial tension [60], where steering forces, estimated from the lateral and normal pressure profiles exerted by the deformed bilayer [30], were added on selected boundary atoms of the protein. Despite this bias, the channel merely opened to $a = 9.4$ Å after 12 ns of simulation, which highlights the limit of MD simulations in the context of probing the gating process. In fact, according to a recent work [62], in all-atom simulations channel opening could be directly observed only when an explicit lateral bias force (in the radially outward direction) was applied to all the five subunits of MscL.

Another artificially accelerated approach is targeted MD (TMD) [63]. In a study by Kong et al. [64], the lipid bilayer membrane was completely ignored and a holonomic constraint was used to drive the opening of an *E. coli*–MscL. The constraining force, however, can be unrealistically large, which makes the TMD useful only in a qualitative sense. In general, the final open structure must be specified in TMD simulations (thus gating is guaranteed), which is another major limitation of this approach. A similar approach was carried out by Bilston and Mylvaganam [65] on a Tb-MscL without an explicit lipid membrane, where the opening of MscL was possible if the membrane tension was above 12 dyn/cm.

During simulations at short timescales, the applied external forces can lead to unrealistic protein structural transitions and protein–lipid interactions. In order to partially circumvent this problem, inspired by the findings from some experiments conducted by Perozo et al. [28], Meyer et al. [66] studied the conformational transitions of *E. coli*–MscL in a precurved lipid membrane. Interesting observations were found through a 9.5-ns simulation, in which major structural rearrangements occurred in the perisplamic loops and extracellular helices. Although the channel radius increment was still limited, this technique attempted to combine bending and equibiaxial tension modes, and the transition of *E. coli*–MscL occurred locally in the absence of global external forces.

Recently, the effect of different lipids was studied in several all-atom simulations of MscL. Elmore and Dougherty [31] carried out simulations with MscL in several lipid bilayers with different head groups and tails. Substituting one lipid head group for another was found to modify protein–lipid interactions, which in turn led to different conformations of MscL. Changing the tail length and therefore the membrane thickness also induced structural changes in the channel. More recently, Debret et al. [67] compared simulations of MscL in DMPE and POPE, two lipids of different tail lengths. The MscL structure remained stable in the POPE simulations, but the transmembrane helices underwent tilts and kinks in the thinner DMPE simulations. These observations confirmed the importance of hydrophobic mismatch in the gating process.

Despite various improvements, all-atom simulations are still computationally intensive and the short simulation time may be inadequate for statistical sampling. Hence, developing coarse-grained models to access longer timescales has become a topic of much interest in the simulation community [68–71]. Most of the efforts in this direction are focused on developing particle-based models in which one bead represents a group of atoms. In the context of MscL, building on their success in developing an effective coarse-grained model for lipids, Yefimov et al. [68] developed a coarse-grained model for MscL based on the transfer of the free energy of amino acids between water and lipid. The gating transition of MscL was successfully observed in the simulation, although it was still obtained at elevated tension (60 mN/m rather than the experimental value of 20 mN/m required to open WT Tb-MscL) and temperature (338 K) values. Moreover, the final open state accessed in the simulation had a pore somewhat smaller than that estimated in the literature (radius of 0.70–0.75 nm vs. 1–2 nm). However, the coarse-grained model did reproduce the experimental observation that the V21D mutant opens at a substantially lower tension than the wild type, confirming the idea that solvation of hydrophobic residues constitutes an important part of the energy barrier for gating. With further improvements being made in the way protein structures are described, the particle-based coarse-graining framework is very promising, especially as a physically transparent way of describing protein–lipid interactions.

Despite its great potential, particle-based coarse-grained models suffer from significant computational cost and it is still not routine to carry out multiple simulations to the millisecond scale, which is often the biologically relevant timescale. Moreover, in the specific context of mechanosensation and mechanotransduction, it is not straightforward to apply particle-based models to study deformations involving large length scales or complex loading modes. These considerations motivated the development of a continuum mechanics–based simulation framework for mechanobiology [16,17] that can potentially bridge multiple length scales and adapt to complex loadings; at the same time, the model can include sufficient molecular details to capture some of the most important characteristics of a specific system. In the MD-decorated finite element method (MDeFEM) framework, biomolecules and their assemblies are modeled as integrated continuum structures that maintain some of the most important structural details (and redundant atomic details are excluded). This is motivated by the observation that the mechanical deformation of a biomolecular system is dictated by the superposition of several of its lowest modes, which can be described well by the collective behavior of its structural motifs (via phenomenological mechanical properties), and most local chemical/atomistic details are not important. Thus, the aim of MDeFEM approach is to efficiently treat deformations at large length scales and complex deformation modes that are inaccessible to conventional MD simulations, while retaining key features from atomistic simulations at various levels of sophistication.

In Section 4.4, I will illustrate the MDeFEM model through discussions of its application to *E. coli*–MscL (Figure 4.2b) [16,17]. In order to keep the focus on key physical principles that govern the gating process, I limit cases to very simple models of the channel in these preliminary studies, bearing in mind that the model can be systematically improved in the future (see Section 4.6). We believe that many of the underlying physical principles that govern the gating of *E. coli*–MscL [17] are applicable to other MS channels also.

4.4 CONTINUUM-BASED APPROACHES: MODEL AND METHODS FOR STUDYING MECHANOSENSITIVE CHANNELS OF LARGE CONDUCTANCE

In Figure 4.1c and d, the top and side views of the continuum model of *E. coli*–MscL are shown [16]; the geometries of all continuum components are measured from the closed state of the homology model [21] (Figure 4.1a and b). Each helix (TM1/TM2/ S1/S2/S3) is modeled as a three-dimensional (3D) elastic cylinder with a diameter of 5 Å (a typical value for the main chain of an α helix). As a first-order approximation the helix is considered homogeneous and isotropic, and the elastic properties remain constants during the gating transition. Due to its potentially important contribution, Pro43 [20] is treated in the helix model with an effective elastic modulus that is different from the remaining part of TM1 helices. The loops are taken to be quasi-one-dimensional elastic springs, whose mechanical properties are also assumed to be residue independent. Effective material properties, such as the nominal Young's modulus and Poisson's ratio of each helix and the spring constant of the loop, are calibrated by matching results of normal mode analyses (with respect to several lowest eigenmodes and frequencies) at the atomistic and continuum levels [16].

At the simplest level, a lipid bilayer can be effectively modeled as a sandwich plate structure (Figure 4.2b) by considering the different roles played by the head and tail regions in transducing mechanical stress [30,31]. Each layer (head group or tail) is assumed to be homogeneous and elastic, whose effective thickness and elastic constants are fitted based on a previous MD study [30]. To host the MS channel, a cavity having a conforming shape is created in the membrane. Based on this continuum model, the assembled continuum structure of an *E. coli*–MscL inside a lipid bilayer is shown in Figure 4.2b.

In MDeFEM, the structural components of biomolecules are integrated together through nonbonded interactions, which represent electrostatic and van der Waals interactions. For simplicity, we do not distinguish these two types of interactions and the total nonbonded energy is assumed to obey an effective pairwise potential function similar to the Lennard–Jones form, that is,

$$E_{\text{int}}(\alpha) = C\left[\frac{n}{m}\left(\frac{d_0}{\alpha}\right)^m - \left(\frac{d_0}{\alpha}\right)^n\right],$$

where d_0 and α are initial equilibrium distance and deformed distance between the two surfaces, respectively; and m and n denote the power indices for repulsive and attractive terms, respectively, with $n < m$ in general. Using this particular functional form is clearly an approximation, although similar kinds of short-range effective interactions have been found appropriate in coarse-grained simulation of other condensed-phase systems including highly polar liquids [72]. Such nonbonded interactions are applied to all relevant pairs of surfaces in the due course of gating; thus, the interaction areas between objects do not change during deformation. The values of the "well depth," C, and d_0, n, and m depend on different pairs of interactions (e.g., interactions between helices and those between a helix and the lipid), and they are calibrated by calculating and matching potential energies at atomistic and continuum levels [16]. Among the nonbonded interactions, the most essential one is protein–lipid interaction; a small perturbation of the corresponding C value has a negligible effect, whereas a significant change (e.g., 50% reduction) of the C parameter leads to more difficult gating.

Parameterization of helices, loops, lipid, and interactions is described by Chen et al. [16]. These parameters are meant to be order-of-magnitude estimates, and more quantitative parameterization (which also includes the functional form of nonbonded interactions) is an essential aspect for refinement in the future (see Section 4.6). The integrated system is meshed with finite elements (see example in Figure 4.1e for the mesh of protein bundle and Figure 4.2b for the elements near the lipid cavity). Commercial software ABAQUS [73] is used for FEM analyses. By taking advantage of the continuum approach, the typical simulation time is only a few hours on a regular workstation having a single central processing unit.

4.5 GATING MECHANISMS OF MECHANOSENSITIVE CHANNELS OF LARGE CONDUCTANCE AND INSIGHTS FOR MECHANOTRANSDUCTION

This Section provides the application of MDeFEM for characterization of the function (e.g. pore opening/closing) of MscL. In particular, this Section demonstrates the effect of loading modes and structural motifs on the mechanotransduction of MscL.

4.5.1 EFFECT OF DIFFERENT LOADING MODES

In many experimental and theoretical studies [9,28,12,38,40,60,62,66], several loading modes have been postulated as the trigger for MscL gating, including dilatational gating (equibiaxial tension) and gating by tilting (axisymmetric bending). These potential gating mechanisms are examined with MDeFEM simulations [17] to explore mechanotransduction pathways under different deformation modes.

4.5.1.1 Gating Behaviors upon Equibiaxial Tension

An equibiaxial strain up to 21% has been applied as a displacement boundary condition on the membrane [17]. Such a large strain is required since the lipid is modeled as a solid slab; despite such bias an appropriate expansion of lipid cavity can be achieved, which is necessary to accommodate the fully opened channel and is appropriate for exploring the important aspects of protein structural transition during gating.

Snapshots of the structural transition of *E. coli*–MscL at intermediate (half-opened) and open states obtained from MDeFEM [17] are given in Figure 4.3; compared with the structural model [21], the states obtained from MDeFEM show good agreement, which is encouraging considering the numerous approximations made in the MDeFEM model (see the following discussion). It is not surprising that with the increase of membrane strain, the lipid cavity expands and the forces are transmitted to the transmembrane helices of the protein structure via nonbonded interactions. Consequently, the transmembrane region undergoes most conformational changes characterized by its radial expansions, and the pore enclosed by the TM1 bundle opens up.

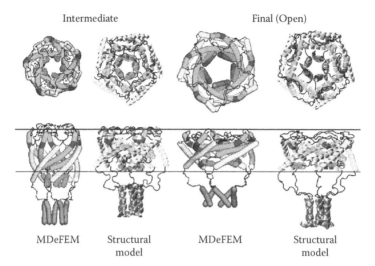

FIGURE 4.3 Snapshots along the gating pathways of a single *Escherichia coli*–mechanosensitive channel of large conductance (*E. coli*–MscL) at half-opened and fully opened states upon equibiaxial tension: Comparisons between the structural model and the molecular dynamics–decorated finite element method simulation (with a maximum membrane strain of 21%) are shown. Gating is primarily realized through the interactions between transmembrane helices and lipid, where the pore enclosed by the TM1 helices is pulled open. Other cytoplasmic helices and loops follow the trajectories of the transmembrane helices.

Besides lateral expansion, visible shrinking in thickness is observed for the transmembrane region, which is correlated with significant tilting of the helices. The longer and more flexible TM1 bends more than TM2 helices. Such significant deformation, which is required to maintain mechanical equilibrium during the gating process [17], remains to be verified using experimental studies at sufficient resolutions (It should be noted that the structural model described by Sukharev and Anishkin [21] also predicted significant helix bending curvature; Figure 4.3).

The MscL pore is defined by TM1 helices; thus the loops connecting TM1 and TM2 helices also impose constraints on the size of the pore. In addition, S1 helices affect pore conformation via interaction with TM1 helices; during gating, the S1 bundle expands in the radial direction and is lifted up toward the transmembrane region, which confirms the swinglike motions of the N-terminus [46]. The expansion of S1 helices, however, is smaller for this model than for the structural model, which might be attributed to neglecting solvation contributions in the current model (see Section 4.6). Being far away from the transmembrane helices, the S3 assembly remains essentially unchanged—this finding is in agreement with the later version of the structural model [21] (as opposed to the previous version [22]), suggesting that S3 helices are less important in terms of their mechanics roles (although they may bear other biochemical functions). The present system is resilient and can recover its closed state when membrane stress is removed.

In Figure 4.4a, the percentage increment of the effective pore radius of *E. coli*–MscL, a, is calculated from MDeFEM simulation as a function of membrane strain

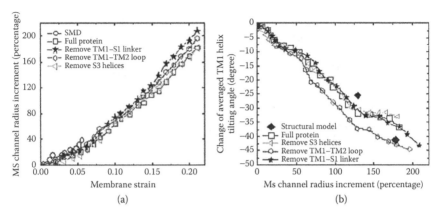

FIGURE 4.4 Predicted behavior of *Escherichia coli*–mechanosensitive channel of large conductance (*E. coli*–MscL) upon equibiaxial tension: (a) Evolution of the effective pore radius (enclosed by TM1 helices) of *E. coli*–MscL versus membrane strain and (b) the change of TM1 helix tilting angle as a function of the effective mechanosensitive channel radius. The results obtained from molecular dynamics (MD)–decorated finite element method simulation are compared with those from steered MD simulation (SMD; at small strains) and the structural model. In addition, various structural motifs are removed to explore the effect of these protein components during gating.

(the open-square curve in Figure 4.4) [17]. Based on elasticity, the relationship between membrane strain and lipid cavity expansion is linear [16], which leads to a monotonic increase in pore radius; the small deviation from perfect linearity in Figure 4.4a is due to the many-body interactions that affect equilibrium. At small strain, the variation of channel radius agrees well between MDeFEM [17] and SMD simulations (the open-circle curve in Figure 4.4a) [60]. This demonstrates that, at least qualitatively, the continuum-based model has a reasonable description for the forces involved in the gating process.

Another parameter characterizing pore shape variation is the averaged tilting angle of TM1 helices (with respect to the membrane plane), which decreases monotonically with pore radius (open-square curve in Figure 4.4b) [17]. Again, the MDeFEM results agree qualitatively with the structural model in the intermediate and open states (solid-diamond symbols) [21]. It is noted that the MDeFEM simulations are based only on the closed structure of MscL (unlike that in many other simulations where the final state must also be given so as to explore the pathways in between [64]), and thus such agreement is quite remarkable considering the preliminary nature of the MDeFEM model. As mentioned in Section 4.3, the mechanistic model of Sukharev, Durell, and Guy [22] differs from that of Perozo and Rees [7] in terms of the rotation of the TM1 helices. Since there is no side chain in the preliminary MDeFEM model, it is difficult to distinguish between the two mechanistic proposals at this stage; this remains an interesting subject for future MDeFEM study that employs more realistic shape for the continuum components.

Although the MDeFEM study demonstrates some promising results and agreements with the structural model, the rather monotonic behaviors of pore radius and helix tilting angle in Figure 4.4 illustrate a main limitation of the current continuum model, in which the effective (free) energy surface is essentially downhill toward the open state in the presence of external load. This is inconsistent with the free energy profile estimated by Sukharev et al. [24], which involves various intermediate states separated by pronounced free energy barriers; moreover, in the open state the channel radius is essentially insensitive to a wide range of tensions in experiments. We believe that further refinements, such as incorporating the effect of solvation forces [49,68,74], may help to make the MDeFEM approach more realistic (see Section 4.6). Indeed, an analysis of gain-of-function mutants [49] suggested that the unfavorable solvation of hydrophobic residues that line the channel pore constitutes a major part of the energy barrier for gating.

4.5.1.2 Gating Behaviors on Bending

Membrane bending is a commonly encountered deformation mode in a flexible cellular structure, and it becomes prominent during cell adhesion/contact. In order to study the pure bending behavior (i.e., decoupling with membrane stretching), a four-point flexure of a circular membrane was used [17]. With respect to Figure 4.5a, one can define the cone angle β using the effective radii of the five TM1 helices at the locations that correspond to the surfaces of the lipid membrane. When the membrane is bent upward the cone angle decreases almost monotonically with bending moment, and the TM1 helices become more upright at the final stage (Figure 4.5a).

Whether pure bending is an effective mode to promote gating can be explored by considering the effective pore radius evolution as a function of bending moment (Figure 4.5b). Despite the wall rotation of the lipid cavity, the average cavity radius throughout cavity thickness remains nearly the same during pure bending; therefore, the overall variation of the channel radius is small. The snapshots of the channel at half and maximum bending moments are also given in Figure 4.5b, where it is shown that despite the change in protein conformation, the transmembrane pore radius is only moderately enlarged. The evolution of pore radius shows a zigzag pathway with bending moment, which is attributed to the many-body interactions among helices. Overall, without the stretching component pure bending is not an effective mode to gate the channel. If excessive bending occurs and is coupled with significant in-plane stretching [9,28,66], the curvature effect becomes important.

4.5.1.3 Insights into Loading Modes versus Mechanotransduction

When an equibiaxial tension is applied on the membrane, gating is realized primarily by the irislike expansion of TM1/TM2 helices in the radial direction, as well as the tilting of the subunits, whose conformation transitions are directly coupled to lipid deformation. The S1 pore is also pulled open, in part due to its nonbonded interactions with the transmembrane helices and in part because of the loop linkers. The conformational transitions of the intermediate and open structures obtained from the MDeFEM simulation are qualitatively similar to the structural models [21]. In addition, the simulation results match with the results of all-atom SMD computations [60] in terms of channel radius evolution at the initial stage of the gating transition (Figure 4.4a). These observations indicate that the gating process is dominated by mechanics principles, including lipid membrane deformation and the deformation or interaction of helices and loops.

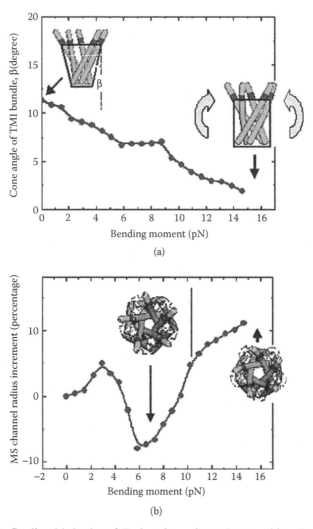

FIGURE 4.5 Predicted behavior of *Escherichia coli*–mechanosensitive channel of large conductance (*E. coli*–MscL) upon axisymmetric pure bending as the membrane is bent upward: (a) Change in the TM1 helix bundle cone angle β as a function of the line-bending moment, with the inserts showing configurations of the TM1 bundle at different instants; and (b) increment of the effective pore radius versus the line-bending moment, with the inserts showing configurations (top view) of the MscL at different instants.

The bending mode has been shown to only slightly affect overall channel radius. Thus, channel gating is relevant to some basic deformation modes (e.g., equibiaxial tension) in the membrane but not others (e.g., pure bending). Other deformation modes have also been studied [17], and from the mechanistic point of view equibiaxial tension is the most efficient way to achieve full gating. When these basic modes are combined, such as bending coupled with tension, the contribution of bending (membrane curvature) can be important for conformational transitions of the protein.

The present study deals with external load acting on the membrane only. However, during gating when the hydrophobic residues of the protein are exposed to water, the solvation force that is generated internally can also play an important role in destabilizing specific conformations and can affect gating [74]. This important contribution is missing in the present MDeFEM approach (see discussions in Section 4.6.1 on future improvements). Thus, the present agreement between MDeFEM and previous experimental and numerical studies may in part be a result of error cancellation.

4.5.2 EFFECTS OF STRUCTURAL MOTIFS

One of the focal goals of biophysical studies of mechanotransduction is to understand the diverse roles played by different structural components, such as the helices and loops in MS channels [21,22]. To that end, one may remove an individual group of structural motifs [17] and explore the change in gating behaviors. Since equibiaxial tension is shown to be the most effective for opening the MS channel, the focus here is on this basic loading mode. The membrane strain is controlled at 21%, which is required for maximum gating for the complete (reference) protein model shown in Figure 4.6a.

In Figure 4.6b [17], upon removal of the loops connecting TM1 and TM2 helices, the constraining effect is reduced and the averaged TM1 tilting angle is decreased by about 10° in the final state. The bending curvature of TM1 and TM2 helices also seems to be increased, which further affects the shapes of S1 and S3 bundles. Thus, the periplasmic loops are moderately important. This observation can be compared with the experimental finding that cutting the periplasmic loops facilitates the gating transition [50,51], although our "loopless" model is an oversimplification of

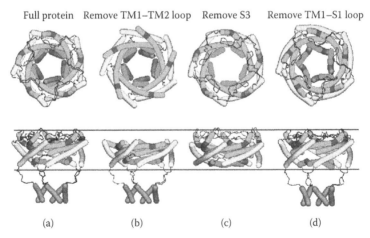

Full protein   Remove TM1–TM2 loop   Remove S3   Remove TM1–S1 loop

(a)   (b)   (c)   (d)

FIGURE 4.6 Effects of protein structural motifs: (a) Full protein, (b) without TM1–TM2 loop, (c) without S3 helix bundle, and (d) without TM1–S1 loop. It is found that moderate structural variations are caused by the removal of the loops, whereas the structural conformation is essentially insensitive to the removal of the S3 bundle. Thus, the continuum simulations show that the S3 bundle plays a relatively minor role during the mechanical gating event whereas the loops constrain gating.

the experimental system, in which the periplasmic loops were cut but not removed. Nevertheless, the qualitative behaviors are consistent.

Removing the S3 helix bundle has an insignificant effect on the deformation of the TM1/TM2/S1 helices (Figure 4.6c) [17], which is consistent with the observations from recent experiments [41]; this is because S3 helices are far away from other transmembrane protein components. When all the structural components except the loops that connect TM1 and S1 helices are kept, the S1 pore becomes distorted, which in turn affects both TM1 and TM2 helices (Figure 4.6d) [17], illustrating the importance of these loops and the S1 bundle. These findings are in qualitative agreement with experimental observations [22].

These trends are quantitatively shown in Figure 4.4a and b. After the S3 helices are removed, both the evolutions of pore radius and the TM1 helix tilting angle are almost identical to those found with the full protein model; as the loops (either TM1–TM2 or S1–TM1 linker) are taken away, the reduced constraining effect leads to a wider pore in the opened state [17]. These results show that during gating, different protein components bear diverse mechanical functions. The loops between TM1 and TM2 helices and the loops between TM1 and S1 helices, for example, may moderately affect the configuration of the channel in the fully open state. The most important helical components are the transmembrane helices, which directly interact with the membrane and sense the forces. These findings [17] are consistent with discussions in previous studies [22,41] and again illustrate the underlying mechanical principles of MscL gating.

4.5.3 Cooperativity of Mechanosensitive Channels

Biological membranes are highly heterogeneous and rich in proteins and other biomolecular components such as polysaccharides. Consider the most fundamental cooperativity problem where two MscLs are spatially close to one another [17]. The configurations of the channel at a membrane strain of 21% as the center-to-center separation between the MscLs is varied are given in Figure 4.7. When the two proteins are separated far apart, they do not affect each other's conformation. When the separation is reduced to 90 Å, the first column of Figure 4.7 shows how the TM1 pore becomes slightly distorted (with respect to the one shown in Figures 4.3 or 4.6a), which implies that the MscL starts to sense the existence of its neighbor. As the two channels further approach each other (with a separation of 60 Å), the second column of Figure 4.7 demonstrates how the TM1 pore becomes more elliptical, indicating that the interchannel interaction is significant.

In order to determine the critical distance at which cooperativity starts to occur, in Figure 4.8 we plot the bias ratio of the pore (the ratio between the shortest and the longest axes) at maximum membrane strain as a function of channel separation. When the two channels are far apart, the bias ratio is 1. The bias ratio decreases gradually with separation, and when the channel center-to-center distance λ falls below about 100 Å, the reduction of the bias ratio becomes more significant; this leads to a critical separation of about 100 Å [17], which is roughly four times the radius of the undeformed lipid cavity (c). This finding is consistent with the simple mechanics analysis based on lipid elastic deformation: According to the plane stress solution [75], the normalized stress concentration factor of a plate containing two circular

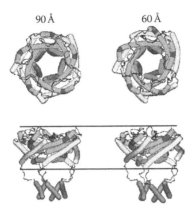

FIGURE 4.7 The interaction between two *Escherichia coli*–mechanosensitive channels of large conductance (*E. coli*–MscLs): The structural configurations of *E. coli*–MscL with a center-to-center separation $\lambda = 90$ and 60 Å (at an equibiaxial membrane strain of 21%). The pore enclosed by TM1 helices becomes increasingly distorted as the channel separation is reduced, indicating stronger magnitude of channel interactions.

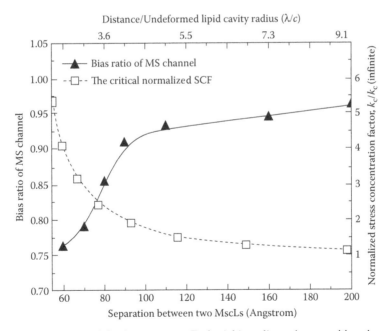

FIGURE 4.8 Cooperativity between two *Escherichia coli*–mechanosensitive channels of large conductance: The bias ratio varies with channel separation, and the results are compared with estimation by elasticity theory (the normalized stress concentration factor, SCF, of an elastic sheet containing two circular holes). The bias ratio is the short axis to long axis ratio of the distorted TM1 pore. The critical separation is about 100 Å, below which the two channels strongly interact with each other.

holes is increased sharply when the distance between these two holes is below about four times the hole radius (shown as the open-square curve in Figure 4.8). This critical separation is close to that identified from the MDeFEM simulation. The critical separation of 100 Å is also consistent with that found by Ursell et al. [55]. Thus, at a phenomenological level, the underlying mechanics principles provide a physical basis for MS channel cooperativity, although the quantitative aspects may depend on other physical and chemical features of the biological membrane and MS channels.

In their study, Ursell et al. [55] used a lipid-centric mechanics model and found that the total membrane energy containing two open channels can be reduced when they are close to each other. It must be noted that the protein deformation free energy and structural distortion were not considered in their model [55]; thus, whether such cooperative gating is realistic remains to be verified. The MDeFEM-based approach [17] further reveals that pore shape can be significantly distorted by channel–channel interactions, a feature that can be observed only with a model that includes structural details of the protein. To test this prediction, a carefully designed channel-recording study can be employed to measure the electric current through each channel as a function of channel density.

4.5.4 LARGE-SCALE SIMULATIONS OF LABORATORY EXPERIMENTS

The advantage of the MDeFEM framework is further demonstrated via simulations of large-scale experiments [17], such as the patch-clamp experiments [24,25,40] that have been instrumental to the mechanistic analysis of mechanotransduction. From the experiment conducted by Sukharev et al. [24], the geometry of a relatively rigid pipette can be measured (Figure 4.9a) and the opening

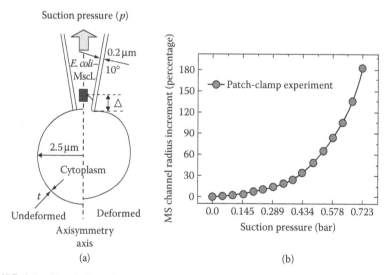

FIGURE 4.9 Simulation of the patch-clamp experiment: (a) Schematic of the experiment and location of the mechanosensitive channel of large conductance in the lipid vesicle; undeformed and deformed configurations of the lipid vesicle are compared. (b) Increment of effective pore radius as a function of suction pressure (until full gating) is shown.

of the pipette is 1 μm. Frictionless contact is assumed between the vesicle and the pipette, which does not correspond to the tight seal condition in many experiments [2,76,77] but can be improved without much difficulty. A change in such boundary conditions can affect the required suction pressure for gating (although the required local membrane strain is still the same). Without losing generality, the lipid vesicle may be modeled as an impermeable shell [40] filled with cytoplasm having a bulk modulus value the same as that of pure water (2.2 GPa) [78]. The averaged size of the liposome used in the patch-clamp experiment conducted by Sukharev et al. [40] was about 5 μm. To facilitate gating as well as to simplify the analysis, one can assume that an *E. coli*–MscL is located at the north pole of the vesicle [17], which leads to an equibiaxial stress field and makes the problem axisymmetric.

Figure 4.9a illustrates the undeformed and deformed liposome configurations. By applying a suction pressure, the top portion of the vesicle membrane is attracted into the pipette and assumes a bulged shape. The stress is also increased nonlinearly with the suction pressure; at about 0.7 bar, local stress near the channel leads to MscL pore opening close to pore opening in full gating (Figure 4.9b). The results also qualitatively agree with those measured from experiments [17,24]. The configuration of the final state of the channel is very close to that upon equibiaxial tension on a flat membrane (Figure 4.3); this is to be expected since the pipette opening is much larger than the protein dimension.

These types of numerical simulations of laboratory experiments at the cellular scale clearly demonstrate the uniqueness of the multiscale approach. The versatile continuum-based simulation framework may explain, guide, and stimulate new experiments, in which the protein location, number, species, and vesicle geometry can be varied. For example, spheroidal or ellipsoidal vesicle geometries can be used, and the resulting membrane tension stress will depend on the principal curvatures. Under such a circumstance, the change of MscL location will lead to different gating behaviors including a distorted open pore; such characteristics can be revealed from carefully designed channel-recording experiments. In addition to patch clamp, alternative loading experiments such as nanoindentation [17,79] can be explored. Finally, the interaction between multiple MS channels in a vesicle can be quite different from the equibiaxial tension case presented in Section 4.5.1.1, since the stress resulting from channel interaction is superimposed with that generated from complex vesicle geometry and loading mode.

4.6 FUTURE OUTLOOK AND SUGGESTED IMPROVEMENTS FOR CONTINUUM MECHANICS FRAMEWORKS

This Section provides the future outlook describing how to improve the MDeFEM model presented in this chapter as well as how to apply the MDeFEM for studying the functions of other biological channels. Section 4.6.1 presents how to enhance the MDeFEM model in order to capture the sophisticated phenomena at atomistic scales, whereas Section 4.6.2 delineates the future applications of MDeFEM for gaining the insights into the underlying mechanisms of mechanotransduction for other biological membranes.

4.6.1 FUTURE IMPROVEMENTS

There is plenty of room for improving the continuum mechanics framework. In essence, the model can be systematically made more realistic by incorporating additional refinements in terms of both structural features of the protein/membrane and simulation protocols. For example, a more sophisticated model of protein can be adopted. In Figure 4.10a, the molecular structure of a subunit (containing TM1/TM2/S1/S2/S3 helices and loops) of *E. coli*–MscL is shown. With the main chain of protein helices treated as a cluster of cylindrical cylinders (as in the present MDeFEM model, Figure 4.1), it is difficult to calculate the contribution of solvation. With respect to Figure 4.10a, real helices have irregular atomic surfaces dominated by side-chain atoms that interact extensively with each other as well as with the surrounding solvent/lipids. Therefore, a more realistic protein model should reflect its surface topology; an example of such an improved model of the MscL subunit is given in Figure 4.10b, which describes the solvent accessible surface area (SASA) [80] with meshed 3D elements. A similar model has been used by Bathe [81] to study low-frequency normal modes of large protein systems. The availability of SASA allows one to couple continuum mechanics and continuum solvation models, for example, the Poisson–Boltzmann (PB) [82], generalized Born (GB) [83], or other implicit methods [84], to describe structural transition of protein under the influence of solvents. It is suspected that the solvation term can cause local lipid membrane bending, thereby reducing the external membrane strain required for gating (which was unrealistically large in the study by Tang et al. [17]).

Besides incorporating the actual morphology of biomolecules such as membrane proteins, more sophisticated parameterization procedures can be adopted. For helices, a critical improvement that is under way is to treat them as heterogeneous (e.g., sequence dependent) and anisotropic. Such improvements are required to study helix

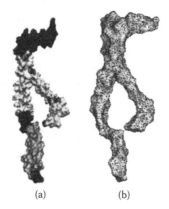

(a)   (b)

FIGURE 4.10 A single subunit (TM1/TM2/S1/S2/S3 helices and loops) in *Escherichia coli*–mechanosensitive channel of large conductance: (a) The solvent accessible surface area in molecular representation and (b) the finite element mesh of the continuum model. Such representation is useful for calculating solvation forces for significantly improving the current continuum model.

kinking, which has been proposed to play an important role in the gating transition of several channels [85]; handling the unfolding events of helices and other secondary structural elements, which have been proposed as part of many large-scale conformational transitions [86], is more challenging at the continuum level and it will be an important subject of research. For lipids, anisotropic properties are necessary to model the membrane in a more realistic fashion: For example, the lipid membrane is unable to bear in-plane shear deformation, and this phenomenological behavior can be modeled using transverse isotropic material description. In addition, local residual stress and curvature, as well as hyperelastic and time-dependent viscoelastic features of lipids, provide ways to avoid excessive strain and stress that is required by the current solid slab membrane model [16]. Although the solid slab lipid model is a major approximation in the continuum-based MDeFEM approach, it is shown that it can nevertheless effectively describe the appropriate expansion of lipid cavity that is needed to accommodate the full gating of MscL upon equibiaxial tension. The elastic model was also able to predict a critical channel separation for cooperativity in qualitative agreement with the study by Ursell et al. [55], despite the fact that the tension is not uniformly distributed in the solid membrane, although stress distribution in the realistic membrane is largely uniform.

The interactions among continuum components can also be described more realistically. Different interaction parameters can be assigned to hydrophilic and hydrophobic surface regions of the protein–lipid, which helps to properly anchor the protein in the membrane. The lateral interaction component can also be incorporated, which can stabilize the system and help create local free energy minima during the gating transition (i.e., making certain intermediate states more stable). Such refinements, along with the incorporation of solvation, can effectively create locally stabilizing wells and barriers and make the continuum model more realistic.

Finally, explicit time dependence in the continuum simulation at a constant temperature can be introduced into the framework of Langevin dynamics as is commonly done in particle-based simulations [87]. Introducing thermal fluctuations is essential to making the gating process more realistic, such as the asymmetry between the motions of different subunits discussed in recent experimental and simulation studies [28,52,62,88]. It is also expected that the application of electric field or external charges can affect the behaviors of membrane proteins [89,90], in addition to mechanical loads.

4.6.2   POTENTIAL APPLICATIONS OF MOLECULAR DYNAMICS–DECORATED FINITE ELEMENT METHOD TO OTHER CHANNEL SYSTEMS

The MDeFEM simulations so far have focused on MscL because of its relatively simple structural topology and extensive experimental background. Once calibrated, the MDeFEM framework can be applied to other MS channels that are less understood than MscL. For example, many transient receptor potential (TRP) channels, which have attracted much attention in recent years due to their diverse roles in important physiological functions [91,92], have mechanosensitivity. Sequence analysis of a specific system analyzed recently [93,94] indicates that the structure is likely similar to that of the voltage-gated potassium Shaker channel. Although the structural response of MscL during gating is very global in nature, the gating transition of the Shaker channel is believed to be dependent on local interactions and conformational transitions [95–97]. In addition, although MscL is mostly MS in nature, many TRP channels are known to be "polymodal" and can be activated by several different forms of stimuli [91,92]. Therefore, it is interesting to explore whether fundamentally different mechanosensing mechanisms are operational in these systems or there are common features, and to identify sequence/structural/energetic features that dictate the variations in the gating characteristics of these systems.

Another MS channel that has attracted much attention is the MS channel of small conductance (MscS). The MscS has homologues in many species and a much wider distribution than MscL. Although an X-ray structure of MscS [98] has been solved, the functional mechanism of this channel remains largely unclear. Indeed, even the identity of the X-ray structure is controversial; the relatively large pore radius led crystallographers [98] to conclude that the structure represents an open state of the channel, whereas the opposite conclusion was reached based on drying transitions observed in MD simulations [99]. More recent MD simulations argue that the X-ray structure is capable of transport once the transmembrane potential is present [100,101], although transmembrane potentials much larger than that found under physiology conditions had to be used to observe transport. The gating process of MscS is more complex than that of MscL in that it is not only activated by membrane tension but also modulated by voltage and has a rather complex inactivation mechanism under sustained stimulus [102–105]. Moreover, a recent study [104] found that the gating behavior of MscS is even sensitive to the rate of stimulation; it is fully responsive to sudden changes in tension but unresponsive to slowly applied tension. These complex features of MscS and the difficulties associated with performing a direct experimental analysis on the interplay between membrane tension and transmembrane voltage make a computational analysis of this system highly worthwhile. Since the opening of MscS is also enhanced by the addition of lysophosphatidylcholine and other amphipaths [106], it is likely that the basic MDeFEM framework established in the MscL study is also applicable, as well as further extensions to a wide variety of membrane proteins.

4.7 CONCLUSION

In this review, recent advances made in the mechanistic understanding of gating transitions in MS channels are summarized as an illustration of multiscale modeling and simulation of membrane proteins. The focus is mainly on insights derived from a continuum mechanics–based approach (MDeFEM) in the study of a widely studied MS channel, the *E. coli*–MscL. Although the model is clearly in its infancy and needs to be improved for quantitative analysis, its initial application has successfully demonstrated that mechanics principles play an important role in governing the conformational response of the MS channel to external mechanical perturbation in the membrane [17]. We believe that many of these mechanical principles are likely to play a role in various membrane-mediated biomechanical processes. The advantage, limitation, and future improvements of the continuum mechanics–based approach are discussed in this chapter. Although having a top-down nature, MDeFEM complements rather than competes with traditional bottom-up all-atom/coarse-grained simulations [68–71]. While the particle-based coarse-grained models are better suited for investigating mechanistic issues involving near-atom scales, such as the role of protein–lipid interactions in the gating of MS channels, the MDeFEM approach is particularly useful for studying processes that involve large-scale deformations and complex mechanistic perturbations. In fact, coarse-grained particle simulations can be used to generate some key properties required in the MDeFEM model as well as to test the reliability of MDeFEM simulations.

With the experience in applying the continuum framework to MscL and continued efforts in refining the versatile MDeFEM approach, similar studies can be made to systems that remain poorly understood, especially those with complex geometry and mechanical loads that are not accessible by standard particle-based simulations. The multiscale framework can also be applied to study the gating mechanisms of different types of ion channels on the application of various external stimuli, such as comparing the gating characteristics of the Shaker potassium channel [91,92] to MscL and the MscS [74] to identify unique structural and energetic features of MS channels. It is envisioned that these studies will lead to further exciting research aimed at uncovering the basic principles of mechanosensation, mechanotransduction, and other mechanobiological processes, as well as extending to the modeling and simulation of a variety of membrane proteins and other systems in mechanobiology.

ACKNOWLEDGMENT

The work involved in this chapter is supported by National Science Foundation (NSF) grants CMS-0407743 and CMMI-0643726.

REFERENCES

1. Kung, C. 2005. A possible unifying principle for mechanosensation. *Nature* 436:647–54.
2. Hamill, O. P., and B. Martinac. 2001. Molecular basis of mechanotransduction in living cells. *Physiol Rev* 81:685–740.
3. Martinac, B. 2004. Mechanosensitive ion channels: Molecules of mechanotransduction. *J Cell Sci* 117:2449–60.
4. Gustin, M. C., X. L. Zhou, B. Martinac, and C. Kung. 1988. A mechanosensitive ion channel in the yeast plasma membrane. *Science* 242:762–5.
5. Sukharev, S., P. Blount, B. Martinac, and C. Kung. 1997. Mechanosensitive channels of Escherichia coli: The MscL gene, protein and activities. *Annu Rev Physiol* 59:633–57.
6. Perozo, E. 2006. Gating prokaryotic mechanosensitive channels. *Nat Rev Mol Cell Biol* 7:109–19.
7. Perozo, E., and D. Rees. 2003. Structure and mechanism in prokaryotic mechanosensitive channels. *Curr Opin Struct Biol* 13:432–42.
8. Wiggins, P., and R. Philips. 2003. Analytical models for mechanotransduction: Gating a mechanosensitive channel. *Proc Natl Acad Sci U S A* 101:4071–6.
9. Wiggins, P., and R. Philips. 2005. Membrane-protein interactions in mechanosensitive channels. *Biophys J* 88:880–902.

10. Ursell, T., K. C. Huang, E. Peterson, and R. Phillips. 2007. Cooperative gating and spatial organization of membrane proteins through elastic interactions. *PLoS Comput Biol* 3:e81.

11. Markin, V. S., and F. Sachs. 2004. Thermodynamics of mechanosensitivity. *Phys Biol* 1:110–24.

12. Turner, M. S., and P. Sens. 2004. Gating-by-tilt of mechanically sensitive membrane channels. *Phys Rev Lett* 93:118103.

13. Perozo, E., A. Kloda, D. M. Cortes, and B. Martinac. 2002. Physical principles underlying the transduction of bilayer deformation forces during mechanosensitive channel gating. *Nat Struct Biol* 9:696–703.

14. Sukharev, S., and A. Anishkin. 2004. Mechanosensitive channels: What can we learn from 'simple' model systems? *Trends Neurosci* 27:345–51.

15. Tang, Y., G. Cao, X. Chen, J. Yoo, A. Yethiraj, and Q. Cui. 2006. A finite element framework for studying the mechanical response of macromolecules: Application to the gating of the mechanosensitive channel MscL. *Biophys J* 91:1248–63.

16. Chen, X., Q. Cui, J. Yoo, Y. Tang, and A. Yethiraj. 2008. Gating mechanisms of the mechanosensitive channels of large conductance part I: Theoretical and numerical framework. *Biophys J* 95:563–80.

17. Tang, Y., J. Yoo, A. Yethiraj, Q. Cui, and X. Chen. 2008. Gating mechanisms of mechanosensitive channels of large conductance part II: Systematic study of conformational transitions. *Biophys J* 95:581–96.

18. Andersen, O. S., and R. E. Koeppe. 2007. Bilayer thickness and membrane protein function: An energetic perspective. *Annu Rev Biophys Biomol Struct* 36:107–30.

19. Martinac, B. 2004. Mechanosensitive ion channels: Molecules of mechanotransduction. *J Cell Sci* 117:2449–60.

20. Sukharev, S., M. Betanzos, C. S. Chiang, and H. R. Guy. 2001. The gating mechanism of the large mechanosensitive channel MscL. *Nature* 409:720–4.

21. Sukharev, S., and A. Anishkin. 2004. Mechanosensitive channels: What can we learn from "simple" model systems? *Trends Neurosci* 27:345–51.

22. Sukharev, S., S. R. Durell, and H. R. Guy. 2001. Structural models of the MscL gating mechanism. *Biophys J* 81:917–36.

23. Chang, G., R. H. Spencer, A. T. Lee, M. T. Barclay, and D. C. Rees. 1998. Structure of the MscL homolog from Mycobacterium tuberculosis: A gated mechanosensitive ion channel. *Science* 282:2220–6.

24. Sukharev, S. I., W. J. Sigurdson, C. Kung, and F. Sachs. 1999. Energetic and spatial parameters for gating of the bacterial large conductance mechanosensitive channel, MscL. *J Gen Physiol* 113:525–39.

25. Sukharev, S. I., P. Blount, B. Martinac, F. R. Blattner, and C. Kung. 1994. A large-conductance mechanosensitive channel in E. coli encoded by MscI alone. *Nature* 368:265–8.

26. Wiggins, P., and R. Philips. 2004. Analytical models for mechanotransduction: Gating a mechanosensitive channel. *Proc Natl Acad Sci U S A* 101:4071–6.

27. Kloda, A., and B. Martinac. 2001. Mechanosensitive channel of Thermoplasma, the cell wall-less Archaea—Cloning and molecular characterization. *Cell Biochem Biophys* 34:321–47.

28. Perozo, E., A. Kloda, D. M. Cortes, and B. Martinac. 2002. Physical principles underlying the transduction of bilayer deformation forces during mechanosensitive channel gating. *Nat Struct Biol* 9:696–703.

29. Moe, P., and P. Blount. 2005. Assessment of potential stimuli for mechano-dependent gating of MscL: Effects of pressure, tension and lipid headgroups. *Biochem* 44:12239–44.

30. Gullingsrud, J., and K. Schulten. 2004. Lipid bilayer pressure profiles and mechanosensitive channel gating. *Biophys J* 86:3496–509.

31. Elmore, D. E., and D. A. Dougherty. 2003. Investigating lipid composition effects on the mechanosensitive channel of large conductance (MscL) using molecular dynamics simulations. *Biophys J* 85:1512–24.

32. Nagle, J. F., and S. Tristram-Nagle. 2000. Structure of lipid bilayers. *Biochim Biophys Acta Rev Biomembr* 1469:159–95.

33. Hamill, O. P., and B. Martinac. 2001. Molecular basis of mechanotransduction in living cells. *Physiol Rev* 81:685–740.

34. Evans, E., and R. Hochmuth. 1978. Mechanical properties of membranes. *Topics membr transp* 10:1–64.

35. Lindahl, E., and O. Edholm. 2000. Spatial and energetic-entropic decomposition of surface tension in lipid bilayers from molecular dynamics simulations. *J Chem Phys* 113:3882–93.

36. Perozo, E. 2006. Gating prokaryotic mechanosensitive channels. *Nat Rev Mol Cell Biol* 7:109–19.

37. Kung, C. 2005. A possible unifying principle for mechanosensation. *Nature* 436:647–54.

38. Markin, V. S., and F. Sachs. 2004. Thermodynamics of mechanosensitivity: Lipid shape, membrane deformation and anesthesia. *Biophys J* 86:370A.

39. Martinac, B., M. Buechner, A. H. Delcour, J. Adler, and C. Kung. 1987. Pressure-sensitive ion channel in Escherichia-coli. *Proc Natl Acad Sci U S A* 84:2297–301.

40. Sukharev, S. I., P. Blount, B. Martinac, and C. Kung. 1997. Mechanosensitive channels of Escherichia coli: The MscL gene, protein, and activities. *Annu Rev Physiol* 59:633–57.

41. Anishkin, A., V. Gendel, N. A. Sharifi, C.-S. Chiang, L. Shirinian, H. R. Guy, and S. Sukharev. 2003. On the conformation of the COOH-terminal domain of the large mechanosensitive channel MscL. *J Gen Physiol* 121:227–44.

42. Steinbacher, S., R. Bass, P. Strop, and D. C. Rees. 2007. Structures of the prokaryotic mechanosensitive channels MscL and MscS. *Curr Top Membr* 58:1–24.

43. Maurer, J. A., D. E. Elmore, D. Clayton, L. Xiong, H. A. Lester, and D. A. Dougherty. 2008. Confirming the revised C-terminal domain of the MscL crystal structure. *Biophys J* 94:4662–7.

44. Kloda, A., A. Ghazi, and B. Martinac. 2006. C-terminal charged cluster of MscL, RKKEE, functions as a pH Sensor. *Biophys J* 90:1992–8.

45. Blount, P., S. I. Sukharev, P. C. Moe, S. K. Nagle, and C. Kung. 1996. Towards an understanding of the structural and functional properties of MscL, a mechanosensitive channel in bacteria. *Biol Cell* 87:1–8.

46. Gu, L. Q., W. H. Liu, and B. Martinac. 1998. Electromechanical coupling model of gating the large mechanosensitive ion channel (MscL) of Escherichia coli by mechanical force. *Biophys J* 74:2889–902.

47. Yoshimura, K., A. Batiza, M. Schroeder, P. Blount, and C. Kung. 1999. Hydrophilicity of a single residue within MscL correlates with increased channel mechanosensitivity. *Biophys J* 77:1960–72.

48. Yoshimura, K., T. Nomura, and M. Sokabe. 2004. Loss-of-function mutations at the rim of the funnel of mechanosensitive channel MscL. *Biophys J* 86:2113–20.

49. Anishkin, A., C. S. Chiang, and S. Sukharev. 2005. Gain of function mutations reveal expanded intermediate states and a sequential action of two gates in mscl. *J Gen Physiol* 125:155–70.

50. Ajouz, B., C. Berrier, M. Besnard, B. Martinac, and A. Ghazi. 2000. Contributions of the different extramembranous domains of the mechanosensitive ion channel MscL to its response to membrane tension. *J Biol Chem* 275:1015–22.

51. Park, K. H., C. Berrier, B. Martinac, and A. Ghazi. 2004. Purification and functional reconstitution of N- and C-halves of the MscL channel. *Biophys J* 86:2129–36.
52. Perozo, E., D. M. Cortes, P. Sompornpisut, A. Kloda, and B. Martinac. 2002. Open channel structure of MscL and the gating mechanism of mechanosensitive channels. *Nature* 418:942–8.
53. Corry, B., P. Rigby, Z. W. Liu, and B. Martinac. 2005. Conformational changes involved in MscL channel gating measured using FRET spectroscopy. *Biophys J* 89:L49–51.
54. Blount, P., M. J. Schroeder, and C. Kung. 1997. Mutations in a bacterial mechanosensitive channel change the cellular response to osmotic stress. *J Biol Chem* 272:32150–7.
55. Ursell, T., K. C. Huang, E. Peterson, and R. Phillips. 2007. Cooperative gating and spatial organization of membrane proteins through elastic interactions. *Plos Comput Biol* 3:803–12.
56. Tirion, M. M. 1996. Low amplitude motions in proteins from a single-parameter atomic analysis. *Phys Rev Lett* 77:1905–8.
57. Valadie, H., J. J. Lacapcre, Y. H. Sanejouand, and C. Etchebest. 2003. Dynamical properties of the MscL of Escherichia coli: A normal mode analysis. *J Mol Biol* 332:657–74.
58. Ikeguchi, M., J. Ueno, M. Sato, and A. Kidera. 2005. Protein structural change upon ligand binding: Linear response theory. *Phys Rev Lett* 94:078102.
59. Karplus, M., and J. A. McCammon. 2002. Molecular dynamics simulations of biomolecules. *Nat Struct Biol* 9:788.
60. Gullingsrud, J., and K. Schulten. 2003. Gating of MscL studied by steered molecular dynamics. *Biophys J* 85:2087–99.
61. Gullingsrud, J., D. Kosztin, and K. Schulten. 2001. Structural determinants of MscL gating studied by molecular dynamics simulations. *Biophys J* 80:2074–81.
62. Jeon, J., and G. A. Voth. 2008. Gating of the mechanosensitive channel protein MscL: The interplay of membrane and protein. *Biophys J* 94:3497–511.
63. Schlitter, J., M. Engels, P. Kruger, E. Jacoby, and A. Wollmer. 1993. Targeted molecular-dynamics simulation of conformational change—application to the T$R transition insulin. *Mol Simul* 10:291–308.
64. Kong, Y. F., Y. F. Shen, T. E. Warth, and J. P. Ma. 2002. Conformational pathways in the gating of Escherichia coli mechanosensitive channel. *Proc Natl Acad Sci U S A* 99:5999–6004.
65. Bilston, L. E., and K. Mylvaganam. 2002. Molecular simulations of the large conductance mechanosensitive (MscL) channel under mechanical loading. *Febs Lett* 512:185–90.
66. Meyer, G. R., J. Gullingsrud, K. Schulten, and B. Martinac. 2006. Molecular dynamics study of MscL interactions with a curved lipid bilayer. *Biophys J* 91:1630–7.
67. Debret, G., H. Valadié, A. M. Stadler, and C. Etchebest. 2008. New insights of membrane environment effects on MscL channel mechanics from theoritical approaches. *Proteins* 71:1183–96.
68. Yefimov, S., E. van der Giessen, P. R. Onck, and S. J. Marrink. 2008. Mechanosensitive membrane channels in action. *Biophys J* 94:2994–3002.
69. Shi, Q., S. Izvekov, and G. A. Voth. 2006. Mixed atomistic and coarse-grained molecular dynamics: Simulation of a membrane-bound ion channel. *J Phys Chem B* 110:15045–8.
70. Marrink, S. J., H. J. Risselada, S. Yefimov, D. P. Tieleman, and A. H. de Vries. 2007. The MARTINI force field: Coarse-grained model for biomolecular simulations. *J Phys Chem B* 111:7812–24.
71. Lopez, C. F., S. O. Nielsen, P. B. Moore, and M. L. Klein. 2004. Understanding nature's design for a nanosyringe. *Proc Natl Acad Sci U S A* 101:4431–4.
72. Izvekov, S., and G. A. Voth. 2005. Multiscale coarse graining of liquid-state systems. *J Chem Phys* 123:134105.

73. ABAQUS. 2004. *Abaqus 6.4 User's Manual.* ABAQUS Inc.: Providence, Rhode Island, USA.

74. Anishkin, A., and C. Kung. 2005. Microbial mechanosensation. *Curr Opin Neurobiol* 15:397–405.

75. Ling, C. B. 1948. On the stresses in a plate containing 2 circular holes. *J Appl Phys* 19:77–81.

76. Hamill, O. P., A. Marty, E. Neher, B. Sakmann, and F. J. Sigworth. 1981. Improved patch-clamp techniques for high-resolution current recording from cells and cell-free membrane patches. *Pflugers Arch* 391:85–100.

77. Sachs, F., and C. E. Morris. 1998. Mechanosensitive ion channels in nonspecialized cells. *Rev Physiol Biochem Pharmacol* 132:1–77.

78. Hartmann, C., and A. Delgado. 2004. Stress and strain in a yeast cell under high hydrostatic pressure. *PAMM* 4:316–7.

79. Gordon, V. D., X. Chen, J. W. Hutchinson, A. R. Bausch, M. Marquez, and D. A. Weitz. 2004. Self-assembled polymer membrane capsules inflated by osmotic pressure. *J Am Chem Soc* 126:14117–22.

80. Sanner, M. 2008. http://www.scripps.edu/sanner/html/msms_home.html.

81. Bathe, M. 2008. A finite element framework for computation of protein normal modes and mechanical response. *Protein Struct Funct Bioinf* 70:1595–609.

82. Baker, N. A., D. Sept, S. Joseph, M. J. Holst, and J. A. McCammon. 2001. Electrostatics of nanosystems: Application to microtubules and the ribosome. *Proc Natl Acad Sci U S A* 98:10037–41.

83. Feig, M., and C. L. Brooks. 2004. Recent advances in the development and application of implicit solvent models in biomolecule simulations. *Curr Opin Struct Biol* 14:217–24.

84. Zhou, Y. C., M. Holst, and J. A. McCammon. 2008. A nonlinear elasticity model of macromolecular conformational change induced by electrostatic forces. *J Math Anal Appl* 340:135–64.

85. Akitake, B., A. Anishkin, N. Liu, and S. Sukharev. 2007. Straightening and sequential buckling of the pore-lining helices define the gating cycle of MscS. *Nat Struct Mol Biol* 14:1141–9.

86. Miyashita, O., J. N. Onichic, and P. G. Wolynes. 2003. Nonlinear elasticity, proteinquakes, and the energy landscapes of functional transitions in proteins. *Proc Natl Acad Sci U S A* 100:12570–5.

87. Zwanzig, R. 2001. *Nonequilibrium Statistical Mechanics.* New York: Oxford University Press.

88. Iscla, I., G. Levin, R. Wray, and P. Blount. 2007. Disulfide trapping the mechanosensitive channel MscL into a gating-transition state. *Biophys J* 92:1224–32.

89. Liu, L., Y. Qiao, and X. Chen. 2008. Pressure-driven water infiltration into carbon nanotube: The effect of applied charges. *Appl Phys Lett* 92:101927.

90. Han, A., X. Chen, and Y. Qiao. 2008. Effects of addition of electrolyte on liquid infiltration in a hydrophobic nanoporous silica gel. *Langmuir* 24:7044–7.

91. Ramsey, I. S., M. Delling, and D. E. Clapham. 2006. An introduction to TRP channels. *Annu Rev Physiol* 68:619–47.

92. Dhaka, A., V. Viswanath, and A. Patapoutian. 2006. TRP ion channels and temperature sensation. *Annu Rev Neurosci* 29:135–61.

93. Zhou, X., A. F. Batiza, S. H. Loukin, C. P. Palmer, C. Kung, and Y. Saimi. 2003. The transient receptor potential channel on the yeast vacuole is mechanosensitive. *Proc Natl Acad Sci U S A* 100:7105–10.

94. Saimi, Y., X. L. Zhou, S. H. Loukin, W. J. Haynes, and C. Kung. 2007. Microbial TRP channels and their mechanosensitivity. *Mechanosensitive Ion Channels Part A* 58:311–27.

95. Tombola, F., M. M. Pathak, and E. Y. Isacoff. 2006. How does voltage open an ion channel. *Annu Rev Cell Dev Biol* 22:23–52.

96. Long, S. B., E. B. Campbell, and R. MacKinnon. 2005. Crystal structure of a mammalian voltage-dependent Shaker family K+ channel. *Science* 309:897–903.

97. Long, S. B., E. B. Campbell, and R. MacKinnon. 2005. Voltage sensor of kv1.2: Structural basis of electromechanical coupling. *Science* 309:903–8.

98. Bass, R. B., P. Strop, M. T. Barclay, and D. C. Rees. 2002. Crystal structure of Escherichia coli MscS, a voltage-modulated and mechanosensitive channel. *Science* 298:1582–7.

99. Anishkin, A., and S. Sukharev. 2004. Water dynamics and dewetting transition in the small mechanosensitive channel MscS. *Biophys J* 86:2883–95.

100. Spronk, S. A., D. E. Elmore, and D. A. Dougherty. 2006. Voltage-dependent hydration and conduction properties of the hydrophobic pore of the mechanosensitive channel of small conductance. *Biophys J* 90:3555–69.

101. Sotomayor, M., V. Vasquez, E. Perozo, and K. Schulten. 2007. Ion conduction through MscS as determined by electrophysiology and simulation. *Biophys J* 92:886–902.

102. Edwards, M. D., I. R. Booth, and S. Miller. 2004. Gating the bacterial mechanosensitive channels: MscS a new paradigm? *Curr Opin Micro* 7:163–7.

103. Edwards, M. D., Y. Li, S. Kim, S. Miller, W. Bartlett, S. Black, S. Dennison, I. Iscla, P. Blount, J. U. Bowie, and I. R. Booth. 2005. Pivotal role of the glycine-rich TM3 helix in gating the MscS mechanosensitive channel. *Nat Struct Mol Biol* 12:113–9.

104. Akitake, B., A. Anishkin, and S. Sukharev. 2005. The "dashpot" mechanism of stretch-dependent gating in MscS. *J Gen Physiol* 125:143–54.

105. Perozo, E. 2006. Gating prokaryotic mechanosensitive channels. *Nat Rev Mol Cell Biol* 7:109–19.

106. Martinac, B., J. Adler, and C. Kung. 1990. Mechanosensitive ion channels of *E. coli* activated by amphipaths. *Nature* 348:261–3.

5 Exploring the Energy Landscape of Biopolymers Using Single-Molecule Force Spectroscopy and Molecular Simulations

Changbong Hyeon

CONTENTS

5.1 INTRODUCTION

The advent of single-molecule (SM) techniques over the past decades has brought a significant impact on the studies of biological systems [1–3]. The spatial and temporal resolutions and a good force control achieved in SM techniques have been used to decipher the microscopic basis of self-assembly processes in biology. Among SM techniques, single-molecule force spectroscopy (SMFS) is adapted not only to stretch biopolymers [4,5] but also to unravel the internal structures and functions of many proteins and nucleic acids [6–10]. By precisely restricting the initial and final conformations onto specific regions of the energy landscape, SMFS also provided a way to probe the collapse or folding dynamics under mechanical control, which fundamentally differs from those under temperature or denaturant control [11,12]. The observables that are usually inaccessible with conventional bulk experiments, for instance, the heterogeneity of dynamic trajectories and intermediate state ensembles, have been measured to provide glimpses of the topography of complex folding landscapes [7,13]. Furthermore, the use of SMFS is being extended to study the function of biological motors [14–20] and cells [21].

Given that foldings of biopolymers are realized through a number of elementary processes, good control over time, length, force, and energy scales are essential to resolve the details of biomolecular self-assembly [22–24]. The ability to control the energy scale within the range of ~k_BT ($k_BT \approx 4.1$ pN·nm at room temperature $T = 300$ K), in particular, allows us to study how biological systems, which are evolved to accommodate the thermal fluctuations, adapt their conformations with versatility to a varying environment. SMFS is an excellent tool to decompose the energy required to disrupt noncovalent bonds, responsible for the stability of biological structures (~$O(1)$ k_BT), into ~piconewton force and ~nanometer length scales. A phenomenological interpretation of bond rupture due to an external force in the context of cell–cell adhesion process and a theoretical estimate of mechanical force associated with the process had already been discussed as early as 1978 by Bell [25]. However, only after the 1990s, with SMFS its experimental realization was achieved [26–30].

Many biological processes, in vivo, are in fact mechanically controlled. The ability to apply pN force to a SM and observe its motion at nanometer scale (or vice versa) has a great significance in molecular biology in that we can elucidate the microscopic and structural origins of a biological process by quantifying both kinetic and thermodynamic properties of biomolecules at SM level and compare them with those from ensemble measurements [31]. Under a constant force condition using the force-clamp method [11,14,23,32] near the transition region, the molecular extension (x) exhibits discrete jumps along basins of attractions as a function of time. This allows the study of the hopping kinetics between the basins of attraction. If the time traces are long enough to sample all the conformations, then we can also construct an equilibrium free-energy profile under tension f [33,34]. The fraction of nativeness $\Phi_N(f)$ or equilibrium constant $K_{eq}(f) = \{1 - \Phi_N(f)\}/\Phi_N(f)$ as a function of f can be accurately measured using calorimetry or denaturant titration in bulk experiments.

Since the first SM force experiment, interplay between theory, simulation, and experiment have affected the experimental design as well as theoretical formulation to interpret results from measurements. A need to understand biomolecular dynamics at SM level further highlights the importance of theoretical background such as polymer physics [35], stochastic theory [36,37] and fluid dynamics. Molecular simulations of SMFS using a simple model provide a number of microscopic insights that cannot be easily gained by experiments alone. This chapter encompasses the force mechanics from the perspectives of theories and molecular simulations. Basic theories for force mechanics and the main simulation technique using the self-organized polymer (SOP) model will be described, followed by a number of findings and predictions for SMFS made through the concerted efforts using force theories and molecular simulations.

5.2 DECIPHERING SINGLE-MOLECULE FORCE SPECTROSCOPY

5.2.1 FORCE-EXTENSION CURVE

Mechanical response of a molecule is expressed with two conjugate variables, force (f) and molecular extension (x), to define a mechanical work ($W = fx$). Force-extension curves (FECs) or the time dependences of $f(t)$ and $x(t)$ are the lowest level data that can infer all the relevant information concerning the mechanical response of biomolecules. For a generic homopolymer whose Flory radius is $R_F \sim N^\nu a$, where N is the number of monomers and a is the size of monomers, the extension x of a polymer should be determined by a comparison between R_F and tensile screening length $\xi_p = k_B T/f$. The force value f determines the parameter $q = R_F/\xi_p$, which satisfies $q \ll 1$ ($q \gg 1$) for small (large) force f. The applied force can be classified into three regimes: (1) For small f, $x \ll R_F$ and $q \ll 1$ are satisfied. Thus, $x \approx \beta R_F^2 f$ is obtained from a scaling argument $x = R_F \Phi(q) \approx R_F q$ since $\Phi(q) \sim q$ if $q \ll 1$; and (2) For an intermediate f ($R_F < x \ll Na$), the shape of globular polymers is distorted to form a string of tensile blobs, where the blob size is given as $\xi_p \sim N_b^\nu$, with N_b being the number of monomers consisting the blobs. The total extension of the string of the tensile blobs under tension is $x \approx \xi_p \times (N/N_b)$, leading to Pincus scaling law $x \sim f^{1/\nu - 1} \sim f^{2/3}$ [38]. Here, note that this scaling law is only observed when $1 \ll (\xi_p/a)^{1/\nu} \ll N$ is ensured [39]. (3) For extremely large f, the chain is fully stretched: $x \approx Na^2\beta f/3$ for an extensible chain and $x \approx Na$ for an inextensible chain.

The scaling argument for biopolymers deviates from that of generic homopolymers with $N \gg 1$ due to the finite size effect and various local and nonlocal interactions [39]. In practice, the persistence length (l_p) and contour length (L) of biopolymers are extracted using a force-extension relation of the wormlike chain (WLC) model, that is, $f = (k_B T/l_p)[(1/4)(1 - x/L)2 - 1/4 + x/L]$ [40], whose energy Hamiltonian H takes into account the bending energy penalty along the polymer chain

$$H/k_B T = \left(l_p/2\right) \int_0^L \left(\partial \mathbf{u}(s)/\partial s\right)^2 ds$$

where $\mathbf{u}(s)$ is the tangential unit vector at position s along the contour [4]. The rips in FEC due to the disruption of internal bonds and a subsequent increase in the contour length from L to $L + \Delta L$ are used to decipher the energetics and the internal structure of proteins and nucleic acids [6,14,23]. The FEC of repeat proteins shows multiple peaks with saw-tooth patterns, suggesting that under tension repeat proteins unfold one domain after another. As a more complicated system, *Tetrahymena* ribozyme with nine subdomains show saw-tooth patterns but with varying peak height and position, requiring careful and laborious tasks of analysis [6].

5.2.2 FORCED UNFOLDING AT CONSTANT FORCE

A phenomenological description of the forced unbinding of adhesive contacts by Bell [25] plays a central role in studying the force-induced dynamics of biomolecules for the last two decades. In the presence of external force f, Bell modified Eyring's transition state theory [41] as follows:

$$k = \kappa \frac{k_B T}{h} \exp\left[-\frac{E^* - \gamma f}{k_B T}\right] \tag{5.1}$$

where k_B is the Boltzmann constant, T is the temperature, h is the Planck constant, and κ is the transmission coefficient. The parameter γ is a characteristic length of the system associated with bond disruption. Under tension f, the activation barrier E^* responsible for a stable bond is reduced to $E^* - \gamma \times f$. The prefactor $k_B T/h$ is the vibrational frequency of a bond due to thermal fluctuation prior to the disruption.

Although the original Bell model correctly describes the stochastic nature of bond disruption, the prefactor $k_B T/h$ fails to capture the physical nature of attempt frequency for the ligand unbinding from catalytic sites or the protein unfolding dynamics under tension, which depends on the shape of potential and viscosity of media. In fact, Eyring's transition state theory is only applicable to chemical reactions in gas phase. A more appropriate theory for the dynamics in condensed media should account for the effect of the solvent's viscosity and conformational diffusion [42]. A one-dimensional reaction coordinate, projected from a multi-dimensional energy landscape, can well represent the dynamics provided that the relaxation time of conformational dynamics along the reaction coordinate is much slower than other degrees of freedom [43]. Under tension f, molecular extension (x), or end-to-end distance (R), is assumed to be a good reaction coordinate. On the one-dimensional reaction coordinate, mean first passage time obeys the following simple differential equation [44]: $\wp_{FP}^*(x)\tau(x) = -1$, where $\wp_{FP}^* = \exp(F_{eff}(x)/k_B T)\partial_x D(x) \exp(-F_{eff}(x)/k_B T)\partial_x$ is the adjoint Fokker–Planck operator. Here, $F_{eff}(x)$ is an effective one-dimensional free energy as a function of a reaction coordinate x, and $D(x)$ is the position-dependent diffusion coefficient. The mean first passage time of a quasiparticle between the interval $a \leq x \leq b$ with reflecting $\partial_x \tau(a) = 0$ and absorbing boundary condition $\tau(b) = 0$ is

$$\tau(x) = \int_x^b dy \exp(F_{eff}(y)/k_B T) \frac{1}{D(y)} \int_a^y dz \exp(-F_{eff}(z)/k_B T) \tag{5.2}$$

In Equation 5.2, the free-energy profile $F_{eff}(x)$ is considered "tilted" by an external force by the amount of fx. As long as the transition barrier, that is, $\Delta F^* = F(x_{ts}) - F(x_b)$, is large enough, the Taylor expansions of the free-energy potential $F(x) - fx$ at the top of the barrier and the bound state position with a saddle point approximation result in the seminal Bell–Kramers equation [36,37]

$$k(f) \approx \frac{\omega_b \omega_{ts}}{2\pi\gamma} \exp[\beta(\Delta F^* - f\Delta x^*)] \tag{5.3}$$

where $\Delta x^* = x_{ts} - x_b$ with x_{ts} and x_b being reaction coordinates at the transition state and initial state (i.e., stable bond), respectively, ω_b and ω_{ts} are the curvatures of the potential, $|\partial_x^2 F(x)|$, at $x = x_b$ and x_{ts}, respectively, and $\gamma = k_B T/Dm$ is a friction coefficient associated with the motion of biomolecules. The experimentally determined speed limit of the folding dynamics (barrierless folding time) of two-state proteins is $\approx (0.1-1)$ μs [42,45,46]. The prefactor $(k_B T/h)^{-1} \approx 0.2$ ps from transition state theory for gas phase should not be used when estimating the barrier height from folding or unfolding kinetics data of biopolymers in condensed phase. A cautionary word is in place. If the barrier height $\Delta F^* - f\Delta x^*$ is comparable to or smaller than $k_B T$, the molecular configuration trapped as a metastable state in the free-energy barrier can move almost freely across the barrier. In fact, the barrier vanishes when f reaches a critical force $f_c \approx \Delta F^*/\Delta x^*$. In this case, there is no separation of timescales between diffusive motion and barrier-crossing event. Hence, the saddle point approximation taken for Equation 5.3 from Equation 5.2 does not hold. Due to thermal noise, the unfolding or rupture event of the system occurs under finite free-energy barrier $(> k_B T)$ before f reaches f_c. For the Bell–Kramers equation to be applicable, f should be always smaller than f_c.

5.2.3 Forced Unfolding at Constant Loading Rate—Dynamic Force Spectroscopy

Although the constant force (force-clamp) experiment is more straightforward for analysis, due to technical reasons many of the force experiments have been performed under a constant loading condition (force-ramp) in which the force is linearly ramped over time [6,47–49]. Dynamic force spectroscopy (DFS) probes the energy landscape of biomolecular complexes by detecting the mechanical response of the molecules. A linearly increasing mechanical force with a rate of $r_f = df/dt$ is exerted on the molecular system until the molecular complex disrupts. On unbinding, the force recorded on the instrument decreases abruptly, thus we can measure the unbinding force of the system of interest (Figure 5.1a). Because of the stochastic nature of an unbinding event, the unbinding force of the molecular complex is not unique, but distributes broadly, defining the unbinding force distribution $P(f)$ (Figure 5.1b).

Under linearly varying force ($f = r_f \times t$), the rate of barrier crossing from bound to unbound state (or from folded to unfolded state) is also time dependent. Hence, the survival probability, that is, the probability to have unbound state (or folded state), at time t is given by

$$S(t) = \exp\left(-\int_0^t d\tau k(\tau)\right)$$

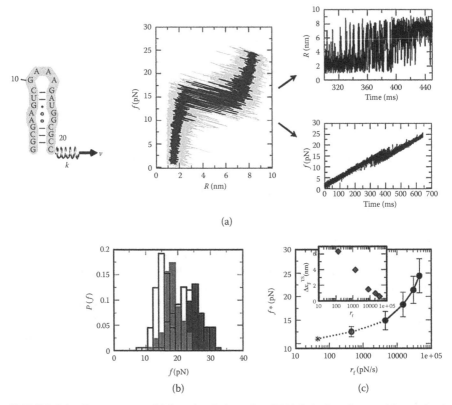

(a)

(b)　　　　　　　　　　　　　　　　　　　(c)

FIGURE 5.1 Force-ramp unfolding simulation of an RNA hairpin using a self-organized polymer model. (a) FEC, $f(R)$ at $r_f = 45$ pN/s ($k = 0.07$ pN/nm, $v = 0.64$ μm/s). $f(R)$ can be decomposed using time t into $R(t)$ and $f(t)$. The experimental setup of the optical tweezer is mimicked by attaching a harmonic spring with strength k. The force is recorded by measuring the extension of this spring. $f(t)$ shows how the force is ramped with time when the system is pulled with constant pulling speed v. The two-state hopping transition in R and f begins when f reaches ≈ 12 pN and ends at $f \approx 17$ pN. (b) Unfolding force distribution from 100 trajectories for each r_f, $r_f = r_f^0$, $r_f^0/3$, $20r_f^0/3$, $10r_f^0$ when $r_f^0 = 4.5 \times 10^3$ pN/s. (c) [log r_f, f^*] plot obtained from (b), which clearly shows a positive curvature. Linear regression at different r_f values provide different Δx^* values. (Adapted from Hyeon, C., and D. Thirumalai., *Biophys J.*, 92:731–43. 2007 With permission.)

Thus, the first passage time distribution is $P(t) = -dS(t)/dt = k(t)S(t)$. Change of variable from t to f leads to an unbinding force distribution.

$$P(f)\frac{1}{r_f}k(f)S(f) = \frac{1}{r_f}k(f)\exp\left[-\int_0^f df' \frac{1}{r_f}k(f')\right] \qquad (5.4)$$

Note that $k(f)$ is an exponentially increasing function of f, whereas $S(f)$ is an exponentially decreasing function of f with greater power at $f \gg 1$, shaping a Gumbel distribution, $P(f) \sim \exp(f - e^f)$, for $k(f) \sim e^f$. The current theoretical issue of deciphering the underlying energy landscape using force hinges on an analysis of $P(f)$ by building not only a physically reasonable but also a mathematically tractable model.

The most probable unfolding force is obtained using $dP(f)/df|_{f=f^*} = 0$.

$$f^* = \frac{k_B T}{\Delta x^*} \log r_f + \frac{k_B T}{\Delta x^*} \log\left(\frac{\Delta x^*}{v_D k_B T \exp(-\beta \Delta F^*)}\right) \qquad (5.5)$$

where $v_D = \omega_b \omega_{ts}/2\pi\gamma$. In conventional DFS theory, Equation 5.5 is used to extract Δx^* and ΔF^* of the underlying one-dimensional free-energy profile associated with force dynamics. For unbinding force f^* to be compatible with the one in the picture of Kramers barrier-crossing dynamics, $f^* < f_c$ should be obeyed as mentioned earlier. The condition $f^* - f_c = (k_B T/\Delta x^*)\log(r_f \Delta x^*/v_D k_B T) < 0$ demands $r_f < r_f^c\ (= v_D k_B T/\Delta x^*)$. For a set of parameters, $k_B T \approx 4$ pN·nm, $\Delta x^* \sim 1$ nm, and $v_D \sim 10^6$ s^{-1}, the critical loading rate is $r_f^c \sim 10^6$ pN/s. The typical loading rate used in force experiments (0.1 pN/s $< r_f < 10^3$ pN/s) is several orders of magnitude smaller than this value. Therefore, in all likelihood unbinding dynamics in typical experimental conditions obey stochastic barrier-crossing dynamics. In contrast, the steered molecular dynamics simulations with all-atom representation [50] typically use $r_f > r_f^c$ due to high computational cost. In such an extreme condition, however, the forced unfolding process can no longer be considered a thermally activated barrier-crossing process. At high $r_f > r_f^c$, it was shown that an average rupture force $<f>$ increases as $r_f^{1/2}$ [51].

It is of particular interest that for a molecular system unfolding through a single free-energy barrier, the force dependence of force-clamp kinetics can be formally expressed with the $P(f)$ from force-ramp experiments as follows [52–54].

$$k(f) = \frac{P(f)/r_f}{1 - \int_0^f d\,f'P(f')/r_f} \qquad (5.6)$$

which is easily shown using the following relation:

$$S(f) = \exp\left(-\int_0^f d\,f'\frac{1}{r_f}k(f')\right) = 1 - \int_0^f d\,f'\frac{1}{r_f}P(f')$$

Technically the two distinct experimental methods are connected through this simple relationship. Therefore, by conducting force-ramp experiments with a sufficiently good statistics to acquire $P(f)$ at varying r_f, we can, in principle, build data for $k(f)$ as in force-clamp experiments.

5.2.4 DEFORMATION OF ENERGY LANDSCAPE UNDER TENSION

The Bell–Kramers equation assumes that an external force changes the free-energy barrier along the reaction coordinate from ΔF^* to $\Delta F^* - f\Delta x^*$ without significantly changing other topology of the energy landscape. A linear regression of both Equations 5.3 and 5.5 provides the characteristic length Δx^* and free-energy barrier

ΔF^* at $f \to 0$. However, in practice, nonlinearity (negative curvature for f vs log $k(f)$ and positive curvature for log r_f vs f^*) is often detected especially when f or r_f is varied over a broad range. Thus, in case f (or r_f) is varied at a large but narrow range of f (or r_f), then substantial errors can arise in the extrapolated values of Δx^* and ΔF^* at zero force; the nonlinear regression will underestimate Δx^* and overestimate ΔF^*. The physical origin of f-dependent Δx^* is found in a complicated molecular response to external force. If the transition state ensemble is broadly spread along the reaction coordinate, then the molecule can adopt diverse structures along the energy barrier with varying f values. Whereas if the transition state ensemble is sharply localized along the reaction coordinate, the nature of the transition state ensemble measured in x-coordinate will be insensitive to the varying f values. Or, more simply, the origin of the moving transition state position can be algebraically explained by plotting the shape of $F_{eff}(x)$ with varying f. Because x_{ts} and x_b are determined from the *force-dependent condition* $F'(x) - f = 0$, all the parameters should be intrinsically f dependent as $\Delta F^*(f)$, $\Delta x^*(f)$, $\omega_{ts}(f)$, and $\omega_b(f)$. By making harmonic approximation of $F(x)$ at $x = x_b$ and $x = x_{ts}$, that is, $F(x) \approx F(x_b) + (1/2)|F''(x_b)|(x - x_b)^2$ and $F(x) \approx F(x_{ts}) + (1/2)|F''(x_{ts})|(x - x_{ts})^2$ and calculating $\Delta x^*(f) = x_{ts}(f) - x_b(f)$ from $F'_{eff}(x) = F'(x) - f = 0$, we can show that

$$\frac{\Delta x^*(f)}{\Delta x^*} = 1 - \chi(f) = 1 - \frac{f}{\Delta x^*}\left(\frac{1}{F''(x_{ts})} + \frac{1}{F''(x_b)}\right) \quad (5.7)$$

Typically for biopolymers under tension, the free-energy profile near the native state minimum is sharp ($|F''(x_b)| \gg 1$). To minimize the difference between $\Delta x^*(f)$ and Δx^* and to make $\chi(f) = 0$, the transition barrier should be sharp for a given $f/\Delta x^*$ value (i.e., $f/\Delta x^*|F''(x_{ts})| \ll 1$). Simulation studies [55–57] in which the free-energy profiles were explicitly computed from thermodynamic considerations alone clearly showed the change of Δx^* when f was varied. Note that the movement of the transition barrier location toward the native state position ($x = x_b$) is consistent with the Hammond postulate [58,59] that explicates the nature of transition state of a simple organic compound when product state is relatively more stabilized than reactant state.

To account for the nonlinear response of biological systems to the force more naturally, Dudko et al. proposed to use an analytically tractable microscopic model for the underlying free-energy profile. For a cubic potential $F(x) = -2\Delta F^*\{(x/\Delta x^*)^3 - (3/2)(x/\Delta x^*)^2\}$ whose distance to the transition state and free-energy barrier are Δx^* and ΔF^*, all the parameters needed for Kramers equation are expressed as a function of f, Δx^*, and ΔF^*; $\omega_{ts}(f)\omega_b(f) = \omega_{ts}\omega_b(1 - f/f_c)^{1/2}$, $\Delta x^*(f) = \Delta x^*(1 - f/f_c)^{1/2}$, and $\Delta F^*(f) = (2/3)\Delta x^*f_c(1 - f/f_c)^{3/2}$, where $f_c \equiv 3\Delta F^*/2\Delta x^*$. Thus, the f-dependent unfolding rate $k(f)$ for cubic potential is given exactly with $v = 2/3$.

$$k(f) = k(0)\left(1 - v\frac{f\Delta x^*}{\Delta F^*}\right)^{1/v-1} \exp\left[-\Delta F^*\left\{\left(1 - v\frac{f\Delta x^*}{\Delta F^*}\right)^{1/v} - 1\right\}\right] \quad (5.8)$$

In fact, with different v values, the same expression with Equation 5.8 is obtained for harmonic cusp potential ($v = 1/2$) and the Bell–Kramers model ($v = 1$). The positive value of v depends on the nature of the underlying potential and could be treated as an adjustable parameter [53]. Consequently, cusp, cubic [52,53,60], or piecewise harmonic potential [61] is suggested as the microscopic model for the underlying free-energy profile. So far, theories for force experiments have been devised mainly the for single barrier picture with a one-dimensional reaction coordinate. Although a two-slope fit using a multiple energy barrier picture was suggested to explain the large curvature observed in DFS data [62], building a multibarrier free-energy profile from one-dimensional information such as a $P(f)$ or [$\log r_f, f^*$] curve is an inverse problem whose answer may not be unique, as is well demonstrated by Derenyi et al. in the context of the forced unfolding over two sequentially located transition barriers [63].

5.3 SELF-ORGANIZED POLYMER MODEL FOR SINGLE-MOLECULE FORCE SPECTROSCOPY

Now, the typical spatial resolution reached in SMFS is a few nanometers, and the dynamics that is probed in SMFS is global rather than local. The details of local dynamics such as the breakage of a particular hydrogen bond or an isomerization of dihedral angle cannot be discerned from FECs or the time trace of molecular extension alone. To gain and provide sufficient insight into molecular simulations in conjunction with SMFS, a simple model with which we can efficiently simulate the forced unfolding dynamics of a large biopolymer at a spatial resolution of SMFS would be of great use. By drastically simplifying the details of local interactions, such as bond angle or dihedral angle along the backbone, which are indeed the major determinants for the dynamics under tension at nanometer scale, we can either gain an acceleration in simulation speed or explore the dynamics of a larger molecule. As is well appreciated in the literature of normal mode analysis, an inclusion of small length scale information does not alter the global dynamics corresponding to a low-frequency mode [64–66]. The SOP model, proposed in this line of thought, is well suited to simulate the forced unfolding dynamics of large biopolymers at the spatial resolution of SMFS. The basic idea of the SOP model is to use the simplest possible energy Hamiltonian to faithfully reproduce the topography of the native fold and to simulate the low-resolution global dynamics of biopolymers of arbitrary size [67–73]. The energy function for biopolymers in the SOP representation is

$$
\begin{aligned}
E_{\text{SOP}}(\{\vec{r}_i\}) &= E_{\text{FENE}} + E_{\text{nb}}^{(\text{att})} + E_{\text{nb}}^{(\text{rep})} \\
&= \sum_{i=1}^{N-1} \frac{k}{2} R_0^2 \log\left[1 - \frac{\left(r_{i,i+1} - r_{i,i+1}^0\right)^2}{R_0^2}\right] \\
&\quad + \sum_{i=1}^{N-3} \sum_{j=i+3}^{N} \varepsilon_h \left[\left(\frac{r_{ij}^0}{r_{ij}}\right)^{12} - 2\left(\frac{r_{ij}^0}{r_{ij}}\right)^6\right] \Delta_{ij} + \sum_{i<j} \varepsilon_1 \left(\frac{\sigma}{r_{ij}}\right)^6 (1 - \Delta_{ij})
\end{aligned}
\tag{5.9}
$$

The first term in Equation 5.9 is the finite extensible nonlinear elastic (FENE) potential for chain connectivity; it has the following parameters: $k = 20$ kcal/mol·Å^2, $R_0 = 0.2$ nm, $r_{i,i+1}$ is the distance between neighboring beads at i and $i + 1$, and $r_{i,i+1}^0$ is the distance in the native structure. The use of FENE potential for backbone connectivity is more advantageous than that of the standard harmonic potential, especially for forced stretching to produce an inextensible behavior of WLC. The Lennard–Jones potential is used to account for interactions that stabilize the native topology. A native contact is defined for bead pairs i and j such that $|i - j| > 2$ and whose distance is less than 8 Å in the native state. We use $\varepsilon_h = 1 \sim 2$ kcal/mol for native pairs, and $\varepsilon_l = 1$ kcal/mol for nonnative pairs. In the current version, we have neglected nonnative attractions. This should not qualitatively affect the results because under tension such interactions are greatly destabilized. To ensure noncrossing of the chain, $(i, i + 2)$ pairs interacted repulsively with $\sigma = 3.8$ Å. There are five parameters in the SOP force field. In principle, the ratio of $\varepsilon_h/\varepsilon_l$ and R_c can be adjusted to obtain realistic values of critical forces. For simplicity, we choose a uniform value of ε_h for all protein constructs. ε_h can be made sequence dependent and ion implicit as $\varepsilon_h \rightarrow \varepsilon_h^{ij}$ if one wants to improve the simulation results. By truncating forces due to Lennard–Jones potential for interacting pairs with $r > 3r_{ij}^0$ or 3σ, the computational cost essentially scales as $\sim O(N)$. We refer to the model as the SOP model because it only uses the polymeric nature of the biomolecules and the crucial topological constraints that arise from the specific fold.

With SOP representation, the dynamics of biopolymers is simulated under force-clamp or force-ramp conditions by solving the equation of motions in the overdamped regime. The position of the ith bead at time $t + h$ is given by

$$x_i(t + h) = x_i(t) + \frac{h}{\varsigma}\left[F_i(t) + \Gamma_i(t) + f_N(t)\right] \tag{5.10}$$

where $F_i(t) = -\nabla_i E_{\text{SOP}}(\{r_i\})$, the x-component of the conformational force acting on the ith bead, and $\Gamma_i(t)$ is a random force selected from a Gaussian white noise distribution

$$P\left[\Gamma_i(t)\right] \sim \exp\left[-\frac{1}{4k_B T\varsigma}\int_0^t d\tau \Gamma_i^2(\tau)\right]$$

Simulation time is converted into physical time by using $\partial_t r \approx a/\tau \sim \varsigma^{-1} \times (k_B T/a)$ and $ma^2/\varepsilon_h = \tau_L^2$.

$$\tau_H \sim \frac{\varsigma a^2}{k_B T} = \left[\frac{\varsigma(\tau_L/m)\varepsilon_h}{k_B T}\right]\tau_L \tag{5.11}$$

When amino acid residue is used as a coarse-grained center, $\varsigma \approx (50 \sim 100)m/\tau_L$, the time step is $h \approx 0.1\tau_L$. For force-clamp condition, a constant force (f_N) is exerted to the Nth bead $f_N(t) = f_N$ with the first bead being fixed, $f_1(t) = 0$. For force-ramp conditions, $f_N(t) = -k(x_N - vt)$ with $f_1(t) = 0$ is used. A harmonic spring with stiffness k is attached to the Nth bead, and the position of a spring is moved with a constant velocity v.

5.4 CHARACTERIZATION OF THE UNFOLDING MECHANICS OF COMPLEX BIOMOLECULES

5.4.1 DECIPHERING THE ENERGY LANDSCAPE OF COMPLEX BIOMOLECULES USING SINGLE-MOLECULE FORCE SPECTROSCOPY

The mechanical response of a molecule becomes more complex with an increasing complexity in the native topology of biomolecules. This section addresses two classes of unfolding scenario for the molecules with complex topology.

First, it is conceivable that a biomolecule with complicated topology in its native state is unraveled via more than two distinct transition state ensembles so that the forced unfolding routes bifurcate [7,74]. In this scenario, the survival probability of a molecule remaining in the native state decays as

$$S_N(t) = \sum_{i=1}^{N} \Phi_i(f) \exp\left(-t/\tau_i(f)\right) \qquad (5.12)$$

where unfolding time along the ith route is given by $\tau_i(f)$ and $\Phi_i(f)$ is the partition factor for ith route with $\Sigma\Phi_i(f) = 1$, both of which are functions of f [42,75]. Mechanical response of a barrel-shaped Green fluorescence protein (GFP), made of 11 β-strands with one α-helix at the N-terminal, is quite intricate, and its unfolding path depends on pulling speed and direction [7,76]. In earlier force experiments on GFP by Rief et al., the intricacy of the GFP forced unfolding is manifested as the indistinguishability of two unfolding routes [77]. After the α-helix is disrupted from the barrel structure due to external force, the second rip is due to the peeling off of β1 or β11. The gains of contour length (ΔL) from the molecular rupture event from β1 and from β11 are, however, identical so that it was impossible to identify the source of the second peak simply by analyzing the FEC [77]. The force simulation using the SOP model suggested a bifurcation into two different routes by showing that 70% of the molecules disrupt from the N-terminal and the remainder of the molecules from the C-terminal. Experimentally, this is confirmed by fixing either the N-terminal or C-terminal direction by introducing a disulfide bridge through mutation [7]. The kinetic partitioning mechanism used for protein and RNA folding [42,78] can be adapted to explain the mechanical behavior of biopolymers.

Second, the reaction coordinate under tension can have sequentially aligned multiple barriers. In comparison to GFP, the forced unfolding experiment on RNase-H was characterized by a peculiar mechanical response [79]. The FEC of RNase-H has a single large rip along the unfolding path but has two rips in the refolding path. This behavior was explained by considering a shape of the free-energy profile in which native (N), intermediate (I), and unfolding (U) states lie sequentially with relatively a high-transition barrier ($\gg k_B T$) between N and I, and a low-transition barrier ($\sim k_B T$) between I and U. On such a free-energy profile, the unfolding to the U state would occur by skipping the I state because the external force that disrupts the N to overcome the first barrier is already larger than the mechanical stability of the I state relative to U. Thus, accessing to the I state from the N state is difficult under an increasing tension while I can be reached from U in the refolding FEC since the free-energy barrier between

I and U is relatively small ($\sim k_B T$). This hypothetical picture was further supported by the force-clamp method. For $I \Leftrightarrow U$, the transition mid-force is $f_{m,I\Leftrightarrow U} \approx 5.5$ pN while the escape from the native basin of attraction (NBA) occurs at $f \approx 15$–20 pN [79]. Even for a molecule hopping between I and U at $f = 5.5$ pN, the molecule can get to the N state. However, once the molecule jumps over the barrier between N and I, being trapped in N, the molecule has little chance to jump back to I within the measurement time.

5.4.2 MEASUREMENT OF ENERGY LANDSCAPE ROUGHNESS

Although the energy landscape of biopolymers is evolutionary tailored such that the potential gradient toward the native state is large enough to drive the biopolymers to their native state, the energetic and topological frustration still remain to render the folding landscape rugged, slowing down the folding processes. To account for the effect of energy landscape roughness in a one-dimensional free-energy profile, $F(x)$ can be effectively decomposed into $F(x) = F_0(x) + F_1(x)$ [80], where $F_0(x)$ is a smooth potential that determines the global shape of the energy landscape, and $F_1(x)$ is the ruggedness that superimposes $F_0(x)$. By taking the spatial average over $F_1(x)$ using $<\exp(\pm\beta F_1(x))>_l = (1/l)\int_0^l dx \cdot \exp(\pm\beta x)$, where l is the ruggedness length scale, the association mean first passage time is altered to

$$\tau(x) \approx \int_x^b dy \exp\left(F_0(y)/k_B T\right)\left\langle \exp\left(\beta F_1(y)\right)\right\rangle_l \frac{1}{D}\int_a^y dz \exp\left(-\frac{F_0(z)}{k_B T}\right)\left\langle \exp\left(-\beta F_1(z)\right)\right\rangle_l$$

By either assuming a Gaussian distribution of the roughness contribution F_1, that is, $P(F_1) \propto \exp(-F_1^2/2\varepsilon^2)$, or simply assuming $\beta F_1 \ll 1$ and $<F_1> = 0$, $<F_1^2> = \varepsilon^2$ and $\beta\varepsilon$ is small, the effective diffusion coefficient can be approximated as $D^* \approx D\exp(-\beta^2\varepsilon^2)$, where D is the bare diffusion constant.

The signature of the roughness of the underlying energy landscape is uniquely reflected in the non-Arrhenius temperature dependence of the unbinding rates. By using mean first passage time with the effective diffusion coefficient with roughness, we can show that the unfolding kinetics for a two-state folder deviates substantially from an Arrhenius behavior as follows:

$$\log k(f,T) = a + \frac{b}{T} - \frac{\varepsilon^2}{T^2} \tag{5.13}$$

where ε^2 is a constant even if the coefficients a and b change under different force and temperature conditions [81]. This relationship suggests that conducting *force-clamp experiments* over the range of temperatures identifies the roughness scale ε. Here, the condition $\varepsilon/\Delta F^* \ll 1$ should be ensured.

To extract the roughness scale, ε, using DFS, a series of DFS experiments should be performed as a function of T and r_f so that reliable unfolding force distributions $P(f)$ and the corresponding f^* value are obtained.

$$f^* \approx \frac{k_B T}{\Delta x^*}\log r_f + \frac{k_B T}{\Delta x^*}\log \frac{\Delta x^*}{v_D k_B T \exp\left(-\Delta F_0^*/k_B T\right)} + \frac{\varepsilon^2}{\Delta x^* k_B T} \tag{5.14}$$

One way of obtaining ε from experimental data is as follows [82]. From the f^* versus log r_f curves at two different temperatures, T_1 and T_2, we can obtain $r_f(T_1)$ and $r_f(T_2)$ for which the f^* values are identical. By equating the right-hand side of the expression in Equation 5.5 at T_1 and T_2, the scale ε can be estimated [81,82] to be

$$\varepsilon^2 \approx \frac{\Delta x^*(T_1)k_B T_1 \times \Delta x^*(T_2)k_B T_2}{\Delta x^*(T_1)k_B T_1 - \Delta x^*(T_2)k_B T_2} \left[\Delta F_0^* \left(\frac{1}{\Delta x^*(T_1)} - \frac{1}{\Delta x^*(T_2)} \right) \right.$$

$$\left. + \frac{k_B T_1}{\Delta x^*(T_1)} \log \frac{r_f(T_1)\Delta x^*(T_1)}{v_D(T_1)k_B T_1} - \frac{k_B T_2}{\Delta x^*(T_2)} \log \frac{r_f(T_2)\Delta x^*(T_2)}{v_D(T_2)k_B T_2} \right] \quad (5.15)$$

Equation 5.15 is used to measure ε for GTPase Ran-Importin β complex ($\varepsilon > 5k_B T$) [82] and transmembrane helices ($\varepsilon \approx 4 - 6\ k_B T$) [83].

5.4.3 Pulling Speed–Dependent Unfolding Pathway

The dynamics of a polymer is extremely intricate due to multiple entangled length scales and timescales. The polymeric nature of RNA and proteins immediately lends itself when a molecule interacts with an external force. A biopolymer adapts its configuration in response to an external stress in a finite amount of relaxation time. In both AFM and laser optical tweezer (LOT) experiments, a force is applied to one end of the chain with the other end being fixed. A finite amount of delay is expected for the tension f to propagate along the backbone of a molecule and through the network of contacts that stabilize the native topology. To understand the effect of finite propagation time of the tension on unfolding dynamics, it is useful to consider a ratio between the loading rate r_f and the rate at which the applied force propagates along polymer chain r_T ($\lambda = r_T/r_f$). r_f is controlled by experiments; r_T most likely depends on the topology of a molecule. Depending on the value of parameter λ, the history of dynamics can be altered qualitatively. If $\lambda \gg 1$, then the applied tension at one end propagates rapidly so that, even prior to the realization of the first rip, force along the chain is uniform. In the opposite limit, if $\lambda \ll 1$, then tension is nonuniformly distributed along the backbone at the moment any rupture event occurs (note that the gradient is formed from $i = 1$ to $i = N$ in the one at the highest loading rate, the top panel of Figure 5.2a). In such a situation, unraveling of RNA begins from a region where the value of local force exceeds the tertiary interactions.

The unfolding simulation using the SOP model of *Azoarcus* ribozyme provides a great insight into the issue of force propagation and r_f-dependent unfolding pathways [67]. The intuitive argument given in Figure 5.2a is clarified by visualizing the change in the pattern of the force propagation for *Azoarcus* ribozyme under three different loading conditions. Alignment of the angles between the bond segment vector ($r_{i,i+1}$) along the force direction can provide the magnitude of force exerted at each position along the backbone. The nonuniformity in the local segmental alignment is most evident at the highest loading rate. The dynamics of the force propagation occurs sequentially from one end of the chain to the other at high r_f. The alignment of the segment along f is more homogenous at lower r_f. These results

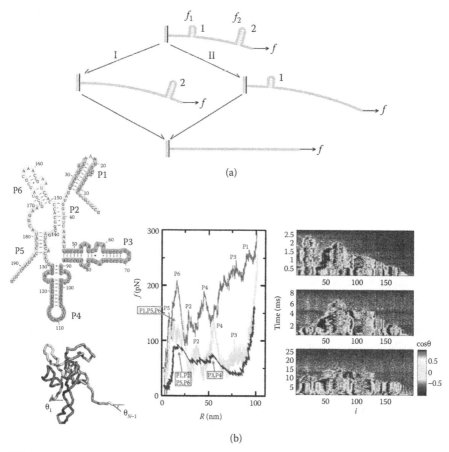

(a)

(b)

FIGURE 5.2 Force propagation. (a) A diagram depicting the pulling speed–dependent unfolding pathway. For a biopolymer consisting of two hairpins 1 and 2 whose barriers associated with forced unfolding are given as ΔF_1^* and ΔF_2^* with $\Delta F_1^* < \Delta F_2^*$, the unfolding will occur through one of the two reaction paths depending on the r_f. For $\lambda \gg 1$, tension will uniformly distribute along the chain, so that $f_1 \approx f_2$. Since $\Delta F_1^* - f_1 \Delta x^* < \Delta F_2^* - f_2 \Delta x^*$, hairpin 1 will unfold before hairpin 2 (pathway I). In contrast, for $\lambda \ll 1$, $f_1 \ll f_2$ leads to $\Delta F_1^* - f_1 \Delta x^* > \Delta F_2^* - f_2 \Delta x^*$, thus hairpin 2 will unfold before hairpin 1 (pathway II). (b) r_f-dependent unfolding pathway simulated with *Azoarcus* ribozyme using self-organized polymer model. FEC (middle panel) shows that unfolding occurs via $N \rightarrow$ [P5] $\rightarrow$ [P6] $\rightarrow$ [P2] $\rightarrow$ [P4] $\rightarrow$ [P3] $\rightarrow$ [P1] at $r_f = 1.2 \times 10^6$ pN/s (top), $N \rightarrow$ [P1, P5, P6] $\rightarrow$ [P2] $\rightarrow$ [P4] $\rightarrow$ [P3] at $r_f = 3.6 \times 10^5$ pN/s (middle), and $N \rightarrow$ [P1, P2, P5, P6] $\rightarrow$ [P3, P4] at $r_f = 1.8 \times 10^4$ pN/s (bottom). From high to low r_f, the unfolding pathways were changed completely. The time evolution of $\cos\theta_i$ ($i = 1$, 2, ..., $N - 1$) at the three r_f visualizes how the tension is being propagated. (Adapted from Hyeon, C., R. I. Dima, and D. Thirumalai., *Structure* 14:1633–45, 2006. With permission.)

highlight an important prediction, which is closely related to polymer dynamics, that the unfolding pathways can drastically change depending on the loading rate r_f. By varying λ (i.e., controlling r_f) from $\lambda \ll 1$ to $\lambda \gg 1$, force experiments will show the dramatic effect of pulling speed dependence on the unfolding dynamics. There may be a dramatic change in unfolding mechanism for two different instruments using different r_f (e.g., LOT and AFM experiments). In addition, predictions of a mechanism for forced unfolding based on all-atom MD simulation [84] should also be treated with caution unless due to topological reasons; the unfolding pathways are robust to large variations in the loading rates regardless of λ value.

5.4.4 EFFECT OF MOLECULAR HANDLES ON THE MEASUREMENT OF HOPPING TRANSITION DYNAMICS

While the idea of SM experiments is to probe the dynamics of an isolated molecule, noise or interference from many possible sources is always a difficult problem to deal with in nanoscale measurements. To accurately measure the dynamics of a test molecule, the interference between molecule and instrument should be minimal. In LOT experiments, dsDNA or DNA–RNA hybrid handles are inserted between the microbead and the test molecule so as to minimize the systematic error due to the microbead–molecule interaction. Unlike force-ramp experiments, issues concerning nonequilibrium relaxation such as the delay of signal or force propagation do not lend themselves in force-clamp experiments as long as the timescale of molecular hopping (τ_{hop}) at given force is greater than the relaxation times of the handle (τ_h), microbead (τ_b), and other part of the instruments (typically, $\tau_{hop} \gg \tau_h$, τ_b). Force-clamp experiments essentially create an equilibrium condition. As a result, tension is uniformly distributed over the handle as well as the test molecule while the molecule hops.

However, a critical issue regarding measurements still remains even at a perfect equilibrium due to *handle fluctuations*. (1) Since the dynamics of the test molecule is measured by monitoring the position of the microbead, to gain an identical signal with the test molecule, the fluctuation of the handle should be minimal. Therefore, we may conclude that *short and stiff* handles (smaller L/l_p) are ideal for precisely sampling the conformational statistics [34] (see Figure 5.3b and caption); and (2). Simulations of hopping dynamics with handles of differing length and flexibility show that the hopping kinetics is less compromised than the true handle-free kinetics when *short and flexible* handles are used [85]. Physically, when a test molecule is sandwiched between handles, the diffusive motion is dynamically pinned. The stiffer the handles, the slower the hopping transition. Thus, *short and flexible* handles are suitable for preserving the true dynamics of the test molecule.

The aforementioned two conditions for (1) precise measurement of thermodynamics, and (2) accurate measurement of true kinetics apparently contradict each other. However, this dilemma can be avoided by simply measuring the folding landscape accurately using *short and stiff* handles as long as the effective diffusion coefficient associated with the reaction coordinate is known. At least, at $f \approx f_{mid}$, the hopping time trace can easily fulfill the ergodic condition, providing a good

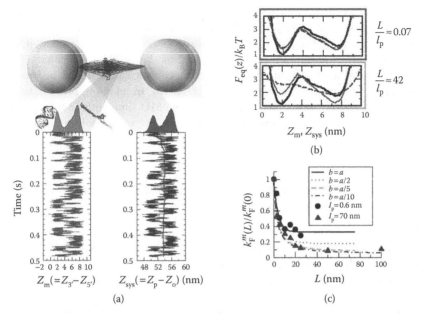

(a)   (b)   (c)

FIGURE 5.3 Effect of handles on measurement and RNA hopping dynamics. (a) Molec-ular simulation of RNA hopping dynamics with two handles of $L = 25$ nm and $l_p = 70$ nm attached to the 5′ and 3′ ends. The illustration was created by using the simulated structures collected every 0.5 milliseconds. An example of the time trace of each component of the system, at $f = 15.4$ pN. z_m and z_{sys} measure the extension dynamics of the RNA hairpin and the handle-RNA-handle system, respectively. The time-averaged value $<z_r(t)> = (1/\tau)\int_0^\tau z_{sys}(t)\,dt$ for the time trace of z_{sys} is shown as the bold line. The histograms of the extension are shown on the top of each column. (b) The free-energy profiles, $F_{eq}(z_{sys})$ from the two ends of the handles attached (dashed line), $F_{eq}(z_m)$ from the 5′ and 3′ ends of RNA (thin solid line), and $F_{eq}^0(z_m)$ from the 5′ and 3′ ends of RNA with no handles attached (thick solid line). When $L/l_p \ll 1$, $F_{eq}(z_{sys}) \approx F_{eq}(z_m) \approx F_{eq}^0(z_m)$. (c) With increasing handle length, the folding rate $k_F^m(L)$ devi-ates from a true folding rate $k_F^m(0)$. This effect is larger than handles with a greater l_p. (Adapted from Hyeon, C., G. Morrison, and D. Thirumalai., *Proc Natl Acad Sci U S A* 105:9604–6, 2005. With permission.)

equilibrium free-energy profile. Furthermore, the molecular extension z (or the end-to-end distance R) used to represent the free-energy profile is indeed a good reac-tion coordinate on which Bell's equation for obtaining the folding rate agrees well with the folding rate one can obtain using a simulation; for a given transition rate from NBA to unfolding basin of attraction (UBA), $k_F(f_{mid})$, the variation of force from $f = f_{mid}$ modifies the rate as $k_F(f_{mid} \pm \delta f) = k_F(f_{mid})\exp(\pm\delta f \cdot \Delta x^*/k_BT)$. Once an accurate free energy $[F(z_m) \approx F^0(z_m) \approx F(z_{sys})$ holds for $L/l_p \ll 1$; see Figure 5.3b and caption] is obtained, one can determine the hopping kinetics by directly calculating the mean first passage time on $F^0(z_m)$. The only unknown, the effective diffusion coefficient on $F^0(z_m)$, can be estimated by mapping the hopping dynamics of the test molecule to the dynamics of a simple analytical model such as the generalized Rouse model (GRM) [86,87].

5.5 FOLDING DYNAMICS ON FORCE QUENCH

Because of its multidimensional nature, the dynamics of biomolecules is sensitive to the condition to which the molecule is imposed. In a rugged energy landscape, the folding rate is not unique. Rather, the folding kinetics can vary greatly depending on initial condition [13]. The difference in the initial condition may be due to denaturant concentration, temperature, or force. For RNA, it is well known that the initial counterion condition can alter the folding route in a drastic fashion [13]. A biopolymer of interest adapts its structure to a given condition and creates a "condition-specific denatured state ensemble (DSE)." The routes to the NBA from the DSE are determined by the shape and ruggedness of the energy landscape due to factors such as side-chain interactions and topological frustrations.

The force-quench refolding dynamics of poly-ubiquitin (poly-Ub) by Fernandez and Li [11] is the first experiment that focused on the folding dynamics of proteins from a fully extended ensemble, in which the dynamics of molecular extension of poly-Ub construct was traced from an initial stretched ensemble prepared at high stretch force $f_S = 122$ pN to a quenched ensemble at low quench force $f_Q = 15$ pN (see Figure 5.4a). The folding trajectory monitored using the molecular extension (R) was at least an order of magnitude slower than the one from bulk measurement and was characterized by continuous transitions divided into at least three (four in [11]) stages: (1) initial reduction of R, (2) long flat plateau, and (3) a cooperative collapse transition at the final stage. Two main questions were immediately raised from the results of the experiment: (1) Why is the folding (collapse) process so slow compared to the one at bulk measurement? (2) What is the nature of the plateau and cooperative transition at the final stage? Regarding point (1), it was suspected that an aggregate was formed between the Ub monomer since the effective concentration is greater than the critical concentration for the aggregate formation [88]. But, this possibility was ruled out since no sign of disrupting the aggregate contacts was observed when the folded poly-Ub construct was restretched; the number of steps indicating the unfolding of the individual Ub domain was consistent with the number of Ub monomers [11].

The immediate interpretation toward the anomalous behavior of refolding trajectories upon force quench is found from the vastly different initial structure ensemble [12,49,57,89,90]. The initial structure ensemble under high tension is fully stretched (stretched state ensemble [SSE]), while the nature of the ensemble for bulk measurements is thermally denatured (thermally denatured ensemble [TDE]).TDE and SSE drastically differ in structural characteristics. The entropy of SSE is smaller than that of TDE ($S_{SSE} < S_{TDE}$) [12,57,90]. Therefore, it is not unusual that the folding kinetics upon force quench is vastly different from the one at the bulk measurement. It is conceivable that the energy landscape explored from these two distinct ensembles vastly differ.

The next more elaborate question is then why the folding rate for force quench is slower. Under force-quench conditions, two driving forces compete each other. One is the free-energy gradient bias toward NBA, while the other is the quench force ($f_Q < f_{mid}$) that resists the collapse process. A free-energy barrier is formed under these two competing forces. Consequently, before making a transition to NBA, the

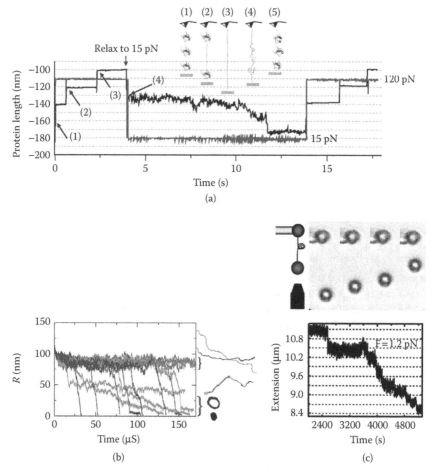

FIGURE 5.4 Refolding (collapse) dynamics of biopolymers upon force quench. (a) Refolding trajectory of poly-Ub generated by atomic force microscopy upon force quench. (Adapted from Fernandez, J. M., and H. Li., *Science* 303:1674–8., 2004. With permission.) (b) The collapse of a semiflexible chain in a poor solvent under force-quench conditions. Toroidal and racquet structures are formed after varying durations of plateau associated with force-induced metastable intermediates. (Adapted from Hyeon, C., G. Morrison, D. L. Pincus, and D. Thirumalai., *Proc Natl Acad Sci U S A.*,106:20288–93, 2009. With permission.) (c) After an addition of multivalent counterions, it takes as long as ≈ 2400 seconds for a λ-DNA to initiate collapse dynamics under a small quench force f_Q = 1.2 pN. (Adapted from Fu, W. B., X. L. Wang, X. H. Zhang, S. Y. Ran, J. Yan, and M. Li., *J Am Chem Soc.*, 128:15040–1, 2006. With permission.)

molecule is trapped in a finite-sized free-energy barrier, forming a metastable intermediate. Due to f_Q, the formation of contact responsible for the collapse process is suppressed. The likelihood to form the metastable intermediate tendency increases with an increasing f_Q, which again increases the refolding time (τ_F). If the time spent for force quench (τ_Q) is too fast, that is, $\tau_Q \ll \tau_F$, producing a nonequilibrated system in which the molecule is trapped in the force-induced metastable intermediate (FIMI), then the plateau in terms of molecular extension x will be observed in the measurement. Compared to TDE, in which the contacts responsible for the collapse are in proximity, the formation of folding (or collapse) nuclei is much more time-consuming for a system trapped in FIMI state at higher f_Q. Thus, the folding route from SSE to NBA differs significantly from the one from TDE to NBA [90].

The FIMI is generic to the force-quench refolding dynamics of any biopolymer, which can be easily tested by applying a tension to a semiflexible polymer that forms a toroid or a racquet structure upon collapse. The energy Hamiltonian is given by

$$
\begin{aligned}
H = \frac{k_s}{2a^2} \sum_{i=1}^{N-1} (r_{i,i+1} - a)^2 + \frac{k_b}{2} \sum_{i=1}^{N-2} (1 - \hat{r} \cdot \hat{r}_{i+1}) \\
+ \varepsilon_{LJ} \sum_{i,j} \left[\left(\frac{a}{r_{ij}} \right)^{12} - 2 \left(\frac{a}{r_{ij}} \right)^6 \right] - f(z_N - z_1)
\end{aligned}
\tag{5.16}
$$

with the parameters $\varepsilon_{LJ} = 1.5k_BT$, $k_s = 2000k_BT$, $N = 200$, $a = 0.6$, and $k_b = 80k_BT$. By abruptly changing $f = f_s = 83$ pN to $f = f_Q = 4$ pN, long plateaus with varying durations are observed in the semiflexible polymer described by the energy Hamiltonian (Figure 5.4b). A LOT experiment showing the difficulty of forming DNA toroids under tension unambiguously supports the argument that FIMI is generic to the collapse dynamics of any polymer under tension [91] (Figure 5.4c).

5.6 CONCLUDING REMARKS

The effort to decipher the energy landscape of a biopolymer by monitoring its mechanical response to an external stress has greatly enriched SM experiments, theories, and molecular simulations associated with the force mechanics of biomolecules and cells. We have also witnessed many instances of successful application of theories and molecular simulations for experimental data analysis [54,81,82,92]. However, as experiments are conducted on more complicated and larger systems among cellular constituents, nontrivial patterns of molecular response are observed more frequently [62,93–95]. Theoretical methods to analyze the mechanical response from multiple barriers and studies on the structural origin of more non-trivial mechanical responses such as the catch-slip bond are still in their infancy. Further studies need to be done beyond force-induced dynamics of a molecular system associated with single free-energy barrier-crossing dynamics on a one-dimensional energy profile [44,92,96]. Interference between a molecule of interest and the instrument itself pose the problem of deconvolution [34,86]. It will be of great interest to see how the perspective of both theory and simulation reveals how hydrodynamic interaction between the subdomains of a biopolymer with complex architecture plays a role during its force-unfolding process. More exciting experiments using mechanical force and breakthroughs in theories and molecular simulation methods are anticipated in the next decade to further reveal the beauty of living systems at the microscopic level.

ACKNOWLEDGMENTS

I am grateful to Dave Thirumalai for intellectually pleasurable discussions and collaborations on theory for force spectroscopy and molecular simulations for the past many years. This work was supported in part by grants from the National Research Foundation of Korea (2010-0000602).

REFERENCES

1. Moy, V. T., E. L. Florin, and H. E. Gaub. 1994. Intermolecular forces and energies between ligands and receptors. *Science* 266:257–9.
2. Ha, T., T. Enderle, D. F. Ogletree, D. S. Chemla, P. Selvin, and S. Weiss. 1996. Probing the interaction between two single molecules: Fluorescence resonance energy transfer between a single donor and a single acceptor. *Proc Natl Acad Sci U S A* 93:6264.
3. Kellermeyer, M. Z., S. B. Smith, H. L. Granzier, and C. Bustamante. 1997. Folding-unfolding transitions in single titin molecules characterized by force-measuring laser tweezer. *Science* 276:1599–600.
4. Bustamante, C., J. F. Marko, E. D. Siggia, and S. Smith. 1994. Entropic elasticity of λ-phase DNA. *Science* 265:1599–1600.
5. Seol, Y., G. M. Skinner, K. Visscher, A. Buhot, and A. Halperin. 2007. Stretching of homopolymeric RNA reveals single-stranded helices and base-stacking. *Phys Rev Lett* 98:158103.
6. Onoa, B., S. Dumont, J. Liphardt, S. B. Smith, I. Tinoco Jr., and C. Bustamante. 2003. Identifying kinetic barriers to mechanical unfolding of the *T. thermophile* ribozyme. *Science* 299:1892–5.
7. Mickler, D., R. I. Dima, H. Dietz, C. Hyeon, D. Thirmalai, and M. Rief. 2007. Revealing the bifurcation in the unfolding pathways of GFP by using single-molecule experiments and simulations. *Proc Natl Acad Sci U S A* 104:20268–73.
8. Rief, M., H. Gautel, F. Oesterhelt, J. M. Fernandez, and H. E. Gaub. 1997. Reversible unfolding of individual titin immunoglobulin domains by AFM. *Science* 276:1109–11.
9. Rief, M., J. Pascual, M. Saraste, and H. E. Gaub. 1999. Single-molecule force spectroscopy of spectrin repeats: Low unfolding forces in helix bundles. *J Mol Biol* 286:553–61.
10. Greenleaf, W. J., K. L. Frieda, D. A. N. Foster, M. T. Woodside, and S. M. Block. 2008. Direct observation of hierarchical folding in single riboswitch aptamers. *Science* 319:630–3.
11. Fernandez, J. M., and H. Li. 2004. Force-clamp spectroscopy monitors the folding trajectory of a single protein. *Science* 303:1674–8.
12. Hyeon, C., G. Morrison, D. L. Pincus, and D. Thirumalai. 2009. Refolding dynamics of stretched biopolymers upon force-quench. *Proc Natl Acad Sci U S A* 106:20288–93.
13. Russell, R., X. Zhuang, H. Babcock, I. Millett, S. Doniach, S. Chu, and D. Herschlag. 2002. Exploring the folding landscape of a structured RNA. *Proc Natl Acad Sci U S A* 99:155–60.
14. Visscher, K., M. J. Schnitzer, and S. M. Block. 1999. Single kinesin molecules studied with a molecular force clamp. *Nature* 400:184–7.
15. Guydosh, N. R., and S. M. Block. 2006. Backsteps induced by nucleotide analogs suggest the front head of kinesin is gated by strain. *Proc Natl Acad Sci U S A* 103:8054–9.
16. Chemal, Y. R., K. Aathavan, J. Michaelis, S. Grimes, P. J. Jardine, D. L. Anderson, and C. Bustamante. 2005. Mechanism of force generation of a viral DNA packaging motor. *Cell* 122:683–92.
17. Hodges, C., L. Bintu, L. Lubkowska, M. Kashlev, and C. Bustamante. 2009. Nucleosomal fluctuations govern the transcription dynamics of RNA polymerase II. *Science* 325:626–8.

18. Coppin, C. M., J. T. Finer, J. A. Spudich, and R. D. Vale. 1996. Detection of sub-8-nm movements of kinesin by high-resolution optical-trap microscopy. *Proc Natl Acad Sci U S A* 93:1913–7.

19. Guydosh, N. R., and S. M. Block. 2009. Direction observation of the binding state of the kinesin head to the microtubule. *Nature* 461:125–8.

20. Asbury, C. L., A. N. Fehr, and S. M. Block. 2003. Kinesin moves by an asymmetric hand-over-hand mechanism. *Science* 302:2130–4.

21. Vogel, V., and M. Sheetz. 2006. Local force and geometry sensing regulate cell functions. *Nat Rev Mol Cell Biol* 7:265–75.

22. Greenleaf, W. J., M. T. Woodside, E. A. Abbondanzieri, and S. M. Block. 2005. Passive all-optical force clamp for high-resolution laser trapping. *Phys Rev Lett* 95:208102.

23. Liphardt, J., B. Onoa, S. B. Smith, I. Tinoco Jr., and C. Bustamante. 2001. Reversible unfolding of single RNA molecules by mechanical force. *Science* 292:733–7.

24. Chen, Y. F., J. N. Milstein, and J. C. Meiners. 2010. Protein-mediated DNA loop formation and breakdown in a fluctuating environment. *Phys Rev Lett* 104:258103.

25. Bell, G. I. 1978. Models for the specific adhesion of cells to cells. *Science* 200:618–27.

26. Tees, D. F. J., O. Coenen, and H. L. Goldsmith. 1993. Interaction forces between red cells agglutinated by antibody: IV. Time and force dependence of break-up. *Biophys J* 65:1318–34.

27. Florin, E. L., V. T. Moy, and H. E. Gaub. 1994. Adhesive forces between individual ligand-receptor pairs. *Science* 264:415–7.

28. Lee, G. U., D. A. Kidwell, and R. J. Colton. 1994. Sensing discrete streptavidin-biotin interactions with atomic force microscopy. *Langmuir* 10:354–7.

29. Hoh, J. H., J. P. Cleveland, C. B. Prater, J. P. Revel, and P. K. Hansma. 1992. Quantized adhesion detected with the atomic force microscope. *J Am Chem Soc* 114:4917–8.

30. Alon, R., D. A. Hammer, and T. A. Springer. 1995. Lifetime of the p-selectin-carbohydrate bond and its response to tensile force in hydrodynamic flow. *Nature* 374:539–42.

31. Tinoco Jr., I. 2004. Force as a useful variable in reactions: Unfolding RNA. *Annu Rev Biophys Biomol Struct* 33:363–85.

32. Block, S. M., C. L. Asbury, J. W. Shaevitz, and M. J. Lang. 2003. Probing the kinesin reaction cycle with a 2D optical force clamp. *Proc Natl Acad Sci U S A* 100:2351–6.

33. Woodside, M. T., W. M. Behnke-Parks, K. Larizadeh, K. Travers, D. Herschlag, and S. M. Block. 2006. Nanomechanical measurements of the sequence-dependent folding landscapes of single nucleotide acid hairpins. *Proc Natl Acad Sci U S A* 103:6190–5.

34. Woodside, M. T., P. C. Anthony, W. M. Behnke-Parks, K. Larizadeh, D. Herschlag, and S. M. Block. 2006. Direct measurement of the full, sequence-dependent folding landscape of a nucleic acid. *Science* 314:1001–4.

35. de Gennes, P. G. 1979. *Scaling Concepts in Polymer Physics*. Ithaca and London: Cornell University Press.

36. Kramers, H. A. 1940. Brownian motion in a field of force and the diffusion model of chemical reaction. *Physica* 7:284–304.

37. Hanggi, P., P. Talkner, and M. Borkovec. 1990. Reaction-rate theory: Fifty years after Kramers. *Rev Mod Phys* 62:251–341.

38. Pincus, P. 1976. Excluded volume effects and stretched polymer chains. *Macromolecules* 9:386–8.

39. Morrison, G., C. Hyeon, N. M. Toan, B. Y. Ha, and D. Thirumalai. 2007. Stretching homopolymers. *Macromolecules* 40:7343–53.

40. Marko, J. F., and E. D. Siggia. 1995. Stretching DNA. *Macromolecules* 28:8759–70.

41. Eyring, H. 1935. The activated complex in chemical reactions. *J Chem Phys* 3:105–15.

42. Thirumalai, D., and C. Hyeon. 2005. RNA and protein folding: Common themes and variations. *Biochemistry* 44:4957–70.

43. Zwanzig, R. 2001. *Nonequilibrium Statistical Mechanics*. New York: Oxford University Press.

44. Hyeon, C., and D. Thirumalai. 2007. Measuring the energy landscape roughness and the transition state location of biomolecules using single-molecule mechanical unfolding experiments. *J Phys Condens Matter* 19:113101.

45. Yang, W. Y., and M. Gruebele. 2003. Folding at the speed limit. *Nature* 423:193–7.

46. Chung, H. S., J. M. Louis, and W. A. Eaton. 2009. Experimental determination of upper bound for transition path times in protein folding from single-molecule photon-by-photon trajectories. *Proc Natl Acad Sci U S A* 106:11839–44.

47. Liphardt, J., S. Dumont, S. B. Smith, I. Tinoco Jr., and C. Bustamante. 2002. Equilibrium information from nonequilibrium measurements in an experimental test of Jarzynski's equality. *Science* 296:1832–5.

48. Marszalek, P. E., H. Lu, H. Li, M. Carrion-Vazquez, A. F. Oberhauser, K. Schulten, and J. M. Fernandez. 1999. Mechanical unfolding intermediates in titin molecules. *Nature* 402:100–3.

49. Fisher, T. E., A. F. Oberhauser, M. Carrion-Vazquez, P. E. Marszalek, and J. M. Fernandez. 1999. The study of protein mechanics with the atomic force microscope. *TIBS* 24:379–84.

50. Lu, H., B. Isralewitz, A. Krammer, V. Vogel, and K. Schulten. 1998. Unfolding of titin immunoglobulin domains by steered molecular dynamics. *Biophys J* 75:662–71.

51. Hummer, G., and A. Szabo. 2003. Kinetics from nonequilibrium single-molecule pulling experiments. *Biophys J* 85:5–15.

52. Dudko, O. K., G. Hummer, and A. Szabo. 2008. Theory, analysis, and interpretation of single-molecule force spectroscopy experiments. *Proc Natl Acad Sci U S A* 105:15755–60.

53. Dudko, O. K., G. Hummer, and A. Szabo. 2006. Intrinsic rates and activation free energies from single-molecule pulling experiments. *Phys Rev Lett* 96:108101.

54. Dudko, O. K., J. Mathe, A. Szabo, A. Meller, and G. Hummer. 2007. Extracting kinetics from single-molecule force spectroscopy: Nanopore unzipping of DNA hairpin. *Biophys J* 92:4188–95.

55. Hyeon, C., and D. Thirumalai. 2005. Mechanical unfolding of RNA hairpins. *Proc Natl Acad Sci U S A* 102:6789–94.

56. Lacks, D. J. 2005. Energy landscape distortions and the mechanical unfolding of proteins. *Biophys J* 88:3494–501.

57. Hyeon, C., and D. Thirumalai. 2006. Forced-unfolding and force-quench refolding of RNA hairpins. *Biophys J* 90:3410–27.

58. Hammond, G. S. 1953. A correlation of reaction rates. *J Am Chem Soc* 77:334–8.

59. Leffler, J. E. 1953. Parameters for the description of transition states. *Science* 117:340–1.

60. Dudko, O. K., A. E. Filippov, J. Klafter, and M. Urbakh. 2003. Beyond the conventional description of dynamic force microscopy of adhesion bonds. *Proc Natl Acad Sci U S A* 100:11378–81.

61. Freund, L. B. 2009. Characterizing the resistance generated by a molecular bond as it is forcibly separated. *Proc Natl Acad Sci U S A* 106:8818–23.

62. Merkel, R., P. Nassoy, A. Leung, K. Ritchie, and E. Evans. 1999. Energy landscapes of receptor-ligand bonds explored with dynamic force microscopy. *Nature* 397:50–3.

63. Derenyi, I., D. Bartolo, and A. Ajdari. 2004. Effects of intermediate bound states in dynamic force spectroscopy. *Biophys J* 86:1263–9.

64. Tirion, M. M. 1996. Large amplitude elastic motions in proteins from a single-parameter atomic analysis. *Phys Rev Lett* 77:1905–8.

65. Bahar, I., and A. J. Rader. 2005. Coarse-grained normal mode analysis in structural biology. *Curr Opin Struct Biol* 15:586–92.

66. Zheng, W., B. R. Brooks, and D. Thirumalai. 2006. Low-frequency normal modes that describe allosteric transitions in biological nanomachines are robust to sequence variations. *Proc Natl Acad Sci U S A* 103:7664–9.

67. Hyeon, C., R. I. Dima, and D. Thirumalai. 2006. Pathways and kinetic barriers in mechanical unfolding and refolding of RNA and proteins. *Structure* 14:1633–45.

68. Hyeon, C., and D. Thirumalai. 2007. Mechanical unfolding of RNA: from hairpins to structures with internal multiloops. *Biophys J* 92:731–43.

69. Hyeon, C., G. H. Lorimer, and D. Thirumalai. 2006. Dynamics of allosteric transitions in GroEL. *Proc Natl Acad Sci U S A* 103:18939–44.

70. Lin, J., and D. Thirumalai. 2008. Relative stability of helices determines the folding landscape of adenine riboswitch aptamers. *J Am Chem Soc* 130:14080–1

71. Chen, J., S. A. Darst, and D. Thirumalai. 2010. Promoter melting triggered by bacterial RNA polymerase occurs in three steps. *Proc Natl Acad Sci U S A* 107:12523–8.

72. Dima, R. I., and H. Joshi. 2008. Probing the origin of tubulin rigidity with molecular simulations. *Proc Natl Acad Sci U S A* 105:15743–8.

73. Hyeon, C., and J. N. Onuchic. 2007. Mechanical control of the directional stepping dynamics of the kinesin motor. *Proc Natl Acad Sci U S A* 104:17382–7.

74. Peng, Q., and H. Li. 2008. Atomic force microscopy reveals parallel mechanical unfolding pathways of T4 lysozyme: Evidence for a kinetic partitioning mechanism. *Proc Natl Acad Sci U S A* 105:1885–90.

75. Zwanzig, R. 1990. Rate processes with dynamical disorder. *Acc Chem Res* 23:148–52.

76. Dietz, H., F. Berkemeier, M. Bertz, and M. Rief. 2006. Anisotropic deformation response of single protein molecules. *Proc Natl Acad Sci U S A* 103:12724–8.

77. Dietz, H., and M. Rief. 2004. Exploring the energy landscape of GFP by single-molecule mechanical experiments. *Proc Natl Acad Sci U S A* 101:16192–7.

78. Thirumalai, D., and S. A. Woodson. 1996. Kinetics of folding of proteins and RNA. *Acc Chem Res* 29:433–9.

79. Cecconi, C., E. A. Shank, C. Bustamante, and S. Marqusee. 2005. Direct observation of three-state folding of a single protein molecule. *Science* 309:2057–60.

80. Zwanzig, R. 1988. Diffusion in rough potential. *Proc Natl Acad Sci U S A* 85:2029–30.

81. Hyeon, C., and D. Thirumalai. 2003. Can energy landscape roughness of proteins and RNA be measured by using mechanical unfolding experiments? *Proc Natl Acad Sci U S A* 100:10249–53.

82. Nevo, R., V. Brumfeld, R. Kapon, P. Hinterdorfer, and Z. Reich. 2005. Direct measurement of protein energy landscape roughness. *EMBO Rep* 6:482.

83. Janovjak, H., H. Knaus, and D. J. Muller. 2007. Transmembrane helices have rough energy surfaces. *J Am Chem Soc* 129:246–7.

84. Lee, E. H., J. Hsin, M. Sotomayor, G. Comellas, and K. Schulten. 2009. Discovery through the computational microscope. *Structure* 17:1295–306.

85. Manosas, M., J. D. Wen, P. T. X. Li, S. B. Smith, C. Bustamante, I. Tinoco Jr., and F. Ritort. 2007. Force unfolding kinetics of RNA using optical tweezers. II. Modeling experiments. *Biophys J* 92:3010–21.

86. Hyeon, C., G. Morrison, and D. Thirumalai. 2005. Force dependent hopping rates of RNA hairpins can be estimated from accurate measurement of the folding landscapes. *Proc Natl Acad Sci U S A* 105:9604–6.

87. Barsegov, V., G. Morrison, and D. Thirumalai. 2008. Role of internal chain dynamics on the rupture kinetics of adhesive contacts. *Phys Rev Lett* 100:248102.

88. Sosnick, T. R. 2004. Comment on "Force-clamp spectroscopy monitors the folding trajectory of a single protein." *Science* 306:411b.

89. Li, M. S., C. K. Hu, D. K. Klimov, and D. Thirumalai. 2006. Multiple stepwise refolding of immunoglobulin I27 upon force quench depends on initial conditions. *Proc Natl Acad Sci U S A* 103:93–8.

90. Hyeon, C., and D. Thirumalai. 2008. Multiple probes are required to explore and control the rugged energy landscape of RNA hairpins. *J Am Chem Soc* 130:1538–9.

91. Fu, W. B., X. L. Wang, X. H. Zhang, S. Y. Ran, J. Yan, and M. Li. 2006. Compaction dynamics of single DNA molecule under tension. *J Am Chem Soc* 128:15040–1.

92. Barsegov, V., and D. Thirumalai. 2005. Dynamics of unbinding of cell adhesion molecules: Transition from catch to slip bonds. *Proc Natl Acad Sci U S A* 102:1835–9.

93. Marshall, B. T., M. Long, J. W. Piper, T. Yago, R. P. McEver, and C. Zhu. 2003. Direct observation of catch bonds involving cell-adhesion molecules. *Nature* 423:190–3.

94. Zhang, Y., S. Sivasanker, W. J. Nelson, and S. Chu. 2009. Resolving cadherin interactions and binding cooperativity at the single-molecule level. *Proc Natl Acad Sci U S A* 106:109–14.

95. Nevo, R., C. Stroh, F. Kienberger, D. Kaftan, V. Brumfeld, M. Elbaum, Z. Reich, and P. Hinterdorfer. 2003. A molecular switch between alternative conformational states in the complex of Ran and importin β1. *Nat Struct Biol* 10:553–7.

96. Suzuki, Y., and O. K. Dudko. 2010. Single-molecule rupture dynamics on multidimensional landscapes. *Phys Rev Lett* 104:048101.

6 Coarse-Grained Modeling of DNA Nanopore Interactions

Yaling Liu, Abhijit Ramachandran, and Qingjiang Guo

CONTENTS

6.1 INTRODUCTION

The concept of personalized medicine has evolved from the fact that the phrase "one size fits all" is no longer valid for the various diseases facing mankind. The human genome project was undertaken to individualize medical needs by determining the genes responsible for their causes. However, the major challenge in genome sequencing apart from the enormous data that is collected in an economical manner to achieve it.

This section provides an overview of gene sequencing using solid-state nanopores which could result in entire genome sequencing in less than $1000 per genome.

6.1.1 OVERVIEW

The cell is the basic structural and functional unit of all known living organisms. It is the smallest unit of an organism that is classified as living and is often called the building block of life [1]. Instructions needed to direct the activities of cells are contained within a chemical, deoxyribonucleic acid (DNA). The DNA of all organisms is made up of the same chemical and physical components. The DNA sequence or genetic sequence is the particular side-by-side arrangement of bases along a DNA strand (e.g., AAGGCTGTAG). This order spells out the exact instructions required to create a particular organism with its own unique traits. The genome is an organism's complete set of DNA. It varies widely in size; for instance, the smallest known genome of a bacterium contains about 600,000 DNA base pairs, whereas human and mouse genomes have some 3 billion DNA base pairs [2].

In 1990, the U.S. Department of Energy (U.S. DOE) and the National Institutes of Health (NIH) developed a plan for sequencing the entire human genome, which resulted in the inauguration of the Human Genome Project (HGP). The ultimate goal of the HGP was to generate a high-quality reference DNA sequence for the human genome's 3 billion base pairs and to identify all human genes [3–5]. This project aimed to help one understand the evolution of human beings, enables the development of personalized medicine, and recognition of the role of environment and heredity in defining the human condition [2]. The HGP (completed in 2003 at the cost of $2.7 billion) took more than 12 years to complete [6]. With the advent of novel sequencing technologies and abundant computational power, the cost and time needed for sequencing a genome has reduced dramatically over the years; now it is ~3¢/base [6]. Yet, this adds up to an enormous amount for 3 billion bases to be sequenced. The challenge of reducing cost and time has furthered the development of sequencing technology. Since the development of the dideoxy sequencing method, which was pioneered by Sanger, Nicklen, and Coulson [7] in 1977, many new technologies have emerged that can massively increase the capacity for de novo DNA sequencing. The major techniques are either electrophoretic methods such as slab gel electrophoresis or non-electrophoretic methods such as pyrosequencing, and/or sequencing by hybridization microarrays. The single-molecule sequencing method that has recently been receiving much attention is a nonelectrophoretic approach that is developed based on the principle of ion channels using nanopores along with single-channel current measurements. The nanopore-based device provides a highly confined space, within which polymers such as DNA and ribonucleic acid (RNA) can be electrophoretically driven, and allows the analysis at high throughput. The unique analytical ability to identify and characterize kilobase (kb)-length polymers without amplification or labeling makes rapid DNA sequencing an inexpensive possibility using nanopores [8]. Thus, despite their material, electrical, and fabrication shortcomings and challenges, nanopore-based devices have the potential to achieve the $1000/genome target.

6.1.2 Principle of Deoxyribonucleic Acid Sequencing

The concept of nanopore-based sequencing of DNA is very simple and straightforward: An electrochemical system is shown in Figure 6.1, where an applied bias voltage drives the ions to flow from anode (−) to cathode (+) through the nanopore. This gives the baseline current value. The DNA carries a −e charge due to its phosphate backbone, and hence it is electrophoretically driven through the nanopore toward the cathode. As it flows through the nanopore DNA blocks the free passage of ions, which causes a dip in the ionic current that can be measured using the electrical circuit. The DNA is forced to pass through a nanopore having a diameter comparable to the size of DNA, and thus it has to pass in a linear mode in a base-by-base fashion instead of remaining coiled. The degree and duration of the ionic current dip is proportional to the structure and length of the molecules (e.g., DNA). This simple, straightforward technique promises a range of analyses with higher throughput than currently used methods.

It is difficult to achieve genome sequencing with such a simplified model. Various attempts to modulate the existing system and obtain a de novo sequencing method are being made. Such a method that successfully demonstrates gene sequencing is described in Section 6.1.3.

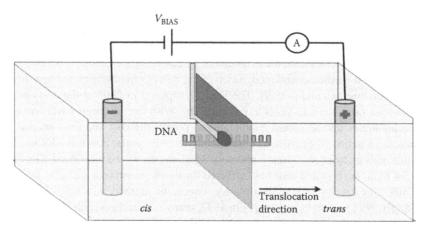

FIGURE 6.1   Electrophoretic translocation of single-stranded deoxyribonucleic acid through a nanopore.

6.1.3 GENOMIC SEQUENCING THROUGH FUNCTIONALIZED NANOPORES

Iqbal, Akin, and Bashir [9] suggested the single-molecule electrophoretic transport measurements of single-stranded DNA (ssDNA) that passes through hairpin-loop (HPL) DNA–functionalized nanopores, which allow gene sequencing. They explained their observations based on the work of Bauer and Nadler [10] who presented theoretical predictions on channel–molecule interactions through nanopores.

The nanopores were coated with HPL DNA molecules (Figure 6.2c). This coating makes the nanopores selective toward short lengths of "target" ssDNA (Figure 6.2d), and leads to transporting perfectly complementary (PC) molecules faster than mismatched ones (MM). Even a single-base mismatch between the probe and the target results in longer translocation pulses and significantly reduces the number of translocation events (Figure 6.2b). The pulses for PC ssDNA were narrower and deeper than those for mismatched ones, indicating facilitated (and faster) transport of PC DNA that interacts with surface-bound hairpin molecules.

6.1.4 CONTENT OF THE CHAPTER

This chapter focuses on fundamental insights into the mechanics of DNA translocation inside biofunctionalized nanopores, which have enabled imparting selectivity to solid-state nanopores. The studies are divided into three sections: (1) interaction between a DNA and a bare nanopore [11], (2) surface properties of a functionalized nanopore, and (3) interaction between a DNA and a functionalized nanopore.

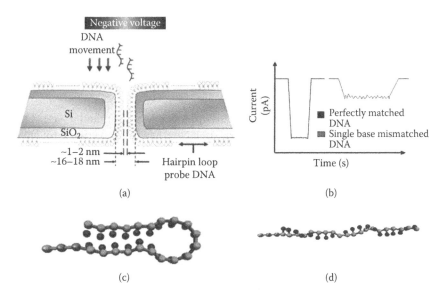

(a)   (b)

(c)   (d)

FIGURE 6.2 (a) Schematic of the biofunctionalized system; (b) schematic to show selectivity achieved due to the difference in translocation times measured using the dip-in ionic current of matched and mismatched deoxyribonucleic acid (DNA); (c) the hairpin-loop DNA structure; and (d) single-stranded DNA structure. (From Iqbal, S. M., D. Akin, and R. Bashir. 2007. *Nat Nanotechnol* 2(4):243–8. With permission.)

The molecular dynamics (MD) of a coarse-grained (CG) system in the simulation is similar to the schematic shown in Figure 6.2a, and ideal selectivity achieved from the system is shown in Figure 6.2b. In order to study the interaction between ssDNA (Figure 6.2d) and HPL-coated nanopore and thereby reveal the underlying molecular mechanisms and process of functionalized nanopore-based DNA sequencing, the translocation processes of an ssDNA with its sequence perfectly matching the probe HPLs and another ssDNA with a completely mismatched sequence are considered for quantitative comparison. Similar functionalization schemes can be used for a variety of ligand–receptor combinations of significant importance, and the solid-state functionalized nanopore can serve as the next generation of sequencing tools, whereas a pore functionalized with a specific probe can be used as a detector of specific nucleotides or biomarkers or both [9].

6.2 REVIEW OF CURRENT DEOXYRIBONUCLEIC ACID SEQUENCING TECHNIQUES

To understand DNA sequencing modalities, the concept on which these technologies are based must be illustrated. This section highlights the basics of DNA and its useful properties, along with a summary of currently available sequencing modalities utilizing DNA properties.

6.2.1 DEOXYRIBONUCLEIC ACID

Deoxyribonucleic acid (DNA) is a nucleic acid that contains the blueprints of genetic instructions used in the development and functioning of all known living organisms. The term nuclein was coined by Miescher in 1869 for substances isolated from the nucleus of a cell and it was later termed as nucleic acid [12]. The DNA has a helical structure as identified by Watson and Crick in 1953 and a diameter of 2 nm [13]. It is a polymer consisting of two long chains made of repeated subunits, called nucleotides, arranged in a linear fashion.

A nucleotide consists of a phosphate group, a pentose sugar, and a base. There are two classes of pentose sugar: (1) ribose (in RNA), and (2) deoxyribose (in DNA). The difference between these two is the presence of a hydrogen atom instead of a hydroxyl group on carbon atom 2 for deoxyribose. Moreover, there are two types of bases: (1) a pyrimidine (thymine [T] and cytosine [C]), which is a heterocyclic aromatic organic compound similar to benzene containing two nitrogen atoms at positions 1 and 3 of the six-member ring; and (2) purines (adenine [A] and guanine [G]), which are more complex fused nitrogenous rings consisting of pyrimidine and imidazole rings [12].

The long chains (strands) forming double-stranded DNA (dsDNA) are antiparallel in nature, that is, they run in opposite directions to each other (5′–3′ and 3′–5′). This is established since the phosphate attaches to the 5′ carbon of the trailing pentose sugar (phosphodiester bond) and the 3′ carbon of the next, resulting in a strand in a 5′→3′ direction, whereas its complementary strand has the 3′→5′ direction. The DNA has a negative charge ($-e$) due to its phosphate backbone, where a free-floating phosphate has a charge of $-2e$ and is represented as PO_4^{-2}. However, the phosphate in DNA is not a free phosphate due to its additional attachment with the sugar, and thus the phosphate in DNA has a charge of $-1e$, (PO_4^{-1}) [14]. Further, the bases of DNA have the property of complementary base pairing, which is achieved with the specific formation of two and three hydrogen bonds between A–T and G–C pairs, respectively (see Figure 6.3b). These hydrogen bonds enable molecular recognition but are weaker than the covalent bonds between sugar and phosphate of the DNA backbone. Base pairing occurs if the complementary bases lie in close proximity to each other. This enables two DNA molecules (complementary to each other) to self-organize and form supramolecular structures without the requirement of energy and/or regulatory helpers. Furthermore, if either of the strands from a dsDNA is removed it can be re-created by using the other as a template, which is an important requirement for basic genetic processes (replication, transcription, and recombination). The processes by which the order of nucleotide bases in a molecule of DNA can be determined are termed DNA sequencing methods [14].

The DNA double helix is very stable due to base pairing between complementary strands and stacking between adjacent bases. The dsDNA can be separated into two single strands of DNA (two ssDNAs) by either increasing the temperature or

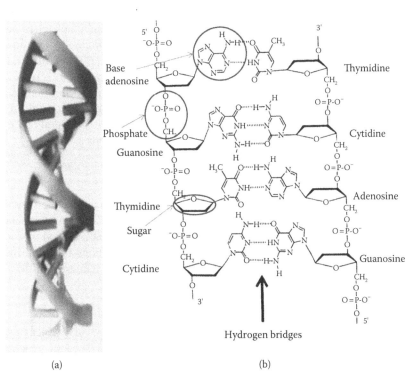

(a)   (b)

FIGURE 6.3 (a) The regular double helix structure of deoxyribonucleic acid (DNA), and (b) the molecular level detail of a DNA strand with complementary base pairing with the formation of the hydrogen bonds. (Reprinted from Wink, M. 2006. *An Introduction to Molecular Biotechnology: Molecular Fundamentals, Methods and Applications in Modern Biotechnology.* Wiley-VCH. With permission.)

changing the salt concentration (pH) of the solution [15]. The process of denaturing the DNA to form single strands by increasing the temperature is termed as melting or "dehybridization of DNA." When the temperature is decreased (i.e., the system is cooled), two dehybridized DNA will reproduce the double helices at regions where sequences of these two DNAs are complementary to each other; this process is called "DNA hybridization." The temperature at which 50% of DNA is present as single strands is termed as "melting temperature" (T_m). The T_m is dependent on the G–C content of DNA in such a way that the higher the percentage of G–C, the higher the temperature needed for dehybridization. This is because the G–C pairs have three hydrogen bonds compared with the two formed between the A–T pairs (see Figure 6.2b). This technique is useful in determining sequence similarity among DNAs of different origin and the amount of sequence repetition within one DNA [15].

Another ability of DNA that is much exploited is replication. For the process of DNA replication, a dsDNA separates into two ssDNAs. These ssDNAs now act as templates for constructing a new dsDNA by utilizing the property of complementary base pairing. For this, a primer (a strand of nucleic acid that serves as a starting point for DNA replication) attaches to the 3′ end of the ssDNA template strand, and the DNA polymerase (family of enzymes that carry out all forms of DNA replication by adding free nucleotides) adds new nucleotides to the existing template strand of DNA from the end where the primer has bound itself. The process is terminated once the new strand is completely synthesized from the template strand, that is, two dsDNAs are obtained. This ascertains that if either strand from a dsDNA is removed, it can be re-created exactly by using the other as a template. This has resulted in the development of a revolutionizing technique renowned as "polymerase chain reaction" (PCR). The PCR is a process to amplify a DNA strand across several orders of magnitude enabling the generation of well above a thousand copies.

6.2.2   DEOXYRIBONUCLEIC ACID SEQUENCING MODALITIES

The HGP was a 13-year effort to obtain the entire human genome for the very first time with a cost of $2.7 billion, which was finished in 2003. The advancement in technologies over the next five years led to the same feat being achieved in a time span of 5 months with a total cost of $1.5 million [16]. Sections 6.2.2.1 through 6.2.2.9 give a general insight into various current DNA sequencing modalities that are technically advanced and significantly reduce the cost.

6.2.2.1   Sequencing Modalities of Deoxyribonucleic Acid

The chain termination sequencing method developed by Sanger and colleagues in 1971 [7] utilizes the base-specific chain terminations in four separate reactions, which correspond to the four different nucleotides (A, T, G, and C) in the DNA composition. This method requires four separate DNA extension reactions. Each reaction contains a template DNA, a short primer, the DNA polymerase, and all four deoxynucleotides (dNTPs→ dATP, dTTP, dGTP, and dCTP), of which one is radioactively labeled for each of the reactions represented by an asterisk (dATP* for the current example). These reactions are also spiked by the corresponding dideoxy-nucleotides (ddNTP→ ddATP, ddTTP, ddGTP, and ddCTP). The dNTPs are added with the nucleotides and they bind to the complementary base, but when the ddNTP is added in connectivity the growth of the chain is terminated. This is because the ddNTP lacks a 3′ hydroxyl group necessary to form the linkage with an incoming nucleotide. As a consequence, the new strand extends until a ddNTP is incorporated. Using a polyacrylamide gel, the radioactive products are separated along the four lanes and scored according to their molecular masses. The location of the strands in the four lanes provides the position of the nucleotide in the strand whereas the specific lane dictates the type of base at that position [17].

The modified Sanger approach utilizes four different fluorescent dyes to label the different ddNTPs instead of radioactively labeling a particular dNTP [18]. The reaction products are then separated electrophoretically in a single glass capillary filled with a polymer. The DNA bands move inside the capillary according to their lengths (masses). Lasers are used to excite the fluorophores at the end of the capillary. The sequence of DNA is determined based on color; each color corresponds to a particular nucleotide.

The Sanger method is still considered the gold standard in sequencing, although it has certain limitations. One of its limitations is the issue of tracking the different strands that terminate at different locations in the gel. But the major limitation is that the cost of this method is $1 per kilobase; thus it would cost $10,000,000 to sequence the entire human genome [19].

6.2.2.2 Capillary Electrophoresis

The need for a faster sequencing device and higher throughput has resulted in the development of capillary array electrophoresis (CAE) [20]. It is designed to separate species based on their size-to-charge ratio in the interior of a small capillary filled with an electrolyte. The principle of CAE is similar to that of slab gel electrophoresis except that the Sanger DNA sequencing fragments are separated on an array of capillaries in CAE; the detection process for both the techniques remains the same. The advantage of CAE is that there is only a single sample and that CAE eliminates tracking problems. In capillary electrophoresis, the fragment migration time is directly related to the number of bases present. The resolution of this system is three base pairs [21].

6.2.2.3 Microfabricated Capillary Arrays

Performance improvements in CAE are being achieved with miniaturization of the current system. This reduces the cost of the devices and separation time and makes more capillaries available in an instrument. Take the T-injector design and cross-T design [22], for instance, where the sample is electrophoresed from left to right such that it fills the T intersection. When the current flow from sample to waste is terminated and a potential is applied across the main cathode and anode, the part of the injected sample of length equal to channel overlap at the T intersection is electrophoresed into the main capillary. This technique provides superior sample control, higher resolution, and shorter separation lengths [23].

6.2.2.4 Fluorescence In Situ Sequencing

Fluorescence in situ sequencing, that is, sequencing by synthesis (SBS), is the most widely used method for four-color DNA sequencing, and it was designed by Smith et al. [24] in 1986 and Prober et al. [25] in 1987. Resolution of an emission signal from a dye-labeled nucleotide into color, with subsequent assignment in the DNA sequence, is the basis for this method. It usually involves the following steps: The DNA to be sequenced is attached onto a solid surface, and the labeled nucleotides with a cleavable chemical group to cap an −OH− group at the 3′ position of the deoxyribose group are added along with the DNA polymerase. The incorporation of these nucleotides will terminate the reaction and the sequence is read from the labels used for the nucleotides. Then the 3−OH− is uncapped and washed from the 3′ position of the deoxyribose group and the sequencing cycle is continued. This method provides a very high throughput since each base is detected while the DNA chain is growing [26,27]. But it has certain limitations such as inefficient excitation of fluorescent dyes, significant spectral overlap, and inefficient collection of emission signals. Attempts to overcome these limitations being made currently include use of fluorescence resonance energy transfer (FRET) to partially address the inefficient excitation problem [28]. Further efforts are being made to utilize fluorescence lifetime [29] and radio frequency modulation [30] to overcome the current deficiencies.

6.2.2.5 Pyrosequencing

Pyrosequencing is a single-nucleotide addition method (SNA) and its name is derived from the pyrophosphate (PPi) that is naturally released when the DNA polymerase incorporates a nucleotide into a duplicate strand. It is a nonfluorescence technique that particularly measures the release of PPi. During DNA synthesis, cycles of four dNTPs are separately dispensed into the reaction mixture iteratively. After each dispensation, the DNA polymerase incorporates the dNTP into the duplicated strand. The PPi is released after each nucleotide is incorporated to the DNA polymerase. The released PPi is then converted into adenosine triphosphate (ATP) by ATP-sulfurylase. This ATP is used to convert luciferase reporter enzyme to oxyluciferin; this conversion causes the emission of light, which is detected using a charge-coupled (CCD) camera. The strength of the light signal is proportional to the number of nucleotides that are incorporated (T, AA, GGG, etc.) into the duplicated strand, and it is recorded as a series of peaks called a "pyrogram." Each signal peak corresponds to the order of complementary nucleotides incorporated and reveals the underlying sequence of the DNA [31,32]. The limitations of pyrosequencing are low throughput and high background noise. Efforts that have been made to improve this limitation include the introduction of dATP-α-S SP isomer and the development of a massively parallel microfluidic sequencing platform (array-based pyrosequencing) [21].

6.2.2.6 Cyclic Reversible Termination

Cyclic reversible termination (CRT) has three stages: (1) incorporation, (2) imaging, and (3) cleaving/unblocking. Reversible terminators are modified nucleotides that terminate DNA strand formation after incorporation since they contain a blocking group at the 3′ end of the ribose group. These terminators are fluorescently labeled and after incorporation enable the reading of bases at each step of the process. Initially, the template strand of DNA is attached onto a substrate and is hybridized to sequencing primers. In the incorporation stage (Figure 6.2b), DNA polymerase incorporates the modified nucleotides (reversible terminators) to the template strand. In the imaging stage that follows the incorporation stage, the DNA sequencing is terminated and the fluorescent nucleotide is imaged. In the last stage, the fluorescent tag is cleaved off and the terminator is unblocked; this allows the DNA sequencing to be continued. The process is repeated until the template strand is completely sequenced [19,21].

6.2.2.7   Sequencing by Ligation

The approach for sequencing by ligation is quite similar to that of the CRT platform except that DNA polymerase is replaced by DNA ligase and the four nucleotides are substituted by fluorescence-labeled oligonucleotide probes [33]. The DNA fragments to be sequenced are amplified by emulsion PCR and captured on beads. These modified beads are then deposited onto a glass slide to form a random array for sequencing. Primers are then added, which hybridize to the adapter sequence on the bead. The four-color dye-labeled oligoprobes compete for ligation to the sequencing primer. Each oligoprobe is eight base long, and the two bases in the middle (fourth and fifth) are encoded. Probe specificity for ligation is achieved by analyzing every fourth and fifth base during ligation. After hybridization, ligation, and detection, the color of the fourth and fifth bases is recorded. Determination of dinucleotide sequences every five bases (e.g., … 4 5 … 9 10 … 14 15 …) is achieved by repeating this process. After a few cycles, the sequencing reaction is reset by denaturing the DNAs, and a new sequencing primer with a one base offset $(n - 1)$ is introduced. By repeating the ligation procedure, dinucleotide sequences every five bases but beginning from $n - 1$ (e.g., … 3 4 … 8 9 … 13 14 …) are acquired. Again the cycle is repeated by providing an offset to the sequencing primer being introduced by 2, 3, and 4 bases $(n - 2, n - 3,$ and $n - 4)$, until complete DNA sequencing is achieved [34].

6.2.2.8   Polony Sequencing

"Polony" sequencing is a method to obtain a colony of DNA that is amplified from a single nucleic acid in situ on a thin polyacrylamide film. The term polony means polymerase colony in molecular biology [35]. The implementation of polony sequencing can be broadly divided into three steps: (1) library construction, (2) template amplification, and (3) sequencing. The polonies are developed by diluting a library of DNA molecules into a mixture that contains PCR reagents and the acrylamide monomer. Distinct spherical polonies are formed by the amplification of the dilute mixture of single-template molecules. Hence, molecules in two distinct polonies are amplicons of different single molecules, but all molecules within a given polony are amplicons of the same single molecule [19].

The amplification primers usually consist of a 5-acrydite modification, which causes it to be covalently attached to the gel matrix. After PCR, the same strand of every double-stranded amplicon is tethered onto the gel. Now the gel is exposed to denaturing conditions, which enables the efficient removal of strands that are unattached. The DNA sequencing is then done using the copy of the strands that are attached onto the gel matrix, and thus the full set of amplified polonies is used in a highly parallel manner for DNA sequencing by synthesis as described in Section 6.2.2.4 [19].

6.2.2.9 Sequencing by Hybridization

Sequencing by hybridization is a well-known concept and is a method to obtain the sequence of a target DNA strand by allowing it to hybridize to the perfectly complementary probe. Microarray technology uses ssDNAs of known genes and attaches them onto a silicon surface/substrate. Each well of this microarray has a unique sequence of the single-stranded probe DNAs attached. The strands that are to be tested are first fluorescently labeled and then applied to the DNA chip. The target DNAs attach to the complementary-tethered (template) strand sequences. The chip is then washed, which causes all the unbound DNAs to be rinsed away, whereas the perfectly adhered target DNAs continue to remain bonded (Figure 6.4) [36]. Finally, the

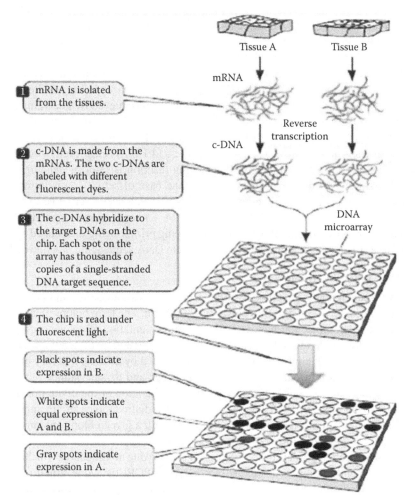

FIGURE 6.4 The deoxyribonucleic acid microarray technology based on the principle of sequencing by hybridization and optical readout measurement. (Reprinted from Hofmann, W. K. 2006. *Gene Expression Profiling by Microarrays: Clinical Implications.* Cambridge University Press. With permission.)

chip is illuminated with ultraviolet light to excite the fluorescence tags and the resulting image captured is shown in Figure 6.4. The output indicates the presence/absence of the target sequences with probed sequences and the intensity of the fluorescence provides the frequency of the probed sequences. The microarray technology that utilizes this technique dominates the current market in terms of sequencing technologies.

6.3 DNA SEQUENCING USING NANOPORES

Nanopore-based DNA sequencing may provide the most economical and reliable means to detect DNA sequencing as compared to all the different modalities mentioned in Section 6.2. This section provides an overview of the technological development in DNA sequencing with nanopores and the challenges for nanopore-based sequencing.

6.3.1 HISTORY OF DEOXYRIBONUCLEIC ACID SEQUENCING USING NANOPORES

The concept of DNA translocation inside nanopores for DNA sequencing relies on a combination of many ideologies. The origin of this concept dates back to 1940, during which Wallace Coulter attempted to standardize the size of paint particles when he ran out of paint. He used his own blood as a substitute and found that even the red blood cells could be detected as they pass through a narrow aperture driven by a pressure difference as they restrict the flow of ions. This technique was termed as "resistive-pulse technique" [37] and was used by Coulter Electronics Company to develop and market instruments for blood cell counting and cell sizing. Deblois and Bean [38] in General Electric (GE) Company worked with nanopores as they used track-etched nanopores, which allowed a detection limit of 60 nm. They also introduced the concept of charged particles driven by electrophoresis through a nanopore [38]. On the organic front, biological nanopores and ion channels were explored by Hodgkin, Huxely, and Hille [39] (http://www.bioetch.com/ion-channels-c-16.html). Ion channels are the transmembrane pores that allow the passage of ions (charged particles) into and out of a cell due to electrochemical gradient. The patch-clamp technique developed on the basis of the voltage clamp method by Nobel Prize winners Neher and Sakaman [40] in the 1970s enables the recording of currents from single ion channels.

Inspired by these combined developments, David Deamer started dreaming about the idea of analyzing DNA with nanopores in 1989 while on a cross-country road trip, and this idea was developed further a few years later. In 1996, the concept of analyzing DNA using nanopores was realized by Deamer and Kasianowicz [41] and this realization was patented with Baldarelli and Church [42], who had independently developed a similar method.

6.3.2 BIOLOGICAL NANOPORES

The organic pore of α-hemolysin is the most extensively used pore for detecting nucleic acids. The pore is formed by the 33-kDa protein secreted by *Staphylococcus aureus*. Figure 6.5 shows a cross-sectional view of the α-hemolysin channel at 1.9-Å resolution as revealed by X-ray crystallography [43,44]. From the *cis* to the *trans* side, the mouth of this channel is measured as 2.6 nm, and there is a vestibule of

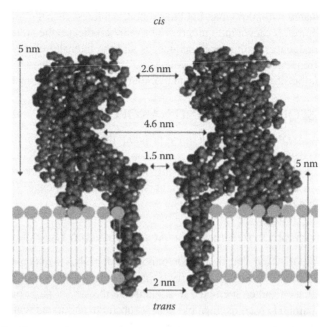

FIGURE 6.5 Cross section of an α-hemolysin channel, which is embedded in a lipid bilayer with channel dimensions. (Reprinted from Branton, D. W. Deamer and M. Akeson. 2000. *Trends in Biotechnology* 18(4):147–151. With permission.)

4.6 nm, which is followed by a limiting aperture of 1.5-nm diameter with a long stem approximately 5 nm in length and an exit diameter of 2 nm.

Kasianowicz et al. [42] in 1996 demonstrated that the single-stranded RNA and DNA molecules could be driven through ion channels in a lipid bilayer membrane under the application of an electric field. During translocation, nucleotides within the polynucleotide must pass through the α-hemolysin pore in a sequential single-file order because the limiting diameter of the pore can accommodate only a single strand of RNA or DNA and/or each polymer, as it passes through the membrane as an extended chain, and it partially blocks the channel. This causes a transient decrease in ionic current whose duration is proportional to the length of the polymer translocation. These channel blockades could be used to measure the polynucleotide length, and with further modifications this might lead to high-speed detection of the sequence of bases of single DNA or RNA molecules (see Figure 6.6) [8]. Polymer translocation through α-hemolysin pores can be achieved under the influence of an applied voltage bias, or in a free motion, or under a mechanical force that is able to pull the DNA in a particular axial direction [45].

The experimental setup used by Kasianowicz was modified by Akeson et al. [46] to allow lower noise levels and lower analysis volume, which is schematically shown in Figure 6.7. They employed this setup to compare the translocation of several RNA homopolymers [poly(C), poly(A), poly(U), and poly(CA)]. They observed characteristic depths and durations for each of the homopolymers. Figure 6.8 indicates the difference in ionic current measurements for comparison between poly(A), poly(C),

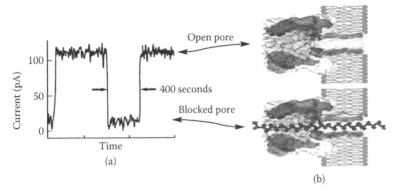

FIGURE 6.6 The schematic shows the principle by which sequencing of deoxyribonucleic acid (DNA) is done using a biological nanopore: (a) Modulation in current measurement with the open pore current when there is no DNA (b; top part) and the dip in current due to the presence of DNA, which blocks the pore (b; bottom part). (Reprinted from Branton, D. et al. 2008. *Nat Biotechnol* 26(10):1146–63. With permission.)

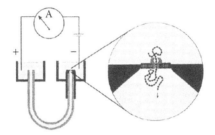

FIGURE 6.7 The improvised electrical setup provided by Akeson et al.: Deoxyribonucleic acid (DNA) polymers are driven through the α-hemolysin channel inserted into the lipid bilayer by an applied voltage of 120 mV. (Reprinted from Akeson, M., D. Branton, J. J. Kasianowicz, E. Brandin, and D. W. Deamer. 1999. *Biophys J* 77(6):3227–33. With permission.)

and poly(CA) homopolymers. The open pore (baseline) ionic current at 120 mV was 126 pA, and each of these homopolymers gave a trait significant to its length and composition. The poly(A) had blockades of 22 ± 6 microseconds per nucleotide and it reduced the current to ~20 pA (85% blockades), whereas poly(C) had blockades of 5 ± 2 microseconds per nucleotide and it reduced the current to ~5 pA (95% blockades). The 35%–55% blockage observed is attributed to partial entry into the pore, and this blockage is not valid for the length-dependent ionic current measurement. Further, for the combined poly(CA) a sequence of 85% and 95% blockages can be observed, which indicates the combination of the nucleotides. But these differences cannot be correlated directly with the nucleotide sequence of RNA as this can be also ascribed to the modified difference in the secondary structures adopted by these homopolymers.

Meller [47–49] extended this knowledge of DNA translocation by studying the translocation of homopolymers and diblock copolymers. Mellar et al. obtained some

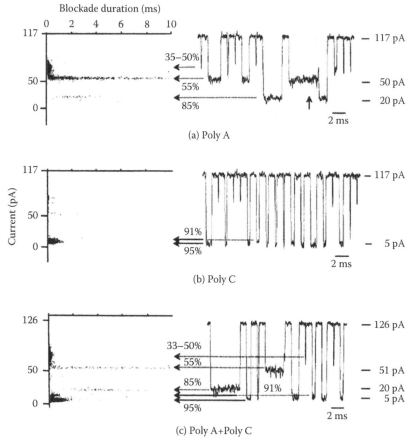

FIGURE 6.8 Different current measurements, which help one detect the different homopolymersbased on the depth of the current dip and its duration: (a) poly A, (b) poly C and (c) 50% poly C and 50% poly A. (Reprinted from Akeson, M., D. Branton, J. J. Kasianowicz, E. Brandin, and D. W. Deamer. 1999. *Biophys J* 77(6):3227–33. With permission.)

statistical results leading to the conclusion that on the basis of well-defined statistical parameters in this case, the current peak value (I_p) and translocation duration peak (t_{p1}) can help one discriminate between polynucleotides of similar length with the nanopores. Furthermore, the translocation duration time t_D alone might help one distinguish between individual polynucleotides of an individual nucleotide basis in some favorable cases. They also concluded that there is strong temperature dependence among the statistical parameters considered. The peak of the translocation time (t_p) of homopolymers and diblock copolymers varies significantly with respect to temperature with the scaling law of $t_p \sim T^{-2}$. At lower temperatures the measurements are highly sensitive, whereas at high temperatures a peak of the translocation time converges. This strong T^{-2} dependence on temperature could lead to the possibility of controlling the translocation speed and enhancing the differences between the numerous types of polymers.

Mathe et al. [50] have identified, on the basis of their MD studies, the translocation process through nanopores to be more complex than the mere passage of bases, concluding that the global orientation of DNA also has a great influence on the DNA–channel interaction. They observed that the 3′–5′-oriented DNA translocation (*trans* to *cis* direction) is two times slower than the poly (dA) strand with the same sequence having a 5′–3′ orientation. This shows that DNA translocation in nanopores also involves tilting of bases and stretching of the ssDNA within the nanopore. Their view is also supported by other simulation studies such as those conducted by Butler and Gundlach [51] and Wang et al. [52]. Specifically, Wang, et al. [52] showed that it is possible to detect and differentiate between all combinations of phosphorylation that are present or absent at either end, based on their finding that phosphorylation affects the ease with which the relevant end of the molecule can enter the pore. It has also been shown that an α-hemolysin pore with 1-M KCL solution and 120-mV potential from *cis* to *trans* results in a current of approximately 90 pA, whereas at the same potential from *trans* to *cis* the current is approximately 120 pA [44].

The driving voltage is a major key parameter that significantly influences the DNA translocation dynamics through nanopores. Henrickson et al. [53] examined the effect of driving voltage on the mechanism by which individual DNA molecules enter nanometer-scale pores. An exponential relation was obtained between the voltage applied and the event frequency. The blockade frequency was found to be proportional to polymer concentration, which increases exponentially with the applied voltage, and the DNA enters the pore more readily through the entrance that has the larger vestibule. Furthermore, a threshold voltage was obtained below which events did not occur due to random diffusion and repulsive forces [53]. The dependence of event frequency on voltage was predicted by Nakane, Akeson, and Marziali [44] and this dependence holds true even at high voltages; the capture rate is in the following form:

$$R = 4 \times C \times D \times a \times P(V) \tag{6.1}$$

where C is the polynucleotide concentration; D is the diffusion coefficient; a is the pore radius; and $P(V)$ is the probability that a polynucleotide, which collides with the pore, undergoes translocation through a pore [54]. $P(V)$ can be estimated based on Monte Carlo simulations. Aksimentiev and coworkers [55–57] have summarized the characteristics of DNA translocation through α-hemolysin pores by using MD, and have mapped the relevant ionic conductance, osmotic permeability, and electrostatic potential. Moreover, there have been attempts to make self-assembled structures using these channels, which act as a cavity to trap organic molecules and transfer such organic molecules from side to side [58].

6.3.3 SOLID-STATE NANOPORES

Although extensive biological pore-based studies have been proved to be useful for some interesting translocation experiments, the usage of biological pores is limited due to their fixed size and sensitivity to environmental conditions such as pH, salt concentration, temperature, and limited lifetime. Fabrication of nanopores from solid-state materials presents obvious advantages over their biological counterparts such as

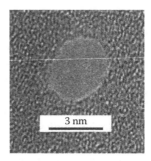

FIGURE 6.9 A transmission electron microscopy image of a 3-nm nanopore fabricated using a focused ion beam technique. (Reprinted from Dekker, C. 2007. *Nat Nanotechnol* 2(4):209–15. With permission.)

high stability, high control of diameter and channel length, adjustable surface properties, and the potential for integration into devices and arrays (see Figure 6.9) [59].

It is only in recent years that advances in nanotechnology have enabled the controlled fabrication of nanopores with diameters as small as 1 nm. Pores of true nanometer dimensions have been fabricated in various ways such as etching a hole into an insulating material [60], fabricating tightly focused e-beams [61], ion beam sculpting [62], or thermal shrinking [63]. Due to the fabrication process, most nanopores have a wide inlet/outlet and a narrow channel in the middle. Besides nanopores, a slitlike nanochannel with one side open has also been fabricated on substrates for DNA diffusion studies [64]. The DNA diffusion in channels of a few hundred nanometers has been reported by Stein, Kruithof, and Dekker [65]; Siwy et al. [66]; and Fan et al. [67]. The state-of-the-art in nanopore development is summarized in the study by Healy, Schiedt, and Morrison [68].

In 2001, Li et al. [62] first reported DNA detection by using a synthetic nanopore. They had observed blockade events for the translocation of 500-bp dsDNA (bp stands for base pair) through the nanopore. Most works to date focus on dsDNA length measurement [69–72]. Each work considers different polymer length and pore diameter and, thus, studies cannot be compared with each other. Li et al. [69] used a 10-nm pore with 3-kb and 10-kb dsDNA strands for comparative study and observed nonoverlapping peaks in the event duration histogram. Mara et al. [70] conducted similar studies, but they utilized a 4-nm pore and shorter dsDNA strands having lengths of 286 bp, 974 bp, and 4126 bp, respectively, and observed the corresponding peaks; but they reported significant overlapping of the peaks. Their observations were partially supported by Storm et al. [72] who conducted studies with 6.6-kb, 9.4-kb, 11.5-kb, 27.5-kb, 48-kb, and 96-kb fragments and found overlapping peaks only for fragments with lengths of 6.6 kb and 9.9 kb and/or fragments with 9.9 kb and 11.5 kb. This result raises concern over the lengthwise resolution of DNA, since peaks due to shorter lengths are completely overlapped by those due to longer lengths.

Heng et al. [71] have utilized MD simulation to study DNA translocation behavior through nanopores having diameters of 3 nm and <3 nm. They found that ssDNA can be distinguished from dsDNA based on the duration and magnitude of the blocking current that is measured when these DNAs translocate through the synthetic nanopores. Researchers have made attempts to investigate the mechanical properties of DNA using electric field–induced translocation of single molecules through a pore with nanometer-scale diameter [67,73]. A threshold of 60 pN is identified for dsDNA translocation through the nanopore [74]. Ho et al. [61] measured the ionic conductance across nanopores as a function of time, bath concentration, and pore diameter. The modulation of pH and its effect on dsDNA translocation through a nanopore has also been studied [60,75,76]. Fologea et al. [75] found that increasing the pH beyond a certain threshold causes the dsDNA to be denatured into ssDNAs. Molecular dynamics studies have enabled the formulation of better insights into this organic nanoscale system. Muthukumar and Kong [77] have shown that the detailed conformation fluctuations of polymer inside the vestibule and beta-barrel compartments of the protein pore plays a significant role in slowing the translocation process. On the other hand, in this study it is interesting to note that no correlation between translocation time and blocked current is observed, which is contrary to experimental results.

When an HPL DNA passes through nanopores, a characteristic pattern for DNA translocation is observed at nucleotide resolutions. Vercoutere et al. [78] observed that when this technique is applied, the current from the baseline initially falls to an intermediate level (termed the shoulder region), remains there for some time (Figure 6.10), and then briefly drops to a lower level before returning sharply to the initial level. Figure 6.10 shows the predicted stages of this specific pattern.

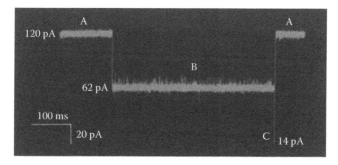

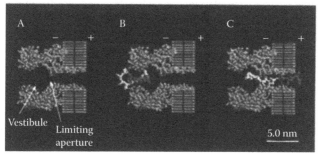

FIGURE 6.10   The characteristic dip in current measurement as the hairpin-loop (HPL) strand translocates through the biological nanopore: There is an initial dip at the point where the HPL blocks the channel, and then it unzips, which causes a deeper dip followed by a sharp rise to the original baseline. (Reprinted from Vercoutere, W., S. Winters-Hilt, H. Olsen, D. Deamer, D. Haussler, and M. Akeson. 2001. *Nat Biotechnol* 19(3):248–52. With permission.)

The hairpin first enters the wide vestibule of the pore, remains there for some time until random thermal fluctuations combine with the electric field–induced force to unzip the hairpin, and then rapidly translocates through the pore.

The energy required to unzip a hairpin increases with its length and so does the durations of the shoulder regions and the degree of blockage. This is because for longer HPL DNAs, large and statistically less frequent thermal fluctuations are required to unzip the DNA. A larger HPL occupies more area of the vestibule and thus displaces more ions. The shoulder region and the blockage depths are highly sensitive to HPL length and readily allow the single base-pair length difference to be identified between the event classes [78–80].

6.3.4 BIOFUNCTIONALIZED NANOPORES

Although "bare" nanopores provide useful information regarding the biophysical properties of DNA, it is difficult to realize the true potential of important developments such as genome sequencing, protein detection, and so forth with bare nanopores. The high speed of polynucleotide translocation through a nanopore is one of the major concerns since the resolution of present-day electronic detection methods limits the amount of information gathered at these high speeds [81]. Reducing the speed of DNA translocation not only helps in gathering more information with reasonable bandwidth requirements on the measurement systems [82] but also allows the detection of individual nucleotides thereby enabling gene sequencing. Researchers have also raised concern over difficulties in wetting the bare nanopores manufactured by regular methods [61,63].

This has prompted researchers to coat bare nanopores with structures that facilitate increased time of DNA within the pores. Siwy et al. [83] embedded a gold nanotube within a mechanically and chemically modified polymeric membrane that allowed the passage of ionic current through the nanopore. The system did not function on transient current pulses due to the entry of the protein, but it worked by using the protein analyte binding to the biochemical molecular recognition agent immobilized at the inlet opening of the nanotube. The protein molecule once it attaches to the inlet of the pore plugs the nanopore effectively due to comparable diameter and completely blocks the flow of ion current, thus leading to its detection [83]. Kohli et al. [84] showed that single-base mismatch selectivity could be achieved with functionalized nanotube membranes. Selective permeation was achieved since the membranes recognize and transport only those DNA strands that are complementary to the transporter strand attached to the inner walls of these nanotubes. They measured flux in a large number of pores in the filter membrane after the simultaneous passage of molecules through it. Thus, this method failed to provide single-molecule translocation signatures. Iqbal, Akin, and Bashir [9] later developed single-molecule electrophoretic devices that enable the measurement of DNA transportation through HPL-functionalized nanopores. Similar functionalization schemes can be used for a variety of ligand–receptor combinations of significance, and the DNA-functionalized solid-state nanopore can serve as the next generation of sequencing tools, whereas a pore functionalized with a specific probe can be used as a detector of specific nucleotides/biomarkers [9]. Wanunu, Sutin, and Meller [85] developed a high-throughput single-molecule method for evaluating small molecule binding to DNA using nanopores of ~3 nm. The regular method to measure the residual current of native DNA is modified and this label-free method estimates the shift in residual ion current that arises from threading of the dye-intercalated DNA molecule.

6.3.5 CHALLENGES FOR NANOPORE SEQUENCING

For the transition of nanopore-based sequencing from the research level to practical molecular diagnostics, improvements are needed in associated technical procedures, robustness, accuracy, and cost [19]. The ionic current blockades caused by ssDNA translocation are not yet sensitive enough to detect individual nucleotides. The minimum length of the pore in experiments, which determine and report the length of ssDNAs on their passage through the nanopores, is 5 nm. This implies that around 10–15 bases are within the nanopore at a time and the resulting ionic blockade is a combined result of these nucleotides [8]. Thus, for an ideal system, distinct electrical signals from the space between the bases are to be obtained. Another challenge is to minimize errors in the results attained; even though the method is sensitive enough, it is known to be computationally intensive and probable to obtain false readouts. For 3 million data points, the 1% false negatives can surely overwhelm the real matches. It must be noted that 25 kb is the maximum length of the DNA that has been analyzed, which implies a high throughput [48]. The key challenge for a nanopore sequencing device is to reduce the speed of DNA translocation in the nanopores, control the translocation, and reduce the fluctuations in translocation kinetics resulting from pore–surface interactions [8]. Another challenge that might be faced by these next-generation sequencing devices is the huge amount of data that will be generated. It is predicted that in the case of personalized medicine for 10% of the U.S. population, such devices would generate 10^5 TB of data in the FASTA format that will need to be searched [19].

6.3.6 SUMMARY

Electrophoretic methods have led the technology of genomic sequencing among all sequencing methods, mainly because of their high throughput (longest readout length). Integration of the technology advances from all the different fields including instrumentation, fabrication, microfluidics, software control, automation, and informatics is the key to developing a robust DNA sequencing platform. The cost per base is substantially lower for these modalities that are being developed in comparison with the current Sanger technology (gold standard). Furthermore, because of the capability to ensure high throughput, decreased sequencing time, and streamline sample preparation, genome centers and commercial enterprises are ready to adopt these new technologies. Most of these new technologies have very short readout lengths. The technologies being developed could surely be modulated for other applications within the genomic world, for instance, single nucleotide polymorphism detection or expression analysis. The fabrication limitations of nanopores, which are being improved, still remain a major bottleneck. DNA sequencing based on nanopores is a potential candidate for bringing about a major revolution in the form of the de novo sequencing modality and promises a bright future for personalized medicine.

6.4 FUNDAMENTALS OF MOLECULAR DYNAMICS

Molecular dynamics (MD) simulations have recently allowed for gaining insights into the dynamics of atoms that comprise a molecule such as DNA and/or proteins by solving the Newtonian equation of motion for atoms that are interacted with each other. This section reviews the fundamentals of MD simulations, particularly in the theoretical frameworks of statistical mechanics and/or classical mechanics.

6.4.1 INTRODUCTION

In nature, a matter is composed of its constituent building blocks such as atoms (or molecules). This has led to the following question: Are the bulk properties of matter a consequence of underlying interactions among the constituent atoms or molecules? Molecular models are used to address such an issue by considering the complicated nature of molecular motions rather than attempting to deduce microscopic behavior directly from experiments. Molecular dynamics simulation is a methodology useful in elucidating the detailed microscopic behavior of atoms comprising a matter, particularly the motion of these atoms; it allows for an insight into how the positions, velocities, and orientations of atoms change with respect to time. As an analogy, MD constitutes the motion pictures of molecules as they move to and fro, turning, twisting, and colliding with one another or with the container.

The theoretical basis for MD embodies contributions from elite scientists like Newton (laws of motion), Laplace (solution to the many-body problem), Babbage (concept of the computer), Euler (relationship between trigonometric functions and the complex exponential function), and Hamilton (equations of motion). A brief overview of the fundamentals of MD simulation is presented in Section 6.4.2.

6.4.2 STATISTICAL MECHANICS

Statistical mechanics forms a link between the energy of a molecular system and the macroscopic thermodynamic functions of the N-body system through rigorous mathematical expressions. It enables us to examine the energetics and mechanisms of conformational changes that take place in proteins and polymers. It serves as a tool to predict macroscopic phenomena from the properties of individual molecules making up a system through time averages over corresponding values in various microstates. For all practical purposes, ensemble averages are considered instead of time averages, since averaging over time (i.e. time-averaging) is a time-consuming process and time averages are equal to ensemble averages for an ergodic system. Statistical mechanics deal with aforementioned ensemble averages:

$$G = \frac{\int G(r1, \dots r_{N_m}) e^{-\beta U(r_1, \dots r_{N_m})} \, dr_1 \dots r_{N_m}}{e^{-\beta U(r_1, \dots r_{N_m})} \, dr_1 \dots r_{N_m}} \tag{6.2}$$

where r_i, ($i = 1, \dots, N$) are the coordinates, $\beta = 1/k_B T$, and k_B is the Boltzmann constant.

An ensemble is a collection of all possible systems that have the same macrostate but different microstates. The different types of ensembles are as follows:

Microcanonical ensemble (NVE): This ensemble denotes a thermodynamic state characterized by a fixed number of atoms, N, a fixed volume, V, and a fixed energy, E. This corresponds to an isolated system.

Canonical ensemble (NVT): This is a collection of all systems whose thermodynamic state is characterized by a fixed number of atoms, N, a fixed volume, V, and a fixed temperature, T.

Isobaric–isothermal ensemble (NPT): This ensemble is characterized by a fixed number of atoms, N, a fixed pressure, P, and a fixed temperature, T.

Grand canonical ensemble (mVT): The thermodynamic state for this ensemble is characterized by a fixed chemical potential, m, a fixed volume, V, and a fixed temperature, T.

Using these ensembles, statistical mechanics is founded on two postulates about the properties of ensembles. The first postulate is that the time average of a dynamical quantity in a macroscopic system is equal to its ensemble average:

$$x = \sum_i P_i x_i \tag{6.3}$$

where x_i is the value of x in microstate i and P_i is the relative probability of this state. The relative probability P_i is equal to the ratio of the number of systems in microstate i to that of systems in the all accessible microstates. Consider a closed box with two white balls and one black ball; if a person inserts his or her hand into this box and blindly removes a ball 99 times, the ratio of black to white balls will be 1:2 (33:66). On the other hand, say 99 hands blindly took one ball each from 99 boxes with the same two white balls and one black ball, again the ratio of black to white balls will be 1:2 (33:66). The former case involves time averaging and corresponds to pressure measurement with a macroscopic measuring instrument, whereas the latter is ensemble averaging. Therefore, time averaging is identical to ensemble averaging (in the ergodic system). The second postulate is that in a microcanonical ensemble, all possible states are equally probable; but this postulate has been extended to canonical and grand canonical ensembles.

$$P_i = P_j \tag{6.4}$$

where i and j are the two microscopic states in a system, and P_i and P_j are their associated probabilities. For two microscopic states i and j in a microcanonical system (fixed NVE), the two ensembles are the same; thus the associated probabilities are equal, $P_i = P_j$. Proof of this postulate can be found in statistical mechanics books [86–88].

6.4.3 CLASSICAL MECHANICS

Statistical mechanics describe the macroscopic behavior of a system using the partition function that is defined as the integration of Boltzmann distributions with respect to all accessible microstates (for details of Boltzmann distribution, refer to the statistical mechanics books, e.g. Refs. 86–88). The total energy of a system comprises the Boltzmann distribution consists of kinetic and potential energies: $E_p + E_k = E_{total}$. Motions of molecules determine their kinetic energy, which can be formulated independent of the configuration of the molecular environment. The potential energy depends on the configuration of the molecular environment; for instance, electrostatic force field constituting the potential energy is strongly dependent on the motion of atoms comprising a system. In general, quantum mechanics is useful to describe the motion of electrons and nucleus, which comprise the atom; this indicates that quantum mechanics model is able to elucidate the detailed mechanism of atomic motion. However, quantum mechanics is not suitable to depict the dynamics of a molecule consisting of a few hundreds due to the computational limitation of quantum mechanics. On the other hand, the classical mechanics is very useful in delineating the motion of a molecule consisting of a few hundred atoms, which leads classical mechanics to be utilized for establishing the foundations of MD simulations [87–90].

6.4.3.1 Newtonian Dynamics

The fundamental of MD simulation is to numerically solve the equation of motion for all atoms that interact with one another

$$F = ma \tag{6.5}$$

where m is the atomic mass, a is an acceleration of an atom, and F is the force acting on an atom. Here, force can be written as the time derivative of momentum:

$$\frac{dp}{dt} = F(x) \tag{6.6}$$

where $p = m\dfrac{dx}{dt}$ is the momentum of the particle in the x direction, and x is a function of time t.

The acceleration is given by

$$a = \frac{d^2x}{dt^2} \tag{6.7}$$

Thus,

$$F(x) = m\frac{d^2x}{dt^2} \tag{6.8}$$

Similar equations can be constructed for y and z directions. For conservative systems, the force $F(x)$ can always be expressed as a derivative of the potential energy $U(x)$ with respect to the position coordinate x:

$$F(x) = -\frac{dU(x)}{dx} \tag{6.9}$$

where $U(x)$ represents the interatomic interaction.

Multiplying both sides by $\dfrac{p}{m} = \dfrac{dx}{dt}$ and integrating over time, we obtain

$$E_1^{kin} + U_1 = E_2^{kin} + U_2 = E^{kin} + U = constant \tag{6.10}$$

From this energy conservation in the viewpoint of classical mechanics, the kinetics energy is represented in the form:

$$E^{\text{kin}} = \frac{p^2}{2m} = \frac{mv^2}{2} \tag{6.11}$$

Thus the total energy (sum of kinetic and potential energies) of a system is constant, that is, it is independent of time t. The aforementioned formulation of classical mechanics is sufficient for a single particle moving along a trajectory described in a Cartesian coordinate system within a conservative force field [91].

6.4.3.2 Hamiltonian Dynamics

Equation 6.5, which is Newton's second law of motion, is independent of time. However, atomic position and forces change with time. Consequently, there should be some function of position and velocities; but this function is constant with respect to time and can be the Hamiltonian H:

$$H(r^{\text{N}}, p^{\text{N}}) = \text{constant} \tag{6.12}$$

where the momentum p_i is defined in terms of velocity such that

$$p_i = m\dot{r}_i \tag{6.13}$$

Total energy (E) is conserved, that is, the combined kinetic and potential energies of molecules in an isolated system is conserved. Therefore, the total energy for an isolated system is considered as a Hamiltonian; then for N spherical atoms, H takes the following form:

$$H(r^{\text{N}}, p^{\text{N}}) = \frac{1}{2m} \sum_i p_i^2 + U(r^N) = E \tag{6.14}$$

where the potential energy U arises from interatomic interactions.

By following a sequence of procedures [88], the Hamiltonian equations of motion are formulated as follows:

$$\frac{\delta H}{\delta p_i} = \frac{p_i}{m} = \dot{r} \text{ and } \frac{\delta H}{\delta p_i} = -\dot{p}_i \tag{6.15}$$

For a system consisting of N atoms, the Hamiltonian equations of motion provides $6N$ first-order differential equations that are equivalent to Newton's $3N$ second-order differential equations.

For a comparison with Newton's second law of motion, the following result can be deduced:

$$F_i = -\frac{\delta H}{\delta r_i} = \frac{\delta U}{\delta r_i} \tag{6.16}$$

This equation provides an insight into the difference between Newtonian and Hamiltonian dynamics. In Newtonian dynamics, motion is the response to an applied force. In Hamiltonian dynamics, forces do not occur explicitly; instead, motion occurs in such a way that it preserves the Hamiltonian function [91].

6.4.3.3 Phase–Space Trajectories

Newtonian and Hamiltonian dynamics help one to realize the ability of MD to describe the trajectories of atomic positions. Hamiltonian or Newtonian dynamics can be extended to include atomic moment in the equation of motion. A phase–space is a $6N$-dimensional hyperspace used for plotting the positions and velocities of N atoms. It consists of the $3N$-dimensional configuration space composed of position vectors and the $3N$-dimensional momentum space that consists of momentum vectors in Cartesian coordinates.

For example, a phase–space trajectory can be explained using a mass–spring system that is isolated from its surroundings. The stiffness of the spring, representing how the spring resists any displacement due to its expansion or compression, is denoted as γ. When the spring is moved to a new position r from its equilibrium position r_0, the displacement of the mass is given by $x = r - r_0$. The potential energy of this system is given by

$$U(x) = \frac{1}{2}\gamma x^2 \tag{6.17}$$

Applying Newton's law of motion, we have

$$F(x) = \dot{p} = m\ddot{x} = -\frac{dU(x)}{dx} \tag{6.18}$$

Therefore, the equation of motion is represented in the form $m\ddot{x} = -\gamma x$, which shows that the force acting on the mass is linearly proportional to its displacement.

To determine the phase–space trajectory for this simple system, we consider the total energy, which is a constant:

$$E = E_k + U = \text{constant} \tag{6.19}$$

where E_k and U are kinetic and potential energies of the mass, respectively.

$$E = \frac{1}{2m}p^2 + \frac{1}{2}\gamma x^2 \tag{6.20}$$

The phase–space for this system is a two-dimensional space composed of a position coordinate x and a momentum coordinate p. Equation 6.20 describes an equation for ellipse; but when $m = \gamma = 1$, this ellipse disintegrates into a circle. The same relation can be obtained using the definition of momentum and two forms of the second law:

$$\dot{x} = \frac{p}{m} \text{ and } \dot{p} = -\gamma x \tag{6.21}$$

Combining the two equations given in Equation 6.21, the phase–space (phase–plane in this case) trajectory is obtained.

$$\frac{\mathrm{d}\,x}{\mathrm{d}\,p} = \frac{p}{m}\left(\frac{-1}{\gamma x}\right) \tag{6.22}$$

Integration of Equation 6.22 results in Equation 6.20. Now, the position and momentum trajectories are attained, which are governed by interaction potentials [91].

6.4.4 Force Fields

A force field provides a depiction of the relative energy or forces of an ensemble for any geometric arrangement of its constituent atoms. This depiction includes the time evolution of bond lengths, bond angles, and torsions, and nonbonded interactions such as van der Waals (vdW) and electrostatic interactions. The interaction potential can be broadly classified into two classes: (1) intramolecular, and (2) intermolecular potentials. Whereas the former describes interactions that arise from bonded structures, the latter describes pair interactions between distant atoms. The total potential energy of the system can be given as follows:

$$U_{\text{Total}} = U_{\text{intramolecular}} + U_{\text{intermolecular}} \tag{6.23}$$

where $U_{\text{intramolecular}}$ and $U_{\text{intermolecular}}$ represent intramolecular interactions (i.e., bonded interactions) and intermolecular interactions (i.e., nonbonded interactions), respectively. Force fields that have been developed are commercially available; for instance, CHARMM [96], AMBER [97], and Cornell [98].

6.4.4.1 Intramolecular (Bonded) Potential

The intramolecular (bonded) potential describes the covalent-bond energy for a molecule, and such intramolecular interactions consist of two-, three-, and four-body interactions of the bonded atoms in the following form:

$$U_{\text{intramolecular}} = U_{\text{bond}} + U_{\text{angle}} + U_{\text{dihedral}} + U_{\text{improper}} \tag{6.24}$$

where U_{bond}, U_{angle}, U_{dihedral}, and U_{improper} represent two-body potential (relevant for covalent-bond stretching), three-body potential (relevant for bending of the bond angle), four-body potential (relevant for twist of the dihedral angle), and four-body improper potential (for twist of the dihedral angle).

6.4.4.1.1 Two-Body Spring Bond Potential

The harmonic vibrational motion between a pair (*i*th and *j*th) of covalently bonded atoms is described by two-body spring bond potential. The schematic illustration is given in Figure 6.11a. The equation for two-body potential is given as follows:

$$U_{\text{bond}} = \sum k(r_{ij} - r_0)^2 \tag{6.25}$$

where the distance between the *i*th and *j*th atoms is given by $r_{ij} = |r_i - r_j|$, r_0 is the equilibrium distance, and k is the spring constant. Alternatively, Morse potential is used for bond breaking:

$$U(r_{ij}) = K_{\text{M}}(e^{-\beta(r_{ij}-r_0)} - 1)^2 \tag{6.26}$$

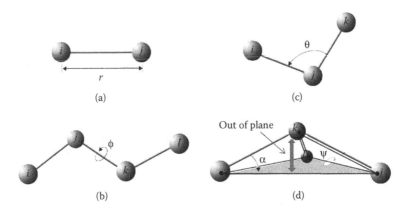

FIGURE 6.11 The (a) two-body (bond), (b) three-body (angle), and (c and d) four-body (dihedral and improper respectively) interactions of covalently bonded atoms: These interactions comprise bonded potential terms. Bond stretching is governed by r, the bond angle term is represented by θ, the dihedral angle is represented by ϕ, and the improper dihedral angle ψ governs the out-of-plane angle α.

where K_M and β are parameters that represent the strength of potential energy and the length-scale relevant for the range of interaction, respectively.

6.4.4.1.2 Three-Body Angular Bond Potential

The angular vibrational motion that occurs between the ith, jth, and kth covalently bonded atoms is described by three-body angular potential (see Figure 6.11b). The angle term depicts how the strain energy required to bend a bond angle changes when it is distorted away from its equilibrium position. The three-body potential can be explicitly written as follows:

$$U_{angle} = \sum k_{\theta_{ijk}} (\theta_{ijk} - \theta_{eq})^2 \qquad (6.27)$$

where θ is the angle formed by the bonds extending between the ith, jth, and kth atoms; and θ_{eq} is the equilibrium angle.

6.4.4.1.3 Four-Body Angular Bond Potential

The four-body torsion angle (dihedral angle) potential describes the angular spring between the planes formed by the first three and last three atoms of the consecutively bonded ith, jth, kth, and lth atoms (see Figure 6.11c). Dihedral or torsion energy for the system describes how the energy of a molecule changes as it undergoes a rotation about one of its bonds. The four-body torsional potential is represented in the following form:

$$U_{dihedral} = \sum \frac{1}{2} k_\phi \left[1 + \cos(n\phi - \delta) \right] \qquad (6.28)$$

where n is periodicity of the angle, ϕ is the dihedral angle, δ is the phase of the angle, and k_ϕ is the force constant. The last term labeled improper dihedral in Equation 6.24

describes the out-of-plane energy. Since the dihedral term alone is not sufficient to maintain the planarity of groups such as sp^2-hybridized carbons in carbonyl groups and aromatic systems, the strain energy for an additional improper dihedral is used and its explicit form is given as follows:

$$U_{improper} = \sum k_\psi (\psi - \psi_0)^2 \tag{6.29}$$

where k_ψ and ψ_0 are the force constant for the energy term and the equilibrium value of the improper dihedral angle, respectively. Another common angle (α) used to define the distortion due to out-of-plane motions is shown in Figure 6.11d [92–94].

6.4.4.2 Intermolecular (Nonbonded) Potential

Nonbonded potential describes the interactions between atoms of different molecules or those between atoms that are not directly bonded together in the same molecule. The overall conformation of the molecular system is defined with the help of these interactions. They are broadly classified as short-range exchange-repulsion and long-range dispersion interactions such as Lennard–Jones (LJ) interactions, electrostatic interactions (elect), and induced or polarized interactions (polar).

$$U_{intermolecular} = U_{elect} + U_{LJ} + U_{polar} \tag{6.30}$$

The intermolecular potential can be also explicitly expressed as follows:

$$U_{intermolecular} = \frac{1}{4\pi \in_0} \sum_{ij-pairs} \frac{q_i q_j}{r_{ij}} + 4\in \left[\left(\frac{\sigma_{ij}}{r} \right)^{12} - \left(\frac{\sigma_{ij}}{r} \right)^6 \right] + \frac{1}{2} \sum \mu_i E_i \tag{6.31}$$

6.4.4.2.1 Lennard–Jones Potential

The LJ potential defines the vdW forces of interactions. It accounts for the weak dipole attraction between distant atoms (dispersion represented by the sixth order term) and the hard-core repulsion (short-range repulsion represented by the twelfth order term), and LJ potential usually excludes pairs of atoms already involved in a bonded term.

$$U_{LJ} = 4\in_{ij} \left[\left(\frac{\sigma_{ij}}{r} \right)^{12} - \left(\frac{\sigma_{ij}}{r} \right)^6 \right] \tag{6.32}$$

where ε is the depth of the potential well, σ is the (finite) distance at which the inter-atomic potential is zero, and r is the distance between atoms. The parameters σ and ε depend on not only two atoms but also the cross-interactions $i \neq j$ between unlike pairs of molecules, and thus they are not so easily obtained. To enable their definitions from parameters of single atoms, certain combination rules can be used such as the Lorentz–Berthelot mixing rules [95], given by

$$\in_{ij} = \sqrt{\in_i \in_j} \text{ and } \sigma_{ij} = \frac{1}{2}\sigma_j + \sigma_j \tag{6.33}$$

where i and j denote the ith and jth atomic species, respectively.

6.4.4.2.2 Polarization Potential

Polarizable force fields provide charge distribution in response to a dielectric environment and have high importance in case of simulations of biological systems. Polarization potential arises from the fact that the charge distribution of a group of molecules is distorted by interactions of the group with its neighbors. If polarizability of the ith atom is denoted as α_i and the field at such an atom is written as $\mathbf{E}_i$ (vector quantity), then the dipole induced at the atom μ_i is as follows:

$$\mu_i = \alpha_i \mathbf{E}_i \qquad (6.34)$$

The total polarization potential in the charged system is given by

$$U_{\text{polar}} = \frac{1}{2} \sum \mu_i \mathbf{E}_i \qquad (6.35)$$

where μ_i is the dipole moment associated with atom i, and $\mathbf{E}_i$ is the electric field experienced at atom i.

6.4.5 PERIODIC BOUNDARY CONDITION

The MD system is relatively small such that the system has around a few thousand atoms. For such small systems, surface effects such as interactions between the atoms and the container wall dominate. A simulation of such a system with solvent will not provide any information regarding the bulk characteristic of the solvent (liquid) but dictates the interaction between liquid and the container walls. To eliminate this surface effect, periodic boundary conditions (PBCs) have to be taken into account.

The schematic shown in Figure 6.12 illustrates a system that is bound free of physical walls. A PBC creates an infinite space-filling array of identical copies of the simulation region. Thus, an atom that leaves the simulation region through a particular bounding face reenters the region through the opposite bounding face. Further, an atom lying within equilibrium distance from the boundary will interact

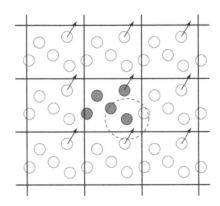

FIGURE 6.12 Periodic boundary condition: As a particle moves out of the simulation box through one of the bounding faces, an image particle moves in to replace it.

with atoms in the adjacent copy of the system or equivalently with atoms near the opposite boundary. This is termed as the "wraparound effect." The wraparound effect has to be accounted for when integrating the equations of motion and while analyzing the results [88].

6.4.6 INTEGRATION EQUATION

The potential energy of a system composed of atoms is critically dependent on the atomic positions of a system. Due to the complicated nature of a molecular system, there is no analytical solution to the equations of motion and hence these equations must be solved numerically. There are many numerical algorithms that are available for integrating the equations of motion [87]. But there are certain criteria to be satisfied such as conservation of energy and momentum, which allows for stable numerical integration over a long time step with computational efficiency. Based on these criteria, two popular algorithms, such as the leapfrog and the Verlet methods, are highlighted due to their accuracy and their ability to obey the energy conservation better than those of higher-order methods; also, their storage requirements are minimal [87].

All integration methods assume that position, velocities, and accelerations can be approximated by a Taylor series expansion:

$$r(t + \partial t) = r(t) + v(t)\partial t + \frac{1}{2}a(t)\partial t^2 + \dots \tag{6.36}$$

$$v(t + \partial t) = v(t) + a(t)\partial t + \frac{1}{2}b(t)\partial t^2 + \dots \tag{6.37}$$

$$a(t + \partial t) = a(t) + b(t)\partial t + \dots \tag{6.38}$$

where r is the position, v is the velocity (the first derivative of position with respect to time), and a is the acceleration (the second derivative of position with respect to time).

6.4.6.1 Verlet Algorithm

The Verlet algorithm uses positions and accelerations at times $t - \delta t$ and t to calculate the new positions at time $t + \delta t$ and uses no explicit velocity. Verlet algorithm is simple and straightforward to be implemented, and it is computationally favorable due to modest storage requirement, whereas it only exhibits the modest precision. The Verlet algorithm is derived from a Taylor expansion:

$$r(t + \partial t) = r(t) + v(t)\partial t + \frac{1}{2}a(t)\partial t^2 \tag{6.39}$$

$$r(t - \partial t) = r(t) - v(t)\partial t + \frac{1}{2}a(t)\partial t^2 \tag{6.40}$$

Combining Equations 6.39 and 6.40, one gets the classical Verlet algorithm:

$$r(t + \partial t) = 2r(t) - r(t - \partial t) + a(t)\partial t^2 \tag{6.41}$$

6.4.6.2 Leapfrog Algorithm

The leapfrog algorithm uses the velocities at time $t + 0.5\delta t$ to calculate the positions, r, at time $t + \delta t$. Thus, the velocities leap over the positions and then the positions leap over the velocities. It is advantageous in that the velocities are explicitly calculated; but the disadvantage of this method is that the velocities are not calculated at the same time as are positions. From a Taylor series expansion, the leapfrog algorithm is explicitly written in the following form:

$$r(t + \partial t) = r(t) + v\left(t + \frac{1}{2}\partial t\right)\partial t \tag{6.42}$$

$$v\left(t + \frac{1}{2}\partial t\right) = v\left(t - \frac{1}{2}\partial t\right) + a(t)\partial t \tag{6.43}$$

The equation for velocity is given by

$$v(t) = \frac{1}{2}\left[v\left(t - \frac{1}{2}\partial t\right) + v\left(t + \frac{1}{2}\partial t\right)\right] \tag{6.44}$$

6.4.7 SUMMARY

The wealth of atomistic details potentially available in MD has made MD a useful and important tool in computational simulations. The rapid growth in computing power and the even greater improvement in the cost:performance ratio has directly benefited MD simulations. Further, the efficient implementation of simulation algorithms on parallel computers has also enhanced the efficiency and accuracy of MD simulations. Molecular dynamics simulations can be applied to fundamental studies for not only understanding diffusion, transport properties, size dependence, or phase transition phenomena, for instance, phase coexistence and order parameters, but also to gain insights into collective behavior such as translational and rotational motion, vibration, dielectric properties, fluid dynamics, biomolecular dynamics, and polymer transport processes to name a few. In recent years, there have been efforts to achieve the simulations at spatial and temporal scales that are beyond the currently available timescales/length scales of classical MD simulations, and although these efforts are still premature they have received much attention in the field as future prospects [87,90].

6.5 ATOMISTIC MODEL SIMULATIONS

In this section, we present the overview an atomistic model that enables the study of interactions between DNA and nanopore. The effect of various parameters, such as nanopore size, applied bias voltage, and pore functionalization, on the DNA translocation dynamics inside a nanopore has been delineated using an atomistic model.

6.5.1 INTRODUCTION

Robust modeling of biomolecular and condensed-phase systems can be attained with the use of atomistic MD simulations [88,99,100]. The ability of MD to dictate the internal motion makes MD an ideal tool to study polymeric systems. There have been many theoretical models to explain chain dynamics, primarily governed by segment chain interactions and connectivity [101–103]. The MD simulations allow one to visualize and gain detailed insights into polymer dynamics, since these simulations include explicit solvents with counterions, reliable force fields, and proper representation of long-range electrostatics. All-atom systems have been successfully implemented to accurately reproduce many experimental results [104–109]. This has prompted the use of this powerful tool to analyze and characterize DNA translocation in bare nanopores. The fundamental understanding of DNA translocation dynamics inside nanopores is important for the future development of lab-on-chip devices enabling biomolecular analysis. However, the interaction between DNA and nanopores is still not well understood because of the small spatial scales of the DNA/nanopore and the dynamic nature of the translocation process.

Studies of electrophoretic transport of DNA in nanochannels have revealed that DNA–channel surface interaction leads to a diffusion rate much lower than that predicted by traditional diffusion theory [110]. Extensive experiments and simulations focus on understanding the translocation of long DNA strands based on the analysis of dips in ionic current due to the blockage of the channel and finding the relationship between the time of dip in the ionic current and the polymer length [47,49,55,73,111–113]. There have also been attempts to understand the stretching of DNA in a nanopore under the application of electric field [74,77], as well as the effect of temperature and DNA sequence on translocation kinetics [114]. These efforts have a common feature in that a DNA with a length scale greater than that of the nanopore and initially located outside the pore is considered. The type and size of the pore influences the electrophoretic mobility for DNA translocation through the pore [70]. It is known that translocation kinetics outside the pore are different from that inside the pore due to ion accumulation near the pore entrances. Meller, Nivon, and Branton [49]. report that in the case of a biological nanopore, a DNA strand with a length scale smaller than that of the pore undergoes faster translocation than one that passes through a smaller pore. Thus, the electrophoretic mobility is higher when length of the DNA strand is comparable to that of the pore. This can be explained on the basis that when the DNA is outside the nanopore, the effective charge on a nanopore is significantly decreased due to the presence of a counterion cloud that moves with the DNA. These counterion cloud charges can be disrupted due to steric effect or the charged inner surfaces of the nanopore [115], and this leads to a change of force experienced by the translocating DNA (t-DNA) and thus to a change in its mobility. A conflicting explanation is electroosmotic flow [116], which also occurs as a consequence of the presence of surface charges in these narrow channels. Chang et al. [63] proposed that the electroosmotic flow is toward the cathode, and thus it opposes DNA translocation causing a reduction in velocity for DNA translocation due to hydrodynamic and electrostatic drag forces. Storm et al. [72] qualitatively validated the power law [117], in case of a polymer, which states that electrophoretic mobility decreases with an increase in DNA length:

$$\mu_e \propto L^{(1-\alpha)} \qquad (6.45)$$

where μ_e is the electrophoretic mobility of DNA translocation, L is the length of the DNA, and $\alpha = 1.27$ is the power law constant. Moreover, in the reported simulation studies higher voltages (compared to experimental voltages) were applied to overcome timescale limitations. This may prevent one from understanding the realistic dynamic behavior of the system since the kinetics could be different at reduced bias voltages compared with higher voltages. This is why Muthukumar and Kong [77] could not find any direct correlation between translocation time and blocked current, which is surprisingly contradictory to the experimental results.

In the absence of surface friction, nanochannels would be a free-solution environment in which DNA molecules could move with length-independent mobility. In addition, due to the entropic effect the stretched DNA relaxes in a nanopore if the electric field is turned off. The importance of channel size is twofold: (1) The channel size constrains the relaxation of the DNA molecule, and (2) The channel surface interacts with DNA bases. Of course, these two effects are coupled together on the premise that higher confinement leads to stronger interaction. Theoretical models of confined polymers predict that hydrodynamic friction coefficients depend on channel diameter and viscosity, although they are independent of the electric field applied [118]. The hydrodynamic friction force on the DNA is given by $f = gv$. Here, v is flow velocity and g is drag coefficient, which is assumed to be in the form of $g = el$ where e is the friction coefficient per unit length and l is the contour length. However, such a friction coefficient ignores the contribution of DNA–nanopore surface interaction. The DNA bases confined in a very narrow channel (<2 nm) are found to be attracted toward the surface, tilting their orientation parallel to the channel surface [50], which leads to a large surface friction. But for a nanochannel with a size greater than 3 nm, DNA tends to fold in various forms and only a small portion of the DNA base pairs come in contact with the nanochannel surface. Therefore, it is necessary to characterize the translocation at different channel-size regimes and to determine the velocity-dependent friction coefficient. In this study, PBCs are applied at the inlet and outlet of the pore. This implies that there is no ion depletion under high voltages and that DNA is always within the pore. This enables the characterization of DNA translocation kinetics inside a nanopore as well as DNA–nanopore interaction without the entrance and exit effects, which eliminate the effects of change of environment.

Nanopores with molecular selectivity are needed to obtain the information of DNA conformation and base pair–level information simultaneously [119]. To provide such selectivity, a simple chemical treatment such as silane coating is insufficient [60,120–122]. Iqbal, Akin, and Bashir [9] have demonstrated that such selectivity can be incorporated by coating tethered DNA hairpin loops on the inner surface of the nanopore. But little is known about the properties of such coated DNA monolayers, which are affected by the size, geometry, and surface curvature of the pore [123]

Other factors that also modulate the tethered DNA orientation are initial orientation, attachments with the wall, applied electric field, and hydrodynamic interactions [45,114]. It is thus important to characterize the interaction between DNA and chemically modified nanopores.

6.5.2 METHOD

The software used to perform fully atomistic scale MD studies is nanoscale MD (NAMD) [124]. Analysis and visualizations are done with visual MD (VMD) [125]. The system used for the simulation consists of a nanopore, a DNA, water molecules, and ions. To perform the simulations, each component of the system has to be built and then integrated to form the entire system.

6.5.2.1 Building the Nanopore

A silicon nitride nanopore is considered, as it is the most commonly used material to fabricate a nanopore in experiments. Simulation conditions have to be matched with those of real experiments to the largest extent possible such that the results could provide better insights into the unknown kinetics of translocation. The pore was built using a single unit of silicon nitride as shown in Figure 6.13a. This unit was then replicated and placed next to the previously existed unit based on the equilibrium distance in order to generate a cubical membrane of silicon nitride as shown in Figure 6.13b. The cubical membrane was then cut into a more convenient geometry to obtain a hexagonal membrane as shown in Figure 6.13c. A hole of desired radius is drilled into this membrane to obtain the final nanopore as shown in Figure 6.13d.

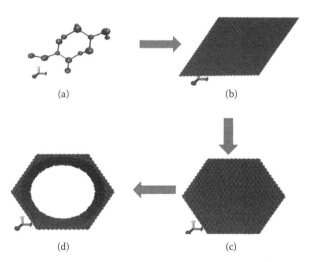

FIGURE 6.13 Formation of nanopore: (a) It begins with a unit cell of silicon nitride; (b) the unit cell is repeated to obtain a membrane; (c) this is followed by obtaining a desired shape for the membrane; and (d) finally the nanopore is obtained.

6.5.2.2 Building the Deoxyribonucleic Acid

In most experimental studies of DNA translocation, the length of the DNA considered is on the order of ~10^3 base pairs (bps), that is, ~1 kbp. Due to computational limitations, atomistic simulations usually consider a DNA with a length of 8–20 bps. An 8-bp dsDNA with a randomly selected sequence is used in this simulation. The pdb (protein data bank) code of dsDNA model is obtained from the Research Collaboratory for Structural Bioinformatics (RCSB) Protein Data Bank [126] (Figure 6.14a). The dsDNA obtained is then converted into two ssDNAs, and the sequence of the ssDNA considered in this chapter is AATTGTGA. The DNA enters the nanopore in a sequential single file, nucleotide by nucleotide, and is stretched using interactive MD (see Figure 6.14) before running the simulation.

The nanopore and the DNA constructed separately from atomistic modeling were then combined and the positon of the DNA was adjusted such that it is in the middle of the pore as shown in Figure 6.15a. This system was then solvated using the automated solvation function of VMD. The solvate function generates water molecules not only inside the nanopore but also outside it. The additional water molecules outside the nanopore and those within 1 Å inside the nanopore were deleted, as illustrated in Figure 6.15b. Ions were added to this system to obtain a molarity of 0.1 M. The psf files (that describe the bonding specifications) needed for the system are constructed using the autopsfgen function of VMD. The CHARMM [96] force field is used for the DNA, nanopore (silicon nitride), and their interactions. The total potential energy comprises bonded energies and nonbonded pair interaction energies:

$$
\begin{aligned}
U &= E_{\text{bond}} + E_{\text{angle}} + E_{\text{dihedral}} + E_{\text{elec}} + E_{\text{vdW}} + E_{\text{constrain}} + E_{\text{other}} \\
&= \sum k_{\text{b}}(b-b_0)^2 + \sum k_{\theta}(\theta-\theta_0)^2 + \sum k_{\chi}[1+\cos(n\chi-\delta)] \\
&\quad + \sum k_{\psi}(\psi-\psi_0)^2 + \sum_{i<j} \frac{q_i q_j}{4\pi\varepsilon_0 r_{ij}} + \sum_{i<j}\left(\frac{A}{r_{ij}^{12}} - \frac{B}{r_{ij}^6}\right) + E_{\text{constrain}} + E_{\text{other}}
\end{aligned}
\tag{6.46}
$$

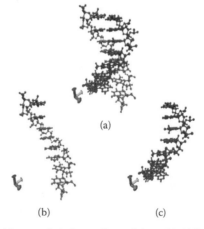

(a)

(b)   (c)

FIGURE 6.14 (a) Double-stranded deoxyribonucleic acid (dsDNA) obtained from the Research Collaboratory for Structural Bioinformatics data bank for obtaining the pdb (coordinate) file: This dsDNA is then used to obtain two single-stranded DNAs (b and c).

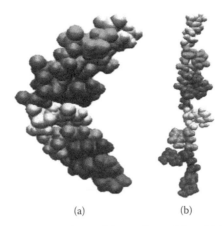

(a)   (b)

FIGURE 6.15 Different configurations of deoxyribonucleic acid (DNA): (a) curled DNA and (b) straightened DNA.

Nonbonded interactions are described using a cutoff distance of 1.2 nm and the LJ potential is smoothly shifted to zero between 1 nm and the cutoff distance. The pair list is updated every step using a cutoff distance of 1.4 nm. Particle-mesh Ewald (PME) is applied for computing long-range electrostatics. The temperature is set to 295 K and a time step of 1 femtosecond is used. The PBCs are applied at the entrance and outlet of the pore. All simulations are performed on the supercomputer cluster POPEL at Pittsburgh Supercomputing Center (PSC) [127]. A 2-nanoseconds simulation based on eight processors takes around 24 hours. The number of atoms in the system varies with the size of the nanopore. A typical simulated system has 14,814 atoms, particularly 8,000 atoms for the nanopore; 6,383 atoms for water; 258 atoms for DNA; and 200 atoms for ions. The timescale accessible with MD simulations is currently limited to a few nanoseconds [56]. To accelerate the translocation events that normally occur at milliseconds, MD simulations are performed at a higher applied voltage than 100–200 mV, which is a typical range in experiments.

6.5.3 RESULTS

To verify the simulation method described in Section 6.5.2, we have considered a benchmark problem, which has been well addressed in literatures. What follows is the characterization of the underlying mechanism of DNA translocation dynamics inside a nanopore as a function of nanopore size, applied bias voltage, and surface functionalization.

6.5.3.1 Benchmark Problem

In order to demonstrate the validity of fully atomistic MD simulations, a benchmark problem is used to reproduce the results obtained by other researchers who experimentally observe the dip-in ionic current when the DNA blocks the ionic flux after entering the nanopore. In the atomistic system, the DNA is already within the pore and the length of the DNA is equivalent to the length of the pore. The PBCs are applied at the inlet and outlet of the pore; thus the process of dip-in ionic current will not be observed for the system results mentioned later in this section.

The cross-sectional views of the two systems under discussion are shown in Figure 6.16a and b. To validate the system, one has to check whether the simulation results are the same as those previously reported in experiments as shown in Figure 6.16a. Figure 6.17 shows the sequence of images as the DNA translocates through the nanopore and their corresponding dip-in ionic current that is being measured as seen in Figure 6.17g. The dip-in ionic current is evident due to the blocking effect of the DNA as it enters the nanopore. The ionic current is measured using the following formula:

$$I\left(t + \frac{\Delta t}{2}\right) = \frac{1}{\Delta t \cdot l_z} \sum_i^N q_i (z_i(t + \Delta t) - z_i(\Delta t)) \qquad (6.47)$$

where z_i and q_i are the z coordinate and charge of the ion i, respectively, and Δt is the simulation time. The rest of the results based on the system are shown in Figure 6.16; the system is useful in elucidating the interaction mechanism of DNA with the nanopore.

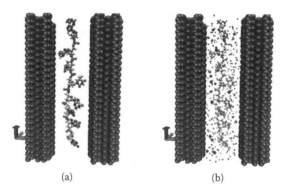

(a)   (b)

FIGURE 6.16 Formation of the final system: (a) The straightened deoxyribonucleic acid is combined with the nanopore and then solvated and ionized to obtain the final system (b).

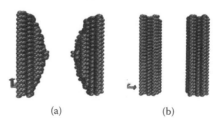

(a)   (b)

FIGURE 6.17 A comparison of the different nanopore configurations: Part (a) allows us to observe the dip in ionic current, whereas part (b) enables us to run simulations at a higher electric field.

6.5.3.2 Nanopore Size–Dependent DNA–Nanopore Interactions

In the absence of surface friction, nanochannels would be a free-solution environment in which DNA molecules would move with length-independent mobility. However, electrophoretic mobility of DNA molecules in slitlike nanochannels has been observed to be dependent on DNA length [110]. This clearly indicates that molecular interactions within the confining walls are significant, and the notion of free-solution electrophoresis breaks down. To understand size-dependent translocation, interaction of DNA with pores of various sizes is studied. A short ssDNA of eight bases (AATTGTGA) is driven by electrophoresis through a nanopore with diameters ranging from 1.5 to 4 nm, as shown in Figure 6.18, where the sequence A, T, G, and C indicate adenosine, thymine, guanine, and cytosine, respectively. The vdW force between the DNA and the nanopore, the translocation velocity of DNA, and the ionic current as a function of pore size are plotted in Figure 6.19a, b, and c, respectively. The water molecules and ions are not shown in the figure for clarity.

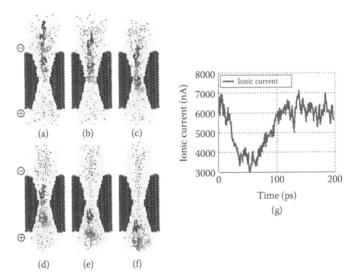

FIGURE 6.18 The deoxyribonucleic acid (DNA) translocation through a nanopore shown through the sequence of images in (a) through (f): The dip in ionic current is measured as the DNA enters the nanopore, and then when it exits the current increases back to the baseline level (g).

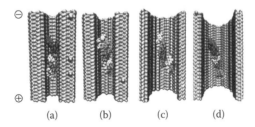

FIGURE 6.19 Deoxyribonucleic acid translocation in pores of different sizes: (a) 1.5 nm; (b) 2 nm; (c) 3 nm; and (d) 4 nm.

As presented in Figure 6.19a, the DNA–nanopore interaction force decreases as pore diameter increases due to the reduced confinement effect. Accordingly, DNA translocation velocity increases with increasing pore size. However, translocation velocity and vdW force do not change linearly with change in size, but they satisfy an exponential curve. There is a gradual increase in translocation velocity as nanopore diameter increases at the beginning. It then reaches a constant limit, indicating that an increase in the pore diameter beyond 3 nm will not affect the ssDNA translocation process anymore. An insight into the relationship between nanopore size and DNA translocation velocity can provide the fundamentals of efficient design of nanopore size for optimal signal yield. The ionic current increases with increasing the pore diameter, and the relation between ionic current and pore diameter obeys a parabolic curve. It should be noted that PBCs are applied at the inlet and outlet of the pore; thus, there is no depletion of ions as is usually observed in nanopore experiments when high voltages are applied. Although experimental data suggests that the translocation time is different for long DNA strands with different sequences [48,56], significant difference was not observed in translocation dynamics for homopolymers poly(dA) or poly(dT) through the nanopores, which might be attributed to the small strand (eight base) of ssDNA used in the simulation. Furthermore, it is expected that there is an optimal pore diameter for a particular voltage bias applied, and this optimal pore diameter would provide sufficient confinement to enhance molecular detection and yet allow the DNA to pass through. Such optimal designs will be explored in future studies.

6.5.3.3 Voltage-Dependent DNA–Nanopore Interaction

Besides nanopore diameter, applied voltage is also a key parameter that determines the DNA translocation speed. In order to consider a balance between confinement and translocation, a nanopore of diameter 2 nm is chosen and various voltages are applied to it. As shown in Figure 6.20, a nearly linear relationship is found between the applied voltage and the translocation velocity (Figure 6.20b), whereas a parabolic trend is observed between applied voltage and vdW force or ionic current (Figure 6.20a and c). Ideally, ionic current should increase linearly with an increase in applied voltage for free ions passing through a nanopore. The

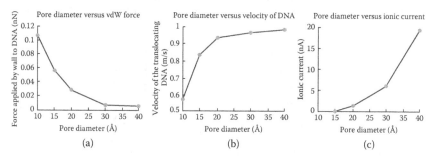

FIGURE 6.20 Deoxyribonucleic acid translocation in pores of different sizes: (a) Van der Waals force; (b) velocity; and (c) ionic current.

nonlinear relationship observed in Figure 6.20c is attributed to the blockage effect of DNA. It should be noted that a high voltage is applied to observe the translocation in a few nanoseconds. To study the translocation at lower voltages, a CG model of DNA has to be developed and this coarse-graining strategy is described later in Section 6.6.

6.5.3.4 Nanopore with Surface Functionalization

Coating nanopores with DNA or other organic molecules such as silanes can make the nanopores more biologically favorable and enables the manipulation of surface charges, hydrophobicity, and chemical functionality. It is important to characterize the surface property, size, and orientation of this chemically modified nanopore surface. The molecule used for coating the nanopore in the simulation is 8-bp ssDNA. The tethered DNAs carry a negative backbone; thus they reorient and are stretched under the applied bias voltage. Characterization of the coating polymer reorientation is essential to predict the effective pore diameter (EPD) under a particular applied voltage, which influences the DNA translocation process. The spacing between attached ssDNA molecules (on nanopore surface) is determined by their radius of gyration. To avoid the peeling off of tethered DNA, the nucleotide of the DNA strand closest to the wall is fixed on the nanopore surface. The charge on each tethered DNA is assumed to be $-e$ due to the phosphate backbone. Under the applied bias voltage, the translocation of ions toward opposite electrodes is observed. Since the coated ssDNA carries a $-e$ charge, these curled and tethered ssDNAs begin to get straightened in the direction of the oppositely charged electrode. The straightened DNA increases the EPD, which is measured as the empty space in the nanopore not occupied by the coated ssDNA. After the system reaches equilibrium, the distances between opposing ssDNA molecules are calculated and averaged to estimate the EPD. There are many factors that contribute to the EPD, for example, the flow of ions, the ionic loop around the DNA, and applied bias voltage. The EPDs are plotted as a function of the applied bias voltage in Figure 6.21. The EPD increases with an increase in the applied electric field strength. The kinetics of DNA translocation in a polymer-coated pore differs from that in a bare nanopore. In a bare nanopore

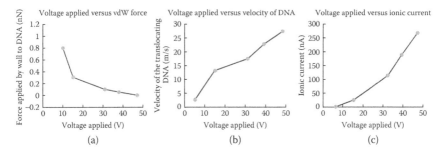

FIGURE 6.21 Deoxyribonucleic acid translocation under different applied voltages: (a) Van der Waals force; (b) velocity; and (c) ionic current. (From Knotts, T. A., N. Rathore, D. C. Schwartz, and J. J. de Pablo. 2007. *J Chem Phys* 126(8):084901. With permission.)

ions inside the nanopore form a double layer, which prevents DNA translocation at a faster speed within the pore. With a tethered DNA coating on the surface, the ion layers disappear and different translocation speeds are expected.

6.5.4 SUMMARY

The DNA translocation speed depends on several physical parameters such as nanopore diameter, electrophoretic bias, and surface coatings. The orientation of the tethered DNA affects the effective nanopore diameter. The effective nanopore diameter can be controlled in two ways: (1) by controlling the bare nanopore size during fabrication, and (2) by controlling the strength of the applied electric field. This model-based system can be used for the optimal design of nanopore systems for DNA/gene sequencing. The major drawback of fully atomistic molecular simulation is the limitation in timescale to a few nanoseconds. In Section 6.6 and its subsections, a coarse-grained DNA model is used to simulate low-voltage DNA translocation with longer timescales.

6.6   COARSE-GRAINED MODELING

To overcome the computational limitation of atomistic MD simulation (such as short timescale accessible with MD), there have been recent efforts to coarsen the atomistic model by reducing the degrees of freedom and simplifying the interatomic potential. This Section presents the overview of coarse-grained model-based simulations of DNA translocation inside nanopore.

6.6.1   INTRODUCTION

Lots of biological phenomena occur over timescales that are well beyond the currently available timescales of atomic-level simulations [128–132]. This challenge leads to the development of coarse-grained (CG) models to overcome the limitations in temporal and spatial scales. Coarse graining is an approach that can be coupled with MD in order to bridge the gap between the atomistic and mesoscopic scales. An all-atom system in which each atom is explicitly represented has Angstrom-level detail and femtosecond timescales, which allows for direct investigations of molecular structure and dynamics such as protein fluctuations. But the major disadvantage of all-atom models is the limited timescale, up to only a few nanoseconds, which implies that all-atom models lack the ability to simulate large systems for a timescale longer than a few nanoseconds. On the other hand, a CG model at mesoscale allows one to observe the average density, charge, or other characteristics of large-scale materials/structures such as self-assembly in biomolecular systems [133]. The schematic shown in Figure 6.22 is a general overview of the temporal and spatial scales that are accessible by current simulation techniques [129]. Thus, coarse graining is an approach that is able to not only reduce the computational complexity but also improve the accessible simulation timescales in such a way that atoms are grouped together into new CG sites. The major challenge of a CG model is that it must be able to reserve the underlying properties of all-atom systems.

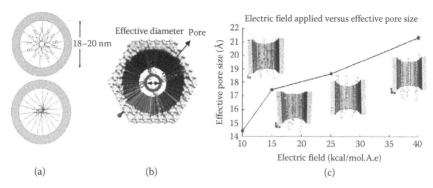

FIGURE 6.22 Single-stranded deoxyribonucleic acids (ssDNAs) coated on a nanopore surface: (a) Radii of gyration for the same length of ssDNA in hairpin and linear conformations, coated inside a pore; (b) effective pore diameter; and (c) DNA reorientation under applied electric fields of various strengths and effective pore diameter as a function of electric field strength.

6.6.1.1 Background

The statistical mechanics formula that describes a CG system is as follows [134]:

$$\exp(-F/k_B T) = C \int dx (\exp[-V(x)/k_B T]) \tag{6.48}$$

$$\exp(-F/k_B T) \simeq C \int dx_{CG} (\exp[-V_{CG}(x_{CG})/k_B T]) \tag{6.49}$$

Equation 6.48 is for the all-atom simulation, where F is the free energy of the system, $V(x)$ is the system's potential energy as a function of x of all atoms, T is the thermodynamic temperature, k_B is the Boltzmann constant, and C is the constant of normalization. Free energy for the CG system shown in Equation 6.49 is represented by x_{CG} and V_{CG} (effective potential). This equation forms the basis for various distribution functions, equilibrium averages and properties, and so forth [134].

6.6.1.2 Approaches to Construct a Coarse-Grained Model

The first step in obtaining a CG model is to convert the all-atom representation into a model having fewer numbers of interaction sites. This is usually done by either a residue-based method or a shape-based approach. In a residue-based method, a single bead is placed at the mass center for a group of atoms in the atomistic model. The CG site formed from the group of atoms (I_i) in the atomistic scale is denoted as R_{Ii}, which is obtained using the linear mapping function M_{R_I} [135]:

$$M_{R_I} = \sum_{i=1}^{n} c_{Ii} r_i \tag{6.50}$$

where c_{Ii} is the constant for the group of atoms that are selected and $c_{Ii} \neq 0$, whereas r_i denotes the coordinates of the group of selected atoms (in most cases, the CG site is located at the mass center of the selected group of atoms). The major advantage of this method is the ability to reconstruct the dynamics of all atoms from the simulation results of a CG model, particularly dynamic information of CG sites. This method has been successfully used in studies of lipids and proteins simulations [136–138].

In case of the shape-based system, a neural network–learning algorithm is used to determine the placement of CG beads. Masses are correlated with clusters of atoms, which are represented by CG beads. This method does not enable the exact reconstruction of the all-atom system, but it helps one understand the targeted properties of structures of interest. Furthermore, this can be used to construct a CG model using electron density maps in case the all-atom model is unavailable. This method has been successfully used in the study of stability and dynamics of viruses [139,140] and rotating bacterium flagellum [141,142].

6.6.1.3   Development of Coarse-Graining Potential

Current approaches to acquire CG potentials are delineated as follows: To begin with, the inverse Monte Carlo technique can reveal the essential physics of a given class of systems and provide qualitative information about a system, but it fails to give quantitatively accurate predictions [143]. The multiscale approach is a bottom-up method, which incorporates force data from atomic MD simulations to construct a CG system. Izvekov et al. [144,145] have established a multiscale CG model, in which the interparticle force field is constructed based on the force matching procedure using force data obtained from the all-atom simulation of the biomolecular system of interest. The canonical (NVT) equilibrium coordinate distribution function for an atomistic model is written as follows [146]:

$$p_r(r^n) = \frac{1}{z_n} e^{-U(r^n)/k_B T} \tag{6.51}$$

where k_b is the Boltzmann constant, T is temperature, $z_n = z(N,V,T) = \int d r^n e^{-U(r^n)/k_b T}$ is the canonical configuration integral, $r^n = [r_1, ..., r_n]$ denotes the Cartesian coordinates for n atoms, and $U(r^n)$ is the potential energy function of these n particles. The canonical equilibrium coordinate function for a CG model obtained from the aforementioned atomistic model is as follows:

$$P_R(R^n) = \frac{1}{Z_n} e^{-U(R^n)/k_B T} \tag{6.52}$$

where $Z_n = Z(N,V,T) = \int d R^n e^{-U(R^n)/k_b T}$ is the canonical configuration integral, $R^n = [R_1, ..., R_n]$ denotes the Cartesian coordinates for the n atoms of the CG model, and $U(R^n)$ is the potential energy function of these n particles. At each CG site, R^n is defined using the mapping equation, Equation 6.50. The details of the procedure can be found in the work by Noid et al. [135]. This multiscale approach has been successfully implemented for the development of CG models for carbohydrates [147], simple and ionic fluids [148], simple and mixed bilayers [149], and peptides [134,150]. This method also takes into account the implicit solvent model [134], which matches with the explicit solvent molecules in the atomistic model and is in accordance with Equation 5.49.

The MARTINI force field–based inversion CG approach, which is suggested by Marrink et al. [153], allows the use of experimental, thermodynamic, and/or average structural properties of the all-atom system to obtain a CG system with similar properties as the all-atom model. An example is the model to reproduce the thermodynamic data between organic and aqueous phases in comparison with experimentally available data. The free energy of partition between organic (oil) and aqueous (water) phases, $\Delta G^{oil/aq}$, is as follows [151]:

$$\Delta G^{oil/aq} = k_b T \ln(\rho_{oil}/\rho_{aq}) \tag{6.53}$$

where ρ_{oil} is the equilibrium density of the organic (oil) phase and ρ_{aq} is the equilibrium density of the aqueous phase of CG particles, respectively. The values of ρ are obtained by long-timescale full atomistic simulations of the two-phase system with small amounts (0.01 M) of target substance dissolved. The major advantage of this approach is that instead of concentrating on the structural details, the target has a broader range without the need for reparameterization of the CG model every time. The system is simple, computationally fast, and flexible enough to study the desired attribute of the system [134]. The MARTINI force field is successfully used to study different types of bilayers elasticity [130], rupture tension, lipid lateral diffusion rates [152], liquid densities [153], and lipid conformations [130]. Figure 6.23b shows

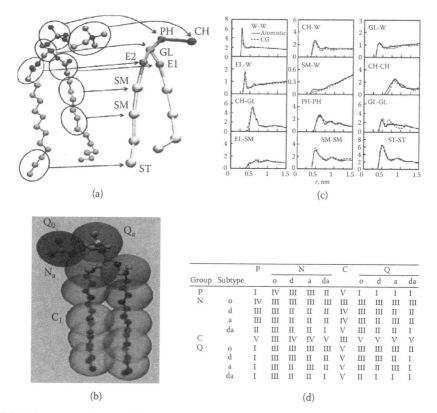

Group	Subtype	P				N				C	Q			
			o	d	a	da	o	d	a	da		o	d	a	da

Group	Subtype	P	N (o)	N (d)	N (a)	N (da)	C	Q (o)	Q (d)	Q (a)	Q (da)
P		I	IV	III	III	II	V	I	I	I	I
N	o	IV	III	III	III	III	III	III	III	III	III
	d	III	III	II	II	II	IV	III	III	II	II
	a	III	III	II	II	II	IV	III	II	III	II
	da	II	III	II	II	I	V	III	II	II	I
C		V	III	IV	IV	V	III	V	V	V	V
Q	o	I	III	III	III	III	V	III	III	III	II
	d	I	III	III	II	II	V	III	III	II	I
	a	I	III	II	III	II	V	III	II	III	I
	da	I	III	II	II	I	V	II	I	I	I

FIGURE 6.23 (a) The DMPC (dimyristoylphosphatidylcholine) coarse-grained (CG) model for multiscale CG modeling: (b) The MARTINI force-field CG model; (c) comparison of force values for the atomistic and CG force data at different intersite potentials; and (d) the interaction potentials used in the case of MARTINI force-field method. (Reprinted from Izvekov, S., and G. A. Voth. 2005. *J Phys Chem B* 109(7):2469–73; Marrink, S. J., A. H. de Vries, and A. E. Mark. 2004. *J Phys Chem B* 108(2):750–60. With permission.)

the CG model obtained by the MARTINI force field such that the CG atom site is placed at the mass center of each subset of four atoms [151]. Figure 6.23d tabulates the general MARTINI force-field potentials representing the interaction potentials [151], which are classified into attractive (I), semiattractive (II), neutral (III), semirepulsive (IV), and repulsive (V), and the groups of polar (P), nonpolar (N), apolar (C), and charged (Q) with subtypes "0" for no hydrogen bonding capabilities present, "d" for groups acting as hydrogen bond donors, "a" for groups acting as hydrogen bond acceptors, and "da" for groups with both donor and acceptor options.

Basically, the method chosen depends on the system under study or the goal of the study. For instance, if the goal is to preserve the general property of the entity of interest, a multiscale approach is preferred.

6.6.1.4 Challenges and Limitations of the Coarse-Grained System

Coarse graining is a promising tool for the future as it enables the understanding of specific attributes of a system by allowing longer timescale simulation in comparison with the short timescales of all-atom models. Coarse graining is still in its developing stages and there are many concerns and challenges to be faced, a few of which are mentioned here. Coarse graining is still not predictive; specifically, the systems are usually built with an idea of what can happen or an intuitive thought of what should happen, which may result in a bias of the CG model. In other words, CG model is generally either based on physical intuition, which implies that it is necessary to have a priori information of the dynamics of a system of interest for the development of CG model. Another concern is the degree of transferability of CG models to other systems and the modifications needed in the thermodynamic conditions. Not all characteristics of a CG model can be transferred, and currently there are no relevant literatures that specify the properties that can or cannot be transferred. Further, with the development of computationally efficient MD algorithms and high-performance central processing units, the timescales accessible with fully atomistic simulations are currently getting increased, which results in a concern that CG modeling might become obsolete if the all-atom model can be simulated with long timescales. Nevertheless, the bright side is that the CG length scale can surely be extended [134].

The CG models are usually built to capture a particular attribute of a system or a particular mechanism. Here, we consider a CG model of DNA, which is able to dictate the hybridization property and a pore–DNA interaction. The CG model is developed based on the residue-based approach and combined with the MARTINI force field.

6.6.2 Coarse-Graining Method

The purpose of the CG method is to obtain a system that is simple and yet sufficiently detailed so as to allow the study of real-time dynamics. The CG model has to be established in such a way that it can dictate the interaction potential between DNA strands. Two of the successfully constructed CG DNA models detailed in the literature are those of Knotts et al. [154] (three-site DNA model) and Schatz and Drukker [155] (two-site DNA model). Descriptions of these models along with a brief outline of the DNA model are delineated in Sections 6.6.2.1 through 6.6.2.5.

6.6.2.1 Three-Site DNA Model

Knotts et al. [154] developed a three-site DNA model that is a mesoscale molecular model. The CG model is able to describe aspects of melting, hybridization, salt effects, the major and minor grooving of DNA, and the mechanical properties of DNA. The groups of atoms replaced by each site of the CG model for a cytosine base are shown in Figure 6.24. The CG site is the mass center of the sugar and phosphate, whereas for the bases the CG site is N1 for purines (adenine and guanine) and N3 for pyrimidines (thymine and cytosine). The DNA system was developed such that

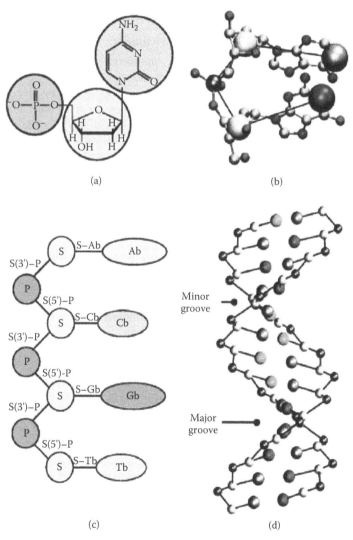

(a) (b)

(c) (d)

FIGURE 6.24 The three-site deoxyribonucleic acid model. (Reprinted from Knotts, T. A., N. Rathore, D. C. Schwartz, and J. J. de Pablo. 2007. *J Chem Phys* 126(8):084901. With permission.)

for an atom with its coordinates (x, y, z), the corresponding atom on the other side is located at $(x, -y, -z)$. Screw symmetry was used to locate successive residues of the DNA with 10 residues per turn and an axial rise of $3.38a°$; thus an atom at (r, φ, z) is adjacent to the atom at $(r, \varphi + 36°, z + 3.38a°)$, which means the DNA is aligned and centered with the z axis [154].

Even though the coordinate geometry file was obtained and the DNA base pairs seemed to form a perfect double helix, it failed to complete the benchmark case. The system was tested to obtain the fraction of the strands melted with respect to temperature. The interstrand potential seemed too high since the strands did not denature completely even at 400 K. Reduction of the interstrand potential inhibited the rehybridization of strands when the temperature was lowered to room temperature.

6.6.2.2 Two-Site Deoxyribonucleic Acid Model

Schatz and Drukker [155] developed the two-site model of DNA based on a simplified backbone–base structure where each nucleotide is symbolized by two sites. One site represents the backbone, that is, it is placed at the combined mass center of sugar and phosphate, whereas the other represents the mass center of the base. Their model was able to not only capture interactions such as hydrogen bond formation and reformation between the strands of DNA, bending and torsions, and the melting transition for DNA decamers containing only A–T or only C–G base pairs, but also assess the importance of the helical structure. The distance between two atoms in two-site DNA was 1.7 Å and screw symmetry was used to place successive DNA residues with 10 residues per turn. The hydrogen bonding sites were at the mass center of the base and were divided into acceptor and donor sites. The hydrogen bonds were formed only if the backbone–donor–acceptor angle was within a defined range [155].

In two-site DNA systems, the charges were not included and even the model parameters were chosen to give reasonable melting behavior only at a salt concentration of 0.1 M. This made their system irrelevant under different salt concentrations. Moreover, modification of the model parameters for each different salt concentration to mimic the correct melting behavior would be impractical. The possible extension to this system is to incorporate charges to the atoms.

6.6.2.3 DNA Model in Simulations

Based on the two-site model, charges are incorporated as described in the CG model of Marrink et al. [130,151–153]. The system is tested at 0.1-M salt concentration for hybridization and melting behavior; however, extension of this work to different salt concentrations is not covered in this chapter as it is an aspect of future work.

In a system similar to atomistic scale system with a silicon nitride pore, tethered DNAs, t-DNA, and ions must be built. All the necessary input files (*.gro, *.itp, *.top, *.index, etc.) for the simulation are obtained using scripts in MATLAB® for the different systems that are built. The nonbonded interactions between particles are described by LJ potential, $E_{\text{nbpair}} = \sum \left(\dfrac{A_{ij}}{r_{ij}^{12}} - \dfrac{B_{ij}}{r_{ij}^{6}} \right)$. Coulombic potential such as $E_{\text{coulombic}} = k \dfrac{(q_1 * q_2)}{r^2}$ is used to define charged-particle interactions. The total potential energy comprises bonded energies and nonbonded pair interaction energies:

$$E_{\text{total}} = E_{\text{bond}} + E_{\text{angle}} + E_{\text{diheral}} + E_{\text{improper}} + E_{\text{nbpair}} + E_{\text{coulombic}}$$

$$= \sum k(r_{ij} - r_0)^2 + \sum k_{\theta_{ijk}}(\theta_{ijk} - \theta_{eq})^2 + \sum \frac{1}{2}k_{\phi}\left[1 + \cos(n\phi - \delta)\right]$$

$$+ \sum k_{\psi}(\psi - \psi_0)^2 + \sum\left(\frac{A_{ij}}{r_{ij}^{12}} - \frac{B_{ij}}{r_{ij}^{6}}\right) + k_q\frac{(q_1 * q_2)}{r^2}$$

(6.54)

where the first four terms describe bonded intersite interactions including bond stretching, bending, and torsion, and the last two terms indicate the pairwise potential, which depicts the nonbonded (nbpair) interactions and the coulombic potential that describes the electrostatic potential. Nonbonded interactions include E_{stack} accounting for the base-stacking phenomena, E_{bp} (which also has a coulombic term) to describe bonding between complementary base pairs, and E_{ex} to dictate excluded volume interactions among the A, T, C, and G bases. All the k subscript parameters are the constants for each term, respectively. The E_{bp} term is critical for modeling the hybridization of an ssDNA with its complementary ssDNA or hairpin loop, and it is parameterized using thermal denaturation experimental data at a fixed salt concentration of ~0.1 M. All parameters used in the CG model can be found in the literature [155]. The bond distance between atoms is 0.47 nm and the angle between backbone sites is 150°, whereas an angle between the base and the phosphate is kept at 180° [123]. The schematic of different models of DNA developed by us are shown in Figure 6.25.

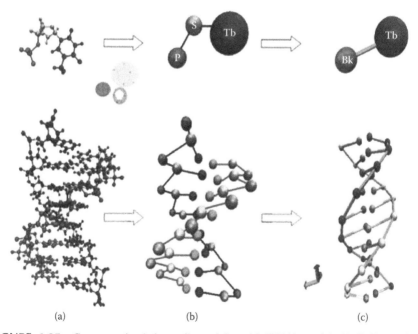

(a)   (b)   (c)

FIGURE 6.25 Coarse-grained deoxyribonucleic acid (DNA) model: (a) Full atomistic scale double-stranded DNA (dsDNA); (b) the three-site dsDNA model; and (c) the two-site DNA model. The first row shows a single nucleotide, whereas the second column shows the entire DNA.

6.6.2.4 Pore–DNA Interaction Potential

As the DNA translocates through a biofunctionalized nanopore, the interaction between DNA and the bare nanopore has to be defined and characterized. This interaction potential will not only be applied between the t-DNA and the nanopore but also between the coated DNAs (c-DNAs) and the nanopore. The interaction potential between the nanopore and the DNA is mapped from the atomistic model into the CG model. The interaction between a DNA base and a bare nanopore surface is dictated as an LJ potential:

$$V_{DN} = 4\varepsilon_{DN}\left(\left(\frac{\sigma_{DN}}{r}\right)^{12} - \left(\frac{\sigma_{DN}}{r}\right)^{6}\right) \tag{6.55}$$

where ε_{DN} is the depth of the potential well, σ_{DN} is the (finite) distance between the DNA base and the nanopore surface at which the potential is zero, and r is the distance between the mass center of the DNA base and the fixed nanopore surface. The vdW energy-density function is given as follows:

$$w_{DN} = \frac{2}{N_0}V_{DN}(r) \tag{6.56}$$

where N_0 is the unit cell of nanopore surface and $V_{DN}(r)$ is the LJ potential function. The total interaction energy (E_{DN}) between the mass center of a DNA nucleotide (q_0) and the total surface of a silicon nitride nanopore (Ω) can be written as the integral equation over the surface:

$$E_{DN} = \int_{\Omega} w_{DN}\left[\|q_0 - x\|\right]d\Omega \tag{6.57}$$

where x is a site on the nanopore surface.

Thus, total vdW interaction energy density is a function of the two LJ parameters (ε_{DN}, σ_{DN}) and can be controlled by modulating the LJ parameters so that they match with the interaction energy from the fully atomistic scale MD simulation. The CHARMM force field is used for the DNA, nanopore (silicon nitride), and their interactions in the fully atomistic scale simulation. The time step is 1 femtosecond with a temperature of 300 K, and a molarity of 0.1 M is used. The discrete force-distance data from the atomistic simulation and from the CG simulation are plotted in Figure 6.26. This process leads to an effective interaction between the CG sites as is present in the underlying all-atom simulation.

If an explicit solvent (water) is used, it increases the number of atoms dramatically and thus the computational demand increases. For this reason, solvent effects are included implicitly into the system by using Langevin dynamics. The solvent bath is represented by Langevin equations of motion using stochastic frictional forces and the velocity Verlet algorithm [156]. The friction constant (ξ) is written as follows [155]:

$$\xi = \frac{4\pi}{m}\eta \cdot r_{eff} \tag{6.58}$$

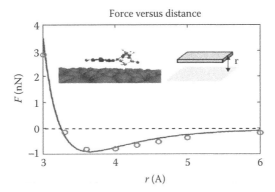

FIGURE 6.26 The deoxyribonucleic acid (DNA) base–surface interaction force–distance functions from the all-atomistic model and from the coarse-grained model. The inset shows a DNA base placed parallel to the surface.

where m is the mass of the CG DNA particles, r_{eff} is the effective hydrodynamic radius considered as 0.5 nm, and η is the solvent viscosity represented in the following form:

$$\eta = \eta_{20} \cdot \exp(-A/B)$$

$$A = 1.37023(T - 20) + 8.36 \times 10^{-4} (T - 20)^2$$

$$B = 109 + T \tag{6.59}$$

where T is the temperature, and $\eta_{20} = 0.93975 \times 10^{-3}$ is the viscosity at 20°C for water.

6.6.2.5 Simulation Details

The CG MD simulations are conducted using GROMACS 4.0 [157]. The electric field applied is in the range of 5–500 mV/nm. The temperature is set to 300 K with a molarity of 0.1 M, and a time step of 0.01 picosecond is used. Nonbonded interactions are defined with cutoff distance of 1.2 nm, and the LJ potential is smoothly shifted to zero between 1 nm and the cutoff distance. The pair list is updated in every step using a cutoff radius of 1.4 nm. Particle-mesh Ewald [158] is applied for long-range electrostatics. To make the system simple, homopolymers are taken into account, and a nanopore is coated with poly(dG), whereas the t-DNA is poly(dC) or poly(dT) depending on the attractive or the neutral case. The tethered and translocating DNA consists of 20 base pairs for better comparative study. The PBCs are applied at the entrance and outlet of the pore. Analysis and visualizations are conducted with VMD. All simulations are performed on the supercomputer cluster Ranger at TACC, which is a part of the TeraGrid resources. A simulation with a timescale of 200 ns using two processors requires a computing time of about 24 hours. The number of atoms in the system varies with the size of the configuration considered. A typical simulation system has 4714 atoms: 2610 atoms for the silicon nitride pore, 1960 atoms of DNA coatings, 40 atoms of DNA (t-DNA), and 144 ions.

6.6.3 BENCHMARK PROBLEM

The system has been validated by considering three benchmark runs in order to ensure that the CG model is able to capture the characteristics of the pure atomistic model.

6.6.3.1 Melting Behavior of DNA

The E_{bp} term, which determines the interaction potential for hybridization between ssDNAs (homopolymers, in this case), is parameterized using thermal denaturation experimental data at a fixed salt concentration of 0.1 M. A 20-bp dsDNA with poly(dGC) is placed in an implicit solvent bath with a temperature of 300 K. A simulation of 100 nanoseconds is run to test if the strands separate at a temperature of 300 K or not; this is done to confirm whether the interstrand potential (E_{bp}) will maintain the hybridized state of the dsDNA.

It was observed that the dsDNA continued to be a double-stranded entity and there was no fraction of melted/separated DNA at 300 K. The next step is to gradually increase the temperature by 5 K after every 10 nanoseconds and this procedure was continued until 400 K; thus the total simulation time is 200 nanoseconds. A script in Tcl (scripting language) [159] is written to measure the distance between the opposite base atoms of complementary strands. The threshold is kept at 4 Å (two-site hydrogen bonds) such that if the separation between the opposite base atoms of the complementary strands is larger than 4 Å, then it is considered as the melted fraction. The graph in Figure 6.27 shows the melting fraction of DNA with gradual increase in temperature. During the initial temperature increments, there is no separation between the two ssDNAs forming the dsDNA, and thus the fraction of melted/separated DNA bases are close to zero. But then the fraction of melted DNA bases increases as the temperature increases; the fraction of melted DNA approaches 1 at temperatures close to 380 K. A reassumed form gives us the value for E_{bp} and the parameters are determined to match with the experimental denaturation data.

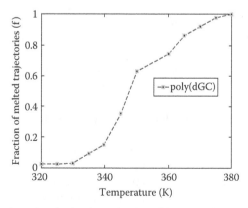

FIGURE 6.27 Fraction of melted trajectories for a 20–base pair double-stranded deoxyribonucleic acid (dsDNA) against the gradual increase of temperature in the system.

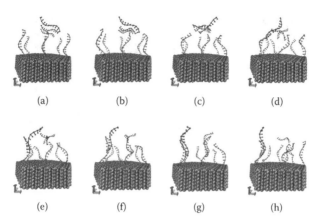

FIGURE 6.28 Brownian hybridization process of free-floating poly(dC) structure with a base-tethered poly(dG) structure, where (a) represents the initial position, (b–g) are different time-points capturing the hybridization process between the complimentary strands, while (h) represents the final hybridized state after a 100ns simulation at 300K.

6.6.3.2 DNA Hybridization

The hybridization of ssDNAs is tested to check the interstrand potential (E_{bp}). The system is set up with 3-ssDNAs poly(dG) having one of their ends attached to the surface while the other 3-ssDNAs poly(dC) are placed in the solvent bath above the substrate. The solvent is implicit with the temperature set at 300 K and the salt concentration at 0.1 M. The fixed base of poly(dG) is constrained on to the pore by applying a force of 10,000 kJ/mol. The system is first equilibrated for 10 nanoseconds to ensure that the strands are in minimized energy formations. After that, a real run of 200 nanoseconds is performed and the image sequence in Figure 6.28 shows that two out of the three poly(dC) ssDNAs have formed hybridized pairs with two of the poly(dG). These results show that the E_{bp} considered is satisfactory.

6.6.3.3 Comparison of a Molecular Dynamics System with a Coarse-Grained System

After the interstrand potentials of DNA are established, a CG system is developed to satisfactorily represent the underlying properties of atomistic scale systems. To ensure that the CG model accurately describes molecular motion, a benchmark case is the comparison between the CG simulation and the all-atom MD simulation. For this, the same formations for both systems are considered as shown in Figure 6.29. Each system consists of two straightened strands of ssDNAs with a length of 20 bps and a circular pore.

The CG system is solvated implicitly through Langevin dynamics while the atomistic scale system is explicitly solvated. A 200-nanoseconds simulation is run with the temperature set at 300 K and the planar electric field applied is 0.1 mV/nm. As shown in Figure 6.29, the DNA displacement-versus-time history under a given electric field is quantitatively comparable between predictions from the CG model and the MD simulation. In particular, the displacement of tethered ssDNA tip is taken into account for the comparison between the CG model and the atomistic scale MD simulation. We can clearly see that both the systems have a similar trend and are

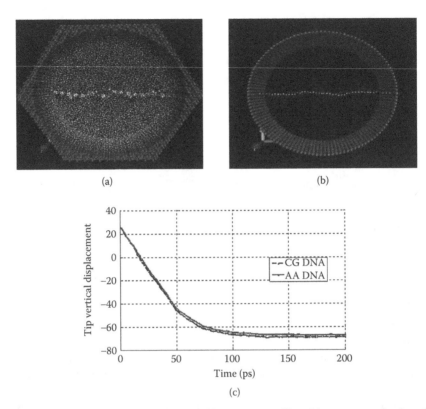

FIGURE 6.29 Comparison between the all-atom system (a) and its coarse-grained model (b): The graph (c) indicates the similarity between the two systems since similar trends are observed in the vertical tip displacement measured for the two systems.

almost identical to each other in terms of the bending of the DNA as indicated by the tips of the two systems.

6.6.4 Results for Biofunctionalized Nanopores

Biofunctionalization is essential to impart selectivity to nanopores. However, this process needs careful characterization to ensure the EPD and translocation kinetics. To understand the effect of biofunctionalization on a nanopore, the parameters to be analyzed are as follows: the type of coating, density of the coating, applied bias voltage, and the EPD. The coating DNA strands have a charge of $-e$, and thus they reorient themselves under the applied bias voltage. The reorientation of the coating molecules leads to an EPD that is different from the original bare pore diameter [11]. A set of simulations is implemented to understand the behavior of these DNA coatings based on their type, circumferential density, and longitudinal density.

6.6.4.1 Effect of Different Coatings

The selectivity of the nanopore can be defined by the type of coatings and their properties under the application of an electric field. Behavior of these coated structures is measured in terms of EPD, whose definition is presented in Figure 6.30.

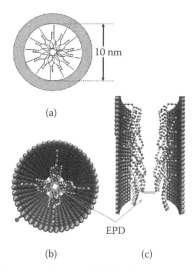

FIGURE 6.30 The effective pore diameter (EPD) schematic (a); top view of the 10-nm EPD (b); and the cross-sectional view of the 10-nm EPD for the coarse-grained model developed (c).

Under an applied bias voltage, the c-DNAs with the $-e$ charge reorient themselves along the electric field direction. These coated structures then begin to bend and this bending depends on their rigidity related to the shape, structure, intrastrand bonds, and so forth. The EPD (shown in Figure 6.30) influences the interaction between the t-DNA and the DNA coatings, as well as the net translocation velocity. To understand the effect of type of coating, HPL coatings are taken into account in comparison with ssDNA coatings for a pore with its actual diameter of 10 nm. The reason for selecting these two types is that when a perfectly matched complementary ssDNA is within a close range, the HPL opens up to form an ssDNA to hybridize with the complementary t-DNA. There is difference in the rigidity of these two structures such that the former (HPL) is more rigid than the latter (ssDNA), which is highly flexible. To characterize their effect on EPDs, a sequence of runs is performed at different electric fields. The probe molecules are to be uniformly coated on the pore surface. The results shown in Figure 6.31 indicate that the HPL-coated nanopore has a smaller EPD compared to the less rigid ssDNA-coated nanopore. This leads to the conclusion that the t-DNA translocating through an HPL-coated nanopore will have more interactions when compared with the ssDNA-coated nanopore. Thus, t-DNA translocates faster in ssDNA-coated pores compared to HPL-coated pores.

6.6.4.2 Effect of Circumferential Density

The number of strands coated circumferentially defines the circumferential density of the coatings. The denser a system, the higher the interaction between the t-DNA and the coatings. To understand the effect of circumferential density, a system of four strands forming a circumferential coating is compared with a system of eight strands. These studies are performed to see the lateral motion of the c-DNAs and to understand if there is a difference in the EPDs obtained for the two systems under consideration. The top views of the two systems are shown in Figure 6.32. The densities

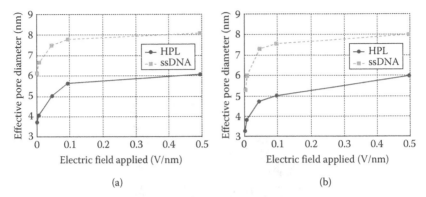

(a)   (b)

FIGURE 6.31 Effective pore diameter (EPD) for two different types of coatings: (a) The EPD obtained when four strands are used for coating; (b) EPD obtained when eight strands are used for coating. Both the results indicate that the EPDs obtained from the hairpin-loop coatings are much smaller than the EPD of the single-stranded deoxyribonucleic acid for nanopores with an original bare pore diameter of 10 nm.

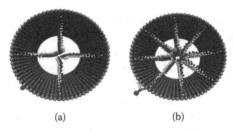

(a)   (b)

FIGURE 6.32 Two systems considered to understand the effect of circumferential density, with density of one strand per 5.0 nm and one strand per 2.5 nm for the four strands circumferentially coated (a) and eight strands circumferentially coated (b) pores, respectively.

of these two systems are one strand per 5.0 nm for four-strand-coated pores and one strand per 2.5 nm for eight-strand-coated pores, respectively.

The results shown in Figure 6.33 suggest a lower EPD for a system with higher circumferential density. All the three graphs show a parabolic trend for the EPDs obtained as the electric field applied is gradually increased. The comparison of results of Figure 6.33c with the other two indicates that the effect of circumferential density is more evidently seen in a nanopore with least coatings in the longitudinal direction. The first two data points show a large difference in the EPDs obtained with different circumferential densities. The difference, however, is reasonable as presented in Figure 6.33a and b, where a consistent parabolic trend is observed. In this case, the translocation velocity of the t-DNA will be higher in the less densely coated nanopore compared to the more densely coated nanopore. In addition, an increase in circumferential density reduces the erroneous results that can be obtained, since the t-DNA may translocate through the path of least resistance without coatings.

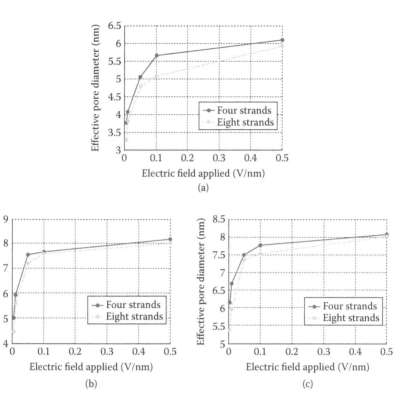

FIGURE 6.33 Impact of circumferential density on effective pore diameter (EPD): The graphs give a comparative analysis of the EPD obtained with four strands coated circumferentially to that with eight strands coated. Different coatings considered are (a) hairpin loop, (b) single-stranded deoxyribonucleic acid (ssDNA) 1 nm apart, and (c) ssDNA 2 nm apart.

6.6.4.3 Effect of Longitudinal Density

The coating density matters not only circumferentially but also longitudinally. To understand the effect of longitudinal density, two ssDNA cases are considered, in which the coatings are placed 1 nm apart compared to 2 nm apart as indicated by Figure 6.34a and b, respectively. The importance of longitudinal density is that interstrand interactions between the coatings increase with an increase in longitudinal density. The lower the longitudinal density, the lesser the interstrand interaction of the coatings and the more the stretching of the coatings under the application of an electric field; this increases the EPD. Cross-sectional views of the two systems are shown in Figure 6.34. The densities of these two systems are one strand per 8.3 nm and one strand per 16.7 nm for the ssDNA-coated pores with longitudinal spacings of 1 nm and 2 nm, respectively.

The results shown in Figure 6.35a and b indicate that at low electric fields, there is significant difference in the EPD obtained from simulation due to the difference in the longitudinal densities of the two systems. As the applied bias voltage increases, the influence of the electric field becomes more apparent compared with interstrand

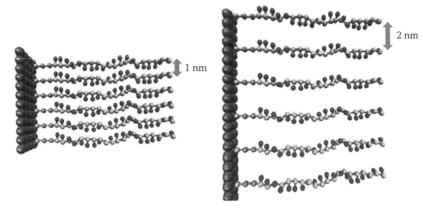

FIGURE 6.34 Variation in the longitudinal density of the systems considered: The strands in part (a) are placed 1 nm apart, whereas in part (b) the strands are placed 2 nm apart.

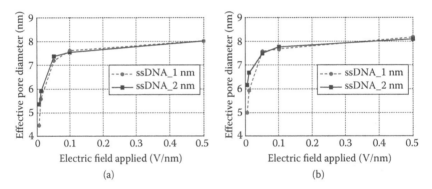

FIGURE 6.35 Effect of longitudinal density on effective pore diameter (EPD): The graphs give a comparative analysis of the EPD obtained with single-stranded deoxyribonucleic acid strands coated 1 nm apart longitudinally to that coated 2 nm apart. Different coatings considered are (a) four-strand circumferential coating and (b) eight-strand circumferential coating.

interactions, which causes the coatings to be stretched to a large extent, overlaying the EPD with different density values at very high electric fields. The velocity of t-DNA will be larger in the less dense system compared to the denser system only for low electric fields. At higher electric fields, there will be no significant difference in the velocity of the t-DNA in the two systems under consideration.

6.6.4.4 Effect of Interaction Potential

Kasianowicz et al. [41] demonstrated that a mere potential difference within the nanopores would not create the selectivity needed to perform gene sequencing. Much more complex procedures are needed to attain this potential difference. The interaction potential of the coatings will surely have a significant influence on the velocity of t-DNA.

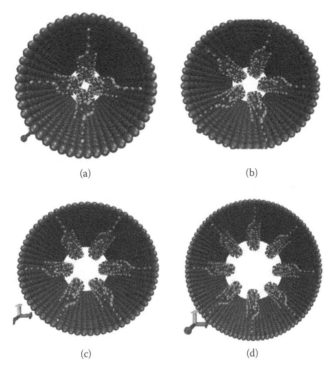

(a) (b)

(c) (d)

FIGURE 6.36 Different effective pore diameters (EPDs): (a) An EPD of 1 nm; (b) 2.8 nm; (c) 4.6 nm; and (d) 81 nm. These are used to test the effect on translocating deoxyribonucleic acid (t-DNA) with attractive and neutral potentials.

In order to understand the effect of the t-DNA potential, the system is developed by considering the final configuration of the c-DNAs in past runs and recoating them to obtain the respective diameters as shown in Figure 6.36. Thus, the stretched coatings are recoated onto the nanopores to obtain an EPD of 1 nm. In addition, it is ensured that irrespective of the types of coating, the initial EPDs will be 1 nm (in the case of Figure 6.36a) for the three different coating systems that are considered. The EPD is also gradually increased by keeping density of the coatings a constant, and as the EPD is increased its initial value for all three coatings is maintained the same. The different cases considered are shown in Figure 6.36. For making the system simple, homopolymers are considered. In order to study the role of interaction potential between translocating DNA and coating DNA on the translocation dynamics, the sequence of DNA used for coating a nanopore is given as "GGGGG CCCCC AAAAA CCCCC", while the sequence of translocating DNA is given as either "CCCCC GGGGG TTTTT GGGGG" (for attraction between translocating DNA and coating DNA) or "AAAAA AAAAA AAAAA AAAAA" (for neutral interaction between translocating DNA and coating DNA).

The results obtained are shown in Figure 6.37, which indicate significant difference in the velocity of the t-DNA in the smallest 1-nm EPD configuration, although in other systems velocities are approximately equal. This is mainly due to the large

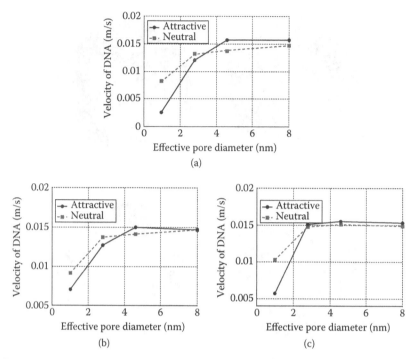

FIGURE 6.37 Influence of interaction potential between the translocating deoxyribonu-cleic acid (t-DNA) and the coating on translocating velocity: The graphs give a comparative analysis of the velocity of the t-DNA for the case of attractive potential system and neutral potential system with different EPD. Different coatings considered are (a) hairpin loop, (b) single-stranded DNA (ssDNA) 1 nm apart, and (c) ssDNA 2 nm apart.

interaction between the t-DNA and the coatings in the smallest system (EPD = 1 nm). The interaction between the t-DNA and the coatings is decreased with increasing EPD, which causes a decrease in the influence of t-DNA potential. Thus there is no specific trend in the velocity obtained, although there is significant difference at the smallest EPD (1 nm). The significant difference in the velocities obtained between the neutral potential t-DNA and the attractive potential t-DNA for the 1-nm EPD shows the impact of the potential on the system if sufficient interactions take place.

In the case of the smallest EPD such as 1 nm, the velocity of the t-DNA under the influence of gradually increasing electric field is observed, which affirms the conclusion from Figure 6.37. The results are illustrated in Figure 6.38, which shows that there is significant difference in the translocation velocities and that the neutral potential t-DNA translocates faster than the t-DNA with attractive potential. The interaction potential becomes less dominant due to an increase in the applied bias voltage. Therefore, a reduced difference of velocities between the t-DNA with neu-tral potential and the attractive potential is observed. The significant differences in velocities based purely on potentials at lower electric fields suggest that modulation of potential can also play a significant role in future nanotechnology applications.

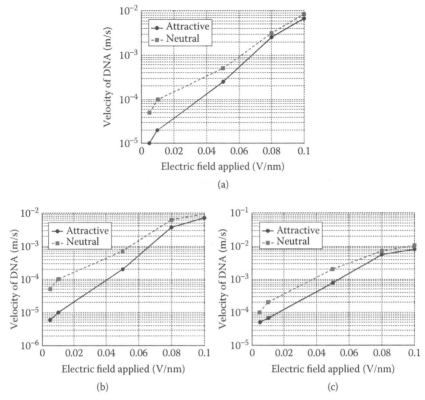

FIGURE 6.38 Influence of interaction potential on deoxyribonucleic acid (DNA) translocation velocity: Each graph gives a comparative analysis of the velocity of the translocating DNA with an attractive potential and that with a neutral potential with an increase in the voltage bias applied. Different coatings considered are (a) hairpin loop, (b) single-stranded DNA (ssDNA) 1 nm apart, and (c) ssDNA 2 nm apart.

This study could also be applied to see the effect of a t-DNA on a repulsive potential and something else, but such aspects do not come within the scope of this chapter.

6.6.4.5 Selectivity

In order to understand the kinetics with which selectivity can be imparted to solid-state nanopores, the actual process of opening up HPL structures into ssDNA and its effect on t-DNA translocation velocity are mimicked after the analysis of different parameters in individual cases separately. Since the hybridization process and the opening of HPL due to the presence of a matched DNA is a complex mechanism, a case is studied to compare an HPL-coated pore with another HPL-coated pore that opens into ssDNA under a voltage bias. The images in Figure 6.39a through f show the translocation process of a mismatched t-DNA that has to translocate through HPL-coated nanopores. The images in Figure 6.39g through l show the translocation process of a matched t-DNA that has to translocate through HPL-coated nanopores, which opens into the ssDNA and thereby increase the EPD and facilitate the translocation process.

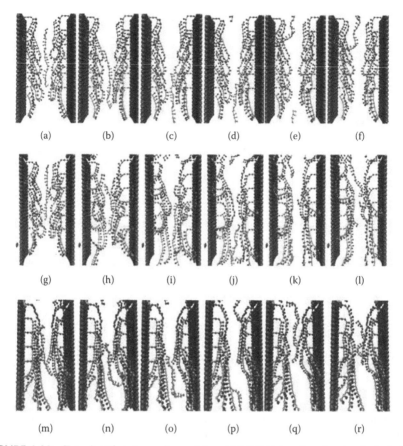

FIGURE 6.39 Translocating deoxyribonucleic acid (t-DNA) translocation in a hairpin loop (HPL)-coated nanopore (a through f), t-DNA translocation in an HPL-coated nanopore that opens into single-stranded DNA (ssDNA) (g through l), and t-DNA translocation in ssDNA (m through r).

It can be understood from the study why the perfectly matched DNA takes less time to translocate through the HPL-coated nanopores compared to the mismatched one. The aforementioned results indicate that t-DNAs should be translocating at a very close range in spite of the difference in potentials unless there is a large inter-strand interaction. Based on the analysis of simulation results as shown in Figure 6.40, there is significant difference between the translocation times of the mismatched and the matched DNA. The matched DNA translocates much faster compared to the mismatched one, since the HPL opens up into the ssDNA and under the influence of the bias applied this increases the EPD; the overall effect causes reduced time of translocation for the matched DNA. This cannot be confirmed until the actual hybridization process between the coated HPL strands and the t-DNA is captured. Based on the results obtained from the current set of simulations (Figure 6.40 and

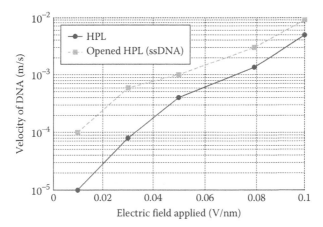

FIGURE 6.40 Effect of biofunctionalization: This graph mimics the actual process of biofunctionalization by giving a comparison with the hairpin-loop (HPL) coatings and shows what happens when the HPL coatings open up into single-stranded deoxyribonucleic acid (ssDNA). The results clearly indicate that the reduced translocation time for the matched DNA is due to the opening of HPL structures into the ssDNA, which increases the effective pore diameter and facilitates translocation.

TABLE 6.1

Velocities for translocating DNA in different kinds of nanopores as a function of electric field

Electric Field (V/nm)	Velocity of DNA in HPL-Coated Nanopores (m/s)	Velocity of DNA in HPL Opening to ssDNAs (m/s)	Percentage of Increase
0.01	1.00E-05	0.0001	900.00
0.05	0.0004	0.001	150.00
0.08	0.00135	0.003	122.22
0.1	0.0048	0.0088	83.33
0.3	0.0348	0.0433	24.42

Table 6.1), there is a notable difference in the translocation velocity. This underscores the fact that the opening up of HPL structures due to the presence of the matched DNA plays a significant role in facilitating its transport. The mismatched DNA has to force its way through HPL coatings, which are more rigid, and this takes a longer translocation time. Furthermore, the percentage increase in the velocity of the DNA shown in Table 6.1 is much larger at smaller bias voltages applied. Even in previous results it is found that the density and coatings play a more influential role in the lower bias voltage range.

6.6.5 SUMMARY

A CG system is developed to understand the kinetics of DNA translocation in a biofunctionalized nanopore. This CG system is first tested with three benchmark cases, which illustrate that the CG system is able to reproduce the results of atomistic simulation. The results are divided into two major divisions; (i) the effect of type and density of coatings on EPD (as a function of electric field), and (ii) role of interaction potential between translocating DNA and DNA used for coating a nanopore in EPD. The parameters under consideration are the type of coatings applied and the circumferential and longitudinal density of the coatings. The EPD increases as the electric field applied (in volts per nanometer) is increased and shows a parabolic trend. The EPD for an HPL-coated nanopore is significantly lower than for the ssDNA-coated nanopore. In the case of density, nanopores circumferentially coated with eight strands had a smaller EPD compared with those coated with four strands. Further, even in the case of a comparative study between ssDNAs longitudinally placed 1 nm apart and ssDNAs longitudinally placed 2 nm apart, it is observed that the EPD is less for the denser ssDNA-coated nanopore placed 1 nm apart. This leads to the following conclusion: the denser the coatings the less the EPD obtained. Another way to reduce EPD is to use different coatings, as it is shown that HPL has a much smaller EPD when compared with the EPD of ssDNA-coated nanopore having the same density.

The next set of results help to understand the effect of t-DNA potential, and it is observed that a t-DNA with an attractive potential to the coatings takes a longer translocation time when compared with a t-DNA that has a semiattractive potential. However, as EPD increases, the difference in velocities becomes negligible and both the t-DNAs have almost equal velocities. Another important observation is the difference between translocation velocities of t-DNA passing through an HPL-coated nanopore and an ssDNA-coated nanopore. The translocation dynamics is slower in the HPL-coated nanopore, whereas it is comparatively faster in the ssDNA-coated nanopore. To understand if selectivity could be imparted in the future, an attempt to capture the intermediate procedure of the HPL opening into ssDNA has to be made with a simpler model. The results support the argument that velocity of t-DNA in the HPL opening into ssDNA system is larger compared to that in the unopened HPL-coated system. This highlights the primary reason why a matched DNA has a lesser translocation time in HPL-coated nanopores: Due to the perfect matching between bases, the coated HPL opens into the ssDNA, which increases the EPD and enables faster translocation of t-DNA, whereas with even a single base mismatch the HPL does not open and it takes much longer for a mismatched DNA to translocate through the nanopore. This shows that selectivity could be imparted into solid-state nanopores; however, to confirm this, the third and final phase of the study, which is to mimic the actual process such that the HPL coatings open only in the presence of the perfectly matched t-DNA and when t-DNA is sufficiently close to the HPL coating, has to be conducted. If successfully demonstrated, this will help the imparting of selectivity to the nanopores. An understanding of this system could then be integrated with real solid-state nanopore devices and bring us closer to achieving the ultimate goal of genome sequencing, whose cost is within $1000.

REFERENCES

1. Alberts, B. 2002. *Molecular Biology of the Cell*. 4th ed. Garland Science: New York, NY, USA.
2. Venter, J. C. 2001. The sequence of the human genome (vol 292, pg 1304, 2001). *Science* 292(5523):1838.
3. Watson, J. D. 1990. The Human Genome Project—past, present, and future. *Science* 248(4951):44–9.
4. Cantor, C. R. 1990. Orchestrating the Human Genome Project. *Science* 248(4951):49–51.
5. Cantor, C. R., K. Tang, J. H. Graber, M. Maloney, D. J. Fu, N. E. Broude, F. Siddiqi, H. Koester, and C. L. Smith. 1997. DNA sequencing after the Human Genome Project. *Nucleosides Nucleotides* 16(5–6):591–8.
6. Service, R. F. 2006. Gene sequencing. The race for the $1000 genome. *Science* 311(5767):1544–6.
7. Sanger, F., S. Nicklen, and A. R. Coulson. 1977. DNA sequencing with chain-terminating inhibitors. *Proc Natl Acad Sci U S A* 74(12):5463–7.
8. Branton, D., D. W. Deamer, A. Marziali, H. Bayley, S. A. Benner, T. Butler, M. Di Ventra et al. 2008. The potential and challenges of nanopore sequencing. *Nat Biotechnol* 26(10):1146–53.
9. Iqbal, S. M., D. Akin, and R. Bashir. 2007. Solid-state nanopore channels with DNA selectivity. *Nat Nanotechnol* 2(4):243–8.
10. Bauer, W. R., and W. Nadler. 2006. Molecular transport through channels and pores: Effects of in-channel interactions and blocking. *Proc Natl Acad Sci U S A* 103(31):11446–51.
11. Ramachandran, A., Y. Liu, W. Asghar, and S. M. Iqbal. 2009. Characterization of DNA-nanopore interactions by molecular dynamics. *Am J Biomed Sci* 1(4):344–351.
12. Gilchrist, T. L. 1997. *Heterocyclic Chemistry*. Eaglewood Cliffs, NJ, USA: Prentice Hall.
13. Watson, J. D., and F. H. C. Crick. 1953. Molecular-structure of nucleic-acids—a structure for deoxyribose nucleic-acid. *Nature* 171:737–8.
14. Wink, M. 2006. *An Introduction to Molecular Biotechnology: Molecular Fundamentals, Methods and Applications in Modern Biotechnology*. Weinheim, Germany: Wiley-VCH.
15. Yakovchuk, P., E. Protozanova, and M. D. Frank-Kamenetskii. 2006. Base-stacking and base-pairing contributions into thermal stability of the DNA double helix (vol 34, pg 564, 2006). *Nucleic Acids Res* 34(3):1082–+.
16. Wheeler, D. A., M. Srinivasan, M. Egholm, Y. Shen, L. Chen, A. McGuire, W. He et al. 2008. The complete genome of an individual by massively parallel DNA sequencing. *Nature* 452(7189):872-U5.
17. Kim, S., H. Tang, and E. R. Mardis. 2007. *Genome Sequencing Technology and Algorithms*. Norwood, MA, USA: Artech House Publishers.
18. Swerdlow, H., and R. Gesteland. 1990. Capillary gel-electrophoresis for rapid, high-resolution DNA sequencing. *Nucleic Acids Res* 18(6):1415–9.
19. Janitz, M. 2008. *Next-Generation Genome Sequencing: Towards Personalized Medicine*. Weinheim, Germany: Wiley-VCH.
20. Huang, X. H. C., M. A. Quesada, and R. A. Mathies. 1992. DNA sequencing using capillary array electrophoresis. *Anal Chem* 64(18):2149–54.
21. Sussman, H. E., and M. A. Smit. 2006. *Genomes*. Cold Spring Harbor, NY, USA: Cold Spring Harbor Laboratory Press.
22. Metzker, M. L. 2005. Emerging technologies in DNA sequencing. *Genome Res* 15(12):1767–76.
23. Xu, F., and Y. Baba. 2005. *DNA Sequencing Studies by Capillary Electrophoresis*. Boca Raton, FL, USA: Encyclopedia of Chromatography CRC Press.

24. Smith, L. M., J. Z. Sanders, R. J. Kaiser, P. Hughes, C. Dodd, C. R. Connell, C. Heiner, S. B. H. Kent, and L. E. Hood. 1986. Fluorescence detection in automated DNA-sequence analysis. *Nature* 321(6071):674–9.

25. Prober, J. M., G. L. Trainor, R. J. Dam, F. W. Hobbs, C. W. Robertson, R. J. Zagursky, A. J. Cocuzza, M. A. Jensen, and K. Baumeister. 1987. A system for rapid DNA sequencing with fluorescent chain-terminating dideoxynucleotides. *Science* 238(4825):336–41.

26. Lewis, E. K., W. C. Haaland, F. Nguyen, D. A. Heller, M. J. Allen, R. R. MacGregor, C. S. Berger et al. 2005. Color-blind fluorescence detection for four-color DNA sequencing. *Proc Natl Acad Sci U S A* 102(15):5346–51.

27. Karger, A. E., J. M. Harris, and R. F. Gesteland. 1991. Multiwavelength fluorescence detection for DNA sequencing using capillary electrophoresis. *Nucleic Acids Res* 19(18):4955–62.

28. Lee, L. G., S. L. Spurgeon, C. R. Heiner, S. C. Benson, B. B. Rosenblum, S. M. Menchen, R. J. Graham, A. Constantinescu, K. G. Upadhya, and J. M. Cassel. 1997. New energy transfer dyes for DNA sequencing. *Nucleic Acids Res* 25(14):2816–22.

29. Zhu, L., W. J. Stryjewski, and S. A. Soper. 2004. Multiplexed fluorescence detection in microfabricated devices with both time-resolved and spectral-discrimination capabilities using near-infrared fluorescence. *Anal Biochem* 330(2):206–18.

30. Alaverdian, L., S. Alaverdian, O. Bilenko, I. Bogdanov, E. Filippova, D. Gavrilov, B. Gorbovitski et al. 2002. A family of novel DNA sequencing instruments based on single-photon detection. *Electrophoresis* 23(16):2804–17.

31. Ronaghi, M., M. Uhlen, and P. Nyren. 1998. A sequencing method based on real-time pyrophosphate. *Science* 281(5375):363–+.

32. Langaee, T., and M. Ronaghi. 2005. Genetic variation analyses by Pyrosequencing. *Mutat Res* 573(1–2):96–102.

33. Brown, R. J. 2007. *Comparative Genomics: Basic and Applied Research.* Boca Raton, FL, USA: CRC Press.

34. Voelkerding, K. V., S. A. Dames, and J. D. Durtschi. 2009. Next-generation sequencing: From basic research to diagnostics. *Clin Chem* 55(4):641–58.

35. Shendure, J., G. J. Porreca, N. B. Reppas, X. X. Lin, J. P. McCutcheon, A. M. Rosenbaum, M. D. Wang, K. Zhang, R. D. Mitra, and G. M. Church. 2005. Accurate multiplex polony sequencing of an evolved bacterial genome. *Science* 309(5741):1728–32.

36. Hofmann, W. K. 2006. *Gene Expression Profiling by Microarrays: Clinical Implications.* Cambridge, UK: Cambridge University Press.

37. Coulter, W. H. 1953. Means for counting particles suspended in a fluid. U.S. Patent Number 2656508.

38. Deblois, R. W., C. P. Bean, and R. K. A. Wesley. 1977. Electrokinetic measurements with submicron particles and pores by resistive pulse technique. *J Colloid Interface Sci* 61(2):323–35.

39. Hille, B. 2001. *Ion Channels of Excitable Membranes.* Sinauer Associates 3rd Casebound edition: Sunderland, MA, USA.

40. Neher, E., and B. Sakmann. 1976. Single-channel currents recorded from membrane of denervated frog muscle fibres. *Nature* 260(5554):799–802.

41. Kasianowicz, J. J., E. Brandin, D. Branton, and D. W. Deamer. 1996. Characterization of individual polynucleotide molecules using a membrane channel. *Proc Natl Acad Sci U S A* 93(24):13770–3.

42. Church, G. M., D. W. Deamer, D. Branton, R. Baldarelli, and J. J. Kasianowicz. 1998. Characterization of individual polymer molecules based on monomer-interface interactions. *U.S. Patent 5795782.*

43. Deamer, D. W., and M. Akeson. 2000. Nanopores and nucleic acids: Prospects for ultra-rapid sequencing. *Trends Biotechnol* 18(4):147–51.

44. Nakane, J. J., M. Akeson, and A. Marziali. 2003. Nanopore sensors for nucleic acid analysis. *J Phys Condens Matter* 15(32):R1365–93.

45. Wang, G. M., and A. C. Sandberg. 2007. Non-equilibrium all-atom molecular dynamics simulations of free and tethered DNA molecules in nanochannel shear flows. *Nanotechnology* 18(13):135702–10.

46. Akeson, M., D. Branton, J. J. Kasianowicz, E. Brandin, and D. W. Deamer. 1999. Microsecond time-scale discrimination among polycytidylic acid, polyadenylic acid, and polyuridylic acid as homopolymers or as segments within single RNA molecules. *Biophys J* 77(6):3227–33.

47. Meller, A. 2003. Dynamics of polynucleotide transport through nanometre-scale pores. *J Phys Condens Matter* 15(17):R581–607.

48. Meller, A., L. Nivon, E. Brandin, J. Golovchenko, and D. Branton. 2000. Rapid nanopore discrimination between single polynucleotide molecules. *Proc Natl Acad Sci U S A* 97(3):1079–84.

49. Meller, A., L. Nivon, and D. Branton. 2001. Voltage-driven DNA translocations through a nanopore. *Phys Rev Lett* 86(15):3435–8.

50. Mathe, J., A. Aksimentiev, D. R. Nelson, K. Schulten, and A. Meller. 2005. Orientation discrimination of single-stranded DNA inside the alpha-hemolysin membrane channel. *Proc Natl Acad Sci U S A* 102(35):12377–82.

51. Butler, T. Z., J. H. Gundlach, and M. A. Troll. 2006. Determination of RNA orientation during translocation through a biological nanopore. *Biophys J* 90(1):190–9.

52. Wang, H., J. E. Dunning, A. P. H. Huang, J. A. Nyamwanda, and D. Branton. 2004. DNA heterogeneity and phosphorylation unveiled by single-molecule electrophoresis. *Proc Natl Acad Sci U S A* 101(37):13472–7.

53. Henrickson, S. E., M. Misakian, B. Robertson, and J. J. Kasianowicz. 2000. Driven DNA transport into an asymmetric nanometer-scale pore. *Phys Rev Lett* 85(14):3057–60.

54. Nakane, J., M. Akeson, and A. Marziali. 2002. Evaluation of nanopores as candidates for electronic analyte detection. *Electrophoresis* 23(16):2592–601.

55. Aksimentiev, A., J. B. Heng, E. R. Cruz-Chu, G. Timp, and K. Schulten. 2005. Microscopic kinetics of DNA translocation through synthetic nanopores. *Biophys J* 88(1):352a.

56. Aksimentiev, A., J. B. Heng, G. Timp, and K. Schulten. 2004. Microscopic kinetics of DNA translocation through synthetic nanopores. *Biophys J* 87(3):2086–97.

57. Aksimentiev, A., and K. Schulten. 2005. Imaging alpha-hemolysin with molecular dynamics: Ionic conductance, osmotic permeability, and the electrostatic potential map. *Biophys J* 88(6):3745–61.

58. Gu, L. Q., S. Cheley, and H. Bayley. 2001. Capture of a single molecule in a nanocavity. *Science* 291(5504):636–40.

59. Dekker, C. 2007. Solid-state nanopores. *Nat Nanotechnol* 2(4):209–15.

60. Smeets, R. M. M., U. F. Keyser, D. Krapf, M. Y. Wu, N. H. Dekker, and C. Dekker. 2006. Salt dependence of ion transport and DNA translocation through solid-state nanopores. *Nano Lett* 6(1):89–95.

61. Ho, C., R. Qiao, J. B. Heng, A. Chatterjee, R. J. Timp, N. R. Aluru, and G. Timp. 2005. Electrolytic transport through a synthetic nanometer-diameter pore. *Proc Natl Acad Sci U S A* 102(30):10445–50.

62. Li, J., D. Stein, C. McMullan, D. Branton, M. J. Aziz, and J. A. Golovchenko. 2001. Ion-beam sculpting at nanometre length scales. *Nature* 412(6843):166–9.

63. Chang, H., F. Kosari, G. Andreadakis, M. A. Alam, G. Vasmatzis, and R. Bashir. 2004. DNA-mediated fluctuations in ionic current through silicon oxide nanopore channels. *Nano Lett* 4(8):1551–6.

64. Bai, J. G., C. L. Chang, J. H. Chung, and K. H. Lee. 2007. Shadow edge lithography for nanoscale patterning and manufacturing. *Nanotechnology* 18(40):405307.

65. Stein, D., M. Kruithof, and C. Dekker. 2004. Surface-charge-governed ion transport in nanofluidic channels. *Phys Rev Lett* 93(3):035901.

66. Siwy, Z., E. Heins, C. C. Harrell, P. Kohli, and C. R. Martin. 2004. Conical-nanotube ion-current rectifiers: The role of surface charge. *J Am Chem Soc* 126(35):10850–1.

67. Fan, R., R. Karnik, M. Yue, D. Y. Li, A. Majumdar, and P. D. Yang. 2005. DNA translocation in inorganic nanotubes. *Nano Lett* 5(9):1633–7.

68. Healy, K., B. Schiedt, and A. P. Morrison. 2007. Solid-state nanopore technologies for nanopore-based DNA analysis. *Nanomedicine* 2(6):875–97.

69. Li, J. L., M. Gershow, D. Stein, E. Brandin, and J. A. Golovchenko. 2003. DNA molecules and configurations in a solid-state nanopore microscope. *Nat Mater* 2(9):611–5.

70. Mara, A., Z. Siwy, C. Trautmann, J. Wan, and F. Kamme. 2004. An asymmetric polymer nanopore for single molecule detection. *Nano Lett* 4(3):497–501.

71. Heng, J. B., C. Ho, T. Kim, R. Timp, A. Aksimentiev, Y. V. Grinkova, S. Sligar, K. Schulten, and G. Timp. 2004. Sizing DNA using a nanometer-diameter pore. *Biophys J* 87(4):2905–11.

72. Storm, A. J., C. Storm, J. H. Chen, H. Zandbergen, J. F. Joanny, and C. Dekker. 2005. Fast DNA translocation through a solid-state nanopore. *Nano Lett* 5(7):1193–7.

73. Heng, J. B., A. Aksimentiev, C. Ho, P. Marks, Y. V. Grinkova, S. Sligar, K. Schulten, and G. Timp. 2006. The electromechanics of DNA in a synthetic nanopore. *Biophys J* 90(3):1098–106.

74. Heng, J. B., A. Aksimentiev, C. Ho, P. Marks, Y. V. Grinkova, S. Sligar, K. Schulten, and G. Timp. 2005. Stretching DNA using the electric field in a synthetic nanopore. *Nano Lett* 5(10):1883–8.

75. Fologea, D., M. Gershow, B. Ledden, D. S. McNabb, J. A. Golovchenko, and J. L. Li. 2005. Detecting single stranded DNA with a solid state nanopore. *Nano Lett* 5(10):1905–9.

76. Yan, H., and B. Q. Xu. 2006. Towards rapid DNA sequencing: Detecting single-stranded DNA with a solid-state nanopore. *Small* 2(3):310–2.

77. Muthukumar, M., and C. Y. Kong. 2006. Simulation of polymer translocation through protein channels. *Proc Natl Acad Sci U S A* 103(14):5273–8.

78. Vercoutere, W., S. Winters-Hilt, H. Olsen, D. Deamer, D. Haussler, and M. Akeson. 2001. Rapid discrimination among individual DNA hairpin molecules at single-nucleotide resolution using an ion channel. *Nat Biotechnol* 19(3):248–52.

79. Mathe, J., H. Visram, V. Viasnoff, Y. Rabin, and A. Meller. 2004. Nanopore unzipping of individual DNA hairpin molecules. *Biophys J* 87(5):3205–12.

80. Meller, A., J. Mathe, V. Viasnoff, and Y. Rabin. 2004. Nanopore unzipping of individual DNA hairpin molecules. *Abstr Pap Am Chem Soc* 228:U301.

81. Healy, K. 2007. Nanopore-based single-molecule DNA analysis. *Nanomedicine* 2(4):459–81.

82. Fologea, D., J. Uplinger, B. Thomas, D. S. McNabb, and J. L. Li. 2005. Slowing DNA translocation in a solid-state nanopore. *Nano Lett* 5(9):1734–7.

83. Siwy, Z., L. Trofin, P. Kohli, L. A. Baker, C. Trautmann, and C. R. Martin. 2005. Protein biosensors based on biofunctionalized conical gold nanotubes. *J Am Chem Soc* 127(14):5000–1.

84. Kohli, P., C. C. Harrell, Z. H. Cao, R. Gasparac, W. H. Tan, and C. R. Martin. 2004. DNA-functionalized nanotube membranes with single-base mismatch selectivity. *Science* 305(5686):984–6.

85. Wanunu, M., J. Sutin, and A. Meller. 2009. DNA profiling using solid-state nanopores: Detection of DNA-binding molecules. *Nano Lett* 9(10):3498–502.

86. McQuarrie, D. A. 2000. *Statistical Mechanics*. Sausalito, CA, USA: University Science Books.

87. Rapaport, D. C. 1996. *The Art of Molecular Dynamics Simulation*. Cambridge, UK: Cambridge University Press.

88. Allen, M. P., and D. J. Tildesley. 1989. *Computer Simulation of Liquids*. Cambridge, UK: Oxford University Press.

89. Field, M. J. 2007. *A Practical Introduction to the Simulation of Molecular Systems*. Cambridge, UK: Cambridge University Press.

90. Lucas, K. 2007. *Molecular Models for Fluids*. Cambridge, UK: Cambridge University Press.

91. Haile, J. M. 1997. *Molecular Dynamics Simulation: Elementary Methods*. Weinheim, Germany: Wiley Professional.

92. Lakhtakia, A. 2004. *Handbook of Nanotechnology: Nanometer Structure Theory, Modeling, and Simulation*. New York, NY, USA: SPIE Publications.

93. Mark, J. E. 2007. *Physical Properties of Polymers Handbook*. Springer.

94. Tersoff, J. 1989. Modeling solid-state chemistry: Interatomic potentials for multicomponent systems. *Phys Rev B* 39(8):5566–8.

95. Boda, D., and D. Henderson. 2008. The effects of deviations from Lorentz-Berthelot rules on the properties of a simple mixture. *Mol Phys* 106(20):2367–70.

96. MacKerell, A. D., D. Bashford, M. Bellott, R. L. Dunbrack, J. D. Evanseck, M. J. Field, S. Fischer et al. 1998. All-atom empirical potential for molecular modeling and dynamics studies of proteins. *J Phys Chem B* 102(18):3586–616.

97. Ponder, J. W., and D. A. Case. 2003. Force fields for protein simulations. *Protein Simul* 66:27–+.

98. Cornell, W. D., P. Cieplak, C. I. Bayly, I. R. Gould, K. M. Merz, D. M. Ferguson, D. C. Spellmeyer, T. Fox, J. W. Caldwell, and P. A. Kollman. 1996. A second generation force field for the simulation of proteins, nucleic acids, and organic molecules (vol 117, pg 5179, 1995). *J Am Chem Soc* 118(9):2309–9.

99. Haile, J. M. 1992. *Molecular Dynamics Simulation: Elementary Methods*. Weinheim, Germany: Wiley Professional.

100. Rapaport, D. C. 2004. *The Art of Molecular Dynamics Simulation*. Cambridge, United Kingdam: Cambridge University Press.

101. Kosztin, D., R. I. Gumport, and K. Schulten. 1999. Probing the role of structural water in a duplex oligodeoxyribonucleotide containing a water-mimicking base analog. *Nucleic Acids Res* 27(17):3550–6.

102. Sung, W., and P. J. Park. 1996. Polymer translocation through a pore in a membrane. *Phys Rev Lett* 77(4):783–6.

103. Lubensky, D. K., and D. R. Nelson. 1999. Driven polymer translocation through a narrow pore. *Biophys J* 77(4):1824–38.

104. Gnanakaran, S., and R. M. Hochstrasser. 1996. Vibrational relaxation of HgI in ethanol: Equilibrium molecular dynamics simulations. *J Chem Phys* 105(9):3486–96.

105. Hofsass, C., E. Lindahl, and O. Edholm. 2003. Molecular dynamics simulations of phospholipid bilayers with cholesterol. *Biophys J* 84(4):2192–206.

106. Amisaki, T., T. Hakoshima, K. Tomita, S. Nishiawa, S. Uesugi, E. Ohtsuka, M. Ikehara, S. Yoneda, and K. Kitamura. 1992. Molecular-dynamics and free-energy perturbation calculations on the mutation of tyrosine-45 to tryptophan in ribonuclease-T-1. *Chem Pharm Bull* 40(5):1303–8.

107. Axelsen, P. H., and F. G. Prendergast. 1989. Molecular-dynamics of tryptophan in ribonuclease-T1. 2. Correlations with fluorescence. *Biophys J* 56(1):43–66.

108. Sands, Z. A., and C. A. Laughton. 2004. Molecular dynamics simulations of DNA using the generalized born solvation model: Quantitative comparisons with explicit solvation results. *J Phys Chem B* 108(28):10113–9.

109. Stefl, R., N. Spackova, I. Berger, J. Koca, and J. Sponer. 2001. Molecular dynamics of DNA quadruplex molecules containing inosine, 6-thioguanine and 6-thiopurine. *Biophys J* 80(1):455–68.

110. Cross, J. D., E. A. Strychalski, and H. G. Craighead. 2007. Size-dependent DNA mobility in nanochannels. *J Appl Phys* 102(2):024701–5.

111. Zhao, X. C., C. M. Payne, P. T. Cummings, and J. W. Lee. 2007. Single-strand DNA molecule translocation through nanoelectrode gaps. *Nanotechnology* 18(42):424018–24.

112. Wanunu, M., J. Sutin, B. McNally, A. Chow, and A. Meller. 2008. DNA translocation governed by interactions with solid-state nanopores. *Biophys J* 95(10):4716–25.

113. Chen, P., J. J. Gu, E. Brandin, Y. R. Kim, Q. Wang, and D. Branton. 2004. Probing single DNA molecule transport using fabricated nanopores. *Nano Lett* 4(11):2293–8.

114. Fyta, M. G., S. Melchionna, E. Kaxiras, and S. Succi. 2006. Multiscale coupling of molecular dynamics and hydrodynamics: Application to DNA translocation through a nanopore. *Multiscale Model Simul* 5(4):1156–73.

115. Sauer-Budge, A. F., J. A. Nyamwanda, D. K. Lubensky, and D. Branton. 2003. Unzipping kinetics of double-stranded DNA in a nanopore. *Phys Rev Lett* 90(23):238101.

116. Apel, P. Y., Y. E. Korchev, Z. Siwy, R. Spohr, and M. Yoshida. 2001. Diode-like single-ion track membrane prepared by electro-stopping. *Nucl Instrum Methods Phys Res B* 184(3):337–46.

117. Clauset, A., C. R. Shalizi, and M. E. J. Newman. 2009. Power-law distributions in empirical data. *SIAM Rev* 51(4):661.

118. Mannion, J. T., C. H. Reccius, J. D. Cross, and H. G. Craighead. 2006. Conformational analysis of single DNA molecules undergoing entropically induced motion in nanochannels. *Biophys J* 90(12):4538–45.

119. Fologea, D., E. Brandin, J. Uplinger, D. Branton, and J. Li. 2007. DNA conformation and base number simultaneously determined in a nanopore. *Electrophoresis* 28(18):3186–92.

120. Kasianowicz, J. J., T. L. Nguyen, and V. M. Stanford. 2006. Enhancing molecular flux through nanopores by means of attractive interactions. *Proc Natl Acad Sci U S A* 103(31):11431–2.

121. Hogg, T., M. Zhang, and R. Yang. 2008. Modeling and analysis of DNA hybridization dynamics at microarray surface in moving fluid. *IEEE Int Conf Rob Autom* 3419–24.

122. Chan, V., D. J. Graves, and S. E. McKenzie. 1995. The biophysics of DNA hybridization with immobilized oligonucleotide probes. *Biophys J* 69(6):2243–55.

123. Hofler, L., and R. E. Gyurcsanyi. 2008. Coarse grained molecular dynamics simulation of electromechanically-gated DNA modified conical nanopores. *Electroanalysis* 20(3):301–7.

124. Kale, L., R. Skeel, M. Bhandarkar, R. Brunner, A. Gursoy, N. Krawetz, J. Phillips, A. Shinozaki, K. Varadarajan, and K. Schulten. 1999. NAMD2: Greater scalability for parallel molecular dynamics. *J Comput Phys* 151(1):283–312.

125. Humphrey, W., A. Dalke, and K. Schulten. 1996. VMD: Visual molecular dynamics. *J Mol Graphics* 14(1):33–8.

126. Research Collaboratory for Structural Bioinformatics (RCSB) protein data bank. http://beta.rcsb.org/pdb/home/home.do

127. Catlett, C., W. E. Allcock, P. Andrews, R. Aydt, R. Bair, N. Balac, B. Banister, T. Barker, M. Bartelt, and P. E. A. Beckman. 2008. TeraGrid: Analysis of organization, system architecture, and middleware enabling new types of applications. HPC and grids in action. In *High Performance Computing and Grids in Action* 16, ed. L. Grandinetti. Amsterdam: IOS Press.

128. Muller, M., K. Katsov, and M. Schick. 2003. Coarse-grained models and collective phenomena in membranes: Computer simulation of membrane fusion. *J Polym Sci B Polym Phys* 41(13):1441–50.

129. Nielsen, S. O., C. F. Lopez, G. Srinivas, and M. L. Klein. 2004. Coarse grain models and the computer simulation of soft materials. *J Phys Condens Matter* 16(15):R481–512.

130. Marrink, S. J., A. H. de Vries, and A. E. Mark. 2004. Coarse grained model for semi-quantitative lipid simulations. *J Phys Chem B* 108(2):750–60.

131. Simons, K., and E. Ikonen. 1997. Functional rafts in cell membranes. *Nature* 387(6633):569–72.

132. Shelley, J. C., M. Y. Shelley, R. C. Reeder, S. Bandyopadhyay, P. B. Moore, and M. L. Klein. 2001. Simulations of phospholipids using a coarse grain model. *J Phys Chem B* 105(40):9785–92.

133. Sales-Pardo, M., R. Guimera, A. A. Moreira, J. Widom, and L. A. N. Amaral. 2005. Mesoscopic modeling for nucleic acid chain dynamics. *Phys Rev E* 71(5):051902.

134. Voth, G. A. 2008. *Coarse-Graining of Condensed Phase and Biomolecular Systems.* Boca Raton, FL, USA: CRC Press.

135. Noid, W. G., P. Liu, Y. Wang, J. W. Chu, G. S. Ayton, S. Izvekov, H. C. Andersen, and G. A. Voth. 2008. The multiscale coarse-graining method. II. Numerical implementation for coarse-grained molecular models. *J Chem Phys* 128(24).

136. Shih, A. Y., A. Arkhipov, P. L. Freddolino, and K. Schulten. 2006. Coarse grained protein-lipid model with application to lipoprotein particles. *J Phys Chem B* 110(8):3674–84.

137. Shih, A. Y., A. Arkhipov, P. L. Freddolino, S. G. Sligar, and K. Schulten. 2007. Assembly of lipids and proteins into lipoprotein particles. *J Phys Chem B* 111(38):11095–104.

138. Shih, A. Y., P. L. Freddolino, A. Arkhipov, and K. Schulten. 2007. Assembly of lipoprotein particles revealed by coarse-grained molecular dynamics simulations. *J Struct Biol* 157(3):579–92.

139. Arkhipov, A., P. L. Freddolino, and K. Schulten. 2006. Stability and dynamics of virus capsids described by coarse-grained modeling. *Structure* 14(12):1767–77.

140. Arkhipov, A. S., P. L. Freddolino, and K. Schulten. 2007. Stability and dynamics of virus capsids described by coarse-grained modeling. *Structure* 14(12):1767–77.

141. Arkhipov, A., P. L. Freddolino, K. Imada, K. Namba, and K. Schulten. 2006. Coarse-grained molecular dynamics simulations of a rotating bacterial flagellum. *Biophys J* 91(12):4589–97.

142. Yin, Y., A. Arkhipov, and K. Schulten. 2009. Simulations of membrane tubulation by lattices of amphiphysin N-BAR domains. *Structure* 17(6):882–92.

143. Stevens, M. J., J. H. Hoh, and T. B. Woolf. 2003. Insights into the molecular mechanism of membrane fusion from simulation: Evidence for the association of splayed tails. *Phys Rev Lett* 91(18):188102.

144. Izvekov, S., A. Violi, and G. A. Voth. 2005. Systematic coarse-graining of nanoparticle interactions in molecular dynamics simulation. *J Phys Chem B* 109(36):17019–24.

145. Izvekov, S., and G. A. Voth. 2005. A multiscale coarse-graining method for biomolecular systems. *J Phys Chem B* 109(7):2469–73.

146. Hansen, J. P., and I. R. McDonald. 1986. *Theory of Simple Liquids.* 3rd ed. Amsterdam, The Netherland: Elsevier.

147. Liu, P., S. Izvekov, and G. A. Voth. 2007. Multiscale coarse-graining of monosaccharides. *J Phys Chem B* 111(39):11566–75.

148. Wang, J. Y., Y. Q. Deng, and B. Roux. 2006. Absolute binding free energy calculations using molecular dynamics simulations with restraining potentials. *Biophys J* 91(8):2798–814.

149. Izvekov, S., and G. A. Voth. 2006. Multiscale coarse-graining of mixed phospholipid/cholesterol bilayers. *J Chem Theory Comput* 2(3):637–48.

150. Liu, P., and G. A. Voth. 2007. Smart resolution replica exchange: An efficient algorithm for exploring complex energy landscapes. *J Chem Phys* 126(4):045106.

151. Marrink, S. J., H. J. Risselada, S. Yefimov, D. P. Tieleman, and A. H. de Vries. 2007. The MARTINI force field: Coarse grained model for biomolecular simulations. *J Phys Chem B* 111(27):7812–24.

152. Marrink, S. J., J. Risselada, and A. E. Mark. 2005. Simulation of gel phase formation and melting in lipid bilayers using a coarse grained model. *Chem Phys Lipids* 135(2):223–44.

153. Marrink, S. J., and A. E. Mark. 2004. Molecular view of hexagonal phase formation in phospholipid membranes. *Biophys J* 87(6):3894–900.
154. Knotts, T. A., N. Rathore, D. C. Schwartz, and J. J. de Pablo. 2007. A coarse grain model for DNA. *J Chem Phys* 126(8):084901.
155. Schatz, G. C., and K. Drukker. 2000. Model for simulating dynamics of DNA denaturation. *Abstr Pap Am Chem Soc* 220:U176–7.
156. Verlet, L. 1967. Computer "experiments" on classical fluids. I. Thermodynamic properties of lennard–jones molecules. *Phys Rev* 159(1):98.
157. Berendsen, H. J. C., D. Vanderspoel, and R. Vandrunen. 1995. Gromacs. A message-passing parallel molecular dynamics implementation. *Comput Phys Commun* 91(1–3):43–56.
158. Essmann, U., L. Perera, M. L. Berkowitz, T. Darden, H. Lee, and L. G. Pedersen. 1995. A smooth particle mesh ewald method. *J Chem Phys* 103(19):8577–93.
159. Ousterhout, J. K. 1994. *Tcl and the Tk Toolkit*. Boston, MA, USA: Addison-Wesley Professional.

7 Mechanical Characterization of Protein Materials

Kilho Eom

CONTENTS

7.1 INTRODUCTION

Several protein materials, including spider silk, [1–4] have recently been highlighted for their remarkable mechanical properties such as high stiffness, high toughness, and superelasticity, albeit they are composed of weak constituents flexible protein domains. One may raise the following question: What makes protein materials that are composed of weak constituents tougher than any other man-made composite? The fundamental insights that can resolve such a question will lead us to establish physical principles, which provide how weak proteins can achieve excellent mechanical properties suitable for performing mechanical functions. For instance, spider silk protein performs notable mechanical functions due to its remarkable mechanical properties such as superelasticity and high toughness. In particular, spider silk has been found to have unique mechanical properties: it possesses both high extensibility and high fracture toughness, which cannot be achieved with engineered materials. Specifically, dragline silk has the elastic modulus of 10 GPa, implying that it is a soft material, while the yield strength is equal to 1.1. GPa, comparable to that of high-tensile steel, and the fracture toughness is equal to 160 MJ/m^3, which is two times larger than that of a strong composite material such as Kevlar [2]. Moreover, spider silk is able to perform its biological functions through its remarkable mechanical properties. The ability of spider silk to capture flying prey is attributed to the superelasticity and high extensibility of spider silk, which enable the conversion of the kinetic energy of the flying prey into heat dissipation, resulting in the capture of the flying prey in the spider silk. This sheds light on the fundamental understanding of the mechanical properties and behaviors of protein materials, which can help in gaining deep insights into the biological functions of protein materials and provide materials scientists with a novel design concept to develop biomimetic materials [3,4].

The mechanical behavior of protein material is closely related to disease expressions. For instance, for patients with Alzheimer's disease, mad cow disease, and so forth, a unique protein fibril was found to be a signature of such diseases [5–7]. In particular, the expression of this protein fibril is attributed to the undesirable aggregation of protein domains, such as β sheet. This assembled fibril, such as islet amyloid polypeptide (IAPP) fibril [8], can replace the specific cells (e.g., insulin-secreting β-cell) in the functional sites (e.g., islet of pancreas), which results in the specific disease expression (e.g., type II diabetes). This indicates that it is essential to understand how protein domains are clustered and aggregated so as to be formed as fibril [7]. Moreover, because it is hypothesized that the replacement of specific cells due to the presence of undesired protein fibrils may be related to the mechanical stability of protein fibrils [9], it is necessary to study the mechanical behaviors of protein fibrils. A recent study [10] has found that a variety of amyloid fibrils that are formed due to assembly of denatured protein possess remarkable mechanical properties such as a bending stiffness that is higher than any other protein materials (see Figure 7.1). This indicates that the mechanical properties of amyloid fibril play a critical role in its biological functions. Moreover, it is of interest to note that the aggregation of weak proteins result in excellent mechanical properties of protein materials (e.g., fibril) composed of weak proteins. The fundamentals of how protein materials are able to achieve their excellent mechanical properties through self-assembly and/or aggregation will

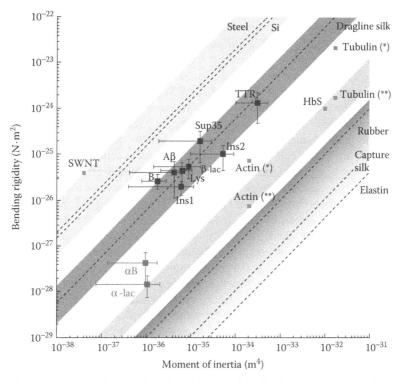

FIGURE 7.1 Mechanical properties of protein materials. Dark region indicates the mechanical properties of various amyloid fibrils, whereas the light-colored regime represents the mechanical properties of other protein materials (e.g., actin, tubulin). Here, * and ** indicates the mechanical properties of actin filaments reported in literatures. The dark, lower region shows the mechanical properties of polymeric materials such as rubber and/or elastin. Here, * and ** indicates the bending rigidities reported in other literatures. The amyloid fibrils possess a larger bending rigidity than that of other protein materials and polymeric materials when the cross-sectional moment of inertia is fixed at a constant value. (Adapted from Knowles, T. P., A. W. Fitzpatrick, S. Meehan, H. R. Mott, M. Vendruscolo, C. M. Dobson, and M. E. Welland. 2007. *Science* 318:1900. With permission.)

provide a design concept for material scientists to construct novel material that can exhibit desired mechanical properties through the assembly of weak constituents [9]. Recent experimental [10] and computational [11] studies have demonstrated that the remarkable mechanical properties of amyloid fibrils are attributed to the intermolecular forces between ladder-shaped β-sheet crystals. Specifically, the hydrogen bonds between the ladder-shaped β-sheet crystals serve as a chemical glue that enhances the mechanical stability of fibril. Moreover, a recent simulation showed that the geometric confinements of β sheets enable the improvement of mechanical properties of protein material composed of β sheets, which is attributed to the hydrogen bonds that act as a glue between β sheets. This suggests that molecular architecture is closely related to the mechanical properties of protein material, which implies that the mechanical properties of protein material are encoded in their molecular architectures.

Mechanical characterization of protein material has been perceived as a novel route to the early diagnosis of diseases such as cancer [12–14]. For instance, a recent experimental study [15] based on the atomic force microscope (AFM) nanoindentation of cartilages obtained from patients with degenerative joint disease, shows that the cartilage tissue of patients with osteoarthritis is more flexible than that of those who do not have osteoarthritis. Recently, AFM nanoindentation has been used to characterize the mechanical properties of viral capsid [16,17], which are dependent on the genetic codes of viral capsid. Moreover, in a recent experimental work [14], a cancerous cell is found to have a coarser cytoskeleton network than a normal cell. This makes the cancer cell more flexible than a normal cell, because the mechanical stiffness of a cell is determined by the cytoskeleton network, which sustains the cell shape. These findings suggest that mechanical characterization of cellular or subcellular materials that consist of proteins can be regarded as a de novo toolkit that enables the early diagnosis of specific diseases. The principle for a mechanics-based diagnosis is ascribed to the hypothesis that the mechanical properties of cellular or subcellular materials are determined by hierarchical structures (i.e., the structure of assembled proteins) that encode the genetic information. This urges us to investigate the structure–property–function relationship for protein materials at various spatial scales ranging from single-molecule resolution to subcellular or cellular level.

Although AFM nanoindentation allows the mechanical characterization of protein materials at various spatial or length scales, it is quite restrictive for gaining the fundamental insights into the structure–property–function relationship of protein material. One of the possible routes to reveal such structure–property–function relation is the computational simulation–based analysis on the mechanical behavior of protein material [18–20]. Specifically, computational simulations based on the atomistic model have enhanced the fundamental insights into the mechanical behavior of proteins [19–21]. For instance, the mechanical unfolding behaviors of various proteins such as muscle protein titin domain [22,23], tubulin dimmer [24], ankyrin [21], and ubiquitin [25] have been well depicted by molecular dynamic (MD) simulations. Recently, Buehler et al. [11] studied the mechanical properties of protein crystals made of β sheets using MD simulations revealing that geometric confinement of β sheet results in the enhancement of the mechanical properties of proteins, which is attributed to the hydrogen bonds that act as a chemical reinforcement (glue) between β sheets. It has been implied that MD simulation may be appropriate for studying the mechanical properties of proteins with respect to their molecular architectures.

Despite its ability to unveil the structure–property–function relationship for proteins, MD simulations are computationally limited to macromolecular structures whose spatial and temporal scales are much larger than those accessible with MD simulations [26]. To resolve the computational restrictions encountered in MD simulations, there have been current attempts [27–33] to develop the coarse-grained (CG) model that enables MD-like simulations of biological macromolecules. The key concept of coarse graining is to reduce the degrees of freedom for the molecular system such that some atoms that are insignificant to describe the dynamics of the molecular system are not taken into account in the CG model. In addition, in the CG model, the complicated potential field prescribed for a molecular system is also simplified as long as such simplified potential is acceptable to depict the dynamics of a molecular system. The details of coarse graining are well summarized in references [27–31].

In this chapter, we briefly review the currently suggested CG models that are able to describe the molecular structure of protein material to gain fundamental insights into the mechanical properties of protein materials. The remainder of the chapter is organized as follows: Section 7.2 reviews the recent advances in atomistic simulations and/or CG model-based simulations of protein materials for studying the role of hierarchy in the mechanical properties of protein materials; Section 7.3 overviews the theoretical model–based studies on the mechanical properties of folded chains to fundamentally study the relationship between folding configuration and the mechanical properties of proteins; Section 7.4 provides the CG model that enables us to gain insight into the mechanical behavior of protein materials in response to applied mechanical strain. It is shown that the CG model of protein materials predicts the elastic properties of protein materials, which are quantitatively comparable to those measured from experiments and/or atomistic simulations; Section 7.5 describes the CG model of amyloid fibril for quantitative understanding of the elastic properties of aggregated amyloid fibrils. It is found that when amyloid fibril exhibits the length of >50 nm, this amyloid fibril becomes mechanically stable. This shows the significance of the length scales of aggregated protein fibril in the mechanical stability of amyloid, which is related to disease expression; and Section 7.6 briefly summarizes the outlook and perspectives.

7.2 SIMULATIONS ON PROTEIN MATERIALS: ROLE OF HIERARCHY

As discussed earlier, some proteins perform excellent mechanical functions and have properties that are much better than expected and/or any other synthetic materials such as composites. To understand why protein materials exhibit remarkable mechanical properties, we should consider the structures of protein materials at various spatial scales ranging from microscopic to macroscopic levels. For instance, to gain insights into why spider silk proteins possess the notable mechanical characteristics such as superelasticity and high fracture toughness, Termonia [34] suggested a theoretical model based on the structure of spider silk, which consist of two phases: (1) β-sheet crystal that is responsible for high fracture toughness, and (2) amorphous matrix that contributes to superelasticity. Despite its ability to capture the mechanical response of spider silk, the phenomenological model by Termonia [34] still lacks the physics on the microscopic properties of the two phases. In particular, a recent study [35] reported that unlike the hypothesis of Termonia, spider silk is composed of β-sheet crystals and well-ordered helices (rather than a random molecular chain resembling the amorphous matrix). Furthermore, a phenomenological model requires a priori information of the elastic properties of two phases—β-sheet crystal and matrix, as a fitting parameter in order for the force-extension curve obtained from the model to be matched with that from experiment. This indicates that the phenomenological model fails to explain why β-sheet crystal exhibits high stiffness although β-sheet crystal is made of weak protein domains. It has been implied that in order to comprehensively understand the mechanical properties of spider silk, we should develop the computational model whose spatial scales range from microscopic to macroscopic scales. Recently, Buehler et al. [36] have considered the CG MD simulations, which allow simulations at large spatial and temporal scales that are inaccessible with conventional MD simulations, to understand the mechanical properties of spider silk. First, they have studied the mechanical properties of β-sheet crystals with their various length scales using atomistic simulations. They have found that geometric confinement of weak protein domains lead to enhancement of mechanical properties, which is attributed to the hydrogen bonds between the weak proteins [37]. In other words, the hydrogen bonds serve as a chemical glue between the protein domains, which result in an increase in the stiffness of the protein materials. This shows the importance of chemical bonds between the protein domains at atomistic scales in the mechanical response of a protein material, at mesoscopic scale, composed of protein domains. They have also considered the theoretical model of dragline spider silk protein such that β-sheet crystals are connected by α helices at atomic scales based on the CG model [38]. Their simulations remarkably show the mechanical response of the model, which resembles the experimentally measured force-extension curve of spider silk. This suggests that the macroscopic mechanical properties of protein materials are highly correlated with the microscopic properties of each constituent such as protein domains.

Another example that shows the significance of understanding the mechanical response of protein materials at various length scales is the formation of plaques that are involved in disease expressions [9]. Specifically, the plaques are formed due to the cross-linking of protofilaments, which are composed of cross-β structures that are made of β-sheet crystals [5–7]. This implies that the mechanical properties of ectopic material related to disease expressions can be fundamentally understood from the physical insights into the mechanical response of the constituents at various length scales ranging from atomistic scales (e.g., β-sheet crystal) to macroscopic scales (e.g., protofilament). Until recently, there have been few attempts to investigate the origin of the mechanical properties of ectopic materials based on the simulations of each constituent at multiple spatial scales [9,18]. Recently, there has been an attempt [39] to theoretically anticipate the mechanical properties of amyloid fibril at short length scales based on the CG model such as the elastic network model (ENM) [40–43]. As elucidated in Section 7.5, the mechanical stability of amyloid fibril is highly correlated with its length scale, implying that the mechanical properties of amyloid fibrils are determined by the molecular architecture of β sheets that are constituents of amyloids.

The above-mentioned examples suggest that the comprehensive understanding of the mechanical properties of biological materials can be made possible by considering the theoretical models and/or simulations at various length scales ranging from microscopic to macroscopic scales. This sheds light on the multiscale simulations and/or CG model–based simulations for understanding the mechanical properties of biological materials in a hierarchical manner. Beyond these two aforementioned examples, there are plenty of examples (e.g., see [9,18,19,44]) that show the important role of hierarchical architecture in the mechanical behavior of biological materials.

7.3 THEORETICAL MODEL OF FOLDED CHAINS: ROLE OF FOLDING TOPOLOGY

Recent studies [25,45,46] have reported that the mechanical properties, that is, mechanical stability, of proteins are highly correlated with their native topology. Specifically, unlike the polymer chain, the protein structure is unique in that the chain is in a folded configuration. The folding topology for the protein structure is maintained by hydrogen bonds between specific residues under physiological conditions. When a mechanical force is applied to the protein structure, the force is transferred from not only the covalent bond but also the nonbonded interactions such as hydrogen bonds. As the force applied to the protein structure increases, the hydrogen bonds share the mechanical force, and at a certain amount of force, the hydrogen bonds start to break, which leads to the unfolding of the folded topology [23]. Recent studies [4,47–50] have shown that the unique mechanical response of a protein is attributed to the unfolding due to mechanical force. In particular, the breakage of hydrogen bonds are well reflected in the saw-tooth-like force-extension curve, in which each peak corresponds to unfolding due to the breakage of hydrogen bonds, which can be experimentally obtained from single-molecule force spectroscopy. This suggests that to gain the physical understanding of the mechanical behavior of a single protein, it is essential to investigate the role of folding topology (i.e., geometric configurations of hydrogen bonds) in the mechanical properties of a single protein.

In this section, we provide the theoretical model of folded protein domains such that a polymer chain is folded and the folded configuration is maintained by hydrogen bonds. We have extensively studied the possible optimal configurations of hydrogen bonds that can maximize the mechanical resistance of a folded chain using an optimization process based on a random search. It is shown that the optimal configuration of hydrogen bonds resembles the parallel strands that are found in mechanically stable and strong proteins, which highlights the significance of the architecture (assembly) of hydrogen bonds in the unique mechanical response of a single protein.

7.3.1 FOLDED CHAIN MODEL

Here, we would like to review the theoretical model of a cross-linked polymer chain, which was developed by Eom et al. [51,52]. We consider a polymer chain that consists of $(N + 1)$ beads that are connected into a chain by N links. We have assumed that the motion of a chain can be described by Gaussian statistics such that the probability distribution for the distance between two monomers i and j is given by [53]

$$P(r_{ij}) = \left(\frac{k}{2\pi k_B T |i - j|} \right)^{3/2} \exp\left[-\frac{k r_{ij}^2}{2 k_B T |i - j|} \right] \tag{7.1}$$

Here, $P(r_{ij})$ is the probability distribution for the distance between two monomers i and j with $r_{ij} = |\mathbf{r}_i - \mathbf{r}_j|$, where $\mathbf{r}_i$ is the position vector of a bead i, k_B is Boltzmann's constant, T is the absolute temperature, and k is the force constant that is related to the mean-square fluctuation for the length of a link, that is, $k = 3k_B T / <|\mathbf{r}_i - \mathbf{r}_j|^2>$, where the angle bracket < > indicates the ensemble average. One possible route to construct a Gaussian polymer chain is to connect the consecutive beads with a harmonic spring with a force constant of k.

To mimic the folding topology of a protein, we have taken into account the cross-links that connect some of the beads. The cross-link that connects the pair of beads is formed in such a way that the pair of beads are connected by a harmonic spring whose force constant is much larger than k (i.e., force constant for Gaussian chain monomer), which is equivalent to zero equilibrium length, that is, $\langle |\mathbf{r}_i - \mathbf{r}_j|^2 \rangle \to 0$. The folding topology is defined as the cross-link list: $C_n = [(i_1, j_1), (i_2, j_2), \dots, (i_n, j_n)]$ (see Figure 7.2).

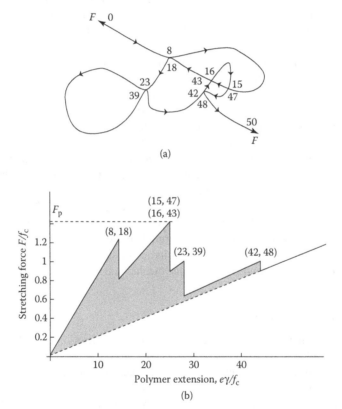

FIGURE 7.2 Simple toy model (coarse-grained elastic model) of cross-linked polymer chain and its mechanical response. (a) shows the cross-linked polymer chain composed of 50 monomers (51 beads), where cross-links are given as [(8, 18), (15, 47), (16, 43), (23, 39), (42, 48)]. The mechanical response of a cross-linked polymer chain is unique in that there are force drops in the force-extension curve [see (b)], which is attributed to the breakage of cross-links. The peak force in the force-extension curve is the force required to break a specific cross-link. The maximum among peak forces is defined as "unfolding force." Moreover, the dissipated energy (excess work) during the mechanical unfolding of cross-links is shown as a shaded region in the force-extension curve. (Adapted from Eom, K., P. C. Li, D. E. Makarov, and G. J. Rodin. 2003. *J Phys Chem B* 107:8730. With permission.)

Now, let us consider the mechanical response of a folded polymer chain to the mechanical extension such that the ends of the folded polymer chain are moved apart from the equilibrium slowly enough that the timescale relevant to pulling is much longer than that of the thermal fluctuations of the polymer chain [51]. In other words, the mechanical strength of the polymer chain is a quasiequilibrium process, which implies that the mechanical response of the polymer to the mechanical stretching is independent of the pulling speed. Since the folded protein domain is denatured upon mechanical stretching, we assume that the mechanical behavior of the folded domain is determined by not only the effective force constant of the polymer chain but also the rupture of cross-links (which resembles the native contact in the protein domain). The bond rupture mechanism is assumed such that a cross-link is denatured when the force it transmits reaches a critical value f_c. When the force the cross-links transfers is below f_c, the force generated by the polymer chain, $F(u)$, is linearly proportional to the extension u due to the assumption that the linear elasticity of the Gaussian polymer chain obeys Hooke's law. As the extension u increases, the bonds (i.e., cross-links) begin to rupture, and as a result, the force-extension curve, that is, $F(u)$, becomes the linearly piecewise linear function of extension u (see Figure 7.2). The peak force observed in the force-extension curve corresponds to the rupture of a single or multiple cross-links, which leads to the abrupt decrease in the force that the polymer chain can bear. Once all cross-links are ruptured, the force-extension curve for the denatured polymer can be described as the slope (i.e., force constant) that is equal to k/N. To characterize the mechanical properties of the folded polymer chain in response to mechanical stretching, we have defined two measurable quantities: (1) excess work W, that is, the dissipated energy due to the rupture of cross-links during the mechanical strength of the polymer chain

$$W = \int_0^{u_{max}} F(v)\,dv - \frac{ku_{max}^2}{2N} \tag{7.2}$$

where u_{max} is the maximally allowable mechanical extension. Here, u_{max} must exceed the mechanical extension at which all cross-links are ruptured; and (2) The second quantity is the unfolding force F_u, which corresponds to the maximum among the peak forces that correspond to the rupture of cross-links. These two quantities W and F_u are the measures of the mechanical resistance of the folded polymer chain to the mechanical stretching. Our aim is to find the folding topology, that is, cross-link configurations C_n, which maximizes these two measures. In other words, for given parameters k, f_c, N, and n, we seek to search the cross-link configuration C_n that maximizes either W or F_u.

Because the mechanical unfolding of a protein domain is generally dependent on the pulling rate [54–57], we now discuss the case in which the folded polymer is stretched with the pulling rate [52]. The pulling rate effect on the mechanical response of a folded protein domain is attributed to the principle that the timescale relevant to mechanical stretching accessible with conventional force spectroscopy such as AFM is comparable to or shorter than that for the thermal fluctuation of a protein. That is, the rupture of native contacts is a stochastic process such that the rupture event is determined by not only the mechanical strength of a native contact but also the competition between timescales relevant to bond rupture and thermal fluctuation of a native contact, respectively. This stochastic nature of bond rupture results in the mechanical response of the folded polymer chain depending on the stochastic bond rupture processes that are correlated with the rate of the mechanical stretching of the polymer chain.

We consider the folded polymer chain whose ends are pulled apart with pulling rate v such that the distance between the ends is given by $u = |\mathbf{r}_N - \mathbf{r}_0| = vt$. We presume that the timescale for the mechanical stretching is slow compared with that for the thermal Brownian motion of the chain. For this case, the value of pulling force $F(t)$ recorded at time t is the force averaged over the thermal motion of the chain. In addition, the timescale for the bond rupture is quantitatively comparable to that of the loading rate so that the bond rupture is reflected into the measurable change in $F(t)$. Specifically, the bond rupture process is described by Bell's theory [58–60], which provides that the probability to observe the rupture of a bond that transmits the force f at the time interval from t to $t + \Delta t$ is given by

$$k[f(t)] \cdot \Delta t = k_0 \exp[f/f_c] \cdot \Delta t \qquad (7.3)$$

One may recall that the probability distribution given by Equation 7.3 is identical with the Arrhenius equation that describes the stochastic nature of a chemical reaction. In other words, the bond rupture process is analogous to the chemical reaction from bonded state to denatured state, a process that operates in a stochastic manner. Here, $k[f(t)]$ is the kinetic rate for the bond rupture process as a function of the force $f(t)$ that a bond transmits, k_0 is the kinetic (reaction) rate for the zero force, and f_c is the critical force. In the viewpoint of Arrhenius-type descriptions, the parameters such as k_0 and f_c are defined as follows [59–62]:

$$k_0 = \frac{D\omega_b\omega_{ts}}{2\pi k_B T} \exp\left(-\frac{\Delta U}{k_B T}\right) \qquad (7.4a)$$

$$f_c = \frac{k_B T}{\Delta x} \qquad (7.4b)$$

where D is the diffusion coefficient of a bond, ΔU is the energy barrier that must be overcome during the bond rupture, ω_b and ω_{ts} represent the angular frequencies for a bond in the bonded configuration and the transition state at which bond rupture occurs, respectively, and Δx is the difference between the reaction coordinates, one of which corresponds to the reaction coordinate at bonded configuration, while the other indicates the reaction coordinate at the transition state. In this work, it is supposed that the bond rupture process happens irreversibly with the timescale of k_0^{-1} even under the zero force. In other words, we assume that the reversible formation of a cross-link is impossible. When the ends of a folded polymer chain are pulled apart with the specified pulling rate, the force-extension curve, that is, $F(u)$, is linearly proportional to the mechanical extension, that is, $u = vt$, before the cross-links undergo the rupture process. When a cross-link is ruptured based on the assumption of the stochastic bond rupture process depicted in Equation 7.3, this rupture event reduces the force that the polymer chain bears, which can be reflected in the abrupt decrease of $F(t)$ in the force-extension curve. The force-extension curve is a piecewise linear function of extension, where the peak in the force-extension curve is a signature of the rupture of cross-links. The folded polymer chain is mechanically stretched with the specified pulling rate until all cross-links are mechanically denatured. The mechanical resistance of the folded polymer chain under the mechanical stretching with the specified rate is defined in terms of the following two measures: (1) an excess work, and (2) an unfolding force. We are aiming at searching the optimal configuration of cross-links that can maximize these two measures for the mechanical resistance of the folded polymer chain, provided the parameters such as k, N, n, k_0, and Δx (or equivalently to f_c) are specified.

7.3.2 QUASISTATIC SIMULATIONS: OPTIMAL CROSS-LINK CONFIGURATIONS

We have analytically found that for $n = 1$ (i.e., a single cross-link) the optimal configuration for the folded polymer chain is given by $C_1 = (i, i + N/2)$, where $i < N/2$, and it is assumed that the bond rupture process is deterministic such that the bond rupture occurs when the force a cross-link transmits reaches the critical value of the force, f_c. For $n = 2$, we have numerically obtained the optimal cross-link configuration, which is in the form of $C_2 = [(i, i + N/2), (i + 1, i + 1 + N/2)]$, through exhaustive searches of all possible configurations. It is shown that the configuration C_2 maximizes both an excess energy and an unfolding force. For $n > 2$, we have performed the random searches to find the optimal cross-link configurations that maximize either an excess work or an unfolding force (Figure 7.3). We describe our major findings on the optimal cross-links configurations as follows: (1) all optimal cross-link configurations maximize both an excess work and an unfolding force simultaneously, and (2) the optimal configurations are in the form of $C_n = [(i_1, j_1), (i_2, j_2), \ldots, (i_n, j_n)]$, where $i_1 < i_2 < \ldots < i_n < j_1 < j_2 < \ldots < j_n$. This optimal configuration found from random searches resembles the parallel strands [63] that are found in the protein domain that typically performs the mechanical functions. This shows that the parallel strands (which are observed in protein domains) optimize the mechanical resistance of protein domains.

To gain deep insights into the structure of optimal folding configurations that are found from random searches, we consider a continuum description of the polymer chain with cross-link configurations. In the continuum limit, the enumeration of beads is assumed to be continuous quantity i, where $0 < i < N$ (here $N \to \infty$). Let us suppose to create a super-cross-link (SCL) configuration such that all n cross-links are placed between two points $(i, i + a)$ and that cross-links are formed in a parallel manner. For this case, when the polymer chain bears the mechanical force F due to mechanical stretching, each cross-link exhibits the force F/n. Then, the optimal configuration that maximizes the excess work is achieved by selecting two points $(x, x + N/2)$, as depicted by the optimal configuration for $n = 1$, that are connected by n consecutive cross-links (i.e., parallel strands).

It is interestingly found that the optimal configurations resemble the SCL configurations except for the constraints prescribed for our chain model, that is, no more than one cross-link can be placed between the pair of beads. Similar to the SCL configuration, the optimal cross-link configurations (i.e., parallel strands) tend to transmit

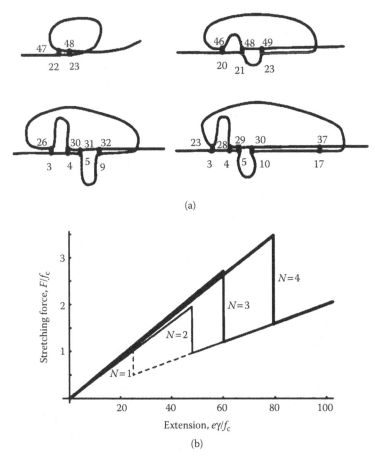

(a)

(b)

FIGURE 7.3 Optimal cross-link configurations and the mechanical response of optimally designed cross-linked polymers. (a) Optimal cross-link configurations for the cross-linked polymers (which consist of 50 monomers) are numerically found from the random searches. Optimal cross-link configurations resemble parallel strands such that cross-links are formed in a parallel manner. (b) Force-extension curves of optimal cross-linked polymer chains. These optimal cross-linked polymer chains can exert the force by distributing the force equally to each cross-link, and when a cross-link is ruptured due to such a force, all cross-links are disrupted (i.e., avalanche-like rupture). (Adapted from Eom, K., P. C. Li, D. E. Makarov, and G. J. Rodin. 2003. *J Phys Chem B* 107:8730. With permission.)

the mechanical load equally to n cross-links. This results in the unique bond rupture mechanism that the breakage of a single bond increases the force that a bond can bear, which leads to the avalanche-like rupture of bonds. These avalanche-like bond rupture mechanisms were found in the optimal cross-link configurations obtained from quasiequilibrium simulations. This is consistent with MD simulation [23] that shows the all-or-none fashion in the breakage of hydrogen bonds (of parallel strand) during the mechanical unfolding of titin immunoglobulin domain. This provides an insight into how a protein structure achieves excellent mechanical resistance using the configurations of chemical bonds (e.g., hydrogen bonds). In particular, a protein structure is generally in the form of a complex, three-dimensional structure, that is, folded structure; this folded structure is formed by hydrogen bonds that stabilize the structure. This folding configuration, especially parallel strands, due to hydrogen bonds is found to play a critical role in the mechanical resistance of a protein such that a mechanical load is effectively transmitted and shared by hydrogen bonds. This is consistent with the optimal cross-link configuration for folded polymer chains.

The optimal cross-link configurations mentioned above are restrictive in that the number of cross-links, n, is much smaller than the possible number of cross-links, that is, $N/2$. For a small number of cross-links, the optimal cross-link configuration is the "parallel strands" that can equally share the mechanical loading. Now, let us consider the case in which the polymer chain is fully folded such that all beads are connected by cross-links whose total number is $N/2$. For computational tractability, we consider the short polymer chain, that is, $N = 19$, such that the total number of cross-links is 10. The optimal cross-link configurations for this case have the following features: (1) all optimal configurations include the subset of three cross-links such as $C_3^* = [(i, i + N/2), (i + 1, i + 3 + N/2), (i + 4, i + 4 + N/2)]$, which is the clamp of parallel strands. The excess work for the clamp without any other cross-links is equal to $W_3^* = 0.79 f_c^2/K$, where K is the stiffness (force constant) of the folded polymer chain before the rupture of any cross-links; (2) By adding any cross-links to the clamp C_3^* one is likely to decrease the excess work (toughness); (3) The maximum of an excess work is given as $W_{max} = 0.93 f_c^2/K$, which corresponds to the configuration $C_{max}^{10} = [(1,15), (2,11), (3,16), (4,14), (5,17), (6,10), (7,9), (8,18), (13,20), (12,19)]$ that also maximizes the unfolding force F_u; and (4) The mean value of excess work for randomly generated cross-link configurations is $<W_{rand}> = 0.35 f_c^2/K$, and only a very small fraction of configurations reaches the toughness comparable to W_{max} (see Figure 7.4). On the other hand, a large fraction of configurations that include the clamp C_3^* approach the toughness close to W_{max} (Figure 7.4). This indicates that the clamp C_3^* (i.e., parallel strand) is a key element that enhances the mechanical resistance of the polymer chain. It is consistent with recent findings that parallel strands formed by hydrogen bonds are responsible for the mechanical toughness of a protein.

7.3.3 RATE EFFECT: IMPLICATIONS FOR SINGLE-MOLECULE PULLING EXPERIMENTS

In Section 7.3.1, the mechanical properties of folded polymer chains are taken into account when the polymer chains are stretched in the regime of a quasiequilibrium process. In other words, earlier simulation ignores the rate effect that is one of the key parameters for single-molecule pulling experiments. However, in general, the accessible loading rate for single-molecule pulling experiments is quite dependent on the loading apparatus; for instance, the loading rate accessible with an optical tweezer is several order of magnitude lower than that accessible with AFM. The loading rate for stretching a biomolecule significantly affects the bond rupture mechanism. As described in Section 7.3.1 (e.g., Equation 7.3), the kinetic rate for bond rupture is critically governed by the loading rate transmitted from a loading apparatus. In particular, the rate for bond rupture, k, is well depicted by Bell's theory [58–60] that

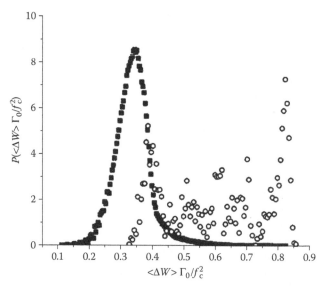

FIGURE 7.4 Probability distributions of excess works for various cross-linked polymer chains. We have considered a polymer chain composed of N monomers (here, $N = 10$) with $N/2$ cross-links, which implies the fully cross-linked polymer chain. We have taken into account two kinds of cross-link formations: (1) cross-link configuration $C_{N/2}$ formed randomly, and (2) $C_{N/2}$ formed in such a way that it contains the subset $C_3 = [(i, i + N/2), (i + 1, i + 3 + N/2), (i + 4, i + 4 + N/2)]$. The probability distribution for the excess work of $C_{N/2}$ configurations generated randomly is shown as a solid rectangular symbol, while the probability distribution for the excess work of $C_{N/2}$, which includes the subset of C_3, is represented as an open circular symbol.

is consistent with the Arrhenius formula [64] (which describes the chemical reaction) used for gaining insight into the energy landscape for folded proteins.

$$k = \chi \exp\left(-\frac{E}{k_B T}\right) \tag{7.5a}$$

where χ is a prefactor, and E is the strain energy for a stretched polymer. When the polymer chain that possess the multiple cross-links (bonds) is pulled apart with the pulling rate v, the strain energy for a bond is given by $E = U - \alpha K v t \cdot \Delta x$, where K is the stiffness of the polymer, α is the dimensionless coefficient, and Δx is the difference between reaction coordinates corresponding to native and denatured states, respectively. Thus, the kinetic rate for the rupture of a specific cross-link is given as follows:

$$k = k_0 \exp\left(\frac{\alpha f}{f_c}\right) = k_0 \exp\left(\frac{\alpha K v t}{f_c}\right) \tag{7.5b}$$

Here, it should be noted that αf is the amount of force that a specific cross-link can bear. For simplicity, let us consider the case of a single cross-link such that a mechanical load f is transmitted to a cross-link, that is, $\alpha = 1$. Now, we introduce the probability function $S(t)$, which is the probability to have an intact cross-link at time t. The decrease rate for $S(t)$ should be balanced by the increase rate for the probability to break a cross-link.

$$-\frac{\mathrm{d}S(t)}{\mathrm{d}t} = S(t) \cdot k(t) \equiv \Phi(t) \tag{7.6}$$

Accordingly, the probability to break a cross-link, $\Phi(t)$, is represented in the form of

$$\Phi(\tau,\theta) = \exp(\theta\tau + \theta^{-1}[1 - \exp(\theta\tau)]) \tag{7.7}$$

where the dimensionless parameters are defined as

$$\tau = k_0 t \tag{7.8a}$$

$$\theta = \frac{\tilde{v}}{1-\tilde{l}} \tag{7.8b}$$

Here, $\tilde{v}$ is the dimensionless pulling rate defined as $\tilde{v} = Kv/k_0 f_c$, and $\tilde{l}$ is defined as $\tilde{l} = l/N$, where a cross-link is formed at $(i, i + l)$, and N is the total number of beads. That is, $\tilde{l}$ indicates the dimensionless length (normalized by the contour length of the polymer chain) for the loop formed by a cross-link. For the rupture of a cross-link, the rupture force (unfolding force) and the excess work are given by

$$F_u(\tau) \equiv f(\tau) = f_c \theta\tau \tag{7.9a}$$

$$W(\tau) = \frac{f_c}{2K}(1-\tilde{l})\tilde{l}\theta^2\tau^2 \tag{7.9b}$$

Consequently, the mean rupture force and the mean excess work can be obtained as follows:

$$\langle F_u \rangle = \int_0^\infty f(\tau)\Phi(\tau,\theta)\,\mathrm{d}\tau \approx f_c \ln\theta = f_c \ln\frac{\tilde{v}}{1-\tilde{l}} \tag{7.10a}$$

$$\langle W \rangle = \int_0^\infty W(\tau)\Phi(\tau,\theta)\,\mathrm{d}\tau \approx \frac{f_c}{2K}\tilde{l}(1-\tilde{l})\ln^2\theta = \frac{f_c}{2K}\tilde{l}(1-\tilde{l})\ln^2\frac{\tilde{v}}{1-\tilde{l}} \tag{7.10b}$$

Equation 7.10 clearly shows that the optimal configuration that maximizes the mean unfolding force or the mean excess work is strongly dependent on the pulling speed v. In particular, the mean unfolding force reaches the maximum when $\tilde{l} \to 1$; the cross-link that connects the ends (i.e., termini) of the chain maximizes the mean unfolding force. On the other hand, the optimal cross-link configuration that maximizes the mean excess work depends on the pulling speed v. For $v \to \infty$, the mean excess work becomes a maximum when $\tilde{l} \to 1/2$; this configuration is identical with the optimal configuration for the case in which rate effect is disregarded (i.e., quasiequilibrium stretching).

Now, we consider the polymer chain that contains more than a single cross-link, where such a polymer chain is pulled apart with pulling speed v. For $n = 3$, we have found that any other configurations cannot exhibit a higher mean unfolding force and mean excess energy than those of the cross-link represented in the form of $C^* = \{(i, i + N\tilde{l}), (i + 1, i + 2 + N\tilde{l}), (i + 3, i + 3 + N\tilde{l})\}$, where the optimal value of $\tilde{l}$ is determined by exhaustive searches with respect to $\tilde{l}$. The mean unfolding force and mean excess energy for C^* with different dimensionless loop lengths $\tilde{l}$ are shown in Figure 7.5. As discussed for the case of $n = 1$, the terminal cross-link (i.e., terminal parallel strand) maximizes the mean unfolding force regardless of the pulling speed. On the other hand, at the low pulling speed, the terminal cross-link maximizes the mean excess energy, whereas the mean excess energy becomes a maximum for the cross-link with $\tilde{l} = 1/2$ at high pulling speed. This shows that the pulling rate plays a role in the mechanical resistance of the folded polymer chain. In particular, at a relatively slow pulling rate, the optimal configuration that maximizes both the mean unfolding force and the mean excess work corresponds to the "terminal parallel strands" [23,25,65–67] that are found in proteins such as ubiquitin that exhibit

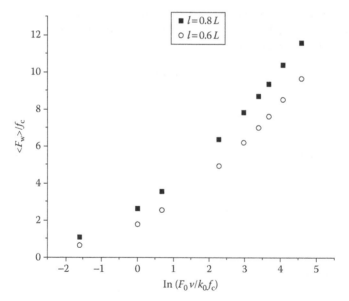

FIGURE 7.5 Pulling rate-dependent unfolding forces for a cross-linked polymer chain. We have considered the unfolding force for a cross-linked polymer with a given cross-link configuration such as $C_3 = [(j, j + l), (j + 1, j + 3 + l), (j + 4, j + 4 + l)]$, where $l = 0.6L$ or $0.8L$ with L being the total number of monomers. It is shown that the unfolding force is linearly proportional to the logarithm of the pulling speed and that the cross-link configuration with $l \to L$ (which resembles the terminal parallel strand) is the optimal configuration that maximizes the unfolding force.

excellent mechanical resilience. Specifically, recent AFM experiments [25,65–67] and/or MD simulations [23,67] showed that some proteins that perform excellent mechanical functions with notable mechanical resistance exhibit the terminal parallel β strands.

The effect of the loading rate on the mechanical resistance of a protein can be elucidated from our simple toy model. Under the physiological conditions, proteins are subjected to a mechanical loading that is different from those of single-molecule experiments and/or MD simulations. Moreover, the timescale at which the protein is stretched is different between experiments and simulations. For example, AFM experiments on the muscle protein titin have shown that an individual immunoglobulin (Ig)-like domain unfolds at forces f ranging from 150 pN to 250 pN depending on the pulling speed v in the range of 0.1 to 10 nm/ms [68,69]. The loading rate can be approximately estimated from $df/dt \sim f/t \sim fv/L$, where L is the contour length and t is time in the order of L/v. With a given L for the Ig domain, the loading rate for the muscle protein titin pulled by AFM is 10^{-9}–10^{-7} N/s. On the other hand, in the experiments that probe the mechanical response of myofibril, the Ig-like domain is subjected to a much lower force (i.e., $f \sim 10$ pN) over the timescale of 1 second to 10 seconds [70]. The loading rate for this case is on the order of 10^{-12}–10^{-11} N/s, which is several orders of magnitude lower than that of AFM pulling experiments. This indicates that the mechanical resistance of a protein is dependent on the loading rate. Our toy model includes the loading rate effect on the mean unfolding force and/or mean excess work. In particular, the loading rate effect is depicted by a dimensionless pulling rate θ given by Equation 7.8b. By rewriting Equation 7.8b, the dimensionless pulling rate θ is in the form of $θ = (df/dt)/k_0 f_c$. Based on the values of $k_0 = 5 \times 10^{-4}$ s and $f_c = 16$ pN, the dimensionless pulling rate for AFM experiments on the muscle protein titin is on the order of 10^5–10^7, while the dimensionless pulling rate for stretching skeleton myofibril is in the range of 10^2–10^3. As anticipated from our toy model, the optimal configuration depends on the dimensionless pulling rate θ, and the mechanical resistance of a folded chain is therefore governed by not only the topology of cross-links but also the loading rate. This suggests a concept that can elucidate how a protein can achieve the optimal mechanical properties that are dependent on the protein's native topology as well as on the loading rate.

7.4 COMPUTATIONAL MODEL OF PROTEIN MATERIALS: ROLE OF PROTEIN TOPOLOGY

Most atomistic simulations and/or theoretical models including the one that was described in Section 7.3 provide an insight into the microscopic view of the mechanical properties of proteins, particularly protein unfolding mechanics. However, the mechanical properties of protein materials at macroscopic scales have not been analyzed based on computational simulations, albeit there have been a few attempts to develop the physical model of protein materials that are formed by the assembly of protein domains. In addition, there have been previous attempts to construct a phenomenological model that can capture the mechanical characteristics of protein materials observed in experiments. For instance, Termonia [34] suggested the phenomenological model of spider silk based on the concept of micromechanics such that, for spider silk, β-sheet crystals are embedded in the amorphous matrix. Despite its ability to capture the experimentally measured mechanical response of spider silk, Termonia's model [34] is inappropriate in that the molecular structure of spider silk, which has recently been found experimentally [35], is quite different from what was conjectured in Termonia's model. In particular, a recent experiment [35] shows that β-sheet crystals are embedded in well-ordered α helices rather than randomly oriented chains. Zhou and Zhang [71] have provided the phenomenological model, which was inspired from AFM single-molecule pulling experiments [72], for spider silk based on the hierarchical model such that spider silk consists of the hierarchical combinations of nonlinear springs. This model requires a priori knowledge of the microscopic mechanical behavior of a nonlinear spring, which was empirically obtained by fitting the experimentally measured mechanical response of a protein from AFM single-molecule experiments. Kasas et al. [73] have suggested the continuum model of microtubules, whose elastic constants were obtained from empirical fitting of the mechanical response predicted from continuum-based simulations to the experimentally observed mechanical behavior. These models [34,71,73] lack the fundamental physics that elucidates how weak chemical bonds (e.g., hydrogen bonds that play a critical role in the formation of three-dimensional folding structure) at atomic scales can tailor the mechanical properties of protein materials at macroscopic scales. This implies that for comprehensive understanding of the mechanical properties of protein materials, we need to develop novel physical model-based simulations that can facilitate the mechanical characterization of proteins at multiple scales ranging from atomic scales (e.g., chemical bonds), to mesoscales (e.g., small protein domain), to macroscopic scales (e.g., spider silk). In this section, we provide the mesoscopic model of protein crystals to characterize and predict the mechanical properties of protein materials.

7.4.1 TYPES OF MODELS

Here, we would like to review our previous simulation techniques [74] based on a physical model (at atomistic scale), which enable the mechanical characterization of protein materials, particularly bulk elastic modulus of protein materials.

7.4.1.1 Mesoscopic Model for Protein Materials

Protein materials, as shown in Figure 7.6, are assumed to be composed of a repeated representative volume element (RVE) that contains the crystallized proteins with a specified space group. We assume that the RVE is gradually extended due to constant, discrete, strain tensor ΔE with the amount of $|\Delta E^0| = 10^{-3}$. Here, it should be noted that the macroscopic strain is equal to the constant strain field applied to the RVE in the context of micromechanics theory [75]. For instance, the strain tensor can be denoted as $\Delta E^0 = e_{xx} \mathbf{j}_x \mathbf{j}_x + e_{yy} \mathbf{j}_y \mathbf{j}_y + e_{zz} \mathbf{j}_z \mathbf{j}_z$, where e_{xx}, e_{yy}, and e_{zz} represent the mechanical strain for x-, y-, and z-directions, respectively, and $\mathbf{j}_x$, $\mathbf{j}_y$, and $\mathbf{j}_z$ indicate the unit vectors for x-, y-, and z-directions, respectively. Here, it is supposed that the RVE is stretched slowly enough that the timescale for strain-induced extension is much longer than that of thermal fluctuation for proteins. This assumption implies that the mechanical

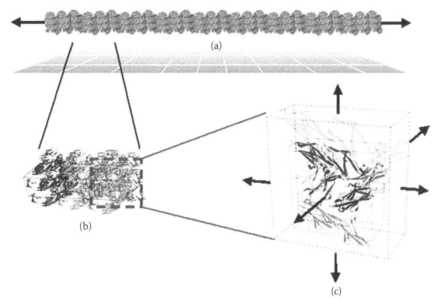

FIGURE 7.6 Mesoscopic model of protein materials. We assume that a protein material at macroscopic scales, shown as (a), consists of representative volume elements (RVEs) in a repetitive manner shown as (b). The enlarged image of the RVE, which includes the protein crystal, is shown in (c). (Adapted from Yoon, G., H.-J. Park, S. Na, and K. Eom. 2009. *J Comput Chem* 30:873. With permission.)

stretching of the RVE obeys the quasiequilibrium process that neglects the stochastic (rate) effect. Once a constant, discrete strain tensor $\Delta \mathbf{E}^0$ is prescribed for the RVE that contains the protein crystal, the displacement vector $\mathbf{u}$ due to strain for a given atomic position $\mathbf{r}$ of a protein is given by $\mathbf{u}(\mathbf{r}) = \Delta \mathbf{E}^0 \cdot \mathbf{r} = e_{xx} R_x \mathbf{j}_x + e_{yy} R_y \mathbf{j}_y + e_{zz} R_z \mathbf{j}_z$, where $\mathbf{r} = R_x \mathbf{j}_x + R_y \mathbf{j}_y + R_z \mathbf{j}_z$. Accordingly, the position vector $\mathbf{r}^*$ for a protein after the application of strain $\Delta \mathbf{E}^0$ to the RVE is given as $\mathbf{r}^* = \mathbf{r} + \mathbf{u}(\mathbf{r})$. Here, it should be noted that a position vector $\mathbf{r}^*$ is not in equilibrium. So we have performed the energy minimization process based on the conjugate gradient method to find the equilibrium position $\mathbf{r}_{eq}$ to ensure the convergence of viral stress.

To compute the mechanical properties of protein materials, it is necessary to evaluate the effective stress for the RVE due to applied mechanical strain $\Delta \mathbf{E}$. The stress tensor $\mathbf{S}(\mathbf{r})$ at position $\mathbf{r}$, which is acquired from application of strain followed by an energy minimization process, can be calculated from viral stress theory [76,77].

$$\mathbf{S}(\mathbf{r}) = \frac{1}{2} \sum_{i=1}^{N} \sum_{j \neq i}^{N} \left[\mathbf{d}_{ij} \otimes \left\{ \frac{1}{d_{ij}} \frac{\partial \Phi(d_{ij})}{\partial d_{ij}} \mathbf{d}_{ij} \right\} \right] \cdot \delta(\mathbf{r} - \mathbf{r}_i) \qquad (7.11)$$

where N is the total number of atoms for proteins in the RVE, $\mathbf{d}_{ij}$ is the distance vector between atom i and j, that is, $\mathbf{d}_{ij} = \mathbf{r}_i - \mathbf{r}_j$ with $\mathbf{r}_i$ being the position vector for the ith atom, d_{ij} is the distance, that is, $d_{ij} = |\mathbf{d}_{ij}|$, $\Phi(d)$ is the interatomic potential as a function of interatomic distance d, and $\delta(r)$ is the Dirac delta function. The overall stress tensor $\mathbf{S}^0$ can be easily estimated as follows [75]:

$$\mathbf{S}^0 \equiv \frac{1}{V}\int_\Omega \mathbf{S}(\mathbf{r})d^3\mathbf{r} = \frac{1}{2V}\sum_{i=1}^{N}\sum_{j\neq i}^{N}\mathbf{d}_{ij} \otimes \left\{ \frac{1}{\mathrm{d}_{ij}}\frac{\partial\Phi(d_{ij})}{\partial d_{ij}}\mathbf{d}_{ij} \right\} \qquad (7.12)$$

Here, V is the total volume of the RVE, and the symbol Ω in the integrand indicates the volume integral with respect to the RVE.

The procedure to obtain the stress–strain relation for protein materials is summarized as follows.

1. The initial RVE is prepared such that the initial conformation of proteins is regarded as a protein crystal as a native conformation deposited in the Protein Data Bank (PDB) in a unit cell (as a RVE). Here, the position vector for proteins at initial conformation is denoted as $\mathbf{r}^0$.
2. A discrete, constant strain field $\Delta\mathbf{E}^0$ is applied to the RVE, so that the displacement vector $\mathbf{u}$ for proteins is given by $\mathbf{u} = \Delta\mathbf{E}^0\cdot\mathbf{r}^0$. Then, the atomic position for proteins in the RVE that is extended by strain is thus given as $\mathbf{r}^* = \mathbf{r}^0 + \mathbf{u}$.
3. In general, the position vector $\mathbf{r}^*$ does not stay in the equilibrium state. The equilibrium position vector $\mathbf{r}_{eq}$ is numerically found from the energy minimization process for an initially provided position vector $\mathbf{r}^*$.
4. The overall viral stress $\mathbf{S}^0$ for the RVE is calculated using Equation 7.12 with a position vector $\mathbf{r} = \mathbf{r}_{eq}$.
5. The initial conformation $\mathbf{r}^0$ is set as $\mathbf{r}_{eq}$, that is, $\mathbf{r}^0 \leftarrow \mathbf{r}_{eq}$.
6. The processes (2)–(5) are repeated until the RVE is extended until the total strain applied to the RVE reaches the prescribed strain field.

In general, the stress–strain relationship for protein materials follows the nonlinear elastic behavior rather than linear elastic characteristics. Thus, to define the elastic properties, we have considered the tangent modulus (in the context of classical elasticity [78]) that can be estimated such as $E = \partial\mathbf{S}^0/\partial e|_{e=0}$, where $\mathbf{e}$ is the total strain applied to the RVE.

7.4.1.2 Interatomic Potential Field: Gō Model and Elastic Network Model

The computational cost to predict a stress–strain relationship for protein materials is strongly dependent on the details of interatomic potential fields as well as the degrees of freedom of a molecular system. We have employed a CG structural model [32] that can capture the dynamic and/or mechanical features of proteins predicted from all atom simulations. In particular, we have used the Gō model [79–81] that describes the protein structure based on the alpha-carbon atoms of a protein backbone chain and that prescribes the coarsened potential field for alpha-carbon atoms. The Gō model presumes that the effective potential field prescribed for alpha-carbon atoms is composed of two major contributions: (1) strain energy for covalent bond stretching, and (2) nonbonded interactions such as Lennard–Jones (LJ) potential. For the Gō model, the interatomic potential prescribed for alpha-carbon atoms is given by

$$\Phi(d_{ij}) = \left[\frac{k_1}{2}\left(d_{ij} - d_{ij}^0\right)^2 + \frac{k_2}{4}\left(d_{ij} - d_{ij}^0\right)^4\right]\delta_{j,i+1} + 4\varepsilon_0\left[\left(\frac{\xi}{d_{ij}}\right)^6 - \left(\frac{\xi}{d_{ij}}\right)^{12}\right](1 - \delta_{j,i+1}) \quad (7.13)$$

where d_{ij} is the interatomic distance between two alpha-carbon atoms i and j, k_1 and k_2 represents the force constant for harmonic and quartic potentials, respectively, ε_0 is the energy depth for nonbonded interactions, ξ is the length scale for nonbonded interactions, superscript 0 indicates the equilibrium state, and δ_{ij} is the Kronecker delta defined as $\delta_{ij} = 1$ if $i = j$; otherwise, $\delta_{ij} = 0$. Here, the parameters are given as $k_1 = 0.15$ kcal/mol·Å^2, $k_2 = 15$ kcal/mol·Å^2, $e_0 = 0.15$ kcal/mol, and $\xi = 5$ Å (for details, see reference [82]). Gō potential is widely used for studying not only protein dynamics such as low-frequency vibrations [79–81] but also protein mechanics such as mechanical unfolding [83–86]. It has been remarkably shown that the protein unfolding mechanics predicted from the Gō model is quantitatively comparable to that observed by single-molecule force spectroscopy.

Tirion [40] provided a more simplified potential field such that nonbonded interactions are approximated as a harmonic potential. Furthermore, the potential field is more simplified such that strain energies for both covalent bond stretching and nonbonded interactions are approximated as a harmonic potential field with an identical force constant. This model is renowned as the ENM [41–43,87–89] (see also Chapter 3) that dictates the short-range harmonic interactions between alpha-carbon atoms within the neighborhood. Specifically, in the ENM, alpha-carbon atoms (which represent the residue) within the neighborhood are connected by a harmonic spring with an identical force constant. Despite its simplicity, the ENM is robust in predicting protein dynamics such as low-frequency vibrations [43,90]. The success of the ENM in analyzing protein dynamics is attributed to the hypothesis that protein dynamics is critically governed by a protein's native topology, but the details of the empirical potential field does not significantly affect the protein's dynamic motion. This hypothesis was supported by an earlier study by Case and Teeter [91], who showed that the low-frequency vibration motion of a protein is insensitive to the details of atomic potential fields. In addition, a recent study by Lu and Ma [92] supports the hypothesis in that the low-frequency vibration motion is well depicted by an ENM whose stiffness matrix is perturbed as long as the perturbation of the stiffness matrix is not significant enough to critically change the protein's native topology. Recently, the ENM has been highlighted for studying the protein unfolding mechanics; a recent study [93] has suggested that the unfolding force anticipated from the ENM with Bell's theory is quantitatively comparable to that observed by single-molecule force spectroscopy. This implies that the ENM may be an appropriate interatomic potential for studying the role of a protein's native topology in the mechanical properties of protein materials. The interatomic potential for the ENM is represented in the following form:

$$\Phi_{ENM}(d_{ij}) = \frac{\gamma}{2}\left(d_{ij} - d_{ij}^0\right)^2 \cdot H\left(r_c - d_{ij}^0\right) \quad (7.14)$$

where γ is a force constant for a harmonic spring that connects two neighboring alpha-carbon atoms, r_c is the cutoff distance that defines the topology of the ENM, and $H(x)$

is the Heaviside unit step function defined as $H(x) = 0$ for $x < 0$; otherwise, $H(x) = 1$. In general, the cutoff distance is empirically determined in the range of 8--~15 Å (for determining the cutoff distance, please refer to reference [41]). The force constant can be determined by fitting the B-factor computed from the ENM to that obtained from experiments (usually, experimental B-factors are available in the PDB).

7.4.1.3 Mechanical Characterization

We consider the protein materials that consist of model protein crystals as summarized in Table 7.1 and their mechanical properties. The model protein crystals are composed of 20--~2000 residues, which are computationally inaccessible with a conventional MD simulation. For mechanical characterization, we take into account the volumetric strain e_V that is applied to the RVE in which the protein crystals reside. Here, the volumetric strain e_V is defined as follows [94]:

$$e_V = \frac{1}{3}(e_{xx} + e_{yy} + e_{zz}) \equiv \frac{1}{3} Tr\left[\mathbf{E}^0\right] \tag{7.15}$$

where $Tr(\mathbf{A})$ is the trace of the matrix $\mathbf{A}$. Once the overall stress tensor $\mathbf{S}^0$ is computed from the viral stress theory given by Equation 7.12, we can calculate the hydrostatic stress (pressure) defined as [94]

$$p = \frac{1}{3}(s_{xx} + s_{yy} + s_{zz}) \equiv \frac{1}{3} Tr\left[\mathbf{S}^0\right] \tag{7.16}$$

where s_{xx}, s_{yy}, and s_{zz} represent the normal stresses in x-, y-, and z-directions, respectively. Here, it should be noted that the RVE is deformed by application of volumetric strain rather than shear strain to compute Young's modulus, which is the axial stiffness, for protein materials. The bulk modulus E_V is defined as the ratio of hydrostatic stress to volumetric strain, that is, $E_V = p/e_V$. With prescribed volumetric strain e_V and hydrostatic stress p calculated from viral stress theory, the bulk modulus for protein materials can be easily computed; then, the elastic modulus (Young's modulus), E, of protein materials can be estimated from the following relation [94]:

$$E_V = \frac{E}{3(1 - 2v)} \tag{7.17}$$

where v is Poisson's ratio of protein materials, which can be computed from the following equation:

$$V_{xy} = -\frac{s_{yy}}{s_{xx}} \tag{7.18}$$

7.4.2 Simulation Results and Discussion

We have studied the mechanical properties, particularly Young's modulus, of protein materials, which consist of protein crystals whose dynamics is depicted by the

TABLE 7.1
Model Protein Crystals for Biological Materials

Protein (PDB)	RVE Parameters			Elastic Modulus (GPa)		Maximum Stress (MPa)	Degree of Fold, Q (%)
	Space Group	No. of Residues	RVE Volume (Å³)	In Silico	In Vitro		
α Helix (1akg)	$P2_12_12_1$	64	$14.6 \times 26.1 \times 29.2$	0.277	N/A	3.82	45
β Sheet (2ona)	P_1	24	$25.8 \times 9.7 \times 15.8$	0.433	N/A	5.69	30
α-Lactalbumin (1hfz)	$P2_1$	982	$38.3 \times 78.6 \times 79.6$	0.186	2	2.45	1.75
Transthyretin (2g5u)	$P2_12_12$	908	$62.2 \times 75.9 \times 134.2$	0.242	5	3.48	4.05
Titin proximal Ig (1g1e)	$P2_12_12_1$	640	$58.6 \times 60.1 \times 77.1$	0.187	N/A	2.36	5.13
Titin distal Ig (1waa)	$P2_12_12_1$	2208	$43.2 \times 85.8 \times 64.7$	0.254	N/A	10.80	1.61
Tubulin (1tub)	$P2_1$	1734	$80 \times 92 \times 90$	0.138	0.1–2.5	1.87	1.13
F-actin rabbit skeletal muscle (1rfq)	$P4_3$	2888	$101.5 \times 101.5 \times 104.2$	0.166	2.2	2.19	1.26
β-Lactoglobulin (1beb)	P_1	312	$37.8 \times 49.5 \times 56.6$	0.166	5	2.37	2.90
Lysozyme (194l)	$P4_32_12_1$	1032	$78.65 \times 78.65 \times 37.76$	0.143	5	7.21	6.77
Fn3 (1fna)	$P2_1$	182	$30.7 \times 35.1 \times 37.7$	0.294	N/A	3.90	9.21
Fn3 (1fmf)	$P2_1$	1472	$64.05 \times 60.67 \times 58.44$	0.239	N/A	3.36	2.43
Fn3 (1fnh)	I222	2152	$68.58 \times 86.29 \times 142.8$	0.189	N/A	2.68	3.33
Fn3 (1ten)	$P4_12_12_1$	712	$49.78 \times 49.78 \times 71.04$	0.307	N/A	4.12	8.81
Titin N-termini (2a38)	P_1	582	$55.41 \times 56.29 \times 74.41$	0.177	N/A	2.40	1.54
Equine cyt c (1hrc)	$P4_3$	416	$58.40 \times 58.40 \times 42.09$	0.365	N/A	4.87	7.47
Ubiquitin (1ubq)	$P2_12_12_1$	304	$50.84 \times 42.77 \times 28.95$	0.532	N/A	7.10	10.2
CspB (1csp)	$P3_22_1$	402	$58.94 \times 58.94 \times 46.45$	0.315	N/A	4.20	11.5

Gō model. Here, the elastic modulus is computed from the bulk modulus, depicted in Equation 7.17, which is calculated from the ratio of hydrostatic stress estimated from viral stress theory to volumetric strain applied to the RVE. The elastic modulus of various protein materials from our simulation is summarized in Table 7.1. First, we consider the microtubule, which plays a mechanical role in sustaining the cell shape, as a model protein material that consists of tubulin dimers. Our simulation predicts that the elastic modulus of a protein material composed of tubulins is $E_{tub} = 0.138$ GPa, which is quantitatively comparable to that obtained from an AFM bending experiment [95] that predicts $E = \sim 0.1$ GPa for microtubules. This indicates that our model is capable of predicting the elastic properties of protein materials *in silico*. However, it should be noted that the elastic properties estimated from experiments are very sensitive to the experimental conditions; for instance, elastic modulus of microtubules obtained from AFM bending experiments is measured differently from that evaluated by nondestructive methods [96] by an order. Moreover, a recent study [95] has shown that the elastic modulus of microtubules is dependent on temperature. In addition, a recent single-molecule imaging experiment [97] shows that the bending rigidity of microtubules is very sensitive to the length of microtubules. Furthermore, to ensure the robustness of our model, we have also taken into account the mechanical properties of titin Ig domains, particularly the Ig distal domain and Ig proximal domain. Our simulation provides that the elastic modulus of the distal domain is larger than that of the proximal domain, which is consistent with the results from single-molecule pulling experiments [98].

Figure 7.7 depicts the stress–strain curves for various protein materials, which are predicted from the mesoscopic model–based simulations. It is interestingly found that the elastic modulus of protein materials is in the range of 0.1–1 GPa, which is consistent with an experimental finding [10] that Young's modulus of protein materials usually ranges between 1 MPa (e.g., elastin) and 10 GPa (e.g., dragline silk). It is also

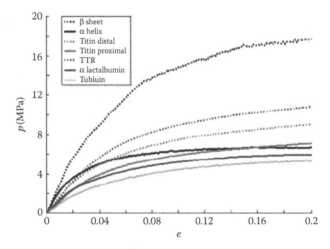

FIGURE 7.7 Stress–strain curves for several protein materials predicted from mesoscopic model based on Gō potential. (Adapted from Yoon, G., H.-J. Park, S. Na, and K. Eom. 2009. *J Comput Chem* 30:873. With permission.)

found that among protein materials, a protein material composed of β sheet exhibits remarkable mechanical resistance such that the elastic modulus and the maximum (yield) stress for β sheet are much larger than that of any other protein materials. This is in agreement with earlier studies [34,51,52], which show that the β-sheet structural motif plays a significant role in stiffening the biological materials.

To gain insights into the role of a protein's native topology on the mechanical properties, we have used the ENM instead of the Gō Model for the interatomic potential to compute the viral stress for two model proteins—α helix and β sheet. Since the ENM assumes that interatomic interaction between alpha-carbon atoms is harmonic potential, the mechanical response of model proteins can be described by piecewise linear elastic behavior (Figure 7.8). It is shown that ENM-based simulation overestimates the elastic modulus of model proteins as compared to Gō Model–based simulation, which implies the importance of anharmonic potential field for accurate predictions on the mechanical properties of protein materials. Nevertheless, ENM-based simulation predicts the mechanical properties of two model proteins, qualitatively comparable to those anticipated from the simulation based on the Gō Model. In particular, ENM-based simulation provides that the elastic modulus of β-sheet-based material is about an order of magnitude larger than that of α-helix-based material. This sheds light on the significance of a protein's native topology in the mechanical properties of protein materials such as elastic modulus and yield stress. In addition, we consider the mechanical properties of protein materials composed of fibronectin III (fn3) domains with different crystallographic structures to further understand the effect of protein's topology on the mechanical properties. It is suggested that a protein material consisting of fn3 domain with a space group of $P4_32_12$ possesses a higher elastic modulus than protein materials composed of fn3 domains with space groups $P2_1$ or $I2\ 2\ 2$. This indicates that the native topology plays a critical role in the mechanical properties of protein materials.

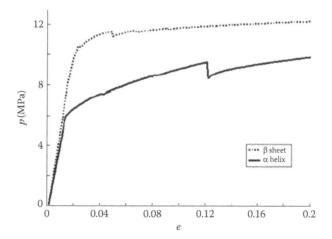

FIGURE 7.8 Stress–strain curves of two protein materials, which are composed of α helices or β sheets, respectively, predicted from the mesoscopic model that uses the elastic network potential. (Adapted from Yoon, G., H.-J. Park, S. Na, and K. Eom. 2009. *J Comput Chem* 30:873. With permission.)

As presented above, it is hypothesized that a protein's native topology is closely related to the mechanical properties of proteins [45,46,51]. In order to support the hypothesis, we have introduced a dimensionless measure Q that represents the degree of protein folding. For a protein composed of N residues, the dimensionless measure Q is defined as $Q = N_c/N_{max}$, where N_c is the number of native contacts and N_{max} is the number of possible, maximum native contacts given as $N_{max} = N(N-1)/2$. Here, the native contact is defined such that if two residues are within the cutoff distance (i.e., 7.5 Å), then these two residues form the native contact. Figure 7.9 shows the relationship between Q and mechanical properties of protein materials such as their elastic modulus and yield stress. It is interestingly shown that the dimensionless measure Q is highly correlated with the elastic modulus of protein materials. For instance,

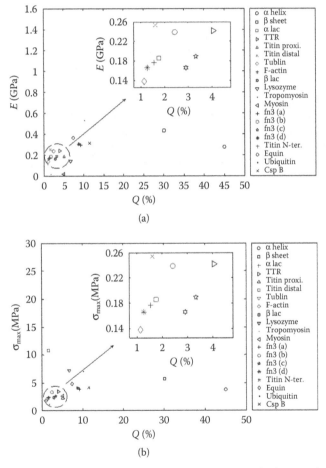

FIGURE 7.9 Relationship between folding topology and mechanical properties for protein materials. (a) Elastic modulus versus degree of fold, and (b) yield stress versus degree of fold are shown. It is found that the more folded the protein crystal, the higher the elastic modulus (and yield stress) it has. (Adapted from Yoon, G., H.-J. Park, S. Na, and K. Eom. 2009. *J Comput Chem* 30:873. With permission.)

the protein materials composed of α helix or β sheets exhibit high elastic modulus due to relatively large Q. On the other hand, some protein materials such as titin Ig domain and/or TTR possess low Q, albeit they exhibit the intermediate value of elastic modulus. These proteins are renowned as mechanical proteins that perform excellent mechanical function due to hydrogen bonding of the β-sheet structural motif. This suggests that hydrogen bonding of β sheet, which exhibits high Q, increases the mechanical stiffness of protein materials. Moreover, we have also studied the correlation between Q and the yield stress of protein materials. Figure 7.9 depicts that the dimensionless measure Q is highly correlated with the yield stress of protein materials. It is shown that β-sheet crystal exhibits high Q as well as high yield stress, whereas α helix possesses high Q and low yield stress. This may be attributed to the fact that the β sheet serves as a nonlinear spring that has high mechanical toughness due to hydrogen bonds, whereas the α helix acts as a helical spring with relatively few hydrogen bonds. In general, the mechanical resistance of proteins is ascribed to the breakage of hydrogen bonds, which results in the unfolding of folded topology.

In this study, it should be noted that the mechanical properties of protein materials, predicted from the mesoscopic model–based simulation, correspond to those of bulk protein materials. In other words, the mechanical properties of protein materials at small scales (i.e., nanoscale) would be different from those of bulk protein materials that can be predicted from our simulation. In particular, this may be attributed to the intermolecular interactions between secondary structures. For instance, as reported in references [99,100], the mechanical properties of intermediate filament composed of α helices are unique in that the unfolding of α helices leads to the generation of β sheets due to intermolecular interactions between two unfolded α helices, which toughens the mechanical resistance of the intermediate filament. This indicates that interaction between the secondary structures (e.g., α helix, β sheet, loop) does also play an important role in the mechanical resistance of protein materials. This intermolecular interaction is not taken into account in our model, which implies the restriction of our model for understanding the mechanical behavior of protein materials at nanoscale. In order to consider this effect, our model has to be modified in such a way that the interaction between the RVE has to be taken into account for simulating the mechanical response of the RVE to obtain the stress–strain relationship for protein materials.

7.5 COMPUTATIONAL MODEL OF PROTEIN FIBRILS: ROLE OF HYDROGEN BONDS

7.5.1 OVERVIEW

In the last decade, it is realized that denatured proteins play an important role in disease expressions in such a way that such proteins act as a catalyst that assists the formation of long fibrils. For instance, it has been found that the expression of Alzheimer's disease is attributed to the presence of undesirable formation of a long amyloid fibril that originates from the self-aggregation of β amyloid peptides [5–7]. A long amyloid fibril was also observed in the pancreas in patients with type II diabetes, and such an amyloid is named "human islet amyloid polypeptide (hIAPP)" [8], which is a key factor associated with insulin-secreting inhibition. In particular, the hIAPP fibril, whose formation is ascribed to hIAPP$_{20-29}$ (SNNFGAILS) as an amyloidogenic core serving as the catalyst in the self-aggregation, can replace the insulin-secreting β cells at the islet of the pancreas. It is implied that the fundamental understanding of the supramolecular structure of amyloid fibrils is essential to gain insights into the self-aggregation patterns that are highly correlated with disease expressions. For this understanding, there have been recent efforts [101–103] to unveil the supramolecular structures of amyloid fibrils using experimental techniques such as solid-state NMR. Moreover, there have been recent attempts [104–107] to theoretically suggest the possible molecular structures of amyloid fibrils using atomistic simulations such as MD simulations. In particular, a recent study [101] provides the eight possible molecular structures of amyloid fibrils based on the hierarchy of multiple ladder-shaped β sheets (i.e., cross-β amyloid).

The ability of amyloid fibrils to replace specific cells (e.g., insulin-secreting β cells) at the specific site (e.g., the islet of pancreas) is attributed to the high mechanical stability of amyloid fibrils in physiological conditions. This implies that it is an a priori requisite to understand the mechanisms of not only the self-aggregation for amyloid formation but also the mechanical stability of amyloid fibrils. This requisite to characterize the mechanical stability of amyloid fibrils has led researchers to study the mechanical stabilities (i.e., properties) of amyloid fibrils. Knowles et al. [10] has found that a variety of amyloid fibrils composed of weak protein domains exhibit excellent mechanical properties such that the elastic stiffness of amyloid fibrils is higher than that of any other protein materials. This highlights the mechanical stability of amyloid fibrils that consist of even weak protein domains. A recent study [108] has reported the bending elastic properties of β-lactoglobulin fibrils measured based on the AFM-based imaging analysis coupled to polymer theory. It is shown that the mechanical properties of β-lactoglobulin are closely related to the structural hierarchy and assembly, particularly helical pitch, fibril length, and fibril thickness (e.g., see Figure 1.6d). Buehler et al. [11] have remarkably shown that, by using atomistic simulations, the geometric confinement of β sheets results in the improvement of the mechanical properties of a protein crystal. This suggests that the hydrogen bonds between β-sheet layers act as a chemical glue between layers, which increases the mechanical stability of a protein crystal. This is consistent with the hypothesis suggested by Knowles et al. [10], who conjectured that intermolecular force between β-sheet layers is a key engineering parameter that determines the mechanical stability of amyloid fibrils. This elucidates the structure–property–function relationship for protein materials, implying that the mechanical stability of a protein material is encoded in the structural hierarchy and self-assembly of a protein material.

Mechanical characterization of the stability of protein materials has been made possible due to recent experimental techniques. For instance, AFM-based nanoindentation and/or force spectroscopy have enabled the mechanical characterization of soft protein materials such as viral capsid [16,17,109], microtubule [95], spider silk protein [72], muscle protein titin molecule [48,68], and amyloid fibrils [10]. Despite its ability to characterize the mechanical stability of amyloid fibrils, the experimental effort [10,108] to measure their mechanical properties is insufficient to unveil the fundamental of how self-aggregation of a mechanically weak protein domain leads to the formation of mechanically stable fibrils. In particular, AFM imaging techniques [108] are unable to visualize the atomistic details of the hierarchical structure of amyloid fibrils. Moreover, the experimentally measured mechanical properties of soft protein materials are significantly dependent on not only the sample preparation [103,110] but also the experimental conditions such as temperature [95]. It is implied that experimental techniques such as AFM-based imaging and/or mechanical tests lack for revealing the structure–property–function relationship for protein materials, particularly amyloid fibrils.

Computational simulations based on atomistic models are one of the possible routes to acquire the structure–property–function relationship for protein materials. As discussed earlier, MD simulation allows us to study the relationship between the mechanical stabilities of protein materials and their structural hierarchy, that is, the structure–property–function relationship for protein materials. However, despite its ability to provide detailed insights into such a relationship, MD simulation does not provide any insight into the structure–function–property relationship for large protein complexes such as long amyloid fibrils due to the limited spatial and temporal scales available for MD simulations [26]. In particular, MD simulation is computationally restrictive in that it can characterize the mechanical stability of amyloid fibrils whose length is at most ~10 nm [111], which is much shorter than that of experimentally observed amyloid fibrils. To overcome this computational limitation, there have been recent efforts [27–33] to develop a CG model, which is able to simulate the dynamics of supramolecular complexes such as long amyloid fibrils. In general, a CG model is constructed in such a way that the degrees of freedom of a molecular system are reduced and the potential field ascribed to the molecular system is also simplified. One of the CG models that have intrigued the biophysics community is the ENM, which was first suggested by Tirion [40] and later by several researchers [41–43,87–89,112–114] (for details of the ENM, see also Chapter 3). The key feature of the ENM is to describe a protein structure based on alpha-carbon atoms as well as to use the simplified harmonic potential field prescribed for alpha-carbon atoms. In particular, the ENM is established in such a way that alpha-carbon atoms within the neighborhood are connected by elastic, harmonic springs with an identical force constant. Despite its simplicity, the ENM is not only computationally efficient but also very robust in predicting the conformational dynamics of large protein complexes. This robustness of the ENM in predicting the protein dynamics is ascribed to the principle that the native topology of a protein structure determines the protein dynamics (e.g., thermal fluctuations) [92] and/or protein mechanics (e.g., mechanical unfolding) [45,46]. This implies that the ENM is appropriate to dictate the key structural feature of a protein material, which is highly correlated with its mechanical properties. This has led researchers to use the ENM for studying the mechanical behavior of protein materials. For instance, Yoon et al. [74] have developed the mesoscopic model of protein materials based on the Gō-like model and/or ENM for characterizing the mechanical properties of protein materials. Golji et al. [115] have reported that the mechanical deformation of α-actinin rod domain, which is computationally predicted from the ENM, is qualitatively comparable to that simulated from MD simulations. Recently, Buehler et al. [39] have studied the mechanical properties of Aβ amyloid fibrils based on an ENM with normal mode analysis (NMA). These studies show that the structural hierarchy (e.g., fibril

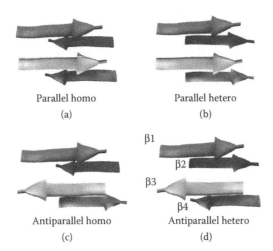

FIGURE 7.10 Four different atomic structures of nano-building blocks for hIAPP amyloid fibrils. (a) Parallel homo (PHM), (b) parallel hetero (PHT), (c) antiparallel homo (AHM), and (d) antiparallel hetero (AHT) configurations are shown. (Reprinted from G. Yoon, J. I. Kim, J. Kwak, S. Na, K. Eom. 2011. *Adv Funct Mater*, in press.)

length) plays a key role in determining the mechanical stability of Aβ amyloid fibril. Deriu et al. have studied the mechanical properties of microtubules based on the ENM. These studies shed light on the ENM as a robust simulation toolkit for studying the mechanical properties of a soft protein material.

In this section, we have studied the mechanical stability of hIAPP fibrils, which takes a functional role on type II diabetes expression, with respect to their structural hierarchies and molecular architectures using NMA with the ENM. Here, we have considered four possible molecular architectures (as shown in Figure 7.10) for hIAPP fibrils, where two molecular architectures are experimentally observed, to study the relationship between the mechanical stability of hIAPP fibrils and the types of their molecular architectures. Moreover, we have also investigated the length-dependent mechanical stability of hIAPP fibrils. It has been found that the mechanical stability of hIAPP fibrils is highly correlated with not only the type of their molecular architecture but also their length. Remarkably, the length-dependent mechanical stability of amyloid fibrils is well elucidated from continuum mechanics theory such as Timoshenko's beam theory.

7.5.2 Types of Models

7.5.2.1 Coarse-Grained Model of Amyloid Fibrils

7.5.2.1.1 Construction of the hIAPP Fibril

hIAPP amyloid fibril is composed of a building block that consists of four identical β sheets whose sequence is given as "NFGAILS." Specifically, the hIAPP fibril is constructed in such a way that a building block is periodically repeated along the fibril axis. The molecular structure of this building block is deposited in the PDB with a PDB code of 2kib that was identified by 2D NMR spectra [103]. Since a recent study [110] has theoretically provided that there could be other molecular structures for such a building block, we have considered four different types of building blocks as shown in Figure 7.10. Here, we denote four β sheets as β_1, β_2, β_3, and β_4, respectively, which constitute the building block of hIAPP fibrils. The building block deposited as 2kib in the PDB is referred to as the "antiparallel hetero" (AHT) configuration, because two β sheets (e.g., β_1 and β_3) form the antiparallel strand while the other two β sheets (e.g., β_1 and β_2) constitute the registered parallel strand. In a similar manner, we have taken into account four possible configurations for the building block of the hIAPP fibril: (1) parallel homo (PHM), (2) parallel hetero (PHT), (3) AHT, and (4) antiparallel homo (AHM) (for details, see Figure 7.10). Based on the experimental observation, we assume that the $hIAPP_{20-29}$ fibril exhibits the twisted structure along the fibril axis. Specifically, an experimental work [103] shows that the helical pitch of the hIAPP fibril is 25.811 nm and 72 β sheets forms one helical pitch of the fibril. Based on this, we have constructed the hIAPP fibril in such a way that a building block is repeated along the fibril axis but rotated 20° about the fibril axis (Figure 7.11).

7.5.2.1.2 Elastic Network Model

For a long amyloid fibril, the full-atomistic model is computationally restrictive due to the large degrees of freedom and complicated potential field. To computationally obtain the natural frequencies (and their corresponding deformation modes) related to the mechanical stability of amyloid fibrils, we have used the CG model that allows for computationally efficient analyses on protein mechanics. Specifically, we have used NMA with the ENM that has been widely employed to characterize the large protein mechanics. As stated earlier, the ENM describes a protein structure based on alpha-carbon atoms in such a way that alpha-carbon atoms in the neighborhood are linked by an elastic, harmonic spring with an identical force constant. With the given structural information of hIAPP fibrils constructed based on a method delineated in Section 7.5.2.1.1, it is straightforward to establish an ENM using alpha-carbon atoms for hIAPP fibrils. With given coordinates of alpha-carbon atoms for hIAPP fibrils, the potential energy V for the ENM (which describes the amyloid structure) is given by [40]

$$V = \sum_{i=1}^{N-1} \sum_{j=i+1}^{N} \frac{\gamma_{ij}}{2} \left[|\mathbf{r}_i - \mathbf{r}_j| - |\mathbf{r}_i^0 - \mathbf{r}_j^0| \right]^2 \cdot H\left(r_c - |\mathbf{r}_i^0 - \mathbf{r}_j^0| \right) \qquad (7.19)$$

where $\mathbf{r}_i$ is the position vector of the ith alpha-carbon atom, γ_{ij} is the force constant for an elastic spring that connects the ith and jth alpha-carbon atoms, r_c is the cutoff distance (which is usually given as $r_c = \sim 10$ Å), superscript 0 indicates the equilibrium conformation (i.e., native conformation), and $H(x)$ is the Heaviside unit step function defined as $H(x) = 0$ if $x < 0$; otherwise $H(x) = 1$. Here, a force constant γ_{ij} is set to be either $\gamma_{ij} = 0.5$ kcal/mol · Å² for nonbonded interactions (i.e., native contact; $j \neq i + 1$) [87] or $\gamma_{ij} = 100$ kcal/mol · Å² for covalent bonds (i.e., backbone chain; $j = i + 1$).

For implementing NMA that provides the natural frequencies (and their corresponding deformation modes) for a long amyloid fibril, we need to compute the stiffness

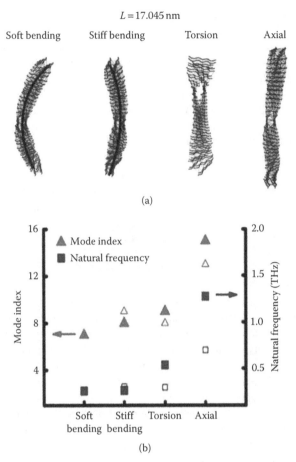

(b)

FIGURE 7.11 Vibrational characteristics on the deformation of amyloid fibrils. (a) Vibrational mode shapes of amyloid fibrils, which are anticipated from ENM-based coarse-graining description, are shown. The solid thick line represents the vibrational mode shape predicted from the continuum elastic beam model. (b) Frequency behaviors (e.g., frequencies and vibrational modes) of amyloid fibril are shown. It is shown that bending deformation modes correspond to the low-frequency mode, while the other two modes such as axial stretching and torsional modes become the high-frequency mode. Filled symbols indicate our simulation results on hIAPP amyloid fibrils, whereas open symbols show the simulation results on Aβ amyloids reported in reference [39]. (Reprinted from G. Yoon, J. I. Kim, J. Kwak, S. Na, K. Eom. 2011. *Adv Funct Mater*, in press.)

matrix based on a given potential energy for the ENM depicted in Equation 7.19. The stiffness matrix **K** consists of 3×3 block matrices $\mathbf{K}_{ij}$ defined as $\mathbf{K}_{ij} = -\partial^2 V(\mathbf{r}_1, \ldots, \mathbf{r}_N)/\partial \mathbf{r}_i \partial \mathbf{r}_j$ given as [116]

$$\mathbf{K}_{ij} = -\left[\gamma H\left(r_c - \left| \mathbf{r}_i^0 - \mathbf{r}_j^0 \right| \right) \frac{\left(\mathbf{r}_i^0 - \mathbf{r}_j^0 \right) \otimes \left(\mathbf{r}_i^0 - \mathbf{r}_j^0 \right)}{\left| \mathbf{r}_i^0 - \mathbf{r}_j^0 \right|^2} \right] (1 - \delta_{ij}) - \delta_{ij} \sum_{l \neq i}^{N} \mathbf{K}_{il} \qquad (7.20)$$

Here, δ_{ij} is the Kronecker delta defined as $\delta_{ij} = 1$ if $i = j$; otherwise, $\delta_{ij} = 0$. Based on the stiffness matrix $\mathbf{K}$, the vibrational motion (i.e., frequencies related to the mechanical properties) of an amyloid fibril can be obtained from NMA such that $\mathbf{Kv} = m\omega^2\mathbf{v}$, where m is the molecular weight of alpha-carbon atom, while ω and $\mathbf{v}$ represents the natural frequency and its corresponding deformation mode, respectively, for ENM descriptions of an amyloid fibril. Here, we excluded the six zero normal modes that correspond to three translational and three rotational rigid body motions. For visualization of the vibrational deformation mode for an amyloid fibril, we introduce the displacement vector for an amyloid fibril in such a way that every alpha-carbon atom moves along the direction of the deformation eigenmode: the position vector of the ith alpha-carbon atom for a deformed amyloid fibril is given by $\mathbf{r}_i^* = \mathbf{r}_i^0 + \alpha_i\mathbf{v}_i^k$, where $\mathbf{r}_i^0$ is the position vector for the ith alpha-carbon atom for an amyloid fibril at native conformation, $\mathbf{v}_i^k$ indicates the directional unit vector (i.e., normalized eigenvector) of the ith alpha-carbon atom at the kth vibrational mode, and α_i is a constant that is determined in such a way that the root mean square distance (RMSD) between native conformation and deformed conformation becomes 1 nm.

7.5.2.2 Continuum Mechanics Model: Euler–Bernoulli Beam Model

As described in Section 7.5.2.1, NMA along with CG descriptions of amyloid fibrils (i.e., ENM-based descriptions) can provide the natural frequencies of an amyloid fibril. Now, to extract the mechanical properties of an amyloid fibril from the measured natural frequencies, we use the continuum mechanics model, especially the Euler–Bernoulli beam model [78,117,118]. In particular, we have described an amyloid fibril as an elastic one-dimensional beam model, because amyloid fibrils exhibit the characteristics such that the transverse dimension (i.e., cross-sectional dimension) is several orders of magnitude smaller than the longitudinal dimension (i.e., length). In the last decade, this simple elastic model (i.e., one-dimensional beam model) is able to capture the nanoscale mechanical characteristics of some nanoscale structures such as nanowires [119], carbon nanotubes [120], and biological filaments (e.g., microtubules) [97], as long as the transverse dimension of the structure is much smaller than the longitudinal dimension by several orders. The mechanical behaviors (i.e., vibration) of a one-dimensional elastic beam model can be described by three major deformation modes such as bending mode, axial stretching mode, and torsional mode. The Euler–Bernoulli beam model depicts these three major deformation modes based on the equations of motion such as [121]

$$E_B I \frac{\partial^4 w(x,t)}{\partial x^4} + \rho A \frac{\partial^2 w(x,t)}{\partial t^2} = 0 \quad \text{for bending mode} \tag{7.21a}$$

$$YA \frac{\partial^2 u(x,t)}{\partial x^2} - \rho A \frac{\partial^2 u(x,t)}{\partial t^2} = 0 \quad \text{for axial stretching mode} \tag{7.21b}$$

$$G_T J \frac{\partial^2 \phi(x,t)}{\partial x^2} - \rho J \frac{\partial^2 \phi(x,t)}{\partial t^2} = 0 \quad \text{for torsional mode} \tag{7.21c}$$

where E_B, Y, and G_T indicate the elastic bending modulus, elastic axial modulus, and torsional shear modulus of an amyloid fibril, respectively, whereas ρ, A, I, and J represent the mass density, cross-sectional area, cross-sectional moment of inertia, and cross-sectional polar moment of inertia, respectively, of an amyloid fibril. The variables $w(x, t)$, $u(x, t)$, and $\phi(x, t)$ represent the transverse displacement (for bending mode), axial displacement (for axial stretching mode), and twist angle (for torsional mode), respectively.

For the vibrational motion of an amyloid fibril, the displacement fields (i.e., transverse displacement, axial displacement, and twist angle) are assumed to be in the form of $w(x, t) = W(x)\exp[j\omega_B t]$, $u(x, t) = U(x)\exp[j\omega_A t]$, and $\phi(x, t) = \Phi(x)\exp[j\omega_T t]$, where ω_B, ω_A, and ω_T indicate the natural frequencies for bending mode, axial stretching mode, and torsional mode, respectively, whereas $W(x)$, $U(x)$, and $\Phi(x)$ are the bending deformation eigenmode, axial stretching eigenmode, and torsional deformation eigenmode, respectively. Consequently, the equations of motion for the vibrational deformation of an amyloid fibril become

$$EI\frac{d^4 W}{dx^4} - \rho A\omega_B^2 W = 0 \quad \text{for bending deformation mode} \tag{7.22a}$$

$$YA\frac{d^2 U}{dx^2} + \rho A\omega_A^2 U = 0 \quad \text{for axial stretching mode} \tag{7.22b}$$

$$G_T J\frac{d^2 \Phi}{dx^2} + \rho J\omega_T^2 \Phi = 0 \quad \text{for torsional deformation mode} \tag{7.22c}$$

The explicit solutions to the ordinary differential equations for each vibrational mode, depicted in Equation 7.22, can be analytically found: $\omega_B^{(n)} = (\lambda_n/L)^2 \sqrt{E_B I/\rho A}$, $\omega_A^{(n)} = (n\pi/L)\sqrt{Y/\rho}$, and $\omega_T^{(n)} = (n\pi/L)\sqrt{G_T/\rho}$, where superscript (n) indicates the nth mode index, and λ_n is the mode index–dependent constant, for example, $\lambda_n = 4.37$. The vibrational mode shapes for each deformation mode are given as

$$W_n(x) = A_n^B\left[\cosh(\lambda_n x/L) + \cos(\lambda_n x/L) - \sigma_n\left\{\sinh(\lambda_n x/L) + \sin(\lambda_n x/L)\right\}\right] \tag{7.23a}$$

$$U_n(x) = A_n^A \cos\frac{n\pi x}{L} \tag{7.23b}$$

$$\Phi_n(x) = A_n^T \cos\frac{n\pi x}{L} \tag{7.23c}$$

where A represents the amplitude for each deformation mode, superscripts B, A, and T indicate the bending mode, axial stretching mode, and torsional mode, respectively, subscript n means the nth mode index, and σ_n is a constant given as $\sigma_n = 0.98$. If the deformation modes for an amyloid fibril, obtained from NMA with the ENM, can be well fitted to those anticipated from the continuum mechanics model depicted as Equation 7.23, then it is straightforward to extract the mechanical properties of an amyloid fibril from the measured fundamental frequencies such as

$$E_B = \frac{\rho A L^4}{\lambda_1^4 I} \left(\omega_1^B \right)^2 \qquad (7.24a)$$

$$Y = \frac{\rho L^2}{\pi^2} \left(\omega_1^A \right)^2 \qquad (7.24b)$$

$$G_T = \frac{\rho L^2}{\pi^2} \left(\omega_1^T \right)^2 \qquad (7.24c)$$

7.5.3 RESULTS

7.5.3.1 Vibrational Characteristics of hIAPP Fibrils

To ensure that a continuum mechanics model (described in Section 7.5.2.2) is able to describe the vibrational deformation mode of an amyloid fibril, we have considered the vibrational modes obtained from the ENM-based NMA simulations. Figure 7.11 shows the major vibrational mode shapes for a hIAPP fibril (in the AHT configuration) with its length of 17.045 nm (equivalent to a single helical pitch). As shown in Figure 7.11, these vibrational mode shapes correspond to the bending, axial stretching, and torsional deformation modes, respectively, which implies that the vibrational deformation of an amyloid fibril can be depicted with a one-dimensional elastic beam model such as the Euler–Bernoulli beam model. Here, we have excluded the deformation modes that correspond to the rigid body motions (i.e., six zero normal modes). The lowest frequency (i.e., seventh mode) is ~0.1 THz, and its corresponding vibrational mode is the bending deformation, where this frequency relevant to the bending mode is quantitatively comparable to that of an Aβ amyloid reported in reference [39]. It is shown that there are two lowest frequency deformation modes that correspond to the two bending modes, which indicates that the hIAPP fibril exhibits anisotropic bending rigidities. This may be attributed to the anisotropic properties of the cross-sectional area. Specifically, there are two major principal cross-sectional moments of inertia I_{max} and I_{min}, which represent the cross-sectional moments of inertia with respect to two principal axes, respectively. As shown in Figure 7.11, the two lowest frequency deformation modes corresponding to the bending modes are well fitted to the theoretical model based on the Euler–Bernoulli beam theory depicted in Equation 7.23. In particular, the bending deformation modes are well dictated using amplitudes such as $A_1^B = 0.03$ nm (for bending with respect to a soft axis) and $A_2^B = 0.01$ nm (for bending about a stiff axis). Because the thermal fluctuations of protein materials are mostly contributed by low-frequency deformation modes [43], the thermal fluctuation motion of an amyloid fibril is mostly attributed to the bending deformation modes corresponding to the lowest frequency normal modes. This is consistent with AFM-based experimental studies [50,122] reporting that the bending rigidity $E_B I$ (related to persistent length L_p) can be measured from the statistical mechanics theory, particularly the wormlike chain model [123,124] that presumes that the thermal fluctuation motion of a fiber (e.g., DNA [125–127], microtubules [97], amyloid fibrils [108]) is ascribed to the fluctuation of bending angles (which results in the bending deformation). On the other hand, high-frequency deformation modes that do not contribute to the thermal fluctuation are depicted as axial stretching and torsional modes, respectively, as presented in Figure 7.11. This elucidates that the principal nanomechanics of an amyloid fibril is well dictated as the bending deformation. Furthermore, it is found that high-frequency deformation modes also include the coupled deformation modes such as coupling between bending and axial stretching modes (or coupling between bending and torsional modes). In this section, we disregard these coupled deformation modes that are high-frequency modes, which are not key eigenmodes that determine the fundamental mechanics (e.g., fluctuation) of an amyloid fibril.

7.5.3.2 Mechanical Properties of hIAPP Amyloid Fibrils

We have theoretically measured the mechanical properties (e.g., bending elastic modulus E_B, axial elastic modulus Y, and torsional shear modulus G_T) of amyloid fibrils using the natural frequencies of amyloid fibrils measured from the CG model (i.e., ENM). Specifically, once the natural frequencies relevant to a specific deformation mode (e.g., bending, axial stretching, and torsional modes, respectively) are calculated from NMA along with the ENM, the mechanical properties of amyloid fibrils are computed from the Euler–Bernoulli beam model depicted in Equation 7.24 that relates the measured natural frequencies to the mechanical properties of amyloid fibrils. Figure 7.12a shows the elastic moduli for each deformation mode (i.e., bending, axial stretching, and torsional modes) of hIAPP fibrils (whose configuration is AHT) with respect to their length. It is found that the bending elastic modulus, E_B, of the hIAPP amyloid fibril is in the range of 7–40 GPa, which is quantitatively comparable to the bending modulus of Aβ amyloid fibrils (in simulation) [39] and other amyloid fibrils (in experiment) [10]. In addition, the axial elastic modulus Y of hIAPP amyloid fibrils is 12 ~ 13 GPa, which is quantitatively comparable to that (Y) computed from MD simulations of Aβ amyloid fibrils [106]. It is also shown that for hIAPP amyloid fibrils, the soft bending elastic modulus (E_B) is much higher than other elastic moduli for axial stretching and torsional modes. This implies that a hIAPP amyloid fibril can resist the mechanical bending deformation in physiological conditions. This suggests that the significant deformation mode for amyloid fibrils is the bending deformation, which is consistent with our finding that the bending deformation plays a significant role in the nanomechanics of amyloid fibrils. In particular, as discussed earlier, the lowest frequency deformation modes that significantly contribute to the thermal fluctuations of amyloid fibrils are the bending deformation modes, while the other deformation modes (i.e., axial stretching and torsional modes) are relatively high-frequency vibrational modes that are not involved in the thermal fluctuations. As shown in Figure 7.12c, as the fibril length increases, the axial stretching and torsional deformation modes become high-frequency vibrational modes, while the bending deformation mode is the lowest frequency normal mode regardless of the fibril length.

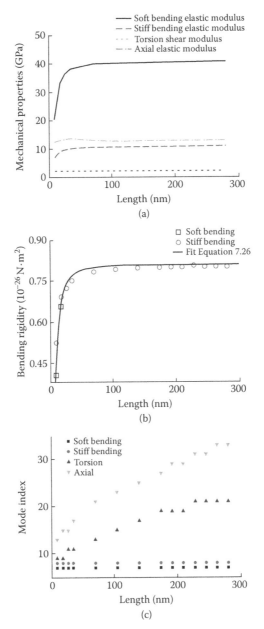

FIGURE 7.12 Mechanical properties of amyloid fibrils with respect to their length. (a) Elastic moduli of amyloid fibrils for each vibrational mode (i.e., bending, axial stretching, and torsional modes) are computed as a function of the fibril length. (b) Bending rigidities of amyloid fibrils are calculated from our simulation with respect to the fibril length. In (a) and (b), the solid line represents the theoretical predictions from the Timoshenko beam model. (c) Mode index (for each vibrational mode) versus the fibril length. It is shown that for longer amyloid fibrils, the axial stretching and torsional modes are the higher-frequency modes. (Reprinted from G. Yoon, J. I. Kim, J. Kwak, S. Na, K. Eom. 2011. *Adv Funct Mater*, in press.)

Moreover, we have also found that the bending elastic modulus E_B is strongly dependent on the fibril length, while the other elastic moduli (e.g., axial elastic modulus Y) are almost independent of the fibril length (see Figure. 7.12a and b). To scrutinize the length-dependent mechanical stabilities of amyloid fibrils, we have considered the bending rigidity $E_B I$ rather than the bending elastic modulus E_B itself because we have anisotropic properties of the cross-sectional shape (i.e., two major cross-sectional moments of inertia, I_{max} and I_{min}). It is revealed that although amyloid fibrils exhibit the anisotropic bending elastic moduli E_B, when the fibril length approaches 70 nm, the bending rigidities with respect to soft and stiff axes are identical with each other, which indicates that the amyloid fibrils with their length of >70 nm possess the isotropic bending rigidity $E_B I$ (Figure 7.12b). This may be ascribed to the geometric shape of amyloid fibrils such that the isotropy of material properties is related to the helically twisted configuration of amyloid fibrils [39]. It is interestingly shown that hIAPP amyloid fibrils have the length-dependent bending rigidity $E_B I$ such that when the fibril length is <70 nm, the bending rigidity significantly depends on the fibril length, and the bending rigidity becomes a steady-state value when the fibril length reaches 70 nm. This length-dependent mechanical property (i.e., bending rigidity) is also found in the case of Aβ amyloids reported in reference [39], which found that the bending rigidity of Aβ amyloid fibrils becomes independent of the fibril length, when the fibril length becomes >200 nm. This length-dependent bending behavior suggests that when a hIAPP fibril formed from aggregation and self-assembly exhibits its length of 70 nm, it has optimum mechanical stability, that is, it can effectively resist the mechanical bending deformations. It is conjectured that the critical length scale for protein aggregation that can improve the mechanical stability is ~70 nm. In other words, when an aggregated hIAPP fibril exhibits a length of ~70 nm, such a fibril exhibits excellent mechanical stability (rigidity) so that it can effectively replace specific cells such as insulin-secreting β cells at a specific functional site such as the islet of the pancreas. We propose that this critical length scale for self-aggregation that remarkably increases the mechanical stability of a fibril should be experimentally validated. The detailed discussion on the length-dependent mechanical property is provided in Section 7.5.4.

7.5.3.3 Bending versus Torsion

We need to consider two deformation modes—bending and torsional deformations—since the torsional rigidity $G_T J$ of a hIAPP fibril is quantitatively comparable to its bending rigidity $E_B I$, albeit the torsional deformation mode is the high-frequency vibrational mode. In other words, the thermal fluctuation motions of amyloid fibrils are not attributed to the torsional deformation mode, which implies that the experiment based on measurement of the thermal fluctuation of biological fibers (e.g., see references [97,108,125]) cannot provide any information on the amyloid's mechanical properties related to the torsional deformation. For quantitative comparison between bending and torsional deformations, we have introduced a dimensionless measure χ defined as $\chi = E_B I / G_T J$. This dimensionless measure describes the competition between two mechanical stabilities corresponding to bending and torsional deformations, respectively. For instance, a biological fiber such as bacterial flagellar hook and filament is found to have the dimensionless measure χ, where $\chi < 1$ [128].

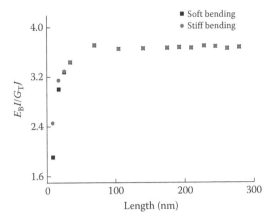

FIGURE 7.13 Bending/twist ratio for amyloid fibrils. As the fibril length increases, the bend-to-twist ratio is significantly increased, which implies that self-assembly and aggregation result in the improvement of mechanical resistance to the bending deformation. (Reprinted from G. Yoon, J. I. Kim, J. Kwak, S. Na, K. Eom. 2011. *Adv Funct Mater*, in press.)

This indicates that the biological fiber can be easily bent rather than twisted in physiological conditions. However, we have interestingly found that unlike the bacterial flagellar hook and filament, amyloid fibrils have a characteristic of $\chi > 1$ (see Figure 7.13), indicating that amyloid fibrils can effectively resist the bending deformation that plays a key role in the mechanics of biological fibers. In particular, as the fibril length increases, χ is significantly amplified. This suggests that the protein aggregation results in the improvement of mechanical stability for the bending deformation. As shown in Figure 7.13, when the fibril length approaches ~70 nm, χ becomes a steady-state value such as $\chi \to$ ~4, which elucidates the significant enhancement of the mechanical resistance of amyloid fibrils to the bending deformation. This value (i.e., $\chi \to$ ~4 when $L \to$ ~70 nm) for amyloid fibrils is much larger than that (e.g., $\chi = 1.3$) for steel solid circular cylinder. This highlights the anomalous mechanical stability of amyloid fibrils with respect to the bending deformations. This clearly demonstrates how protein aggregation gives rise to the enhancement of the mechanical stability of amyloid fibrils under physiological conditions.

7.5.3.4 Role of Molecular Architectures in Mechanical Properties

As stated earlier with Figure 7.10, there could be a few possible molecular structures for a building block that constitutes the amyloid fibrils: (1) PHM, (2) PHT, (3) AHT, and (4) AHM. This has led us to study the mechanical properties of amyloid fibrils whose building blocks are in these configurations, respectively, to gain a fundamental insight into the role of atomistic structures in the mechanical stability of amyloid fibrils. Figure 7.14 shows the mechanical properties of amyloid fibrils with respect to the atomistic structures of their building blocks as a function of the fibril length. It is interestingly found that the bending elastic moduli for hIAPP fibrils are in the order of PHT < PHM < AHT < AHM. For instance, the bending elastic modulus for a hIAPP fibril composed of a building block in the form of PHT is ~38 GPa, which is smaller than that (i.e., ~42 GPa) in the form of AHM, at the fibril length of $L = 280$ nm.

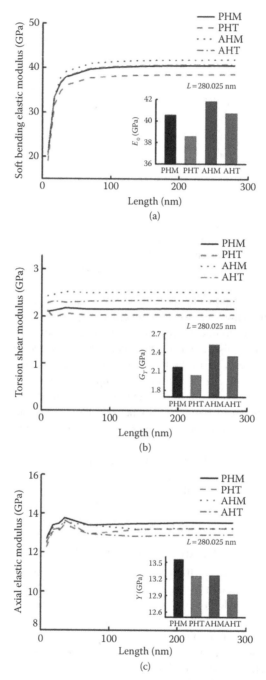

FIGURE 7.14 Mechanical properties of amyloid fibrils with respect to the atomic structures of the building block composing the amyloid: (a) Bending elastic modulus, (b) torsional shear modulus, and (c) axial elastic modulus. (Reprinted from G. Yoon, J. I. Kim, J. Kwak, S. Na, K. Eom. 2011. *Adv Funct Mater*, in press.)

This indicates that the stacking of β sheets in an antiparallel manner plays a key role in the mechanical stability of amyloid fibrils, particularly their excellent mechanical resistance to the bending deformation. This can be elucidated from our finding from simulation. In particular, when all four β sheets are stacked in a parallel manner (i.e., PHT configuration), the bending rigidity becomes the minimal value, while the maximum value of bending rigidity is found for AHM, where all four β sheets are stacked in an antiparallel manner. The intermediate values of bending rigidity were found for PHM and AHT, where two β sheets are stacked in an antiparallel manner while the other two β sheets are laid in a parallel manner. Our finding of the role of the stacking manner in the mechanical stability of β-sheet crystal–based materials is consistent with a previous study [11], which has shown that β-sheet crystals formed by stacking β strands in an antiparallel manner is responsible for the mechanical stiffness (toughness) that governs the excellent mechanical resistance of spider silk protein. This highlights the fact that the stacking of β sheets in an antiparallel manner is an efficient route for enhancing the mechanical stability of amyloid fibrils, especially their remarkable mechanical resistance to the bending deformation.

For gaining deep insights into the structure–property relation for amyloid fibrils, we have considered the contact order (C) that quantitatively describes how a protein structure is folded. In other words, a contact order quantifies how many hydrogen bonds (which play a role in protein folding) a protein structure exhibits. It is shown that the contact order ($C = 0.37$) for β sheets stacked in an antiparallel manner is larger than that ($C = 0.35$) for β sheets stacked in a parallel manner. This clearly demonstrates that the stacking manner, which is related to mechanical stability, is highly correlated with the number of hydrogen bonds. It may imply that the mechanical stability of protein materials is highly correlated with the stacking manner that is directly related to the number (or configuration) of hydrogen bonds. This is consistent with a recent suggestion that the mechanical properties of protein materials are determined by how the hydrogen bonds between β strands are geometrically confined. Furthermore, it is also found that stacking β sheets in an antiparallel manner results in the improvement of the torsional shear modulus G_T, which is larger than that for stacking β sheets in a parallel manner. In summary, the antiparallel stacking of β sheets is an effective method to enhance the mechanical stability of amyloid fibrils with respect to bending and torsional deformations.

However, unlike the case of elastic bending properties, it is interesting in that the axial elastic moduli Y for amyloids composed of building blocks (in their different configurations) are in the order of AHT < PHT ≈ AHM < PHM. This indicates that a β-sheet crystal formed by stacking β strands in a parallel manner is able to effectively resist the axial stretching deformation. In other words, when a force is applied perpendicular to the plane formed by β sheets, the β-sheet crystal formed by parallel stacking exhibits better mechanical resistance than that constructed by antiparallel stacking. This suggests that for the design of β-sheet crystal–based materials that can possess remarkable axial elastic stiffness, the stacking of β strands in a parallel manner results in high mechanical toughness to the axial deformation. This finding is consistent with previous studies [23,25,51,52,63] reporting that parallel strands (which can be formed by parallel stacking of β sheets) are a mechanically efficient clamp when a force is applied to the termini of a protein domain.

Our finding on two different features of stacking manner–dependent mechanical properties for bending and axial stretching deformations suggests that the effective design of protein materials is determined by not only the stacking manner but also the loading mode. The development of protein-based material that can be mechanically resistant to the bending deformation could be made possible using antiparallel stacking. On the other hand, the effective design of protein material with a high mechanical stability for axial stretching deformation can be developed using parallel stacking. This shows how protein materials can achieve excellent mechanical stability related to their biological functions based on their atomistic structural designs.

7.5.4 LENGTH-DEPENDENT MECHANICAL PROPERTIES

In Section 7.5, we have studied the mechanical stability of amyloid fibrils, which play an important role in disease expressions such as type II diabetes, using ENM-based NMA. It is shown that the vibrational modes of amyloid fibrils have been well described by continuum mechanics models such as the classical Euler–Bernoulli beam model. We have interestingly found that the bending rigidity of amyloid fibrils is strongly dependent on the fibril length. This length-dependent bending property of amyloid fibrils cannot be elucidated from the Euler–Bernoulli beam theory because such a theory ignores the shear effects that significantly affect the bending deformation of a fiber to possess its low aspect ratio.

To understand the length-dependent bending properties of amyloid fibrils, we have employed the Timoshenko beam model [118] that accounts for the shear effects on the bending deformation. This model has been successfully used to elucidate the length-dependent bending rigidities of biological materials such as microtubules and β-sheet crystals. In particular, when a force F is applied to the middle point of a fibril whose two ends are clamped on the substrate, the deflection δ of the fibril is attributed to not only the bending deformation but also the shear deformation.

$$\delta = \frac{FL^3}{aE_B^0 I} + \frac{cFL}{bG_S A} \tag{7.25}$$

where E_B^0 and G_S represent the length-independent, intrinsic elastic bending modulus and shear modulus of an amyloid fibril, respectively, L, I, and A are the length, cross-sectional moment of inertia, and cross-sectional area of an amyloid fibril, respectively, while a and b are constants that depend on the boundary condition, and c is a shear coefficient that depends on the cross-sectional shape (e.g., $c = 3/2$ for a rectangular cross-sectional shape). Since we have a relationship between deflection δ and bending elastic modulus E_B as $\delta = FL^3/aE_B I$, the bending elastic modulus E_B is represented in the form

$$E_B = E_B^0 \left(1 + \frac{a}{b} \frac{cE_B^0 I}{G_S AL^2} \right)^{-1} \tag{7.26}$$

Equation 7.26 shows the dependence of the elastic bending modulus of an amyloid fibril on its length. This model is able to capture the length-dependent bending properties of amyloid fibrils, which are computed from ENM-based NMA (see Figure 7.12a). In particular, it is found that a hIAPP fibril has the intrinsic properties such as $E_B^0 = 0.8 \times 10^{-26} \, \text{N} \cdot \text{m}^2$ and $G_S = 1.1$ GPa. To quantify the role of shear effects on the bending stability of an amyloid fibril, we have introduced a dimensionless measure ε defined as $\varepsilon = (a/b)\left(cE_B^0 I/G_S AL^2\right)$, which indicates how the bending deformation is contributed to by shear effects. Based on this dimensionless measure, we have found that for the fibril with a length of >70 nm, ε is given as $\varepsilon < 1\%$, which indicates that the shear effect becomes insignificant in the mechanical deformation of an amyloid fibril when its length reaches ~70 nm. This suggests that the critical length scale, at which an amyloid fibril becomes mechanically stable, can be determined based on two intrinsic material properties such as E_B^0 and G_S.

In conclusion, we have provided the fundamental mechanisms on how the mechanical stability of amyloid fibrils could be optimized. Specifically, the mechanical stability of an amyloid fibril is determined by the atomistic structural design such as the stacking manner, and the stacking manner–based optimal design is also dependent on the loading mode. Remarkably, we have shown that the mechanical stability of an amyloid fibril is highly correlated with the fibril length in such a way that the mechanical stiffness of the amyloid increases as the fibril length increases, which makes it clear that the protein aggregation is an efficient route for enhancing the mechanical stability. This length-dependent bending behavior has been well described by the Timoshenko beam model. It is suggested that by using the Timoshenko model the critical length scale at which the mechanical stability is achieved can be determined from two design parameters E_B^0 and G_S.

7.6 OUTLOOK AND PERSPECTIVES

In this chapter, we have provided the recent advances in CG modeling techniques that allow the mechanical characterization of protein materials across multiple length scales ranging from atomic scale, to single-molecule scale, to mesoscale. In particular, we have first described the CG elastic chain model that can capture the mechanics of a single-protein domain, particularly protein unfolding mechanics. This CG model provides fundamental insights into the relationship between the native topology and the mechanical unfolding characteristics. Second, we have suggested the mesoscale model, which allows for quantitative predictions on the mechanical properties of protein crystals at mesoscales, based on the Go-like model (or ENM) coupled with the micromechanics model. This model clearly elucidates that the mechanical properties of protein crystals at mesoscale are highly correlated with the native topology of protein domains. Finally, we have described the ENM-based CG modeling of amyloid fibrils, which play an important role in disease expressions due to their excellent mechanical properties.

We should note that each CG model has its own advantages and disadvantages. In particular, the CG elastic chain model is still not able to mimic the experimentally or computationally observed mechanical unfolding behavior of a highly packed, macromolecular protein domain (i.e., globular domain), albeit the elastic chain model is able to roughly provide the insights into the role of cross-link (bond) topology on the mechanical properties of cross-linked polymers such as small proteins. Second, the micromechanics-based mesoscale model is not able to accurately capture the mechanical deformation behavior of large protein complexes (or protein materials) because this model neglects the intermolecular interactions between secondary structures (or interactions between protein domains). As presented in reference [24], the nanomechanical behavior of an α/β dimer comprising a microtubule is different from that of a single α or β monomer itself, which implies the significant role of the intermolecular interactions in the mechanical deformation of proteins. Moreover, as described in references [99,100], the mechanical properties of protein materials are determined by the configuration of intermolecular interactions between the secondary structures.

As compared to the previously mentioned two models—the elastic chain model and micromechanics-based mesoscopic model—the ENM-based CG modeling techniques are suitable to study the mechanical behavior of large protein complexes (or protein material) such as amyloid fibril. However, our CG model can only be applied to describe the single amyloid fiber whose length is at most ~300 nm, which is relatively shorter than that of fibril experimentally observed in patients. To characterize the experimentally observed amyloid fibrils, our CG model has to be extended to describe the longer fibril structure as well as multiple hierarchical structures ranging from single amyloid fibril, protofilament (bundle of amyloid fibrils), even up to plaques formed by the aggregation of protofilament. The de novo computational techniques (e.g., techniques that allow the computationally efficient diagonalization of the Hessian) have to be developed to apply NMA to the amyloid materials at multiple hierarchies. Moreover, our CG descriptions of the mechanical properties of amyloid fibrils are based on NMA (i.e., harmonic approximation), which implies that our coarse-graining techniques can only provide the linear elastic material properties, although the mechanical deformation of protein materials cannot be described using only the linear elastic model. To overcome this restriction, we need to improve our coarse-graining techniques by introducing MD-like simulations that can describe the deformation of the atomic structure due to the mechanical force.

In summary, we conclude that our coarse-graining techniques may be ad hoc models, so there is room for researchers including ourselves to develop the physical models that can capture the characteristics of protein materials across multiple length scales and/or multiple hierarchies. Nevertheless, our coarse-graining techniques may provide further guidance in the development of generic coarse-graining techniques that are capable of quantitatively describing the mechanical stability of various protein materials at multiple scales.

ACKNOWLEDGMENT

We gratefully acknowledge the financial support from the National Research Foundation of Korea (NRF) under Grant No. NRF-2010-0026223.

REFERENCES

1. Shao, Z., and F. Vollrath. 2002. Surprising strength of of silkworm silk. *Nature* 418:741.
2. Gosline, J., P. Guerette, C. Ortlepp, and K. Savage. 1999. The mechanical design of spider silks: from fibroin sequence to mechanical function. *J Exp Biol* 202:3295.

3. Porter, D., and F. Vollrath. 2009. Silk as a biomimetic Ideal for structural polymers. *Adv Mater* 21:487.

4. Li, H. 2008. Mechanical engineering of elastomeric proteins: Toward designing new protein building blocks for biomaterials. *Adv Funct Mater* 18:2643.

5. Miller, Y., B. Ma, and R. Nussinov. 2010. Polymorphism in alzheimer A-beta amyloid organization reflects conformational selection in a rugged energy landscape. *Chem Rev* 110:4280.

6. Merlini, G., and V. Bellotti. 2003. Molecular mechanisms of amyloidosis. *N Engl J Med* 349:583.

7. Pepys, M. B. 2006. Amyloidosis. *Annu Rev Med* 57:223.

8. Hoppener, J. W. M., B. Ahren, and C. J. M. Lips. 2000. Islet amyloid and type two diabetes mellitus. *N Engl J Med* 343:411.

9. Buehler, M. J., and Y. C. Yung. 2009. Deformation and failure of protein materials in physiologically extreme conditions and disease. *Nat Mater* 8:175.

10. Knowles, T. P., A. W. Fitzpatrick, S. Meehan, H. R. Mott, M. Vendruscolo, C. M. Dobson, and M. E. Welland. 2007. Role of intermolecular forces in defining material properties of protein nanofibrils. *Science* 318:1900.

11. Keten, S., Z. Xu, B. Ihle, and M. J. Buehler. 2010. Nanoconfinement controls stiffness, strength and mechanical toughness of beta-sheet crystals in silk. *Nat Mater* 9:359.

12. Suresh, S. 2007. Biomechanics and biophysics of cancer cells. *Acta Mater* 55:3989.

13. Suresh, S. 2007. Nanomedicine: Elastic clues in cancer detection. *Nat Nanotechnol* 2:748.

14. Cross, S. E., Y.-S. Jin, J. Rao, and J. K. Gimzewski. 2007. Nanomechanical analysis of cells from cancer patients. *Nat Nanotechnol* 2:780.

15. Stolz, M., R. Gottardi, R. Raiteri, S. Miot, I. Martin, R. Imer, U. Staufer et al. 2009. Early detection of aging cartilage and osteoarthritis in mice and patient samples using atomic force microscopy. *Nat Nanotechnol* 4:186.

16. Roos, W. H., R. Bruinsma, and G. J. L. Wuite. 2010. Physical virology. *Nat Phys* 6:733.

17. Michel, J. P., I. L. Ivanovska, M. M. Gibbons, W. S. Klug, C. M. Knobler, G. J. L. Wuite, and C. F. Schmidt. 2006. Nanoindentation studies of full and empty viral capsids and the effects of capsid protein mutations on elasticity and strength. *Proc Natl Acad Sci U S A* 103:6184.

18. Vaziri, A., and A. Gopinath. 2008. Cell and biomolecular mechanics in silico. *Nat Mater* 7:15.

19. Buehler, M. J., S. Keten, and T. Ackbarow. 2008. Theoretical and computational hierarchical nanomechanics of protein materials: Deformation and fracture. *Prog Mater Sci* 53:1101.

20. Buehler, M. J., and S. Keten. 2010. Colloquium: Failure of molecules, bones, and the Earth itself. *Rev Mod Phys* 82:1459.

21. Sotomayor, M., and K. Schulten. 2007. Single-molecule experiments in vitro and in silico. *Science* 316:1144.

22. Gao, M., M. Wilmanns, and K. Schulten. 2002. Steered molecular dynamics studies of titin I1 domain unfolding. *Biophys J* 83:3435.

23. Lu, H., B. Isralewitz, A. Krammer, V. Vogel, and K. Schulten. 1998. Unfolding of titin immunoglobulin domains by steered molecular dynamics simulation. *Biophys J* 75:662.

24. Dima, R. I., and H. Joshi. 2008. Probing the origin of tubulin rigidity with molecular simulations. *Proc Natl Acad Sci U S A* 105:15743.

25. Carrion-Vazquez, M., H. Li, H. Lu, P. E. Marszalek, A. F. Oberhauser, and J. M. Fernandez. 2003. The mechanical stability of ubiquitin is linkage dependent. *Nat Struct Biol* 10:738.

26. Elber, R. 2005. Long-timescale simulation methods. *Curr Opin Struct Biol* 15:151.

27. Sherwood, P., B. R. Brooks, and M. S. P. Sansom. 2008. Multiscale methods for macromolecular simulations. *Curr Opin Struct Biol* 18:630.

28. Tozzini, V. 2005. Coarse-grained models for proteins. *Curr Opin Struct Biol* 15:144.

29. Ayton, G. S., W. G. Noid, and G. A. Voth. 2007. Multiscale modeling of biomolecular systems: in serial and in parallel. *Curr Opin Struct Biol* 17:192.

30. Rader, A. J. 2010. Coarse-grained models: getting more with less. *Curr Opin Pharmacol* 10:753.

31. Voth, G. A. 2009. *Coarse-Graining of Condensed Phase and Biomolecular Systems.* Boca Raton, FL: CRC Press.

32. Eom, K., G. Yoon, J.-I. Kim, and S. Na. 2010. Coarse-grained elastic models of protein structures for understanding their mechanics and dynamics. *J Comput Theor Nanosci* 7:1210.

33. Trylska, J. 2010. Coarse-grained models to study dynamics of nanoscale biomolecules and their applications to the ribosome. *J Phys Condens Matter* 22:453101.

34. Termonia, Y. 1994. Molecular modeling of spider silk elasticity. *Macromolecules* 27:7378.

35. van Beek, J. D., S. Hess, F. Vollrath, and B. H. Meier. 2002. The molecular structure of spider dragline silk: Folding and orientation of the protein backbone. *Proc Natl Acad Sci U S A* 99:10266.

36. Nova, A., S. Keten, N. M. Pugno, A. Redaelli, and M. J. Buehler. 2010. Molecular and nanostructural mechanisms of deformation, strength and toughness of spider silk fibrils. *Nano Lett* 10:2626.

37. Buehler, M. J. 2010. Tu(r)ning weakness to strength. *Nano Today* 5:379.

38. Keten, S., and M. J. Buehler. 2010. Nanostructure and molecular mechanics of spider dragline silk protein assemblies. *J R Soc Interface* 7:1709.

39. Xu, Z. P., R. Paparcone, and M. J. Buehler. 2010. Alzheimer's A-beta(1-40) amyloid fibrils feature size-dependent mechanical properties. *Biophys J* 98:2053.

40. Tirion, M. M. 1996. Large amplitude elastic motions in proteins from a single-parameter, atomic analysis. *Phys Rev Lett* 77:1905.

41. Atilgan, A. R., S. R. Durell, R. L. Jernigan, M. C. Demirel, O. Keskin, and I. Bahar. 2001. Anisotropy of fluctuation dynamics of proteins with an elastic network model. *Biophys J* 80:505.

42. Haliloglu, T., I. Bahar, and B. Erman. 1997. Gaussian dynamics of folded proteins. *Phys Rev Lett* 79:3090.

43. Bahar, I., A. R. Atilgan, M. C. Demirel, and B. Erman. 1998. Vibrational dynamics of folded proteins: Significance of slow and fast motions in relation to function and stability. *Phys Rev Lett* 80:2733.

44. Launey, M. E., M. J. Buehler, and R. O. Ritchie. 2010. On the mechanistic origins of toughness in bone. *Annu Rev Mater Res* 40:25.

45. Paci, E., and M. Karplus. 2000. Unfolding of proteins by external forces and temperature: The importance of topology and energetics. *Proc Natl Acad Sci U S A* 97:6521.

46. Klimov, D. K., and D. Thirumalai. 2000. Native topology determines force-induced unfolding pathways in globular proteins. *Proc Natl Acad Sci U S A* 97:7254.

47. Forman, J. R., and J. Clarke. 2007. Mechanical unfolding of proteins: insights into biology, structure and folding. *Curr Opin Struct Biol* 17:58.

48. Tskhovrebova, L., J. Trinick, J. A. Sleep, and R. M. Simmons. 1997. Elasticity and unfolding of single molecules of the giant muscle protein titin. *Nature* 387:308.

49. Marszalek, P. E., H. Lu, H. B. Li, M. Carrion-Vazquez, A. F. Oberhauser, K. Schulten, and J. M. Fernandez. 1999. Mechanical unfolding intermediates in titin modules. *Nature* 402:100.

50. Strick, T. R., M. N. Dessinges, G. Charvin, N. H. Dekker, J. F. Allemand, D. Bensimon, and V. Croquette. 2003. Stretching of macromolecules and proteins. *Rep Prog Phys* 66:1.

51. Eom, K., P. C. Li, D. E. Makarov, and G. J. Rodin. 2003. Relationship between the mechanical properties and topology of cross-linked polymer molecules: Parallel strands maximize the strength of model polymers and protein domains. *J Phys Chem B* 107:8730.

52. Eom, K., D. E. Makarov, and G. J. Rodin. 2005. Theoretical studies of the kinetics of mechanical unfolding of cross-linked polymer chains and their implications for single-molecule pulling experiments. *Phys Rev E* 71:021904.

53. Doi, M., and S. F. Edwards. 1986. *The Theory of Polymer Dynamics*. New York: Oxford University Press.

54. Hummer, G., and A. Szabo. 2003. Kinetics from nonequilibrium single-molecule pulling experiments. *Biophys J* 85:5.

55. Sheng, Y.-J., S. Jiang, and H.-K. Tsao. 2005. Forced Kramers escape in single-molecule pulling experiments. *J Chem Phys* 123:09112.

56. Dudko, O. K., G. Hummer, and A. Szabo. 2006. Intrinsic rates and activation free energies from single-molecule pulling experiments. *Phys Rev Lett* 96:108101.

57. Dudko, O. K., G. Hummer, and A. Szabo. 2008. Theory, analysis, and interpretation of single-molecule force spectroscopy experiments. *Proc Natl Acad Sci U S A* 105:15755.

58. Bell, G. I. 1978. Models for the specific adhesion of cells to cell. *Science* 200:618.

59. Evans, E., and K. Ritchie. 1997. Dynamic strength of molecular adhesion bonds. *Biophys J* 72:1541.

60. Lin, H.-J., H.-Y. Chen, Y.-J. Sheng, and H.-K. Tsao. 2007. Bell's expression and the generalized garg form for forced dissociation of a biomolecular complex. *Phys Rev Lett* 98:088304.

61. Garg, A. 1995. Escape-field distribution for escape from a metastable potential well subject to a steadily increasing bias field. *Phys Rev B* 51:15592.

62. Kramers, H. A. 1940. Brownian motion in a field of force and the diffusion model of chemical reactions. *Physica* 7:284.

63. Rohs, R., C. Etchebest, and R. Lavery. 1999. Unraveling proteins: A molecular mechanics study. *Biophys J* 76:2760.

64. Frauenfelder, H., S. G. Sligar, and P. G. Wolynes. 1991. The energy landscapes and motions of proteins. *Science* 254:1598.

65. Best, R. B., B. Li, A. Steward, V. Daggett, and J. Clarke. 2001. Can non-mechanical proteins withstand force? Stretching barnase by atomic force microscopy and molecular dynamics simulation. *Biophys J* 81:2344.

66. Brockwell, D. J., G. S. Beddard, E. Paci, D. K. West, P. D. Olmsted, D. A. Smith, and S. E. Radford. 2005. Mechanically unfolding the small, topologically simple protein L. *Biophys J* 89:506.

67. Brockwell, D. J., E. Paci, R. C. Zinober, G. S. Beddard, P. D. Olmsted, D. A. Smith, R. N. Perham, and S. E. Radford. 2003. Pulling geometry defines the mechanical resistance of a b-sheet protein. *Nat Struct Biol* 10:731.

68. Rief, M., M. Gautel, F. Oesterhelt, J. M. Fernandez, and H. E. Gaub. 1997. Reversible unfolding of individual titin immunoglobulin domains by AFM. *Science* 276:1109.

69. Rief, M., J. M. Fernandez, and H. E. Gaub. 1998. Elastically coupled two-level systems as a model for biopolymer extensibility. *Phys Rev Lett* 81:4764.

70. Minajeva, A., M. Kulke, J. M. Fernandez, and W. A. Linke. 2001. Unfolding of titin domains explains the viscoelastic behavior of skeletal myofibrils. *Biophys J* 80:1442.

71. Zhou, H., and Y. Zhang. 2005. Hierarchical chain model of spider capture silk elasticity. *Phys Rev Lett* 94:028104.

72. Becker, N., E. Oroudjev, S. Mutz, J. P. Cleveland, P. K. Hansma, C. Y. Hayashi, D. E. Makarov, and H. G. Hansma. 2003. Molecular nanosprings in spider capture-silk threads. *Nat Mater* 2:278.

73. Kasas, S., A. Kis, B. M. Riederer, L. Forro, G. Dietler, and S. Catsicas. 2004. Mechanical properties of microtubules explored using the finite elements method. *ChemPhysChem* 5:252.

74. Yoon, G., H.-J. Park, S. Na, and K. Eom. 2009. Mesoscopic model for mechanical characterization of biological protein materials. *J Comput Chem* 30:873.

75. Nemat-Nasser, S., and M. Hori. 1999. *Micromechanics: Overall Properties of Heterogeneous Materials*. Amsterdam, North Holland: Elsevier Publisher.

76. Zhou, M. 2003. A new look at the atomistic virial stress: on continuum-molecular system equivalence. *Proc Roy Soc Lond A* 459:2347.

77. Andia, P. C., F. Costanzo, and G. L. Gray. 2006. A classical mechanics approach to the determination of the stress-strain response of particle systems. *Modell Simul Mater Sci Eng* 14:741.

78. Landau, L. D., and E. M. Lifshitz. 1986. *Theory of Elasticity*. St. Louis: Butterworth-Heinemann.

79. Ueda, Y., and N. Go. 1976. Theory of large-amplitude conformational fluctuations in native globular proteins—Independent fluctuating site model. *Int J Pept Protein Res* 8:551.

80. Suezaki, Y., and N. Go. 1975. Breathing mode of conformational fluctuations in globular proteins. *Int J Pept Protein Res* 7:333.

81. Noguti, T., and N. Go. 1982. Collective variable description of small-amplitude conformational fluctuations in a globular protein. *Nature* 296:776.

82. Weiner, S. J., P. A. Kollman, D. A. Case, U. C. Singh, C. Ghio, G. Alagona, S. Profeta, and P. Weiner. 1984. A new force field for molecular mechanical simulation of nucleic acids and proteins. *J Am Chem Soc* 106:765.

83. Cieplak, M., T. X. Hoang, and M. O. Robbins. 2002. Folding and stretching in a Go-like model of titin. *Proteins Struct Funct Genet* 49:114.

84. Cieplak, M., A. Pastore, and T. X. Hoang. 2005. Mechanical properties of the domains of titin in a Go-like model. *J Chem Phys* 122:054906.

85. Sikora, M., J. I. Sulkowska, and M. Cieplak. 2009. Mechanical strength of 17,134 model proteins and cysteine slipknots. *PLoS Comput Biol* 5:e1000547.

86. Sulkowska, J. I., and M. Cieplak. 2008. Selection of optimal variants of Go-like models of proteins through studies of stretching. *Biophys J* 95:3174.

87. Eom, K., S.-C. Baek, J.-H. Ahn, and S. Na. 2007. Coarse-graining of protein structures for normal mode studies. *J Comput Chem* 28:1400.

88. Jang, H., S. Na, and K. Eom. 2009. Multiscale network model for large protein dynamics. *J Chem Phys* 131:245106.

89. Kim, J. I., S. Na, and K. Eom. 2011. Domain decomposition-based structural condensation of large protein structures for understanding their conformational dynamics. *J Comput Chem* 32:161.

90. Tama, F., and Y. H. Sanejouand. 2001. Conformational change of proteins arising from normal mode calculations. *Protein Eng* 14:1.

91. Teeter, M. M., and D. A. Case. 1990. Harmonic and quasiharmonic descriptions of crambin. *J Phys Chem* 94:8091.

92. Lu, M. Y., and J. P. Ma. 2005. The role of shape in determining molecular motions. *Biophys J* 89:2395.

93. Dietz, H., and M. Rief. 2008. Elastic bond network model for protein unfolding mechanics. *Phys Rev Lett* 100:098101.

94. Gould, P. L. 1983. *Introduction to Linear Elasticity*. New York, NY: Springer-Verlag.

95. Kis, A., S. Kasas, B. Babic, A. J. Kulik, W. Beno, G. A. D. Briggs, C. Schonenberger, S. Catsicas, and L. Forro. Nanomechanics of microtubules. 2002. *Phys Rev Lett* 89:248101.

96. Wagner, O., J. Zinke, P. Dancker, W. Grill, and J. Bereiter-Hahn. 1999. Viscoelastic properties of f-actin, microtubules, f-actin/alpha-actinin, and f-actin/hexokinase determined in microliter volumes with a novel nondestructive method. *Biophys J* 76:2784

97. Pampaloni, F., G. Lattanzi, A. Jonas, T. Surrey, E. Frey, and E.-L. Florin. 2006. Thermal fluctuations of grafted microtubules provide evidence of a length-dependent persistence length. *Proc Natl Acad Sci U S A* 103:10248.

98. Tskhovrebova, L., and J. Trinick. 2004. Properties of titin immunoglobulin and fibronectin-3 domains. *J Biol Chem* 279:46351.

99. Qin, Z., and M. J. Buehler. 2010. Molecular dynamics simulation of the alpha-helix to beta-sheet transition in coiled protein filaments: Evidence for a critical filament length scale. *Phys Rev Lett* 104:198304.

100. Qin, Z., L. Kreplak, and M. J. Buehler. 2009. Hierarchical structure controls nanomechanical properties of vimentin intermediate filaments. *PLoS One* 4:e7294.

101. Sawaya, M. R., S. Sambashivan, R. Nelson, M. I. Ivanova, S. A. Sievers, M. I. Apostol, M. J. Thompson et al. 2007. Atomic structures of amyloid cross-beta spines reveal varied steric zippers. *Nature* 447:453.

102. Tycko, R. 2003. Insights into the amyloid folding problem from solid-state NMR. *Biochemistry* 42:3151.

103. Nielsen, J. T., M. Bjerring, M. D. Jeppesen, R. O. Pedersen, J. M. Pedersen, K. L. Hein, T. Vosegaard, T. Skrydstrup, D. E. Otzen, and N. C. Nielsen. 2009. Unique identification of supramolecular structures in amyloid fibrils by solid-state NMR spectroscopy. *Angew Chem Int Ed* 48:2118.

104. Jahn, T. R., O. S. Makin, K. L. Morris, K. E. Marshall, P. Tian, P. Sikorski, and L. C. Serpell. 2010. The common architecture of cross-beta amyloid. *J Mol Biol* 395:717.

105. Paparcone, R., and M. J. Buehler. 2009. Microscale structural model of Alzheimer A-beta(1-40) amyloid fibril. *Appl Phys Lett* 94:243904.

106. Paparcone, R., M. A. Pires, and M. J. Buehler. 2010. Mutations alter the geometry and mechanical properties of Alzheimer A-beta(1-40) amyloid fibrils. *Biochemistry* 49:8967.

107. Berryman, J. T., S. E. Radford, and S. A. Harris. 2009. Thermodynamic description of polymorphism in Q- and N-Rich peptide aggregates revealed by atomistic simulation. *Biophys J* 97:1.

108. Adamcik, J., J.-M. Jung, J. Flakowski, P. De Los Rios, G. Dietler, and R. Mezzenga. 2010. Understanding amyloid aggregation by statistical analysis of atomic force microscopy images. *Nat Nanotechnol* 5:423.

109. Roos, W. H., M. M. Gibbons, A. Arkhipov, C. Uetrecht, N. R. Watts, P. T. Wingfield, A. C. Steven et al. 2010. Squeezing protein shells: How continuum elastic models, molecular dynamics simulations, and experiments coalesce at the nanoscale. *Biophys J* 99:1175.

110. Madine, J., E. Jack, P. G. Stockley, S. E. Radford, L. C. Serpell, and D. A. Middleton. 2008. Structural insights into the polymorphism of amyloid-like fibrils formed by region 20–29 of amylin revealed by solid-state NMR and X-ray fiber diffraction. *J Am Chem Soc* 130:14990.

111. Paparcone, R., and M. J. Buehler. 2011. Failure of A-beta(1-40) amyloid fibrils under tensile loading. *Biomaterials* 32:3367.

112. Yang, L., G. Song, and R. L. Jernigan. 2009. Protein elastic network models and the ranges of cooperativity. *Proc Natl Acad Sci U S A* 106:12347.

113. Kim, M. K., R. L. Jernigan, and G. S. Chirikjian. 2005. Rigid-cluster models of conformational transitions in macromolecular machines and assemblies. *Biophys J* 89:43.

114. Lu, M., and J. Ma. 2008. A minimalist network model for coarse-grained normal mode analysis and its application to biomolecular x-ray crystallography. *Proc Natl Acad Sci U S A* 105:15358.

115. Golji, J., R. Collins, and M. R. K. Mofrad. 2009. Molecular mechanics of the alpha-Actinin Rod Domain: Bending, torsional, and extensional behavior. *PLoS Comput Biol* 5:e1000389.

116. Kim, J.-I., S. Na, and K. Eom. 2009. Large protein dynamics described by hierarchical-component mode synthesis. *J Chem Theor Comput* 5:1931.

117. Gere, J. M. 2003. *Mechanics of Materials.* Belmont, CA: Thomson Learning.

118. Timoshenko, S. P., and J. N. Goodier. 1970. *Theory of Elasticity.* New York, NY: McGraw-Hill.

119. He, J., and C. M. Lilley. 2008. Surface effect on the elastic behavior of static bending nanowires. *Nano Lett* 8:1798.

120. Dai, M. D., K. Eom, and C.-W. Kim. 2009. Nanomechanical mass detection using non-linear oscillations. *Appl Phys Lett* 95:203104.

121. Meirovitch, L. 1967. *Analytical Methods in Vibrations.* New York: Macmillan.

122. Strick, T., J. F. Allemand, V. Croquette, and D. Bensimon. 2000. Twisting and stretching single DNA molecules. *Prog Biophys Mol Biol* 74:115.

123. Yamakawa, H., and M. Fujii. 1973. Wormlike chains near the rod limit: Path integral in the WKB approximation. *J Chem Phys* 59:6641.

124. Marko, J. F., and E. D. Siggia. 1995. Stretching DNA. *Macromolecules* 28:8759.

125. Wiggins, P. A., T. van der Heijden, F. Moreno-Herrero, A. Spakowitz, R. Phillips, J. Widom, C. Dekker, and P. C. Nelson. 2006. High flexibility of DNA on short length scales probed by atomic force microscopy. *Nat Nanotechnol* 1:137.

126. Mazur, A. K. 2007. Wormlike chain theory and bending of short DNA. *Phys Rev Lett* 98:218102.

127. Rivetti, C., M. Guthold, and C. Bustamante. 1996. Scanning force microscopy of DNA deposited onto mica: Equilibration versus kinetic trapping studied by statistical polymer chain analysis. *J Mol Biol* 264:919.

128. Flynn, T. C., and J. Ma. 2004. Theoretical analysis of twist/bend ratio and mechanical moduli of bacterial flagellar hook and filament. *Biophys J* 86:3204.

Section II

Simulations in Nanoscience and Nanotechnology

8 Nature's Flexible and Tough Armor

Geometric and Size Effects of Diatom-Inspired Nanoscale Glass

Andre P. Garcia, Dipanjan Sen, and Markus J. Buehler

CONTENTS

8.1 INTRODUCTION: THE DIATOM

Since the dawn of civilization, nature has been a source of countless inspiration. The impetus for discovery in many notable minds of the past, like Newton, da Vinci, and Darwin, has centered around nature and its fascinating characteristics. With the

273

unfolding of time, nature has left its indelible impact in the design of synthetic structures. Presently, the field of biomimetics and bioinspiration has come into the age of the nanoscale, offering an impressive array of inventions that redefine conventional thought. In this chapter a systematic analysis of the mechanical properties of diatom inspired nanoscale mineralized structures is undertaken. As the future foundation of synthetic materials may be influenced by mechanisms and principles of the nanoworld, this type of research could serve as a guiding light and motivation for the use of nanoscale minerals in conventional design paradigms and applications, turning the weakness of minerals—their extreme brittleness—into strength, to realize strong but also tough and lightweight materials and structures. The awe, magnitude, and beauty of the nanoworld is aptly embodied by the following quote:

"To see a world in a grain of sand, And a heaven in a wild flower, Hold infinity in the palm of your hand, And eternity in an hour ..."

William Blake [1]

Invisible to the naked eye and yet unconsciously implemented in structures throughout human history, diatoms have served as a silent backbone to human civilization and continue to play a role in improving the compressive strength of cement, carbon sequestration, manufacture of water filters, and as a source of oil [2–8]. Several impressive arrays of diatom types and applications are shown in Figures 8.1 and 8.2. But how can diatoms play such an important role in a wide variety of applications? The answer may lie at the nanoscale. Diatoms are micrometer-sized photosynthetic algae with silicified, porous shells. Generally, these pores and surrounding walls have nanometer to micrometer dimensions, serve to protect the organism, and sustain multiple biological functions. The structure of the

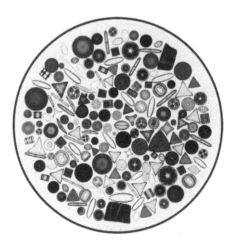

FIGURE 8.1 Marine diatoms that represent the wide array of diatom species, which reach up to approximately 100,000: Common diatom morphologies are circular, triangular, quadrilateral, and elliptical. Diatoms generally range from 2–2000 μm in overall size [90]. (Reprinted from Wipeter. Circle of diatoms on a slide. 2009. http://en.wikipedia.org/wiki/Diatom. With permission.)

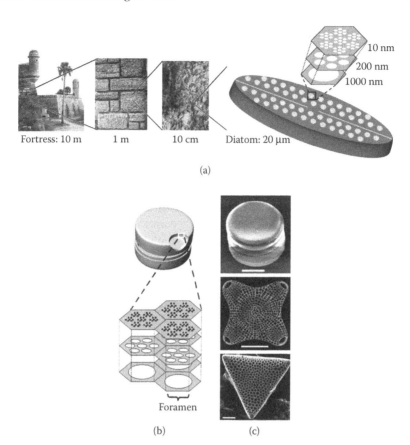

(a)

(b)   (c)

FIGURE 8.2 Hierarchical structure of diatoms, showing their porous silica structure: (a) Merger of structure and material in engineering design. (b) Schematic of the centric diatom frustule, showing the three porous silica layers lying along a hexagonal grid. (c) Atomic force microscope and scanning electron microscope images of various diatom species, revealing their porous structural makeup. Scale bar is 10 μm. (Reprinted from Losic, D., J. G. Mitchell, and N. H. Voelcker. 2009. *Adv Mater* 21:2947–58. With permission.)

cell wall is ornate and ordered, resembling images from a kaleidoscope, with a nanoporous hierarchical structure. Examples of mechanical protection include the prevention of virus penetration, crushing from some predator's mandibles, and the ability to survive once the diatom has been ingested by certain species [9–11]. Diatoms are found in both aquatic and terrestrial environments, with the ability to live in certain man made structures, such as in the walls of the Castillo de San Marcos, a Spanish built fortress located in St. Augustine, Florida, and shown in Figure 8.2(a). Interestingly, the coquina stone found in this fortress is extremely resilient in absorbing impacts such that cannon balls sink into it, rather than shattering or puncturing it, attesting to its overall ductile response, although its constituents are brittle materials.

Interestingly, biological systems such as collagen [12–14] and intermediate filaments [15–18] also share the hierarchical structural makeup of diatoms, presenting a universal design paradigm of biology that has been shown to turn weaknesses to strengths [19–21]. The reason for nature's universal implementation of hierarchical structures might lie in the intricate interplay between the triumvirate of process, structure, and property, which are all contingent on the requirement of survival of biological species such as defense under extreme conditions and under severe limitations of material quality and quantity. Because of the unique mechanical properties of diatoms, a second look at biomineralization is necessary to incorporate the governing role of genetics in the formation of structures. Several primary steps have already been taken in the form of genetic sequencing of certain diatoms, allowing a better understanding of which proteins are responsible for the intricate nanoscale shapes seen on the frustules [22]. The primary material that comprises diatom shells, silica, is also the most abundant material on Earth's crust; this serves as a cornerstone in the efficient mass production of lightweight functional materials, devices, and machines from diatoms (see Figures 8.3 and 8.4).

One attractive quality of diatoms and algae is their ability to sequester carbon and serve as an alternative energy source for human beings. For example, algae bioreactors are used to reduce carbon dioxide (CO_2) concentrations from power plant emissions. The Massachusetts Institute of Technology installed air-lift reactors on the roof of their power plant and observed 82.3% and 50.1% efficiencies for the removal of CO_2 for sunny and cloudy days, respectively [23]. Indeed, these efficiencies are significant and point to the feasibility of using algae air-lift reactors to sequester carbon from power plant flue gases.

Diatoms can also serve as viable sources for mass production, since they double in population about every day [24]. The most common form of diatom reproduction is through cell division. As a consequence of cell division, the new half of a diatom shell is smaller than its counterpart. This feature results in a continual reduction of the cell size of diatoms, and after reaching a minimum size the diatom sexually reproduces by forming an auxospore, thereby creating a larger cell [9].

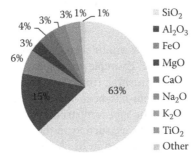

FIGURE 8.3 Abundance of silica: Silica is an extremely abundant mineral, comprising 63% of the Earth's crust. Materials made of silica are generally very brittle, such as glass, and thus have limited structural applications. However, biomineralizing proteins found in diatoms transform silica into a functional material by improving its toughness and ductility, giving rise to groundbreaking structural applications. (Adapted from Taylor, S. R. 1964. *Geochim Cosmochim Acta* 28(8):1273–85.)

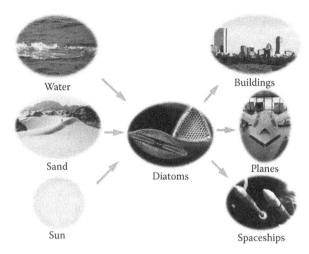

FIGURE 8.4 How to turn abundant materials into functional materials: By incorporating biomineralizing proteins found in diatoms, a new paradigm of large-scale construction can be achieved with nanoscale resolution, high toughness, and low weight. As diatomaceous proteins require resources that are in abundance, such as water, sand, and light, they offer an opportunity for efficient and cost-effective fabrication of structures such as buildings, planes, spaceships, and microelectromechanical systems. (Courtesy of Andre P. Garcia. Sand image credit: Luca Galuzzi [www.galuzzi.it]. Diatoms image credit: Mel Pollinger [nyms .org/Gallery.htm]. Spaceships image credit: National Aeronautics and Space Administration.)

8.2 MECHANICAL PROPERTIES OF DIATOMS AND NANOSTRUCTURES SIMILAR TO DIATOMS

Only a few experimental data have been collected on the mechanical properties of diatom shells, called frustules. Hamm et al. [25] used a glass needle to load and break diatom frustules in order to probe their mechanical response at failure, as shown in Figure 8.5. Their findings for the *F. kerguelensis* species revealed an elastic modulus of 22.4 GPa and a maximum stress along the costae of ~0.6 GPa in tension and ~0.7 GPa in compression. A comparison of mechanical properties between different structural materials and diatom regions is shown in Table 8.1. Other studies [26] have used atomic force microscope (AFM) nanoindentation to study nanoscale material properties of the porous frustule layers of diatoms, identifying pore sizes in the order of several tens of nanometers at the smallest levels of the hierarchy with ultrathin silica walls on the order of several nanometers. For *Coscinodiscus* sp. it was found that the porosity increased from the outer membrane to the inner membrane. The pore sizes were 45 ± 9, 192 ± 35, and 1150 ± 130 nm for the cribellum, cribrum, and areola layers, respectively. Of the three layers, the cribellum had the lowest hardness and elastic modulus, 0.076 ± 0.034 GPa and 3.40 ± 1.35 GPa, respectively, whereas the areola had the highest, 0.53 ± 0.13 GPa and 15.61 ± 5.13 GPa, respectively [26,27]. Researchers observed that the variation of mechanical properties between frustule layers could be influenced by pore size, pore distance, porosity, and different biomineralization processes.

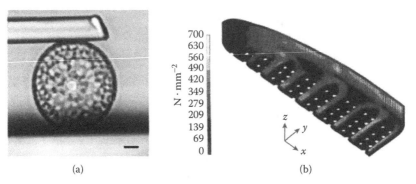

(a)   (b)

FIGURE 8.5 (a) Glass needle compressive tests of live single cells of *Thalassiosira punctigera*. Scale bar is 10 μm. (b) Finite element calculation of *Fragilariopsis kerguelensis* frustules, showing the equivalent von Mises stress as a function of the pressure (total force is 750 μN) on the girdle region. This finite element model simulates the glass needle crush test. The stress distribution within the frustule is homogeneous, and thus statically favorable. (Reprinted from Hamm, C. E., R. Merkel, O. Springer, P. Jurkojc, C. Maier, K. Prechtel, and V. Smetacek. 2003. *Nature* 421(6925):841–3. With permission.)

TABLE 8.1
Comparison of Structural Materials and Diatom Regions

Materials and Diatom Regions	Strength (GPa)	Young's Modulus (GPa)
Cement (compression)	20–55	10–35
Steel	0.2–1	200
Aluminum	0.1–0.2	70
Copper	0.2–0.4	110
Silicon	0.2	180
Silica	0.04–0.06	80–90
Cribellum	0.042–0.11	2.05–4.75
Cribrum	0.02–0.24	0.86–2.54
Areola layers	0.4–0.66	10.48–20.74

The mechanical properties of diatoms and links to associated structural features have already been discussed in the literature. For instance, the raphid diatom has a raphe, which is a split in the frustule along its length with blunted ends; it helps to reduce stress concentrations [10,25]. Further, many diatoms share a hexagonal pattern for the frustules, which feature a high moment of inertia, due to the large distance between the hexagon edge and the hexagon centroid. A high moment of inertia increases the capacity of the structure to withstand deformation [25]. The material that makes up the frustules is also important for the flexibility and toughness of diatoms. For example, during the initial stages of frustules' growth, silaffins and polyamines proteins coprecipitate with silicic acid, which is found in the aqueous environment, forming a composite organic material [28–32]. The cross-sectional shape of the outermost layer, the cribellum, is a shallow dome with the tip pointing outward and the cribrum forming another dome pointing inward, similar to eggshells or hillocks. Connecting the base of the cribrum and the areola is a wall-type structure reminiscent of an I beam. The lower base of the wall forms the surface of the areola. This I-beam shape is very common in macroscale engineering applications, such as in steel I beams used in the construction of buildings, and offers an extremely efficient design due to its high moment of inertia, resulting in increased bending stiffness and shear resistance. From a structural point of view, domes are used in engineering design due to their capacity to distribute compressive loads along walls. Overall, a review of earlier structural and mechanical analyses reveals that diatoms are fascinating nanoscale structures that incorporate multiple engineering design concepts, perhaps resulting in an optimization of their mechanical stability.

Other studies of the mechanical properties of brittle materials under extreme conditions of geometric confinement, in certain nanowires, revealed that by decreasing cross-sectional diameter, a material becomes stronger and more capable of undergoing significantly large deformation before breaking [33–35]. Experimental studies on silica nanowires of widths ranging from 230–800 nm revealed that fracture stress is influenced by specimen size; however, the modulus is not affected by size [36]. This study also determined that the modulus stayed constant at around 68 GPa, whereas the fracture stress varied from 8.77 to 6.35 GPa for the smallest to the largest silica wire widths, respectively. Ni, Li, and Gao [37] experimentally determined that modulus is independent of diameter for amorphous silica nanowires ranging from 50 to 100 nm in diameter. On the other hand, Silva et al. [38] found increasing stiffness for lower diameters by performing molecular dynamics (MD) simulations on amorphous silica nanowires. The fracture toughness of bulk vitreous silica was determined to be approximately $0.8 \, MPa \cdot m^{1/2}$ [39]. The failure mechanisms occurring within the process zone at the crack tip in amorphous silica glass was studied by Bonamy et al. [40]. By implementing AFM experiments and MD simulations, the authors found that nucleation and growth of cavities are the dominant mechanism for failure within the process zone. Integrated silicon circuits have been manufactured with a wavy structural layout and 1.7 μm thickness, and they can be elongated up to 10% [41]. Other studies revealed highly ductile amorphous silica nanowires, by using a taper-drawing process. With diameters on the order of 20 nm and highly smooth surfaces, nanowires achieve extreme flexibility such that ropelike twists and spiral coils are realized [42]. However, although these synthetic nanostructures offer attractive mechanical properties, they are generally unfeasible for mass production due to the need for complex and expensive manufacturing. For this reason, a second look at harnessing biomineralization is necessary. The initial steps have already been taken by genetic sequencing of certain diatoms, allowing a better understanding of what proteins are responsible for the intricate nanoscale shapes seen on frustules [22].

As the ability to actually synthetically create and manipulate these magnificent structures increases, so does the necessity to understand the fundamental mechanical properties they are endowed with. In light of this, our recent work focuses on a range of mineralized nanostructures found in diatoms, from which we extracted simple generic models that reflect their hierarchical structural makeup. We use mesoscale and atomistic simulations to probe the mechanical response and failure mechanisms of these structures based on a computational materials science approach that provides a powerful bottom-up material description [43]. We hypothesize that the nanoporous geometry of frustules is the key to providing enhanced toughness, ductility, and strength to the constituting material, although the material itself is inherently brittle. Thereby, the formation of nanostructured geometry, such as the one that includes thin, geometrically confined structures of brittle elements, may play a crucial role in understanding the material's behavior.

8.3 COMPUTATIONAL METHODS

An overview of atomistic simulations and the reactive force field, ReaxFF [44], is provided in Sections 8.3.1 through 8.3.3. The evolution of the atomic theory is discussed, with an overview of models and simulation techniques. A description of the theoretical framework, applications, and limitations are also given.

8.3.1 ATOMISTIC MODELING

The inception of the atomic theory dates back to approximately 400 BC, when the Greek scholars Democritus and Leukippus stated that the universe is made up of empty space and indivisible particles called atoms [45]. They captured a compelling description [46]: "Atoms, divergent in form, propel themselves through their separation from the infinite into the great vacuum by means of their mutual resistance and a tremulous, swinging motion. Here gathered, they form one vortex where, by dashing together and revolving round in all sorts of ways, the like are separated off with the like" (p. 309).

Indeed this represents a striking resemblance to the actual behavior of atoms in terms of their motion, attraction, and repulsion. This description was a harbinger to the more complex atomistic models of the twentieth century. It was not until 1957 that one of the first MD simulations was reported [47]. The fundamental principles of MD are illustrated in Figure 8.6. Each atom is simplified to a point representation, with trajectories in conjunction with the interactions between other atoms through

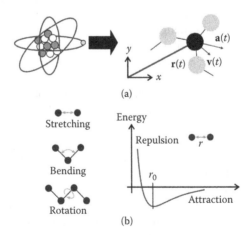

(a)

(b)

FIGURE 8.6 Molecular dynamics description: (a) Atoms are represented as points that have a certain position, velocity, and acceleration. (b) Each atom has an energy landscape and different bond terms, such as stretching, bending, and rotation between those atoms. (Reprinted from Buehler, M. J., and S. Keten. *Rev Mod Phys* 82(2):1459.)

interatomic potentials. By incorporating the Newtonian equation of motion, $F = ma$, and thermodynamic parameters, a description of an atom's position $\mathbf{r}(t)$, velocity $\mathbf{v}(t)$, and acceleration $\mathbf{a}(t)$ are obtained.

Molecular dynamics offers a powerful elucidation of atomistic phenomenon and in many respects allows pioneering advances in understanding nanoscale material properties, deformation mechanisms, and even biological processes. In retrospect, the ability to describe atomic interactions has evolved in part thanks to a symbiotic relationship between computational hardware and parallelized simulation methods. With the dawn of the computer age signified by mass production of the Universal Automatic Computer (UNIVAC) in the 1950s, MD studies entered a new era. For example, the first MD simulation, reported in 1957 [47], studied the phase transformation of a 32-particle system, proceeding at about 300 collisions per hour on the UNIVAC. The amazing development of computational processing allows the simulation of millions of atoms with near quantum detail today [48].

Within MD, the trajectories of atoms are generated by applying ensembles such as the microcanonical (NVE) and canonical (NVT) ensembles [49]. In the NVE ensemble, volume, energy, and the number of moles are constant. The exchange between the kinetic and potential energies describes the path of atoms. In the NVT ensemble, volume, temperature, and the number of moles are constant. The energy created by straining the material is removed with a Berendsen thermostat. The Berendsen thermostat allows the velocities of atoms to rescale toward the desired temperature of 300 K. The velocity Verlet algorithm is used to update the velocity of atoms. The flowchart shown in Figure 8.7 lists the main steps comprising MD simulations.

8.3.2 Mechanical Analysis

The virial stress [50] is calculated as follows:

$$\sigma_{IJ} = \frac{\sum_k^N m_k v_{k_I} v_{k_J}}{V} + \frac{\sum_k^N r_{k_I} \cdot f_{k_J}}{V} \tag{8.1}$$

where I and $J = x, y, z$; and $N, m, r, f,$ and V are the number of atoms, mass of an atom, velocity, position, force, and volume of total system, respectively.

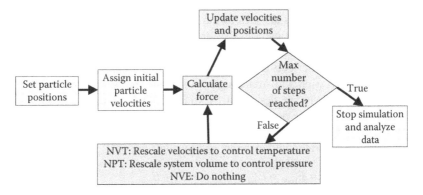

FIGURE 8.7   Flowchart showing the general methodology of molecular dynamics simulation.

The engineering strain is defined as follows:

$$\varepsilon = \frac{\Delta L_y}{L_y} \tag{8.2}$$

where L_y is the initial length of the specimen and ΔL_y is the change in length along the deformation direction (which is the y coordinate). The stress–strain curve is then used to determine the elastic modulus E, where $\sigma = \sigma_{22}$ (tensile stress in the loading direction):

$$E = \frac{\partial \sigma}{\partial \varepsilon} \approx \frac{\Delta \sigma}{\Delta \varepsilon} \tag{8.3}$$

Once stress–strain curves are determined, their integral is taken in order to determine toughness:

$$E_v = \int_0^{\varepsilon_f} \sigma(\varepsilon)\, d\varepsilon \tag{8.4}$$

where E_v and ε_f are energy per unit volume and strain at failure, respectively. A higher toughness indicates a greater ability of the material to adsorb energy due to stresses before failure (resulting, for instance, in a large fracture process zone). The von Mises stress σ_v is calculated as follows:

$$\sigma_v = \sqrt{\frac{(\sigma_{xx} - \sigma_{yy})^2 + (\sigma_{yy} - \sigma_{zz})^2 + (\sigma_{xx} - \sigma_{zz})^2 + 6(\sigma_{xy}^2 + \sigma_{yz}^2 + \sigma_{xz}^2)}{2}} \tag{8.5}$$

where σ_{II} ($I = x, y, z$) are normal components of the stress tensor; and σ_{xy}, σ_{yz}, and σ_{xz} are shear components of the stress tensor. Equation 8.5 is applied for von Mises stress in plotting the atomic stresses throughout the loading event.

8.3.3 REACTIVE FORCE FIELD: REAXFF

Over the past several decades, molecular simulation tools have expanded to take various principles and forms, as shown in Figure 8.8. Many incorporate MD schemes with various types of potentials such as the Tersoff–Brenner potential. At the heart of MD are potential energy functions (i.e., empirical force field [EFF]) [51,52], which depend on the location of atoms and their respective energy contributions. Bond interactions can be described as springs (i.e., as simple harmonic equations) that describe the compression and stretching of bonds and the bending of bond angles. Nonbonded interactions are described by van der Waals potential functions and Coulomb interactions. The potential functions that are defined within EFF are fitted against experimental or quantum chemical data (i.e., training sets) [52]. The total system energy is separated into different energy contributions:

$$\begin{aligned} E_{system} &= E_{bond} + E_{over} + E_{under} + E_{lp} + E_{val} \\ &\quad + E_{pen} + E_{tor} + E_{conj} + E_{vdWaals} + E_{Coulomb} \end{aligned} \tag{8.6}$$

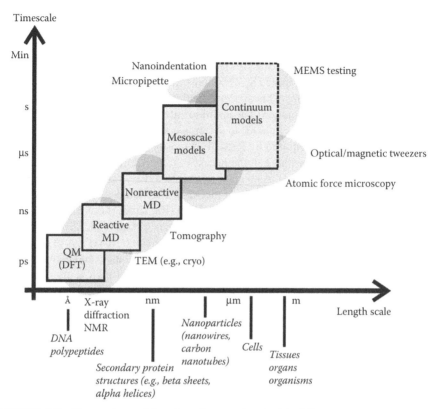

FIGURE 8.8 Timescales and length scales associated with different computational tools used in modeling materials: One important concept lies in the coupling of different computational and experimental tools that enable scaling up in terms of system size and time. For example, certain reactive molecular dynamics (MD), such as ReaxFF, utilize principles derived from quantum mechanics and experiments. Continuum models reach the largest timescales and length scales, and can incorporate atomistic results. Experimental techniques, such as those shown surrounding the shaded ellipses, enable the validation of and provide data for different computational methods. In this chapter, the method that is used is ReaxFF, a type of reactive MD. (Reprinted from Buehler, M. J., and Y. C. Yung. 2009. *Nat Mater* 8(3):175–88.)

Since the force field is defined for the specific range of chemistry for which it was developed, it should be applied only to systems similar to the training sets [52,53]. Reactive chemical systems are not properly defined by EFF methods since the shape of the potential function that defines the bond length–bond energy relationship makes it difficult to accurately determine a parameter value near the dissociation limit [52]. For instance, REBO, the EFF method developed by Brenner, allows the dynamic simulation of reactions in systems having more than 100 atoms. Nevertheless, the use of REBO is limited because it excludes nonbonded interactions and is based on a relatively small training set [52].

The development of the ReaxFF [44] (reactive force field) provides a more accurate view of complex chemical systems (specifically for systems undergoing a large strain or hyperelasticity) and therefore offers an alternative to Tersoff-type potentials. The force field ReaxFF is founded on quantum mechanical (QM) principles, such as the discrete Fourier transform, and defines the atomic interactions with QM accuracy [44]. It has been shown to more accurately describe the material behaviors (e.g., the bond breaking and formation process) of nonmetals (C, O, H, N), metals (Cu, Al, Mg, Ni, Pt), semiconductors (Si), mixed Si–O systems, silica–water interfaces [52,54–61], and biological materials [62]. This aspect is critical in describing the properties of materials under large deformation. The ab initio calculations are quite computationally expensive and thus limit ReaxFF to systems containing a few thousand atoms. The computational time required by ReaxFF is mostly determined by the complexity of mathematical expressions and the necessity to perform a charge equilibration (QEq) at each iteration [63,64]. As a comparison, ReaxFF is approximately several orders of magnitude faster than quantum mechanics techniques based on first principle (ab initio) methods. However, ReaxFF is 20–100 times more expensive computationally than simple empirical force fields such as CHARMM, DREIDING, or covalent-type Tersoff's potentials [64]. The basic idea behind ReaxFF is to modulate bond properties (i.e., bond order, bond energy, and bond distance) and thereby properly dissociate the bonds to separated atoms [44]. The valence angle terms are bond-order dependent, as a result of which their energy contributions disappear when the bonds break. Upon bond dissociation the bond terms all smoothly become zero, and therefore no energy discontinuities appear during the reaction [44]. The ReaxFF force field also describes the nonbonded interactions between atoms by incorporating Coulomb and van der Waals potentials. Indeed, ReaxFF serves as a link between QM and empirical force fields [62,64].

8.4 HOW HIERARCHY, SIZE, AND SHAPE AFFECT THE MECHANICAL PROPERTIES OF NANOSCALE GLASS

As diatom structures are relatively complex, reaching multiple degrees of hierarchy and size, revealing the intricate interplay between structure and mechanical response requires delicate and systematic analysis. In order to more comprehensively understand a diatom's mechanical properties, hierarchical silica structures, a common feature observed in diatoms, are now examined. The mechanical properties of diatoms are quite attractive, allowing very high toughness, which is a quality that is indeed attributed to the complex hierarchical topology resembling honeycombs within honeycombs. In Sections 8.4.1 through 8.4.3, two structures resembling that in Figure 8.2b [65] are discussed.

8.4.1 Reactive Force Field Simulation: Revealing the Toughening Mechanisms of Diatoms and Tailoring Their Mechanical Properties

A full atomistic representation of two distinct silica geometries found in diatoms is discussed here, hierarchical meshes and foil structures, similar to those found in many diatoms, such as Bacillariophyceae and *Ellerbeckia arenaria*, respectively. By carrying out a series of MD simulations with the first principles–based reactive force field ReaxFF, it is shown that when concurrent mechanisms occur, such as shearing and crack arrest, toughness is optimally enhanced. This occurs, for example, when structures encompass two nanoscale levels of hierarchy: (1) an array of thin-walled foil silica structures, and (2) a hierarchical arrangement of foil elements into a porous silica mesh structure. For wavy silica, unfolding mechanisms are achieved by increasing amplitude and allow for greater ductility. Furthermore, these deformation mechanisms are governed by the size and shape of the structure.

The extreme ductility of certain diatom species and communities is another fascinating attribute, especially since diatoms mostly comprise amorphous silica, which is a typically brittle constituent. A particularly interesting colonial diatom is *E. arenaria* because the species live in waterfalls and are thus able to resist significant and continual mechanical stress. These colonies are also able to elastically stretch up to about 33% [66,67]. Two possible reasons for this extreme mechanical response are the intricate shape of the cell wall and the organic coating called "mucilage" surrounding the cell wall surface of such species. In this section, a new geometry is analyzed because of its presence in certain diatom species that are able to elongate and resist extreme mechanical stress from the environment. We, therefore, focus on the corrugated, wavy shape found along the sides of *E. arenaria* (see Figure 8.9a) and propose that this particular shape is essential to providing flexibility as well as combining high strength and toughness [68].

As a model system we consider a structural design comprised of α-quartz crystals. One structure is a foil or an infinitely tall thin wall, whereas the other is a mesh comprised of interlocking silica foils, thus forming uniform and ordered rectangular voids. Since the simulation box is periodic, the foil structure can be thought of as a periodic array of thin foils having a spacing s equivalent to that of the mesh structure. Figure 8.10 shows the geometries considered here. Both the mesh and foil structures are exposed to free surfaces. The foil has free surfaces parallel to the y axis, and the mesh has free surfaces along the x and y axes. All mesh structures have the same void dimension of 76 Å × 30 Å. The only parameter varied here is the wall width w, which ranges from 5 to 72 Å for both foil and mesh (see Figure 8.10). The number of atoms varies from ~750 to ~17,000 for the smallest- to the largest-width silica systems. The largest simulation cell has dimensions of 151 Å × 196 Å × 8.5 Å in the x, y, and z directions. A relaxation is performed for 80 picoseconds to a 31-Å foil structure with free surfaces along the x and y axes. It was found that there is negligible average stress after relaxation. This result could be explained since surface reconstruction does not take place.

The next structure we considered is a foil or an infinitely tall thin wall with varying amplitude and width, resembling a wave, as shown in Figure 8.9b. Since the simulation box is periodic, the foil structure can be thought of as an array of waves having a spacing equivalent to the peak-to-peak amplitude. The z axis has no free surface, and the structure can be described as infinitely tall. Figure 8.9c shows the geometries considered here. All wave structures have an equivalent wavelength of 63.5 Å. The only parameters varied here are the wall width w and amplitude A, which range from 20 to 120 Å and from 0 to 60 Å, respectively (see Figure 8.9c). The number of atoms varies from ~650 to ~7000 for the smallest- to the largest-width

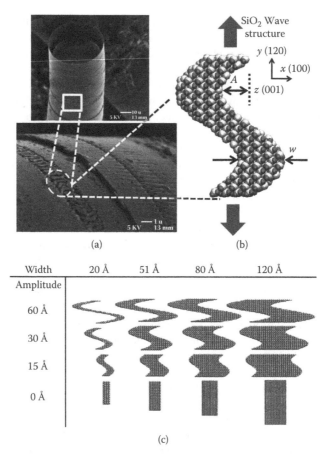

FIGURE 8.9 (a) A colonial diatom, *Ellerbeckia arenaria*, which lives in waterfalls and contains girdle bands with intricate patterns. Specifically, a wave shape can be seen, which might be an important contribution to an elastic response of approximately 33%, as observed through atomic force microscope experiments. (b) Initial geometry of a bioinspired silica structure used in our simulations, illustrating the wall width and amplitude (*w* and *A*, with definitions indicated in the structure). (c) Initial geometry of all wave structures considered in this chapter, illustrating the range of variation in amplitude and width. (Reprinted from Gebeshuber, I. C., J. H. Kindt, J. B. Thompson, Y. Del Amo, H. Stachelberger, M. Brzezinski, G. D. Stucky, D. E. Morse, and P. K. Hansma. 2004. *J Microsc Oxford* 214:101. With permission.)

silica systems. For the wave structures, the largest simulation cell has dimensions of 177 Å × 63.5 Å × 8.5 Å in the *x*, *y*, and *z* directions.

We begin our analysis by focusing on the effect of changing the wall width on the mechanical properties of two levels of hierarchy (see Figure 8.10b) for the two geometries considered, that is, foil and mesh. As shown in Figure 8.10, the wall widths are varied between *w* = 5 Å and *w* = 72 Å for both the foil and mesh structures. From a structural hierarchy perspective of materials design, the first order of hierarchy consists of a foil oriented in such a way that only two free surfaces are in the *x* direction. The second order of hierarchy comprises a mesh of interlocking foils with voids and

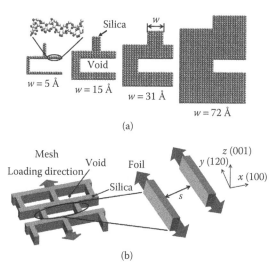

(a)

(b)

FIGURE 8.10 Geometry of the hierarchical bioinspired silica structure, and the setup used in our simulations: (a) Initial geometry of mesh structures considered, illustrating the wall width (w; definition indicated in one of the structures) variation in the geometry. Inset shows a detailed view of the relaxed surface structure. Wall widths are varied for the foil as in the mesh structure. (b) Three-dimensional schematic of the silica mesh structure (shown on left), with periodic boundaries along the x, y, and z directions. On the right, the geometry of the silica foil array structure is shown, and it has periodic boundaries along the z and y directions with free surfaces only along the x axis. The spacing s between foils is equal to that in the corresponding mesh structure. The crystallographic orientation is the same for all silica structures considered. The lowest level of hierarchy represented here is the foil, and the highest is the mesh structure. The arrows indicate tensile load applied uniformly along the structure. (Reprinted from Garcia A., D. Sen, M. J. Buehler, Hierarchical silica nanostructures inspired by diatom algae yield superior deformability, toughness and strength. *Metallurgical and Materials Transactions A*, DOI: 10.1007/s11661-010-0477-y, 2011.)

free surfaces in both x and y directions. The z axis has no free surface, and thus the structure can be described as consisting of infinitely tall walls, or foils.

Figure 8.11a shows the stress–strain response of the foil structure for varying values of w, and Figure 8.11b shows the same data for the mesh geometry. Both the foil and mesh structures show an increase in deformation in the plastic regime, a lower modulus, and lower maximum stress with decreasing wall width w. However, silica mesh structures have a much greater plastic regime than silica foil. The first important observation made here is that although silica is considered a brittle material, it is possible to transform silica into a ductile system for small (nanoscale) wall widths, which reaches maximum failure strains of 90% and 120% for the silica foil and the mesh structures, respectively. In a similar fashion to foil structures, the silica mesh also shows an increased modulus and maximum stress for larger wall widths. The plastic regime in the mesh structure decreases with increasing wall width, albeit showing a less severe drop in this case than in the case of the foil structure. Interestingly, the maximum tensile stress is reached at roughly the same strain of 34% for the silica foil for all $w \geq 15$ Å (see Figure 8.11). Another important

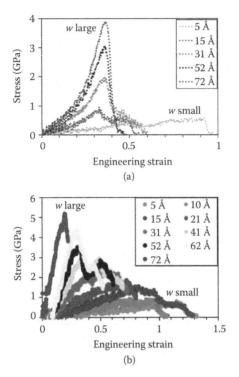

FIGURE 8.11 Stress–strain graph of silica foil (a) and mesh structure (b), for all sizes (wall widths range from $w = 5$ to 72 Å): (a) For smaller wall widths, there is a greater plastic regime, lower maximum stress, and lower modulus. Thus, due to the lowering of the wall width of the structure (w), the system behaves in a very ductile fashion and sustains very large deformations of up to 115%. The strain hardening region observed for $w = 52$ Å is the result of a temporary crack arrest due to the stretching and shearing of a secondary horizontal ligament. The crack continues its path once the secondary ligament stops deforming. (b) For wall widths more than 15 Å, there exists a plastic regime of about 1% to 5%. The greatest deformation is obtained for the smallest wall width of 5 Å. Failure mechanisms are characterized by void formations near the center or edge of the foil, which then coalesces with other voids until the structure is no longer intact. Cracks are not observed for the foil, whereas cracks occur in some mesh structures. The reason for the stiffening of wall widths >15 Å is nonlinear elasticity within the core, along with a Poisson effect as seen in earlier studies. (Reprinted from Garcia A., D. Sen, M. J. Buehler. Hierarchical silica nanostructures inspired by diatom algae yield superior deformability, toughness and strength. *Metallurgical and Materials Transactions A*, DOI: 10.1007/s11661-010-0477-y, 2011.)

difference between the foil and mesh structure lies in an increase in elastic modulus with increasing wall width, which is gradual for the foil structure and sharp for the mesh structure, respectively. Furthermore, the foil structures show a stiffening effect at $w > 15$ Å. Previous studies also have shown either stiffening or softening effects, which are affected by the orientation of the loading (since silica is anisotropic). For example, silica nanorods modeled using the Beest-Kramer-Santen (BKS) and Tsuneyuki-Tsukada-Aoki-Matsui (TTAM) potentials showed a softening effect

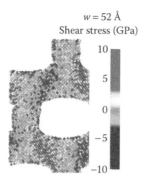

$w = 52$ Å
Shear stress (GPa)

FIGURE 8.12 Shear stress σ_{xy} taken at maximum stress for a system with wall width $w = 52$ Å: High regions of shear stress form a diagonal pattern and suggest possible areas for deformation to occur. Multiple sets of the periodic cell are shown so that the stress pattern can be clearly seen. In order to improve image clarity, only those stress values associated with silicon atoms within the silica system are shown. (From Garcia A., D. Sen, M. J. Buehler. Hierarchical silica nanostructures inspired by diatom algae yield superior deformability, toughness and strength. *Metallurgical and Materials Transactions A*, DOI: 10.1007/s11661-010-0477-y, 2011.)

when under tension [69,70]. Another study conducted ab initio simulations using the projector augmented-wave method and found dependence of pressure on the bulk modulus of silica [71]. Experimental measurements of silver nanowires loaded in tension along the [0 1 1] direction found a stiffening behavior [72]. Copper nanowires also show either stiffening or softening depending on crystallographic orientation, as shown by earlier analyses using the embedded atom model potential [73].

High levels of shear stress also manifest in a diagonal pattern as shown in Figure 8.12 for $w > 31$ Å, which facilitate possible regions of void nucleation and the start of failure. Once crack failure starts, stress is concentrated around the fracture process zone, as shown in Figure 8.13. The deformation mechanisms observed for varying wall widths are dramatic in that they correlate with the peaking of toughness, and they can be summarized as follows: For $w > 62$ Å and $w < 21$ Å, the dominant failure mechanisms are brittle crack propagation and beading of structures to thin atom chains, respectively. However, for 21 Å $\leq w \leq 62$ Å, crack propagation and shear mechanisms occur in a competing fashion and result in increased toughness. For example, for a structure with $w = 52$ Å, one can observe a crack arrest phenomenon and a corresponding increase of tensile stress from 2 to 2.8 GPa, which is due to shear mechanisms occurring within the junctions of the mesh elsewhere (see Figure 8.14).

The analysis discussed in the preceding paragraph explains the remarkable stress–strain response of mesh structures with thin wall widths, as shown in Figure 8.11b. The explanation for this lies in the geometric pattern; large deformations can be accommodated by the mesh by changing from a rectangular pattern to a hexagonal one at large strains (see, for example, Figures 8.12 and 8.13), specifically for wall widths below 31 Å. One fundamental reason for achieving very large strains without failure is due to the increasingly homogeneous distribution of stresses and the geometric transformation from rectangular to a hexagonal shape; which is attained primarily for meshes with smaller wall widths.

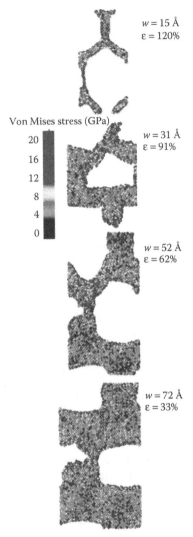

FIGURE 8.13 Von Mises stress field during failure for different mesh wall widths (the strain value at which the snapshot was taken is indicated in the plot): For systems with wall widths of 15 and 31 Å, necking and formation of beaded molecular structure is observed. At widths of 52 Å and larger, cracks initiate at free surfaces, near to the corners of the pore. For $w \geq 31$ Å, we observe the formation of voids within the sample, specifically within regions surrounding the failure process zone. The failure mechanism remains similar for $w \geq 52$ Å and is characterized by a structural change from a rectangular to a hexagonal shape. An analogy to deformation in macroscopic plastic hinges can be drawn to describe the mechanism for accommodating large deformations. For larger systems, however, the failure mode is consistent crack propagation, an effect that has been confirmed to exist for varying strain rates. In order to improve image clarity, only the stress values associated with silicon atoms within the silica system are shown. (From Garcia A., D. Sen, M. J. Buehler. Hierarchical silica nanostructures inspired by diatom algae yield superior deformability, toughness and strength. *Metallurgical and Materials Transactions A*, DOI: 10.1007/s11661-010-0477-y, 2011.)

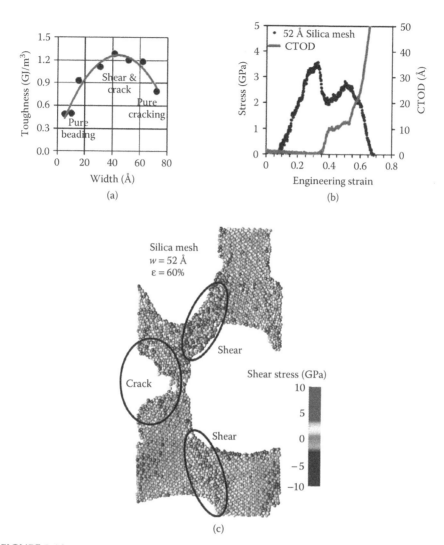

FIGURE 8.14 (a) Toughness map with corresponding failure mechanism for the silica mesh. (b) Toughening and stiffening mechanisms caused by the competing mechanisms of shear and crack formation. The crack tip opening displacement (CTOD) measurement reveals crack arrest and is plotted against the corresponding stress–strain data, which reveals how well correlated both the mechanisms are. (c) Locations of shear and crack formation. For purposes of clarity, only silicon atoms are shown. (From Garcia A., D. Sen, M. J. Buehler. Hierarchical silica nanostructures inspired by diatom algae yield superior deformability, toughness and strength. *Metallurgical and Materials Transactions A*, DOI: 10.1007/s11661-010-0477-y, 2011.)

In Figure 8.15, the effect of wall width variations and hierarchy levels on mechanical properties—the plastic regime, toughness, maximum stress, and ductility—is summarized. In all the structures considered here, the plastic regime increases with decreasing w. The maximum stress and modulus both increase with wall width, and ductility increases for smaller wall widths. For the largest wall width in silica and

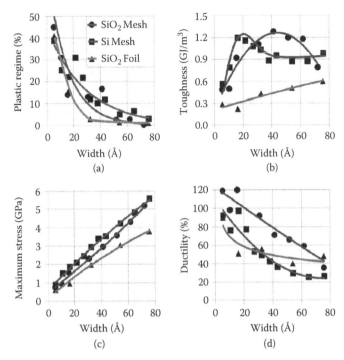

FIGURE 8.15 Comparison between silica mesh, silicon mesh, and silica foil structures showing the effect of wall width on (a) plastic regime, (b) toughness, (c) maximum stress, and (d) ductility: It is found that both the maximum stress and modulus increase with wall width. Ductility generally increases for smaller wall widths. The plastic regime is estimated by measuring the length of the linear plateau region, which is associated with constant stress. In silicon, for wall widths larger than 27 Å, toughness plateaus at around at 9×10^8 J/m³. Between 16 and 27 Å a sharp increase in toughness is observed, with a maximum at ~16 Å. Below 16 Å the toughness drops to around 7×10^8 J/m³, denoting an inverse trend. In the silica mesh, the highest toughness is observed for wall widths of 41 Å, reaching values of ~1.29×10^9 J/m³. Although silica foil generally increases in toughness with the wall width, it has lower toughness when compared with mesh structures. Thus, by increasing the level of hierarchy higher toughness, stress, and modulus can be achieved. (From Garcia A., D. Sen, M. J. Buehler. Hierarchical silica nanostructures inspired by diatom algae yield superior deformability, toughness and strength. *Metallurgical and Materials Transactions A*, DOI: 10.1007/s11661-010-0477-y, 2011.)

silicon structures, the range of ductility is between ~30% and 50%. The greatest ductility is observed for the silica mesh having the smallest wall width of 5 Å, of value 120%. Silica foil shows a gradual increase in modulus with width, with a maximum of 6.7 GPa for $w = 72$ Å, whereas mesh structures show a sharp increase in modulus, reaching 36 and 29 GPa for silicon and silica meshes, respectively. The effect of hierarchy on toughness is quite striking because the foil does not show size-dependent toughness peaking response, as was observed for meshes, because it has a consistently lower toughness than meshes. For example, the maximum toughness values observed for silicon and silica meshes are 1.20×10^9 J/m³ and 1.29×10^9 J/m³, respectively, whereas the maximum toughness for silica foil is only 0.60×10^9 J/m³. The reason for the greater toughness observed in the higher hierarchy of meshes lies in competing mechanisms of shear and crack, wherein crack arrest is achieved either through shearing of another foil subcomponent in the mesh structure (as observed in the silica mesh, $w = 52$ Å) or through the simultaneous cracking of different regions (as observed in the silicon mesh, $w = 43$ Å). These competing mechanisms are enabled through the hierarchical assembly of foil elements into a mesh structure and cannot be achieved in unit foil structures alone. This result demonstrates that inclusion of higher levels of hierarchy is beneficial in improving the mechanical properties and deformability of silica structures.

The greatest ductility is achieved for wavy structures with high amplitudes and low widths, whereas low amplitudes and high widths yield the greatest strength. Structures possessing the highest toughness have a width and amplitude bounded by regions of highest ductility and stress. Interestingly, it is a balance between the two geometric parameters (A and w) and a combination of deformation mechanisms (unfolding, shearing, and cracking) that result in the highest toughness. Furthermore, this concept of geometric effects on stress–strain behavior is displayed in Figure 8.16.

An analogy to protein structures can be drawn wherein sacrificial bonds and hidden lengths are responsible for enhanced toughness [19,74,75]. These sacrificial bonds are weaker than the carbon backbone but stronger than van der Waals or hydrogen bonds, and they allow for sawtooth-shaped force–extension curves. Sulfate bonds are a good example of sacrificial bonds in systems containing DOPA, a common amino acid found in biological adhesives. As each sacrificial bond breaks energy is released in the form of heat, and a regional unfolding, or uncoiling of a hidden length

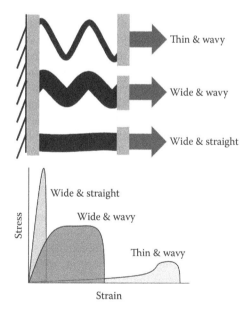

FIGURE 8.16 Schematic of stress–strain response for tensile deformation of different morphologies of silica waves: Thin and wavy structures provide the greatest ductility, whereas wide and straight structures provide high values of stress at the cost of ductility. However, when combining wavy and wide morphologies, significant toughness is gained. (Reprinted from Sen, D., and M. J. Buehler. *Int J Appl Mech* (in press).)

segment, occurs. This process is repeated until all the sacrificial bonds are broken and the structure is completely unfolded. Only then will the carbon backbone break, resulting in the highest peak of stress. The interplay of multiple failure zones and deformation mechanisms is strikingly similar to that found in silica wave structures. A link is made between the sacrificial bonds found in proteins and the shearing mechanism found in a silica wave. The catastrophic carbon backbone failure is also analogous to cracking in a silica system. When integrating these multiple mechanisms, a universal concept of enhancing toughness is achieved.

8.4.2 SURFACE RECONSTRUCTION MECHANISMS

An important consideration in the behavior of material properties is their structural arrangement at the nanoscale. In the case of crystals, degrees of amorphization are usually present as well as surface reconstruction. Surface atoms rearrange and shift in their spacing, and such relocations can affect even multiple layers at the surface. In the case of silicon, for example, the fcc lattice along the [1 0 0] surface reconstructs to a 1×1 square array. By cleaving silicon along other surfaces, different patterns emerge on the surface, such as the hexagonal shape of the 7×7 reconstruction [76].

Surface reconstruction may have some effect on mechanical behavior. For example, an MD study reported a stiffening of silicon nanowires once the surfaces were reconstructed [77]. For a 1.05-nm thick nanowire, the modulus increased from approximately 150 to 160 GPa once the [0 0 1] surface was reconstructed. The authors attributed this stiffening effect to bond saturation as reconstruction takes place. Another study used ReaxFF and observed the reconstruction of a zinc oxide surface after 300 picoseconds at 700 K [78]. In yet another study, it was shown that a total timescale of 240 picoseconds was required to obtain a reconstructed silicon surface by an annealing process with the Tersoff potential and a total system size of 308 atoms [79]. Although our ReaxFF-based approach can capture surface reconstruction in principle, it is computationally expensive to capture this reconstruction due to the associated timescale and system size and thus this was not observed in our simulations. Indeed, for small systems such as systems with 1000 atoms, ReaxFF could simulate the aforementioned timescales quite well. However, for systems having more than 10,000 atoms, ReaxFF would be very expensive computationally. The investigation of surface reconstruction and its effects on the mechanics of hierarchical silica structures can be an interesting subject for future studies.

In summary, we ascribe these magnificent improvements in the mechanical properties of mesh structures to two competing atomistic mechanisms of deformation: (1) shear, and (2) brittle crack propagation. In the toughest silica mesh, for example, a crack is arrested by the shearing of another strut, which consequently stiffens the system (see Figure 8.14). Interestingly, a size-dependent peaking of toughness was observed in both silicon and silica meshes, with a maximum toughness of 1.29×10^9 J/m^3 corresponding to $w = 41$ Å for the silica mesh (see Figure 8.15b). The increased toughness of these nanostructures also make them viable candidates for impact-resistant lightweight structures, and this should be tested in further detail. Another powerful concept derived from this study is the ability to transform a brittle material into a ductile one by simply manipulating the geometry of a constituent structure to resemble that of ordered nanopores, or a mesh. The reasons for the high ductility observed are threefold: (1) a homogeneous distribution of surface stress throughout the entire structure, (2) occurrence of a conformational change from rectangular to hexagonal pores, and (3) presence of competing mechanisms of shear and crack arrest.

In a similar fashion to structural materials found in nature, such as bone, nacre, or diatom shells, it seems that hierarchical architecture is a cornerstone of nanomaterial design that allows for enhanced mechanical properties. More importantly, the concept of hierarchical structures made of the same materials (as demonstrated in this section) is fundamental to fully realizing the enormous potential of nanomaterial design. Therefore, by introducing structural hierarchies a weakness can be turned into a strength, that is, an intrinsically strong but brittle material can be made exceedingly tough, strong, and ductile. The fact that a similar behavior was found in silicon [43] suggests that this may indeed be a generic design concept that could be used for many materials. The observation that a weakness can be turned into a strength is also reminiscent of similar behaviors in hydrogen bonds, which are by themselves highly brittle and weak but can reach extreme levels of toughness and strength once arranged in particular hierarchical patterns [80].

8.4.3 MESOSCALE SIMULATION OF DIATOM MORPHOLOGY: TURNING WEAKNESSES INTO STRENGTHS

Atomistic modeling and simulation, albeit quite powerful in elucidating mechanisms with atomistic detail and based solely on QM input, has limitations with respect to accessible length scales and timescales. To probe the mechanical behavior of silica structures at the micron length scale, a mesoscale spring-lattice network model is developed, as recently reported by Sen and Buehler [81], and the multiscale simulation paradigm outlined in Figure 8.8 is realized. The model is derived from (as well as validated against) full atomistic simulation results. Spring-lattice network models operate at the micron length scale and are thus able to capture elasticity, plasticity, and fracture phenomena at much larger length scales than full atomistic models. Yet, the coarser grained model is directly derived from atomistic simulation results and as such provides a direct link to the fundamental chemical scale of a material without any need to introduce any empirical parameter. The details of the method and its representative results are described in this section.

As an application we consider a nanocomposite of bulk silica and nanoporous silica, in which the nanoporous phase represents the material discussed in Section 8.4.1, that is, a constituent that by itself has an underlying structure. The combination of these two structures to form a nanocomposite, therefore, represents the creation of another level of structural hierarchy and thus mimics the kind of structures seen in diatoms (see, e.g., Figure 8.2a). Since the bulk structure represents bulk silica it is stiff and brittle, whereas the nanoporous structure is soft and ductile and features great extensibility (as explained in detail in Section 8.4.1). The stress–strain law for the mesoscale model is hyperelastic in order to fit the constitutive behavior of nanosilica and bulk silica under tensile load, as shown in Figure 8.17.

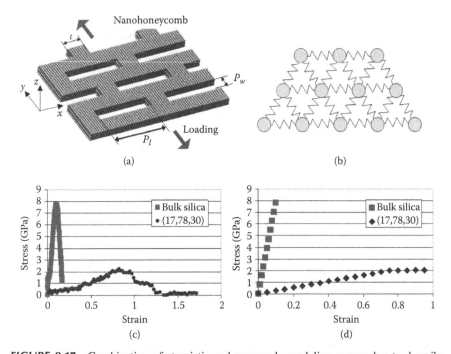

FIGURE 8.17 Combination of atomistic and mesoscale modeling approaches to describe the material from the nano- to the microscale: Parameters for the mesoscale model are derived from constitutive behavior at the nanoscale obtained using atomistic simulations. (a) Geometry of the nanohoneycomb used as the building block for composite structures; (b) a section of the triangular mesh mesoscale particle-spring model setup; (c) stress–strain curves obtained from atomistic simulations of a nanohoneycomb structure and for bulk silica with a crack of the same size as the pores in the nanohoneycomb. The legend defines the nanohoneycomb structure, which is shown as [$t\ p_l\ p_w$] parameters for the structure (numerical values given in angstroms). The bulk silica structure shows purely brittle fracture, whereas the nanohoneycomb structure shows ductile fracture; and (d) Behavior of the mesoscale triangular mesh lattice fitted to this constitutive behavior (the agreement with the full atomistic result depicted in part (c) is evident and provides direct validation of the mesoscale model). (Reprinted from Sen, D., and M. J. Buehler. *Int J Appl Mech* (in press).)

Parameters for the mesoscale model are derived from constitutive behavior at the nanoscale obtained using atomistic simulations. Appropriate potential formulations are obtained by identifying appropriate interparticle potential functions, for which details are reported by Sen and Buehler [81]. Specifically, the hyperelastic spring potential simulates and models atomistic results for the nanohoneycomb as elastic–perfectly plastic behavior and bulk silica as elastic–brittle behavior.

Next, sharp-edged cracks in all materials are created and they are loaded under quasistatic mode I loading. Loading is carried out by stepped edge displacement boundary conditions and relaxing the global positions of all material particles using a conjugate gradient energy minimization scheme [82]. The initial crack size is 3.9–5.4 µm, and crack initiation is identified by the advance of the crack front at a particular loading strain value. Figure 8.18a shows representative overall stress–strain response for

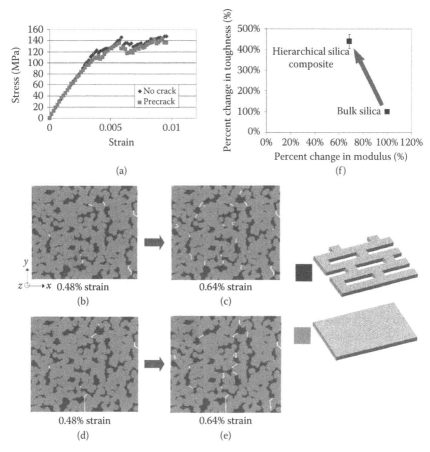

FIGURE 8.18 Stress–strain curves for (a) bulk silica reinforcing composite structures, with and without the presence of a precrack. The near identical response shows the flaw tolerance behavior for these structures to precracks of the given size. The structures show multiple cracking throughout the material, and this is reflected in the stress–strain curve as a gradual loss of stiffness of the material as the number and size of the multiple cracks grow. (b) Crack pathways for composite structures with the nanohoneycomb structure as the matrix and brittle silica as the reinforcing fiber phase. The volume fraction of the silica phase is 76%. (b) and (c) show fracture starting from a material with no precrack, and (d) and (e) show fracture progress in the same material with a precrack present. In some cases it is observed that the precrack propagates for a small distance but does not propagate through the sample, and other smaller cracks are initiated throughout the sample. These multiple small cracks determine the stress–strain response of the structure. The structure is thus flaw tolerant to precracks of these sizes, and fracture stress and behavior are almost independent of the size of the precrack. This is reflected in the stress–strain curve for the stress–strain response of structures with no defects and cracked structures (shown in (a)). (f) The design of the hierarchical silica composite improves the toughness of bulk silica significantly (~4.4 times) while compromising the stiffness only slightly (~70% of bulk). This points toward the use of hierarchies along with a single design material to improve undesirable mechanical properties significantly (low toughness here) without compromising on the desirable ones (high stiffness here). (Reprinted from Sen, D., and M. J. Buehler. *Int J Appl Mech* (in press).)

composite structures with brittle matrix morphologies, with and without the edge crack. The brittle fiber structures show multiple microcracking sites (throughout the sample) under tensile load. The presence of precracks is not seen to affect this phenomenon, and the material fails by the growth and coalescence of several microcracks, as shown in Figure 8.18b, c, d, and e. This phenomenon is also captured by the absence of any response on the stress–strain curve to the presence of precracks of a certain size and below (Figure 8.17) and can be classified as a defect-tolerant state. Fracture mechanics formulations cannot be used for such material microstructures and crack sizes, and damage mechanics that deals with the evolution of damage with applied load, for example, in the form of diffuse microcracking, must be used to characterize the failure response. Diffuse cracking also affects the elastic modulus (slope of the stress–strain curve) of the material, as seen in Figure 8.17, and the modulus is lowered at higher strain values when the microcracks start to increase in size and number.

The improvement in toughness while retaining the stiffness of bulk silica is achieved by designing small regions of nanoporous geometry within the bulk phase; which points to a design methodology for obtaining stiff and tough materials out of an inherently brittle material (e.g., silica) (Figure 8.18f). The design philosophy can be summed up as the use of the single material silica, which is traditionally considered an undesirable (mechanically inferior) structural material due to its brittleness, arranged in a hierarchical pattern with substructures that go down to nanoscale dimensions. This enables one to obtain highly functional materials that show enhanced toughness and stiffness. Use of the design paradigm outlined here leads to substantial increase in the design space for brittle materials, without the need to add additional materials, solely by geometrical design via the utilization of hierarchical structures.

8.5   CONCLUSION

The ability to improve multiple mechanical properties, such as toughness, strength, and ductility, is extremely important when designing nanoscale structures, such as nanoelectromechanical systems (NEMS). Altering the mechanical properties of two of the most brittle and abundant minerals on Earth's surface, silica and silicon, allows a new window of opportunity for humanity to create applications and reinvent materials. The ability to massively transform the mechanical properties, such as from brittle to ductile or from weak to tough, through geometric alterations at the nanoscale is another profound discovery that will undoubtedly introduce a new paradigm in the way materials are designed and applied. Indeed, the culminating goal of materials design is to maintain environmental sustainability, infrastructure superiority, multifunctional capacity, and economic feasibility. Nanoscale materials implemented through design and fabrication concepts found in nature, such as in diatom algae, bone, and sea sponges, hold the promise of providing these advantages.

8.5.1   SUMMARY OF KEY CONCEPTS PRESENTED IN THIS CHAPTER

Revealing the intricate interplay between structure and mechanical response of diatoms requires delicate and systematic analysis. This chapter establishes a framework on size effects, hierarchy, and shape, based on the mechanical response of silicon and silica structures. The main findings are summarized as follows:

- Silicon and silica meshes yield highly tunable mechanical properties through alterations of their width between 5 and 76 Å, as shown in Figure 8.15. A region of optimum toughness is observed to lie between $w = 20$ Å and $w = 50$ Å. Crack arrest and shear mechanisms are the reasons for the high toughness observed, as shown in Figure 8.14.
- Hierarchical structures such as meshes have superior toughness and ductility when compared to bulk silica, foils, as shown in Figure 8.11. Foils do not exhibit the multiple mechanisms of failure that are observed in meshes.
- By incorporating small regions of nanoporous geometry within the bulk phase, stiff and tough materials can be obtained from an inherently brittle material (silica), as shown in Figure 8.18. The hierarchical silica composite shows significantly improved toughness over bulk silica (≈ 4.4 times), although it compromises on stiffness slightly ($\sim 70\%$ of bulk).

8.5.2 DISCUSSION AND FUTURE WORK

Several challenges remain in the form of fabricating and more accurately modeling diatom inspired structures. For example, biomineralization from self-assembling proteins, which guide silica precipitation, has been studied [28,83–85]. However, synthesis of complex hierarchical structures still remains a challenge. The recent determination of certain diatom genetic sequences will further the understanding of accurately controlling and fabricating silica structures. In modeling these systems, a key consideration is the effect of surface reconstruction on mechanical properties. However, surface reconstruction occurs on timescales that are intractable for many atomistic methodologies, such as ReaxFF or quantum-based approaches.

Future research could be geared toward atomistic simulations on the deformation and failure of different morphologies found in diatom species. Moreover, mineralized structures are found in many other biological systems, such as deep-sea sponges [86], which could be studied using a similar molecular approach. Another important step is attaining a greater convergence between actual diatom frustules and those that are modeled. Key challenges include reaching greater size scales, incorporating organic material, amorphization, and surface termination. The size-scale issue can be generally overcome with coarse graining or utilizing massive supercomputers on the order of hundreds of central processing units with ReaxFF. With larger systems, more complex shapes can be modeled, such as incorporating different shapes throughout the z axis. Perhaps a more complicated challenge is the addition of organic material, such as proteins, within the silica structure. The existing ReaxFF force field that models both organics and silica is limited to glyoxal, and does not encompass nitrogen bonds; glyoxal is an element found in many organic structures, such as collagen [87]. Once an adequate force field is developed, proteins such as silaffins and collagen could be added to silica. As mentioned in previous studies, proteins within diatoms are found in their adhesives and they enable self-assembly

[88]. Amorphization of silica is another critical concept that must be explored, since it could affect mechanical properties and is found in diatoms. Another avenue for further research is surface termination, as it occurs when silica is exposed to water and does affect the mechanical response of silica structures. Future simulations of surface-terminated structures can encompass larger systems in order to more fully capture the effect at different length scales.

REFERENCES

1. Gilchrist, A., and A. B. Gilchrist. 1880. *Life of William Blake, with Selections from His Poems and Other Writings*. London: Macmillan and co.
2. Losic, D., J. G. Mitchell, and N. H. Voelcker. 2009. Diatomaceous lessons in nanotechnology and advanced materials. *Adv Mater* 21:2947–58.
3. Fragoulis, D., M. G. Stamatakis, D. Papageorgiou, and E. Chaniotakis. 2005. The physical and mechanical properties of composite cements manufactured with calcareous and clayey Greek diatomite mixtures. *Cem Concr Compos* 27(2):205–9.
4. Chatterji, S. 1978. Mechanism of the $CaCl_2$ attack on Portland cement concrete. *Cem Concr Res* 8(4):461–7.
5. Fenical, W. 1983. *Plants: The Potentials for Extracting Protein, Medicines, and Other Useful Chemicals: Workshop Proceedings*. Washington, DC: U.S. Gov. Print. Off.
6. Allen, J. T., L. Brown, R. Sanders, C. Mark Moore, A. Mustard, S. Fielding, M. Lucas et al. 2005. Diatom carbon export enhanced by silicate upwelling in the northeast Atlantic. *Nature* 437(7059):728–32.
7. Litchman, E., C. A. Klausmeier, and K. Yoshiyama. 2009. Contrasting size evolution in marine and freshwater diatoms. *Proc Natl Acad Sci* 106(8):2665–70.
8. Ramachandra, T. V., D. M. Mahapatra, B. Karthick, and R. Gordon. 2009. Milking diatoms for sustainable energy: Biochemical engineering versus gasoline-secreting diatom solar panels. *Ind Eng Chem Res* 48(19):8769–88.
9. Raven, J. A., and A. M. Waite. 2003. The evolution of silicification in diatoms: Inescapable sinking and sinking as escape? *Tansley Rev* 162:45–61.
10. Hamm, C. E., and V. Smetacek. 2007. Armor: Why, when, and how. In *Evolution of Primary Producers in the Sea*, ed. P. G. Falkowski, and A. H. Knoll, 311–32. Boston: Elsevier.
11. Fowler, S. W., and N. S. Fisher. 1983. Viability of marine phytoplankton in zooplankton fecal pellets. *Deep Sea Res* 39(9):963–9.
12. Fratzl, P., and R. Weinkamer. 2007. Nature's hierarchical materials. *Prog Mater Sci* 52(8):1263–334.
13. Fratzl, P. 2008. *Collagen: Structure and Mechanics*. New York: Springer.
14. Gupta, H. S., J. Seto, W. Wagermaier, P. Zaslansky, P. Boesecke, and P. Fratzl. 2006. Cooperative deformation of mineral and collagen in bone at the nanoscale. *Proc Natl Acad Sci U S A* 103:17741–6.
15. Kreplak, L., and D. Fudge. 2007. Biomechanical properties of intermediate filaments: From tissues to single filaments and back. *BioEssays* 29(1):26–35.
16. Guzmán, C., S. Jeney, L. Kreplak, S. Kasas, A. J. Kulik, U. Aebi, and L. Forró. 2006. Exploring the mechanical properties of single vimentin intermediate filaments by atomic force microscopy. *J Mol Biol* 360(3):623–30.
17. Strelkov, S. V., L. Kreplak, H. Herrmann, and U. Aebi. 2004. Intermediate Filament Protein Structure Determination. In *Methods in cell biology*, ed. B. M. Omary, and P. A. Coulombe, 78:25–43. Basel, Switzerland: Academic Press.

18. Qin, Z., L. Kreplak, and M. J. Buehler. 2009. Hierarchical structure controls nanomechanical properties of vimentin intermediate filaments. *PLoS One* 4(10):e7294.
19. Buehler, M. J., and Y. C. Yung. 2009. Deformation and failure of protein materials in physiologically extreme conditions and disease. *Nat Mater* 8(3):175–88.
20. Buehler, M. J. 2010. Nanomaterials: Strength in numbers. *Nat Nano* 5(3):172–4.
21. Buehler, M. J. 2010. Turning weakness to strength. *Nano Today* 5:379–383.
22. Armbrust, E. V., J. A. Berges, C. Bowler, B. R. Green, D. Martinez, N. H. Putnam, S. Zhou et al. 2004. The genome of the diatom Thalassiosira pseudonana: Ecology, evolution, and metabolism. *Science* 306(5693):79–86.
23. Vunjak-Novakovic, G., Y. Kim, X. Wu, I. Berzin, and J. C. Merchuk. 2005. Air-lift bioreactors for algal growth on Flue gas: Mathematical modeling and pilot-plant studies. *Ind Eng Chem Res* 44(16):6154–63.
24. Rivkin, R. B. 1986. Radioisotopic method for measuring cell division rates of individual species of diatoms from natural populations. *Appl Environ Microbiol* 51(4):769–75.
25. Hamm, C. E., R. Merkel, O. Springer, P. Jurkojc, C. Maier, K. Prechtel, and V. Smetacek. 2003. Architecture and material properties of diatom shells provide effective mechanical protection. *Nature* 421(6925):841–3.
26. Losic, D., K. Short, J. G. Mitchell, R. Lal, and N. H. Voelcker. 2007. AFM nanoindentations of diatom biosilica surfaces. *Langmuir* 23(9):5014–21.
27. Losic, D., R. J. Pillar, T. Dilger, J. G. Mitchell, and N. H. Voelcker. 2007. Atomic force microscopy (AFM) characterisation of the porous silica nanostructure of two centric diatoms. *J Porous Mater* 14:61–69.
28. Kroger, N. 2007. Prescribing diatom morphology: Toward genetic engineering of biological nanomaterials. *Curr Opin Chem Biol* 11(6):662–9.
29. Sumper, M. 2002. A phase separation model for the nanopatterning of diatom biosilica. *Science* 295(5564):2430–3.
30. Schroder, H. C., X. Wang, W. Tremel, H. Ushijima, and W. E. Muller. 2008. Biofabrication of biosilica-glass by living organisms. *Nat Prod Rep* 25(3):455–74.
31. Hildebrand, M. 2008. Diatoms, biomineralization processes, and genomics. *Chem Rev* 108(11):4855–74.
32. Poulsen, N., and N. Kroger. 2004. Silica morphogenesis by alternative processing of silaffins in the diatom Thalassiosira pseudonana. *J Biol Chem* 279(41):42993–9.
33. Park, H. S. 2009. Quantifying the size-dependent effect of the residual surface stress on the resonant frequencies of silicon nanowires if finite deformation kinematics are considered. *Nanotechnology* 20(11):115701.
34. Kim, T. Y., S. S. Han, and H. M. Lee. 2004. Nanomechanical behavior of beta-SiC nanowire in tension: Molecular dynamics simulations. *Mater Trans* 45(5):1442–9.
35. Chuang, T.-j. 2006. On the tensile strength of a solid nanowire. *Nanomechanics of Materials and Structures*, ed. T.-J. Chuang, P. M. Anderson, M.-K. Wu, and S. Hsieh. Dordrecht, Netherlands, Springer:67–78.
36. Namazu, T., and Y. Isono. 2003. Quasi-static bending test of nano-scale SiO2 wire at intermediate temperatures using AFM-based technique. *Sens Actuators A* 104(1):78–85.
37. Ni, H., X. Li, and H. Gao. 2006. Elastic modulus of amorphous SiO_2 nanowires. *Appl Phys Lett* 88(4):043108-3.
38. Silva, E. C., L. Tong, S. Yip, and K. J. Van Vliet. 2006. Size effects on the stiffness of silica nanowires. *Small* 2(2):239–243.
39. Lucas, J. P., N. R. Moody, S. L. Robinson, J. Hanrock, and R. Q. Hwang. 1995. Determining fracture toughness of vitreous silica glass. *Scr Metall Mater* 32(5):743–8.
40. Bonamy, D., S. Prades, C. Rountree, L. Ponson, D. Dalmas, E. Bouchaud, K. Ravi-Chandar, and C. Guillot. 2006. Nanoscale damage during fracture in silica glass. *Int J Fract* 140(1):3–14.

41. Kim, D.-H., J.-H. Ahn, W. M. Choi, H.-S. Kim, T.-H. Kim, J. Song, Y. Y. Huang, Z. Liu, C. Lu, and J. A. Rogers. 2008. Stretchable and foldable silicon integrated circuits. *Science* 320(5875):507–11.

42. Tong, L., J. Lou, Z. Ye, G. T. Svacha, and E. Mazur. 2005. Self-modulated taper drawing of silica nanowires. *Nanotechnology* 16(9):1445.

43. Garcia, A. P., and M. J. Buehler. 2010. Bioinspired nanoporous silicon provides great toughness at great deformability. *Comput Mater Sci* 48(2):303–9.

44. van Duin, A. C. T., S. Dasgupta, F. Lorant, and W. A. Goddard. 2001. ReaxFF: A reactive force field for hydrocarbons. *J Phys Chem A* 105:9396.

45. Stewart, B., and P. G. Tait. 1875. *The Unseen Universe: Or, Physical Speculations on a Future State.* New York: Macmillan.

46. Hegel, G. W. F., E. S. Haldane, and F. H. Simson. 1892. Lectures on the history of philosophy. London, K. Paul, Trench, Trübner, & Co.

47. Alder, B. J., and T. E. Wainwright. 1957. Phase transition for a hard sphere system. *J Chem Phys* 27(5):1208–9.

48. Nakano, A., R. K. Kalia, K.-i. Nomura, A. Sharma, P. Vashishta, F. Shimojo, A. C. T. van Duin, W. A. Goddard, R. Biswas, and D. Srivastava. 2007. A divide-and-conquer/cellular-decomposition framework for million-to-billion atom simulations of chemical reactions. *Comput Mater Sci* 38(4):642–52.

49. Berendsen, H. J. C., J. P. M. Postma, W. F. van Gunsteren, A. DiNola, and J. R. Haak. 1984. Molecular dynamics with coupling to an external bath. *J Chem Phys* 81(8):3684–90.

50. Zimmerman, J. A., E. B. Webb III, J. J. Hoyt, R. E. Jones, P. A. Klein, and D. J. Bammann. 2004. Calculation of stress in atomistic simulation. *Model Simul Mater Sci Eng* 12(4):S319.

51. Duin, A. C. T., K. Nielson, W. Q. Deng, J. Oxgaard, and W. A. Goddard. 2004. Application of ReaxFF reactive force fields to transition metal catalyzed nanotube formation. *Abstr Pap Am Chem Soc* 227:U1031.

52. Duin, A. C. T. 2002. *ReaxFF User Manual.* Pasadena, CA: California Institute of Technology.

53. Sutmann, G. 2002. John von Neumann Institute for Computing. *Classical Mol Dyn* 10:211–54.

54. Sen, D., and M. J. Buehler. 2007. Chemical complexity in mechanical deformation of metals. *Int J Multiscale Comput Eng* 5:181–202.

55. van Duin, A. C. T., A. Strachan, S. Stewman, Q. Zhang, X. Xu, and W. A. Goddard. 2003. ReaxFF$_{SiO}$: Reactive force field for silicon and silicon oxide systems. *J Phys Chem A* 107:3803–11.

56. Strachan, A., A. C. T. van Duin, D. Chakraborty, S. Dasgupta, and W. A. Goddard. 2003. Shock waves in high-energy materials: The initial chemical events in nitramine RDX. *Phys Rev Lett* 91(9):098301.

57. Nielson, K. D., A. C. T. van Duin, J. Oxgaard, W. Deng, and W. A. Goddard. 2005. Development of the ReaxFF reactive force field for describing transition metal catalyzed reactions with application to the initial stages of the catalytic formation of carbon nanotubes. *J Phys Chem A* 109(3):493–9.

58. Buehler, M. J., A. C. T. van Duin, and W. A. Goddard. 2006. Multi-paradigm modeling of dynamical crack propagation in silicon using the ReaxFF reactive force field. *Phys Rev Lett* 96(9):095505.

59. Cheung, S., W. Q. Deng, A. C. T. van Duin, and W. A. Goddard. 2005. ReaxFF$_{MgH}$ reactive force field for magnesium hydride systems. *J Phys Chem A* 109(5):851–9.

60. Chenoweth, K., S. Cheung, A. C. T. van Duin, W. A. Goddard, and E. M. Kober. 2005. Simulations on the thermal decomposition of a poly(dimethylsiloxane) polymer using the ReaxFF reactive force field. *J Am Chem Soc* 127(19):7192–202.

61. Fogarty, J. C., H. M. Aktulga, A. Y. Grama, A. C. T. van Duin, and S. A. Pandit. 2010. A reactive molecular dynamics simulation of the silica-water interface. *J Chem Phys* 132(17):174704–10.

62. Buehler, M. J. 2006. Large-scale hierarchical molecular modeling of nano-structured biological materials. *J Comput Theor Nanosci* 3(5):603–23.

63. Rappe, A. K., and W. A. Goddard. 1991. Charge equilibration for molecular dynamics simulations. *J Phys Chem A* 95(8):3358–63.

64. Buehler, M. J., J. Dodson, A. C. T. van Duin, P. Meulbroek, and W. A. Goddard. 2006. The Computational Materials Design Facility (CMDF): A powerful frame-work for multiparadigm multi-scale simulations. *Mater Res Soc Symp Proc* 894 0894-LL03-03.1.

65. Garcia, A., D. Sen, and M. J. Buehler. 2011. Hierarchical silica nanostructures inspired by diatom algae yield superior deformability, toughness, and strength. *Metall Mater Trans A*, DOI: 10.1007/s11661-010-0477-y.

66. Gebeshuber, I. C., H. Stachelberger, and M. Drack. 2005. Diatom bionanotribology-biological surfaces in relative motion: Their design, friction, adhesion, lubrication and wear. *J Nanosci Nanotechnol* 5(1):79–87.

67. Gebeshuber, I. C., J. H. Kindt, J. B. Thompson, Y. Del Amo, H. Stachelberger, M. Brzezinski, G. D. Stucky, D. E. Morse, and P. K. Hansma. 2004. Atomic force microscopy study of living diatoms in ambient conditions. (vol 212, pg 292, 2003). *J Microsc Oxford* 214:101.

68. Garcia, A. P., N. Pugno, and M. J. Buehler. Superductile, wavy silica nanostructures inspired by diatom algae. *Advanced Engineering Materials* 13.

69. Zhu, T., J. Li, S. Yip, R. J. Bartlett, S. B. Trickey, and N. H. de Leeuw. 2003. Deformation and fracture of a SiO_2 nanorod. *Mol Simul* 29(10):671–6.

70. Silva, E., J. Li, D. Liao, S. Subramanian, T. Zhu, and S. Yip. 2006. Atomic scale chemo-mechanics of silica: Nano-rod deformation and water reaction. *J Comput Aided Mater Des* 13(1):135–59.

71. Kimizuka, H., S. Ogata, and Y. Shibutani. 2007. Atomistic characterization of structural and elastic properties of auxetic crystalline SiO_2. *Phys Status Solidi B* 244(3):900–9.

72. Wu, B., A. Heidelberg, J. J. Boland, J. E. Sader, X. Sun, and Y. Li. 2006. Microstructure-hardened silver nanowires. *Nano Lett* 6(3):468–72.

73. Liang, H., M. Upmanyu, and H. Huang. 2005. Size-dependent elasticity of nanowires: Nonlinear effects. *Phys Rev B* 71(24):241403.

74. Fantner, G. E., E. Oroudjev, G. Schitter, L. S. Golde, P. Thurner, M. M. Finch, P. Turner et al. 2006. Sacrificial bonds and hidden length: Unraveling molecular mesostructures in tough materials. *Biophys J* 90(4):1411–8.

75. Smith, B. L., T. E. Schaffer, M. Viani, J. B. Thompson, N. A. Frederick, J. Kindt, A. Belcher, G. D. Stucky, D. E. Morse, and P. K. Hansma. 1999. Molecular mechanistic origin of the toughness of natural adhesives, fibres and composites. *Nature* 399(6738):761–3.

76. Brommer, K. D., M. Needels, B. Larson, and J. D. Joannopoulos. 1992. Ab initio theory of the Si(111)-(7 × 7) surface reconstruction: A challenge for massively parallel computation. *Phys Rev Lett* 68(9):1355.

77. Zhang, W. W., Q. A. Huang, H. Yu, and L. B. Lu. 2009. Size-dependent elasticity of silicon nanowires. In *Micro and Nano Technology—1st International Conference of Chinese Society of Micro/Nano Technology(Csmnt)*, ed. X. Wang, 315–9. Stafa-Zurich: Trans Tech Publications Ltd.

78. Raymand, D., A. C. T. van Duin, M. Baudin, and K. Hermansson. 2008. A reactive force field (ReaxFF) for zinc oxide. *Surf Sci* 602(5):1020–31.

79. Liu, W., K. Zhang, H. Xiao, L. Meng, J. Li, G. M. Stocks, and J. Zhong. 2007. Surface reconstruction and core distortion of silicon and germanium nanowires. *Nanotechnology* 18(21):215703.

80. Keten, S., Z. Xu, B. Ihle, and M. J. Buehler. 2010. Nanoconfinement controls stiffness, strength and mechanical toughness of beta-sheet crystals in silk. *Nat Mater* 9(4):359–67.

81. Sen, D., and M. J. Buehler. 2010. Atomistically-informed mesoscale model of deformation and failure of bioinspired hiearchical silica nanocomposites. *Int J Appl Mech* 2(4):699–717.

82. Polyak, B. T. 1969. The conjugate gradient method in extremal problems* 1. *USSR Comput Math Math Phys* 9(4):94–112.

83. Kroger, N., and N. Poulsen. 2008. Diatoms-from cell wall biogenesis to nanotechnology. *Annu Rev Genet* 42:83–107.

84. Sumper, M., and G. Lehmann. 2006. Silica pattern formation in diatoms: Species-specific polyamine biosynthesis. *Chembiochem* 7(9):1419–27.

85. Sumper, M., and E. Brunner. 2006. Learning from diatoms: Nature's tools for the production of nanostructured silica. *Adv Funct Mater* 16(1):17–26.

86. Aizenberg, J., J. C. Weaver, M. S. Thanawala, V. C. Sundar, D. E. Morse, and P. Fratzl. 2005. Skeleton of euplectella sp.: Structural hierarchy from the nanoscale to the macroscale. *Science* 309(5732):275–8.

87. Kramer, R. Z., J. Bella, B. Brodsky, and H. M. Berman. 2001. The crystal and molecular structure of a collagen-like peptide with a biologically relevant sequence. *J Mol Biol* 311(1):131–47.

88. Dugdale, T. M., R. Dagastine, A. Chiovitti, and R. Wetherbee. 2006. Diatom adhesive mucilage contains distinct supramolecular assemblies of a single modular protein. *Biophys J* 90(8):2987–93.

89. Ashby, M. F., and I. Books24x. 2005. Materials selection in mechanical design, 3rd ed. Retrieved August, 11, 2010, from http://www.books24x7.com/marc.asp?bookid=25372.

90. Tomas, C. R., G. R. Hasle, E. E. Syvertsen, K. A. Steidinger, and K. Tangen. 1996. Marine diatoms. In *Identifying Marine Diatoms and Dinoflagellates*, ed. C. R. Tomas, 5–385. San Diego: Academic Press.

91. Wipeter. 2009. Circle of diatoms on a slide. http://en.wikipedia.org/wiki/Diatom. Retrieved September 6, 2010.

92. Buehler, M. J., and T. Ackbarow. 2007. Fracture mechanics of protein materials. *Mater Today* 10(9):46–58.

93. Taylor, S. R. 1964. Abundance of chemical elements in the continental crust: A new table. *Geochim Cosmochim Acta* 28(8):1273–85.

94. Garcia, A. P. 2010. Hierarchical and size dependent mechanical properties of silica and silicon nanostructures inspired by diatom algae. In *Civil and Environmental Engineering*. Master of Science: 91. Cambridge: Massachusetts Institute of Technology.

95. Buehler, M. J., and S. Keten. Colloquium: Failure of molecules, bones, and the Earth itself. *Rev Mod Phys* 82(2):1459.

9 Resonant Theranostics
A New Nanobiotechnological Method for Cancer Treatment Using X-Ray Spectroscopy of Nanoparticles

Sultana N. Nahar, Anil K. Pradhan,
and Maximiliano Montenegro

CONTENTS

9.1 INTRODUCTION

In biomedicine, cancer-related screening, diagnostic workup, image-guided biopsy and therapy delivery, and all X-ray sources such as simulators, linear accelerators, and computed tomography (CT) scanners generate broadband radiation. High-energy radiation doses are needed for sufficient tissue penetration, and higher exposure is needed for linear absorption. Hainfeld et al. (2004) found that irradiation with gold nanoparticles embedded in the malignant tumor is more effective than direct irradiation. Since X-rays interact more efficiently with high-Z elements, the nanoparticles are made of heavy high-Z elements that are *not* abundant in living tissues such as C, O, and Fe. They are also chosen to be nontoxic after injecting into the body and designed with antigens that seek out antibodies produced in the tumor. Some of the high-Z elements in medical applications are bromine, iodine, gadolinium, platinum, and gold, which are used in compounds that do not react unfavorably in the body. High-energy irradiating X-rays interact mainly with heavy elements, while lighter elements remain inactive because of low absorptivity. The typical size of a nanoparticle varies from a few nanometers to a few tens of nanometers and hence can penetrate vascular cells of typical sizes <30 nm or so. The high-energy X-rays have an additional advantage that they can penetrate the tissue before much loss of energy and reach the embedded nanoparticles.

Among the radiosensitizing agents commonly used in medicine are bromo-deoxyuridine (BUdR) or iododeoxyuridine (IUdR) that contain atoms of moderately heavy elements. Increasingly, the application of nanobiotechnology entails nanoparticles made of high-Z elements such as platinum and gold. Although the background X-ray absorption cross sections are large for ionization of outer electron shells, they are small for inner-shell electrons. But the inner-shell electrons are more likely to be ionized at high X-ray energies needed for deep penetration into the body tissue.

Gold nanoparticles have been extensively taken into account in nanobiotechnology applications, especially for in vivo cancer treatment because of their interaction with high-energy X-rays and nontoxicity. Laboratory experiments, using gold nanoparticles injected into mice tumors and then irradiated with high-energy 140–250 keV X-rays, have shown considerable reduction in tumor sizes (Hainfeld et al. 2004). The K-shell ionization energy of gold is 80.73 keV. Therefore, a 250-kVp X-ray source with output energies up to 250 keV is capable of ionizing all inner n-shells and subshells of the gold atom, with L-, M-, N-, O-, P-shell ionizations approximately at 12–14, 3.4–2.2, 0.11–0.9, 0.01–0.11, 0.0.009 keV, respectively.

In an effort to narrow down the radiation bandwidth, and to enable sufficient penetration, our earlier investigations (Pradhan et al. 2009; Montenegro et al. 2009) have focused on K-shell ionization of heavy elements (Nahar, Pradhan, and Sur 2008). Sharp edges in photoionization cross sections at threshold energies of various electronic shells are well known. Hence, energies at the K-shell ionization, or somewhat above, are expected to initiate enhanced emission of electrons and photons through the Auger process. However, this enhancement has not been achieved (Larson et al. 1989). We discuss the reason for this in Section 9.2 and describe a method that targets the energy bands in which such *resonant* enhancements are possible.

9.2 RESONANT THERANOSTICS

We have investigated a novel approach "resonant theranostics" (RT) (Pradhan et al. 2007; Silver, Pradhan, and Yu 2008; Pradhan et al. 2009; Montenegro et al. 2009) for more precise radiation therapies and diagnostics with reduced harmful exposure. RT proposes the use of a monochromatic X-ray source to irradiate heavy-element (high-Z) nanoparticles as shown in Figure 9.1. Tumors could be doped with these nanoparticles of heavy elements. The irradiating X-rays will be focused at resonant frequencies on the nanoparticles. We will illustrate that there are resonant energies at which X-ray absorption probability increases by orders of magnitude. Through accurate atomic calculations, we can determine these energy ranges and the strength of these resonances (e.g., for platinum and uranium, see Nahar, Pradhan, and Lim 2011). Hence, the incident X-rays need to be not only monochromatic but also preselected for maximum interaction cross sections.

Use of narrow-band resonant energies in diagnostics spectroscopy can provide more detailed microscopic and accurate information in contrast to broadband imaging. Also, spectroscopically targeted radiation should be far more efficient with reduced exposure. In Section 9.2.1, we discuss some basics of producing the monochromatic X-rays. Focusing the impinging X-rays to resonant energies will require more elaborate details to understand the physical processes, and these will be discussed in Section 9.4. Numerical simulations are then performed to see the effect of the atomic/molecular/biological processes in the body tissue along potential pathways.

9.2.1 PRODUCTION OF MONOCHROMATIC X-RAYS

In a typical X-ray machine, a beam of electrons is accelerated across the potential difference between the cathode and the anode and strikes a high-Z target such as tungsten, producing Bremsstrahlung radiation at all energies from zero to the peak value of the potential. The typical shape of the emitted Bremsstrahlung spectrum

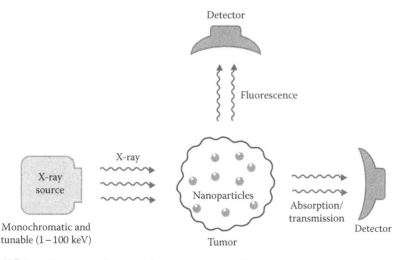

FIGURE 9.1 Schematic diagram of resonant theranostics.

shows a maximum around one-third of the output peak voltage (expressed as kVp or MVp). For example, a conventional high-energy linear accelerator (LINAC) with a peak voltage of 6 MVp produces radiation up to 6 MeV, but with a broad maximum around 2 MeV. Although there are a number of accelerator facilities around the world to produce X-rays via synchrotron radiation, they are expensive and impractical for common usage.

One of the main ideas of RT is to produce monochromatic X-rays using generally available X-ray machines. Bremsstrahlung radiation from a conventional X-ray source may be directed toward a high-Z target, rotated at a selected angle. Inner-shell ionizations, followed by radiative decays from outer shells, would produce X-ray fluorescence at monochromatic energies. For instance, the intensity of the monochromatic K_α radiation from partial conversion of Bremsstrahlung radiation that causes K-shell ionizations can be estimated as follows:

$$I(K_\alpha) \sim N(X) \int_{E \geq E_k}^{E(\text{kVp})} f_B \sigma_K(E) \, dE$$

where $N(X)$ is number density of atoms of element X, f_B is the Bremsstrahlung flux distribution in energy, and σ_K is the K-shell photoionization cross section. It may be assumed to a good approximation that each K-shell ionization leads to the ejection of a K_α photon. The K-fluorescence yield ω_K may be obtained from the branching ratio as

$$\omega_K = A_r(L - K)/[A_r(L - K) + A_a(L)]$$

where $A_r(L - K)$ is the radiative decay rate from L to K shell and A_a is the autoionization decay rate, respectively. For high-Z elements such as platinum and uranium, the K-fluorescence yield ($\omega_K > 0.95$) approaches unity. Hence, all photons from the Bremsstrahlung source above the K-shell ionization energy, $E > E_K$, may be *converted* into monochromatic K_α radiation with high efficiency. Then, monochromatic deposition of X-ray energy may also be localized using high-Z nanoparticles. The RT scheme predicts considerable production of electron ejections and photon emissions via the Auger process and secondary Coster–Kronig and super-Coster–Kronig branching transitions (the relevant atomic physics is described in the textbook "Atomic Astrophysics and Spectroscopy," Pradhan and Nahar [2011]).

9.3   X-RAY INTERACTION WITH NANOPARTICLES

X-ray interaction with high-Z atoms implies inner-shell ionization, absorption, and emission via Auger processes. Although high-Z elements are now commonly used in nanobiotechnology, their detailed physical properties are relatively unknown either experimentally or theoretically. To establish the conceptual framework of the RT technology, it is important to understand the precise interaction of X-rays as a function of incident energy and the atomic structure of the target. We illustrate that resonances are formed during X-ray–nanoparticle interaction at energies *below* the K-shell ionization energy and that these resonances represent the enhancement of absorption coefficients by orders of magnitude.

9.3.1 Radiative Processes of Photoabsorption, Photoemission, and Opacity

Interaction between an atomic species X^{+Z} of charge Z and an X-ray photon (hv) can involve the following processes.

Photoexcitation: An electron in an atomic or a molecular system, X^{+Z}, absorbs the photon and jumps to a higher excited level while remaining in the atomic or molecular system:

$$X^{+Z} + hv \rightarrow X^{+Z*}$$

The asterisk (*) denotes an excited state. The oscillator strength (f) represents the strength of the transition. Deexcitation (inverse of excitation) occurs as an excited electron drops down to a lower state, typically to the ground state by emitting a photon. The atomic parameter for this process is the radiative decay rate or the Einstein A-coefficient.

Photoionization/photodissociation/photoelectric effect: An electron absorbs a photon and leaves the atom or molecule:

$$X^{+Z} + hv \rightarrow X^{+Z+1} + e$$

This direct process gives the *background* photoionization. It can also occur via an intermediate doubly excited autoionizing state, leading to autoionization (AI) when the electron goes free, or to dielectronic recombination (DR) when a photon is emitted with capture of the interacting electron, that is,

$$X^{+Z+1} + e \leftarrow \rightarrow (X^{+Z**}) \leftarrow \rightarrow (X^{+Z} + hv) \text{ (DR)}$$

or

$$\leftarrow \rightarrow X^{+Z+1} + e \text{ (AI)}$$

For photoionization, we calculate cross section σ_{PI}. The intermediate doubly excited state (double asterisks), in which two electrons are in excited levels, manifests itself in a resonance in the cross section.

Collision excitation: An atom or a molecule goes to an excited state by the impact of a free electron and later decays by emitting a photon.

$$X^{+Z} + e \rightarrow X^{+Z*} + e'$$

This collision process can also occur through an intermediate doubly excited autoionizing state, that is, a resonance. The atomic parameters for the process are the collision strength and the collision cross section.

9.3.2 Auger Process and Coster–Kronig Cascades

Photoionization through ejection of an inner-shell electron creates a hole in the electronic structure of an atomic system and leads to the Auger effect. The Auger effect is a

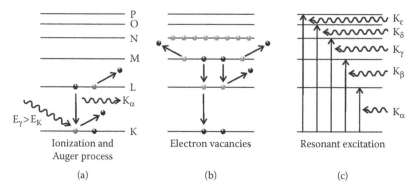

Ionization and   Electron vacancies   Resonant excitation
Auger process

(a)   (b)   (c)

FIGURE 9.2 Schematic diagrams of (a) Auger process, (b) Coster–Kronig cascades, and (c) resonant photoexcitation. (From Pradhan, A. K., S. N. Nahar, M. Montenegro, Y. Yu, H. L. Hang, C. Sur, M. Mrozik, and R. M. Pitzer. 2009. *J Phys Chem A* 113:12356–63. With permission.)

downward transition of an upper level electron to the lower inner-shell hole when the transition energy is released as a photon.

The Auger process and related cascades are illustrated in Figure 9.2. Figure 9.2a shows a $1s$ electron in the K-shell ejected by absorption of an X-ray photon of energy higher than the K-shell ionization energy, $E_\gamma > E_K$, creating a hole or vacancy. An L-shell electron drops down filling the hole, and the transition energy between the K-shell and the L-shell is released as a photon. Figure 9.2b illustrates the Coster–Kronig cascade effect, which causes ejection of more electrons. The emitted photon in the Auger process may knock out another L-shell electron, resulting in two holes in the L-shell. The holes may then be filled up by the decay of two electrons of the higher level, say from the M-shell, along with emissions of two photons. Hence with each decay the number of holes, and thereby the emitted electrons and photons, may double. This cascading process, known as the Coster–Kronig process, will continue until the vacancies reach the outermost level. This can result in up to 20 or more electrons released in a single K-shell ionization of a high-Z atom with electrons up to O and P shells, such as in gold. Most of these electrons are at low energies since they are ejected from high-lying outer shells. The *reverse* process of resonant photoexcitation (Figure 9.2c) is discussed in Subsection 9.4.2.

9.4   PARAMETERS FOR ATOMIC PROCESSES

The parameters for these atomic processes can be obtained from atomic structure calculations (e.g., Nahar, Pradhan, and Sur 2008), or from the more elaborate R-matrix method (e.g., Nahar et al. 2003). The wave function, ψ_B for the bound state and ψ_F for the continuum or free state, and energies can be obtained by solving the Schrodinger equation:

$$H\psi = E\psi$$

We employ the relativistic Breit-Pauli (BP) approximation to calculate the parameters. The BP Hamiltonian H_{BP} is written as

$$H_{BP} = H_{NR} + H_{mass} + H_{Dar} + H_{so}$$

$$+ \frac{1}{2} \sum_{i \ne j}^{N} [g_{ij}(\text{so}+\text{so}') + g_{ij}(\text{ss}') + g_{ij}(\text{css}') + g_{ij}(\text{d}) + g_{ij}(\text{oo}')]$$

where the nonrelativistic Hamiltonian is

$$H_{NR} = \sum_{i} \left[\nabla^2 - \frac{2Z}{r_i} + \sum_{j>i} \frac{2}{r_{ij}} \right]$$

The one-body correction terms are

$$H_{mass} = -\frac{\alpha^2}{4} \sum_{i} p_i^4, \quad H_{Dar} = -\frac{\alpha^2}{4} \sum_{i} \nabla^2 \left(\frac{Z}{r_i} \right), \quad H_{so} = \alpha^2 \sum_{i} \frac{Z}{r_i^3} l_i \cdot s_i$$

where l_i and s_i are orbital angular momentum and spin angular momentum of individual electrons. The rest are two-body correction terms. Of them, the most important part is the Breit interaction term.

$$H_B = \frac{1}{2} \sum_{i \ne j}^{N} [g_{ij}(\text{so}+\text{so}') + g_{ij}(\text{ss}')]$$

In these equations, s stands for spin, s' for other spin, o for orbit, o' for other orbit, c for contraction, and d for Darwin. In the present calculations, terms up to the Breit interaction are included. The last three weak terms are neglected. In the Breit interaction term

$$g_{ij}(\text{so}+\text{so}') = -\alpha^2 \left[\left(\frac{r_{ij}}{r_{ij}^3} \times p_i \right) \cdot (s_i + 2s_j) + \left(\frac{r_{ij}}{r_{ij}^3} \times p_j \right) \cdot (s_j + 2s_i) \right]$$

$$g(\text{ss}') = 2\alpha^2 \left[\frac{s_i s_j}{r_{ij}^3} - 3 \frac{(s_i r_{ij})(s_j r_{ij})}{r_{ij}^5} \right]$$

Substitution of wave function expansion, which is a linear combination of configuration state functions, in the Schrodinger equation yields bound states ψ_B when $E < 0$, and free or continuum states ψ_F when $E > 0$. More details of the theory can be found in the study by Nahar (2006) and the study by Pradhan and Nahar (2011).

Due to electromagnetic interaction, the bound–bound transition matrix is given by $<\psi_B \| D \| \psi_B>$, where $D = \Sigma_i(r_i)$ is the dipole operator of the atomic system of N-electrons (the $\|$ notation represents a reduced matrix element). For the bound-free photoionization, the transition matrix is $<\psi_F \| D \| \psi_B>$. These matrix elements can be reduced to give the desired quantity the line strength S as follows:

$$S = \left| \left\langle \psi_f \left| \sum_{j=1}^{N} r_j \right| \psi_i \right\rangle \right|^2$$

The oscillator strength f_{ij} and radiative decay rate A_{ji} for the bound–bound transition can be obtained from the line strength as

$$f_{ij} = [E_{ji}/(3g_i)]S, \ A_{ji}(s^{-1}) = [0.8032 \times 10^{10} E_{ji}^3/(3g_j)]S$$

where g_i and g_j are statistical weight factors of states i and j, respectively. The photoionization cross section for bound-free transitions can be obtained from the corresponding line strength as follows:

$$\sigma_{PI}(\nu) = [4\pi^2 \alpha a_o^2 E(Ry)]/(3g_i)S$$

where α is the fine structure constant, a_o is the Bohr radius in centimeters, E is the photon energy in Ry, and g_i is the statistical weight factor of the initial ion.

We may refer to the resonant transitions to doubly excited states above the ionization threshold as "Auger line strengths." The K-shell transitions ($1s$-np), such as K_α, K_β, K_γ, with $n = 2, 3, 4, \ldots$, are examples of these resonant transitions. They appear at energies of corresponding transitions; their cross sections can be obtained from their Auger line strengths and using the above equation for σ_{PI}.

The other useful quantity for propagation of radiation in a medium is the mass absorption or attenuation coefficient κ. Since κ is related to σ_{PI}, it may be expressed in terms of units of mass as follows:

$$\kappa(\nu) = \sigma_{PI}/(uW_A)$$

where u is atomic mass unit, amu = 1.66054e-24 g, and W_A is the atomic weight of the element being photoionized. It is also the same quantity as the plasma opacity for interaction cross section per density, cm^2/g.

The κ derived from oscillator strengths correspond to a single-line resonance. However, a resonance in a medium has a structure spread over an energy width. Broadened resonant structures of various K-shell transitions in photoionization cross sections can be obtained from Auger line strengths by convolving the resonant cross sections over a normalized Gaussian function $\varphi(\upsilon)$ as (e.g., Seaton et al. 1994)

$$\sigma_{PI}(\nu) \ \varphi(\upsilon) \quad \text{where } \int \varphi(\upsilon)d\nu = 1$$

$\varphi(\upsilon)$ can be a Gaussian function of form

$$\varphi(E = h\upsilon) = [1/\sqrt{(2\pi)}] \ \Delta E \exp[-E^2/(2\Delta E^2)]$$

where ΔE is the width of the resonant profile.

9.4.1 TRANSMITTED INTENSITY OF X-RAYS

The attenuation of radiation through a plasma depends on temperature, density, and the component material. Quantitatively the radiation intensity decreases exponentially and can be described as (NIST)

$$I(\nu) = I_0(\nu)\exp[-(\sigma/\rho)x] = I_0(\nu)\exp[-\kappa(\nu)x]$$

where I_0 is the incident intensity, σ is the photoionization cross section, ρ is the density of the plasma, and x is the depth.

9.4.2 INCIDENT FLUX FOR RESONANT ABSORPTION

With vacancies in the L-shell, it is possible to have the inverse of the Auger decay, that is, a K-shell electron may be excited upward to fill up the L-shell hole under an external photon field, as illustrated in Figure 9.2c. A photon can induce the K-electron to go through the resonant K_α transition and create a hole in the K-shell. Such a condition will present a competition between the downward decay and the upward excitation. The process can be represented as the Einstein A-coefficient for radiative decay and the B-coefficient for photon absorption (e.g., Pradhan and Nahar 2011)

$$A_{ji}(\nu) + B_{ji}(\nu)\,\rho_{ij}(\nu) = B_{ij}(\nu)\,\rho_{ij}(\nu)$$

for any two-level system $i \rightarrow j$, transition frequency ν, and radiation field density $\rho_{ij}(\nu)$. In the special case of no downward transitions into the K-shell, because it may be filled, there will be no stimulated emission. For such a situation, we can rewrite the B and A coefficients including only radiative transitions from higher n-shells to fill the L-shell vacancy, also competing with upward K-shell excitations. Given the incident photon flux Φ at the resonant energy of K_α, we can write (Pradhan et al. 2009)

$$\Phi(K_\alpha)g_K\,B(K_\alpha) = \sum_{n_i>2} g_i\,A[n_i\,(S_iL_iJ_i) \rightarrow 2(SLJ)]$$

where $n_i(S_iL_iJ_i)$ refers to specific fine structure levels of an n-shell, for example, $2(S_iL_iJ_i)$ refers to the L-shell levels, and g_K and g_i are the statistical weights of the K-shell and excited levels i, respectively. The above condition suggests a *critical photon density* Φ_c could be defined such that any flux greater than

$$\Phi_c(K_\alpha) = \left\{\sum_{n_i>2} g_i\,A\big[n_i(S_iL_iJ_i) \rightarrow 2(SLJ)\big]\right\}/g_K\,B(K_\alpha)$$

could initiate the resonant absorption for upward excitation to fill an L-shell hole.

The radiative decay coefficients for a high-Z element, such as gold and platinum are very large (Nahar, Pradhan, and Sur 2008; Nahar, Pradhan, and Lim 2010), and any vacancy in the L-shell is likely to be filled on a relatively short timescale by a radiative transition from a higher shell, given transition array coefficients $A(M \rightarrow L)$ of approximately 10^{14-15} s^{-1} and $A(L \rightarrow K)$ of approximately 10^{15-16} s^{-1}. Since the resonant absorption $B(K_\alpha)$ coefficients are related to the $A(K_\alpha)$ coefficients by the factor $[c^3/(8\pi h\nu^3)]$ (e.g., Pradhan and Nahar 2011), we find that for high-energy X-ray transitions, the $B(K_\alpha)$ are of the same order of magnitude as the $A(M \rightarrow L)$. Therefore, the RT mechanism also invokes the fact that X-ray irradiation of high-Z atoms results in Auger decays of photon emission and electron ejections creating multiple electron vacancies. These vacancies may be filled either by radiative decays from higher electronic shells or by excitations from the K-shell at resonant energies by an external X-ray source. Calculations of $\Phi_c(K_\alpha)$ require a large number of A and B coefficients, photon fluency rates, and photoabsorption cross sections for high-Z atoms to be computed, as described in Section 9.5.

9.5   X-RAY-INDUCED RADIATIVE TRANSITION RATES AND CROSS SECTIONS IN BROMINE AND GOLD

We aim at studying the physical processes due to irradiation of X-rays at resonant energies by high-Z material embedded in the body tissue. The radiosensitizing agents commonly used in medicine are BUdR or IUdR that contain heavy elements bromine and iodine, respectively. For BUdR, irradiating high-energy X-rays will interact mainly with bromine ($Z = 35$), while remaining transparent to lighter elements H, N, O. The incident energy should be more than the K-shell ionization energy of 13.5 keV of bromine because lighter elements have K-shell ionization energies $E < 0.5$ keV and the photoionization cross sections decrease as E^{-3}.

Let a neutral bromine atom with ground configuration $1s^2 2s^2 2p^6 3s^2 3p^6 3d^{10} 4s^2 4p^5$ undergo L-shell ionization by an external X-ray source, with the residual configuration with a $2p$-hole: $1s^2 2s^2 \mathbf{2p^5} 3s^2 3p^6 3d^{10} 4s^2 4p^5$. This state lies above the bound state energies and is a short-lived autoionizing one that decays quickly. As explained earlier, such decay may introduce a resonance in the absorption or ionization cross section. The external radiation can induce a ($1s$-$2p$) transition such that a K-shell electron absorbs an X-ray photon and jumps into the L-shell hole. In the inverse situation, bromine is ionized through a K-shell electron, and an L-shell electron may decay down to the K-shell, with emission of a photon. The line strength for the $1s$-$2p$ transition is the same for both the processes. But other quantities such as oscillator strength and the radiative decay rate can differ because of their dependence on kinematical factors and energy.

The resonances due to $1s$-np (where $n = 2, 3, 4, …$) transitions lie below the $1s$ or K-shell ionization energy. Table 9.1 presents oscillator strengths and corresponding cross sections and photoabsorption coefficients for $1s$-$2p$, $1s$-$3p$, and $1s$-$4p$ transitions for the singly ionized bromine, bromine II (Nahar et al. 2010). For atomic or molecular ions, many transitions are possible if electron vacancies exist in subshells, thereby allowing for strong dipole upward excitation transitions by incident photons. Radiative attenuation coefficients, or plasma opacities, are greatly enhanced due to such resonant line transitions, which will be illustrated in Section 9.6. Because of the formation of various states with the same configuration, the number of transitions could be large. For example, there are 30 different transitions among states arising from configurations of $2p^5$ and $1s$ while the rest of the orbital occupations remain the same in bromine II. Table 9.1 lists all 30 transitions for $1s$-$2p$ transitions, while only the total transition parameters are given for $1s$-$3p$ and $1s$-$4p$ transitions. The table shows the decrease in total f, cross section (CS), and absorption coefficient κ as n increases. These transition parameters were computed through configuration interaction calculations in the relativistic BP approximation using the atomic structure code SUPERSTRUCTURE (SS) (Eissner, Jones, and Nussbaumer 1974, Nahar et al. 2003) and PRCSS (e.g., Nahar 2006).

TABLE 9.1

Transition Parameters f_{ij}, S, A_{ji}, Cross Section σ_{PI}, and Absorption Coefficient κ for Resonant 1s-2p, 1s-3p, and 1s-4p Transitions in Unfilled np Subshells of Bromine II (Ground: $1s^2 2s^2 2p^6 3s^2 3p^6 3d^{10} 4s^2 4p^4$). LS Multiplets Are from the Same Spin-Multiplicity Transitions

Z	Ne	slpc:i	slpc:j	g_i	g_j	wl (Å)	E (keV)	E_i (Ry)	E_j (Ry)	f_{ij}	S	A_{ji} (s − 1)	σ_{PI} (Mb)
								$C_i(9)=1s^2 2s^2 2p^5 \ldots 4p^5$, $C_j(11)=1s^2 2s^2 2p^6 \ldots 4p^5$					
35	34	3Se 9	3Po11	3	5	1.033	12.002	118.37	1000.86	1.66E-02	1.69E-04	6.24E+13	1.34E-01
35	34	3De 9	3Po11	7	5	1.033	12.002	118.37	1000.86	1.03E-01	2.45E-03	9.02E+14	8.30E-01
35	34	3Pe 9	3Po11	5	5	1.033	12.002	118.40	1000.86	4.68E-02	7.96E-04	2.93E+14	3.78E-01
35	34	3De 9	3Po11	3	5	1.037	11.956	121.85	1000.86	8.62E-02	8.82E-04	3.21E+14	6.95E-01
35	34	3De 9	3Po11	5	5	1.037	11.956	121.86	1000.86	5.13E-02	8.76E-04	3.19E+14	4.14E-01
35	34	3Pe 9	3Po11	3	5	1.037	11.956	121.87	1000.86	2.63E-07	2.69E-09	9.79E+08	2.12E-06
35	34	1Pe 9	1Po11	3	3	1.033	12.002	118.43	1000.87	6.76E-03	6.89E-05	4.23E+13	5.45E-02
35	34	1De 9	1Po11	5	3	1.033	12.002	118.43	1000.87	5.86E-05	9.97E-07	6.11E+11	4.73E-04
35	34	1Se 9	1Po11	1	3	1.033	12.002	118.44	1000.87	9.76E-02	3.32E-04	2.04E+14	7.87E-01
35	34	3Se 9	3Po11	3	1	1.033	12.002	118.37	1000.89	1.28E-03	1.31E-05	2.41E+13	1.04E-02
35	34	3De 9	3Po11	3	1	1.037	11.956	121.85	1000.89	8.00E-07	8.19E-09	1.49E+10	6.45E-06
35	34	3Pe 9	3Po11	3	1	1.037	11.956	121.87	1000.89	3.44E-02	3.52E-04	6.40E+14	2.77E-01
35	34	3Se 9	3Po11	5	3	1.033	12.002	118.37	1000.90	8.83E-03	9.00E-05	5.52E+13	7.12E-02
35	34	3Pe 9	3Po11	3	3	1.033	12.002	118.40	1000.90	1.83E-03	3.10E-05	1.90E+13	1.47E-02
35	34	3De 9	3Po11	3	3	1.037	11.956	121.85	1000.90	8.91E-04	9.12E-06	5.53E+12	7.19E-03
35	34	3De 9	3Po11	5	3	1.037	11.956	121.86	1000.90	3.16E-03	5.39E-05	3.27E+13	2.55E-02
35	34	3Pe 9	3Po11	3	3	1.037	11.956	121.87	1000.90	6.49E-02	6.64E-04	4.03E+14	5.23E-01
35	34	3Pe 9	3Po11	3	3	1.037	11.956	121.93	1000.90	9.78E-02	3.34E-04	2.02E+14	7.89E-01
35	34	1Pe 9	3Po11	3	5	1.033	12.002	118.43	1000.86	3.65E-04	3.72E-06	1.37E+12	2.94E-03

(Continued)

TABLE 9.1 (Continued)

Transition Parameters f_{ij}, S, A_{ji}, Cross Section σ_{PI}, and Absorption Coefficient κ for Resonant 1s-2p, 1s-3p, and 1s-4p Transitions in Unfilled np Subshells of Bromine II (Ground: $1s^2 2s^2 2p^6 3s^2 3p^6 3d^{10} 4s^2 4p^4$). LS Multiplets Are from the Same Spin-Multiplicity Transitions

$C_i(9) = 1s^2 2s^2 2p^5 \ldots 4p^5$

Z	Ne	slpc:i	slpc:j	g_i	g_j	wl (Å)	E (keV)	E_i (Ry)	E_j (Ry)	f_{ij}	S	A_{ji} (s − 1)	σ_{PI} (Mb)
35	34	1De 9	3Po11	5	5	1.033	12.002	118.43	1000.86	4.91E-03	8.35E-05	3.07E+13	3.96E-02
35	34	3Se 9	1Po11	3	3	1.033	12.002	118.37	1000.87	7.63E-02	7.78E-04	4.77E+14	6.15E-01
35	34	3Pe 9	1Po11	5	3	1.033	12.002	118.40	1000.87	5.44E-02	9.24E-04	5.67E+14	4.38E-01
35	34	3De 9	1Po11	3	3	1.037	11.956	121.85	1000.87	1.61E-02	1.65E-04	1.00E+14	1.30E-01
35	34	3De 9	1Po11	5	3	1.037	11.956	121.86	1000.87	4.87E-02	8.31E-04	5.03E+14	3.92E-01
35	34	3Pe 9	1Po11	3	3	1.037	11.956	121.87	1000.87	3.89E-03	3.98E-05	2.41E+13	3.14E-02
35	34	3Pe 9	1Po11	1	3	1.037	11.956	121.93	1000.87	5.39E-03	1.84E-05	1.12E+13	4.35E-02
35	34	1Pe 9	3Po11	3	3	1.033	12.002	118.43	1000.89	6.74E-02	6.87E-04	1.26E+15	5.43E-01
35	34	1Pe 9	3Po11	3	1	1.033	12.002	118.43	1000.90	2.85E-02	2.91E-04	1.78E+14	2.30E-01
35	34	1De 9	3Po11	5	3	1.033	12.002	118.43	1000.90	9.80E-02	1.67E-03	1.02E+15	7.91E-01
35	34	1Se 9	3Po11	1	3	1.033	12.002	118.44	1000.90	5.38E-03	1.83E-05	1.12E+13	4.34E-02
LS						880.91	11.985	119.97	1000.87	6.97E-01	7.12E-03	1.45E+15	5.62E+00

Total number of trans = 30 E(keV) = 11.985   f, σ_{PI}, κ =   1.03E+00   8.31E+00   6.26E+04

$C_i(7) = 1s^2 2s^2 2p^6 3s^2 3p^5 \ldots 4p^5$, $C_j(11) = 1s^2 2s^2 2p^6 \ldots 4p^5$

Z	Ne	slpc:i	slpc:j	g_i	g_j	wl (Å)	E (keV)	E_i (Ry)	E_j (Ry)	f_{ij}	S	A_{ji} (s − 1)	σ_{PI} (Mb)
LS						986.38	13.420	14.49	1000.87	1.44E-02	6.56E-04	1.87E+14	1.16E-01

Total number of trans = 30 E(keV) = 13.420   f, CS, Absrp cf =   8.00E-02   6.45E-01   4.86E+03

$C_i(11) = 1s^2 2s^2 2p^6 3s^2 3p^6 \ldots 4p^4$, $C_j(11) = 1s^2 2s^2 2p^6 \ldots 4p^5$

Z	Ne	slpc:i	slpc:j	g_i	g_j	wl (Å)	E (keV)	E_i (Ry)	E_j (Ry)	f_{ij}	S	A_{ji} (s − 1)	σ_{PI} (Mb)
LS						1000.83	13.617	0.04	1000.88	2.54E-05	6.85E-07	2.04E+11	2.05E-04

Total number of trans = 14 E(keV) = 13.617   f, CS, Absrp cf =   1.34E-04   1.08E-03   8.13E+00

TABLE 9.2
Averaged Energies E_{res} and Resonant Photoionization Cross Sections σ_{res} for K_α (1s-2p) Transitions in Gold Ions, from Hydrogen-Like to Fluorine-Like

Ion Core	Transition Array	No. of Transitions	$\langle E(K_\alpha) \rangle$ (keV)	$\langle \sigma_{res}(K_\alpha) \rangle$ (Mb)
F-like	$1s^2 2s^2 2p^5 \ldots 1s2s^2 2p^6$	2	68.324	0.99
O-like	$1s^2 2s^2 2p^5 \ldots 1s2s^2 2p^6$	14	68.713	4.10
N-like	$1s^2 2s^2 2p^5 \ldots 1s2s^2 2p^6$	35	68.943	5.17
C-like	$1s^2 2s^2 2p^5 \ldots 1s2s^2 2p^6$	35	69.136	8.63
B-like	$1s^2 2s^2 2p^5 \ldots 1s2s^2 2p^6$	14	68.938	3.48
Be-like	$1s^2 2s^2 \ldots 1s2s^2 2p$	2	68.889	3.47
Li-like	$1s^2 2s \ldots 1s2s2p$	6	68.893	2.82
He-like	$1s^2 \ldots 1s2p$	2	68.703	3.93
H-like	$1s \ldots 2p$	2	69.663	1.58

Source: Pradhan, A. K., S. N. Nahar, M. Montenegro, Y. Yu, H. L. Hang, C. Sur, M. Mrozik, and R. M. Pitzer. 2009. *J Phys Chem A* 113:12356–63. With permission.

We also present similar transition rates for 1s-2p transitions in gold ions ($Z = 79$, $M = 196.97$ a.u.), from hydrogen-like to fluorine-like ions, in Table 9.2 (Nahar, Pradhan, and Sur 2008; Pradhan et al. 2009). These highly charged ions illustrate all possible transitions among the levels with all possible vacancies in K- and L-shells. We consider the energy region from K-shell excitations up to the K-edge. This refers to the transitions $K \to L, M, N, O, P, \ldots$ or ΔE ($n = 1 \to 2, 3, 4, 5, \ldots$), and up to the 1s or K-shell ionization energy. The set of resonant K-shell transitions then refers to K_α, K_β, K_γ, K_δ, K_ε, etc. They represent the strongest resonances, and the level of accuracy in the SUPERSTRUCTURE calculations is sufficient for demonstrating the efficacy of monochromatic radiation absorption via these resonances. These yield cross sections that are orders of magnitude larger than that of the background, as shown in Section 9.6. Transition strengths are computed for all possible transitions among fine structure levels *SLJ* of the K- and L-shells. The number of transitions varies widely for each ion from the H-like to F-like state.

Line or resonance absorption strengths of these ions were used to incorporate broadening profiles for K_α resonances. We computed the resonant photoabsorption cross sections, according to Seaton et al. (1994), as follows:

$$\sigma(K-L_i) = [4\pi^2 \alpha a_0^2 E(K-L_i)]/(3g_K)\, S(K-L_i)\varphi(\nu)$$

where L_i refers to the upper level in the open (ionized) L-shell, $E(K-L_i)$ is the corresponding energy, $S(K-L_i)$ is the line strength, and $g_K = 1$ is the initial level statistical weight. The profile factor $\varphi(\nu)$ depends on the plasma temperature and density and is normalized to unity for each resonance complex, K_α, K_β, K_γ, etc., as follows:

$$\int_{\Delta \nu res} \varphi(\upsilon) d\nu = 1$$

Table 9.2 presents the transition arrays, number of resonances N_T, the average K_α energy $E(K_\alpha)$, and cross section $<\sigma_{res}>$. $<\sigma_{res}>$ was obtained from the total $\Sigma_i \sigma(K-L_i, E)$ cross section averaged over the K_α energy range $\Delta E_{K\alpha}$ (given by Nahar, Pradhan, and Sur 2008). Each transition corresponds to a specific K_α resonance. Table 9.2 gives a rough estimate of the magnitude of the averaged absorption for each ion. These resonance strengths at energies below the K-shell ionization threshold contribute photoabsorption, as discussed in Section 9.6.

9.6 PHOTOABSORPTION COEFFICIENTS OF BROMINE AND GOLD

We study X-ray spectroscopy of BUdR through that of bromine as the active element in the compound. As shown in Table 9.1, bromine emits or absorbs X-rays in the narrow energy range of ~12–14 keV in the ($1s$-np) core transitions. The background photoionization, without autoionizing resonances, typically decreases with increase in energy. However, as the energy reaches an inner-shell ionization threshold, the background cross section jumps, but again falls beyond the ionization edge. Figure 9.3 shows background (the curve showing K- and L-edges) photoabsorption coefficient κ, which is the photoionization cross sections per gram of bromine (NIST). Edges in κ correspond to ionization jumps at various K, L, M (sub)shell energies, with the K-shell ionization edge at about 13.5 keV.

Enhanced ionization implies enhanced emission of Auger electrons. Earlier studies have focused on X-ray absorption for ionization at energies of the K-shell ionization threshold, and just above, where the cross sections rise over the background. Figure 9.3 shows that although there is a K-edge jump, the magnitude is much smaller, by orders of magnitude, than the cross sections at lower energies corresponding to higher shell thresholds. This is the main reason for the low probability of any observable enhancement of emission of electrons. On the other hand, the autoionizing resonances due to $1s$-np transitions lying below the K-shell ionization threshold form high-peak resonances (sharp lines), where absorption cross sections are orders of magnitude higher than that at the K-shell ionization edge.

Auger processes induced by ionization of the inner-most K-shell in bromine, in the radiosensitizing agent BUdR, are not sufficiently strong. They do not lead to significant enhancement in the emission of electrons and photons to break up DNA strands in malignant cells (Larson et al. 1989). But now we invoke the RT mechanism activated via these resonances below the K-shell ionization threshold, particularly at energies for the $K_\alpha(1s$-$2p)$ transitions in heavier elements such as gold. We assume that K-shell ionization can lead to vacancies in higher electronic shells due to the Auger process. Thus, an incident monoenergetic K_α beam of X-rays with sufficient intensity may affect the inverse process, that is, excite the K-shell electrons into the L-shell. Similarly, monochromatic excitation into higher shells may also be possible through resonant transitions $K_\beta(1s$-$3p)$, $K_\gamma(1s$-$4p)$, etc.

The calculation of averaged plasma attenuation coefficients requires photoabsorption cross sections as a function of incident photon energy. Using the transition rates for the K_α resonances calculated by Nahar, Pradhan, and Sur (2008), we computed

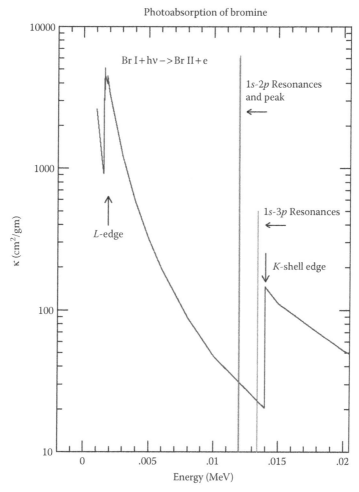

FIGURE 9.3 Photoabsorption attenuation coefficient κ of neutral bromine illustrating resonance effects. The curve is the background cross sections (without resonances) (data from
NIST website http://www.nist.gov), where the rise in κ at various energies correspond to
ionization edges of the *K* and *L* (sub)shells. The sharp lines correspond to energies and resonance peaks due to 1*s*-2*p* and 1*s*-3*p*, 1*s*-4*p* transitions (3*p* and 4*p* lie together). (From Nahar,
S. N., Y. Luo, I. Le, A. K. Pradhan, E. Chowdhury, and R. Pitzer. 2010. In WF06, *Abstracts
of 65th International Symposium on Molecular Spectroscopy*, Columbus, OH, June 21–25, p.
197. With permission.)

the resonant photoabsorption cross sections, broadened with an assumed beam width
and other broadening effects in the target. Similar convolution was carried out for
all individual transitions within higher complexes, such as K_β, K_γ, K_δ, and K_ε, corresponding to transitions to the *M*, *N*, *O*, and *P* shells, respectively. We adopted a normalized Gaussian function φ with an arbitrary full width half maximum (FWHM)
of 100 eV, and computed resonant and nonresonant mass attenuation coefficients κ
for different gold ions (Pradhan et al. 2009). These mass absorption coefficients are

plotted in Figure 9.4. The background photoionization cross sections including the K-edge for various ionic states were computed using the relativistic distorted wave approximation (Pradhan et al. 2009).

Figure 9.4 presents the monochromatic photoabsorption coefficients of gold ions below the K-edge, 80.729 keV. The ionization stages in various panels correspond to highly ionized L-shell states, from hydrogen- to fluorine-like, that is,

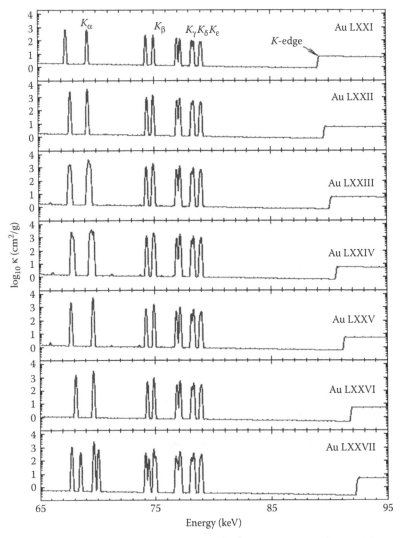

FIGURE 9.4 Total mass attenuation coefficients κ (cm²/g) for gold ions from Au LXXVII (hydrogen-like) to Au LXXI (fluorine-like). X-rays in the range 67–79 keV are absorbed in the high-peak resonances below the K-shell ionization threshold. The resonance peaks due to $1s$-np transitions in the core (Table 9.2) are orders of magnitude higher than at the K-edge jumps. (From Pradhan, A. K., S. N. Nahar, M. Montenegro, Y. Yu, H. L. Hang, C. Sur, M. Mrozik, and R. M. Pitzer. 2009. *J Phys Chem A* 113:12356–63. With permission.)

ions isoelectric with H, He, Li, Be, B, C, N, O, and F-like sequences. More enhanced X-ray photoabsorption occurs via a multitude of strong dipole transition arrays ($1s$-np). The K-shell resonance complexes are found to enhance total absorption by large factors up to 1000 at the corresponding energies relative to the background. In all panels in Figure 9.4, it can be seen that the effects of K-shell absorption in all ions, in much of the energy region from the K_α resonance complex up to the K-edge (68.7–80.7 keV), could cause considerable X-ray extinction. The total absorption by complexes K_β, K_γ, K_δ, etc., decreases as n^{-3}. However, it may be noted (and not explored before) that these resonant cross sections are orders of magnitude higher than that of the K-ionization. Thus, compared to resonant energies, continuum X-ray opacity in the high-energy regime is considerably smaller.

The above findings can be applied to neutral gold. When the K-shell is ionized to create one- or two-electron vacancies in the L-shell by an external source, then the absorption coefficient for K_α resonant excitation can be taken to be the same as for F-like and O-like gold ions, respectively (Table 9.1). The reason for the approximation is that transition strengths for deep inner-shell transitions, especially the $K\rightarrow L$, are largely unaffected by outer-shell electron correlations and influenced mainly by nuclear charge screening. The total K_α oscillator strengths are approximately independent of Z along a given isoelectronic sequence.

Since there is a strong dependence of $\kappa(\nu)$ on each ionization stage capable of K-shell excitation, the attenuation coefficient can be written as follows:

$$\kappa(\nu, K_\alpha) = \frac{1}{uW_A} \frac{\sum_j w_j \sum_i \sigma_{res}(\nu, K \rightarrow L_{ji})}{\sum_j w_j}$$

where w_j is the ionization fraction for an ion core j in the L-shell. For instance, we may perform simulations of X-ray propagation assuming equipartition between only one and two L-shell vacancies, that is, excitation of a plasma ionized up to F-like and O-like ions of gold.

It may also be noted that the resonant absorption peaks for gold ions are significantly higher than those of bromine due to the large difference between the nuclear charge of gold and bromine. Gold has higher Z, 79, than that of bromine ($Z = 35$), and hence the background cross sections, which show $1/Z^2$ dependence, are lower for gold than bromine. But the oscillator strengths do not depend on Z and are of the same order for both elements. Hence, the background-to-resonant peak ratio of gold is much larger than that of bromine.

9.7 TEST OF RESONANT THERANOSTICS USING NUMERICAL SIMULATIONS

Radiation propagation in a medium involves several other processes in addition to those mentioned above. One is "Compton scattering" where the photon transfers some energy to an electron.

$$h\nu + X^{+Z} \leftrightarrow X^{+Z*} + h\nu'$$

This process could be significant for light elements in the body, given high-energy X-ray irradiation. In addition, there are secondary electron collision processes that could be as follows:

1. Ionization: $e + X^{+Z} \rightarrow e' + e'' + X^{+Z+1}$
2. Excitation: $e + X^{+Z} \rightarrow e' + X^{+Z*}$

Hence, in a numerical simulation, these processes need to be considered. We now seek to model the physical processes due to irradiation of X-rays and passage through body tissue interspersed with layers of high-Z material. Our aim is to study the enhanced X-ray interactions owing to the presence of high-Z material and at resonant energies as opposed to the nonresonant background. Modeling of photon transport in biological tissue numerically with Monte Carlo simulations is common and can provide accurate predictions compared to handling a difficult radiation transfer equation or diffusion theory. Monte Carlo simulations can keep track of many physical quantities simultaneously, with any desired spatial and temporal resolution, making it a powerful tool. The simulations can also be made accurate by increasing the number of photons traced. Hence, these methods are standard for simulated measurements of photon transport for many biomedical applications.

Although flexible, Monte Carlo modeling is rigorous since it is statistical and therefore requires significant computation time to achieve precision. In this method, transportation of photons is expressed by probability distributions, which describe the step size of the photon propagation between sites of photon–tissue interaction and the angles of deflection in the photon's trajectory when a scattering event occurs. The required parameters needed are the absorption coefficient, the scattering coefficient, the scattering phase function, and so on. As the photon interacts with the medium, it deposits energies due to absorption and is scattered to other parts of the medium. Any number of variables can be incorporated along the way, depending on the interest of a particular application. Each photon packet will repeatedly undergo a number of steps until it is terminated, reflected, or transmitted.

Compared to other simulations for biological systems, relatively fewer codes are capable of describing effects of radiation in microscopical entities, such as cellular structures and the DNA molecule, of size of the order of a few nanometers or tens of nanometers. However, few general Monte Carlo codes have the capability of describing particle interactions at these scales. The Monte Carlo simulation code Geant4 (Agostinelli et al. 2003) is enough to consider the radiation transport and high-energy X-rays needed to model RT. It is an open-source code developed with an object-oriented design with the possibility to implement or modify any physical process without changing other parts of the code. Its versatility has enabled its application to medical physics.

9.7.1 Monte Carlo Simulations Using Geant4

We have adopted the Monte Carlo code Geant4 version 9.2 for the simulations of X-ray absorption by gold nanoparticles at resonant energies (Montenegro et al. 2009). The physical phenomena were not heretofore explored. The Geant4 package treats atoms as neutral, and photoelectric cross sections do not include electronic shell-ionization effects. We have modified the code such that it can include the resonant cross sections in addition to its own input data for various processes. Geant4 has an extension package known as "low-energy electromagnetic processes," which is the product of a wider project called Geant4-DNA, which addresses specifically the extension of Geant4 to simulate radiation effects at the cellular and DNA scales, and has been widely tested. It includes the Auger process. It is this package that has been modified to consider resonant enhancement and fluorescent yields that may be used for monochromatic imaging and diagnostics, as well as therapy. However, the standard version of Geant4 we have modified to accommodate resonant structures does not include shell structures in an atomic system and hence is incapable of treating the Coster–Kronig cascade process.

For numerical simulations, we perform a test for a tumor measuring $2 \times 5 \times 5$ cm^3 located 10 cm below the skin and assume that it is doped with a thin layer of gold nanoparticles. Since the muscular tissue has a density of 1.06 g/mL similar to that of water (1.0 g/mL), the numerical experiments were carried out in a phantom of water $15 \times 5 \times 5$ cm^3 with a thin layer of gold nanoparticles of 2-cm thickness with 0.1 mm/g at 10 cm inside from the surface. The phantom setup is shown in Figure 9.5a. Monochromatic X-rays of various energies are passed separately through the water into the thin layer of gold nanoparticles.

We assumed preionization of gold atoms by another X-ray beam at energy above the K-edge energy of ~81 keV to create electron vacancies (as illustrated in Figure 9.2). The simulations considered resonances for the preionized atomic core of gold ions Au LXXI (or F-like with one-electron vacancy) and Au LXXII (or O-like with two-electron vacancies). We assume equipartition between these two ion cores for simplification of the calculation of the resonant cross sections and reduced the cross sections to be just the average value of both ions, for example, an estimate of

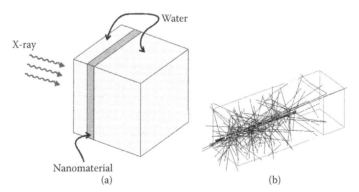

Nanomaterial
(a)   (b)

FIGURE 9.5   (a) Geometry of the Geant4 simulations. The water phantom models a tumor (nanomaterial) 10 cm inside normal tissue. The monochromatic beam collides perpendicular to the square face of the water block. (b) Snapshot of 68-keV X-ray photons propagating inside the tissue. Black lines show the photon trajectories in the phantom. The tumor is embedded with gold nanoparticles at 5 mg. Note how only a small fraction of photons are able to cross the region up to the gold nanoparticles.

~2.5 megabarns (1 MB = 10^{-18} cm^2) at ~68 keV. Outside the resonant energies of these ions, the gold nanoparticles were assumed neutral.

To see the resonant effect, we passed three different monochromatic X-rays into the phantom: (1) a beam with a beam energy corresponding to the averaged K_α energy at 68 keV, which is the energy difference between the L ($n = 2$) and the K ($n = 1$) shells, (2) an 82-keV beam, with energy just above the ionization energy of the K-shell electrons at 81 KeV, and (3) a high-energy beam of 2 MeV, which corresponds to the maximum intensity of the Bremsstrahlung radiation distribution from a LINAC with 6-MVp peak voltage X-ray source used in many medical facilities. A "snapshot" of the propagating beam at 68 keV is shown in Figure 9.5b. It shows that there are strong photon interactions being deflected or absorbed, reducing drastically the number of available photons to reach the tumor or beyond.

9.7.2 RESONANT EFFECTS ON MONTE CARLO SIMULATIONS

The X-ray irradiation of gold ions results in Auger decays of photon emission and electron ejections, creating multiple electron vacancies. These vacancies may be filled either by radiative decays from higher electron shells or by excitations from the K-shell at resonant energies by an external X-ray source. One main objective was to observe the predicted enhancement in absorption of X-rays at 68 keV resonant energy by gold nanoparticles as this would provide a new insight into unsuccessful experimental searches to find such enhancement at and around the K-shell ionization energy of ~81 keV. The absorption spectra from the Monte Carlo simulations on extinction of X-rays with depth from the surface are shown in Figure 9.6. The darker curves represent absorption by the body muscle without the nanoparticles and the lighter curves that with the nanoparticles. The incident X-ray energies are high for any resonance for the water molecules. Hence, these molecules experience only Compton scattering with small cross sections and produce some low-energy electrons. The rise in the lighter curves at the entrance of the tumor, 10 cm inside, in each panel shows the energy deposited in the nanoparticles. Figure 9.6 shows that resonant excitations via K_α, K_β, etc., transitions result in a considerable enhancement in localized X-ray energy deposition at the layer with nanoparticles compared with nonresonant processes and energies. The top panel shows that the absorption peak at 68 keV is considerably higher, ~10.6 keV/mm, compared to the nonresonant background and complete absorption of radiation within the layer.

The enhanced absorption at 68 keV is expected to lead to enhanced emission of photons and electrons to kill the tumor cells. The peak in the middle panel, ~0.38 keV/mm for 82 keV X-rays, is relatively small, which results in no visible rise in secondary electron production. The lowest panel shows an X-ray absorption peak of ~4 keV/mm for 2 MeV X-rays. This peak is higher, although lower than that of 68 keV, because of higher penetration of the beam into the tissue. Both the 68 keV and the 82 keV X-ray beams in the top and middle panels could retain only about a quarter of their intensity before reaching the tumor, whereas the 2 MeV beam in the bottom panel has little interaction with the tissue, and the deposited energy is reduced by only about 20% before reaching the tumor. This is the primary reason for the use of high-energy beams, to ensure that the tumor receives a sufficiently

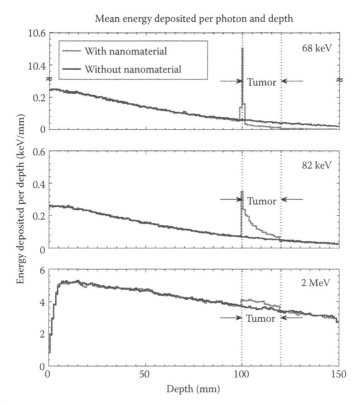

FIGURE 9.6 X-ray energy deposition by depth across the phantom at three X-ray beams used: 68 keV (averaged K_α resonant energy), 82 keV (K-edge energy), and 2 MeV (high energy common in clinical usage). The region between 100 and 120 mm represents the tumor, and it can be embedded with gold nanoparticles at 5 mg/mL (lighter curve) or water (darker curve). The presence of gold nanoparticles increases the energy deposited at the tumor. The highest absorption is at 68 keV, more than 25 times higher than that at 82 keV.

high-radiation dose. The downside of this approach is that not only the tumor but also the normal tissue along the beam path receives high dosage.

The simulations also show that with higher X-ray absorption a larger number of secondary Auger electrons are produced, as illustrated in Figure 9.7. Again, the darker curve represents the water or body tissue, while the lighter curve represents electron production by the water; the gold nanoparticles are assumed to be placed at 10 cm from the surface. While a small rise in electron production is noted for 82 keV and at 2 MeV (middle and the lowest panels), a high peak in the electron counts is seen at the resonant energy of 68 keV, assuming existing L-shell vacancies created by a preionization X-ray beam. A larger number of electrons, by more than an order of magnitude, are then produced at 68 keV compared with that at 82 keV. Consistent with deeper penetration by high-energy X-rays, the high-energy beam at 2 MeV produced more electrons than that at 82 keV, but still much smaller than that at 68 keV.

The photoelectrons produced by the X-ray irradiation should trigger the breakage of DNA strands and, consequently, destruction of cancerous cells. The

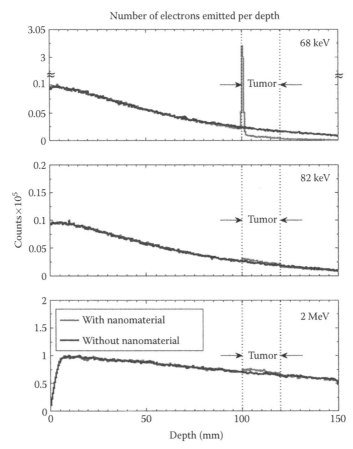

FIGURE 9.7 Electron production with depth at the three X-ray beam energies 68 keV, 82 keV, and 2 MeV. Auger production of electron peaks at 68 keV are higher by more than an order of magnitude than at 82 keV, and over five times that at 2 MeV.

low-energy photoelectrons produced in the tumor are absorbed in the immediate vicinity. The mean free paths of electrons produced from gold nanoparticles are 47, 46, and 52 μm for the 68 keV, 82 keV, and 2 MeV beams, respectively, compared with the 1-mm mean free path of electrons produced in the water portion of the phantom.

In contrast to electron emissions, photon production is relatively low since most of the input energy from the incident X-ray beams goes into the production of Auger electrons and relatively little to secondary photons via fluorescence. Figure 9.8 shows production of photons with depth from the three different X-ray energies. Darker curve represents photon production from water, and the lighter curve represents photon production with gold nanoparticles. The number of photons produced by the 68-keV X-ray beam shows a clear surge at the tumor site, while the production is almost invisible for the 82 keV and 2 MeV beams.

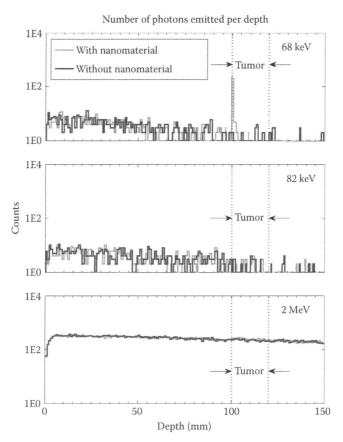

FIGURE 9.8 Photon emission with depth inside the phantom at three X-ray beams: 68 keV, 82 keV, and 2 MeV. Only 68 keV shows a surge in photon production.

9.8   RADIATION DOSE ENHANCEMENT FACTOR

The radiation dose enhancement factor (DEF) is defined as the ratio of the average radiation dose absorbed by the tumor when it is loaded with a contrast medium or agent (e.g., iodine) to the dose absorbed without the agent. We consider the nanoparticle solution as the contrasting medium. These values can be obtained from Figure 9.6 by integrating over the energy absorbed by the tumor with and without nanoparticles, using the same phantom under both conditions. The ratio of the mean total energy absorbed per photon by the tumor under the two conditions gives the DEF.

The simulations also assumed different concentrations of the gold nanoparticles from 0 to 50 mg/mL. DEFs were obtained by irradiation at the three energies: 68 keV, 82 keV, and 2 MeV. The phantom was divided longitudinally into sections of 1 mm, and the energy deposition, as well as particle generation, at each slice was recorded. A monochromatic linear beam, with a circular cross section of 3-mm diameter, was aimed at the phantom perpendicular to the square face far from the tumor region. Each simulation had 500,000 events, a value that generates error fluctuations of less than 0.1%.

The results are plotted in Figure 9.9. The DEFs obtained for the resonant X-ray beam of 68 keV are almost an order of magnitude greater, even at low concentrations, than those calculated at higher energies and concentrations. At 68 keV, the DEF at a concentration of 5 mg/mL was 11.02 and increased slowly to 11.7 for a concentration of 50 mg/mL. This indicates that low concentration is sufficient to achieve high DEF for resonant energies. On the other hand, the DEFs for the 82-keV beam increased steadily from about 1 to achieve the high DEFs obtained in the resonant case. For the 2 MeV case, we obtained low but slowly increasing values of DEFs as the concentration increased, confirming our observation in Figures 9.6 and 9.7. It is possible to obtain an enhanced dosage with high-energy beams. But the enhancement is quite small and increases slowly with nanoparticle concentrations. The behavior shown in these curves is in agreement with the calculations done by Solberg, Iwamoto, and Norman (1992), from which they found that DEF with a 250-keV beam was smaller, and had a weaker concentration dependence, than that with a 140-keV beam.

The almost flat behavior for the DEF at 68 keV can be explained by the large mass attenuation factor obtained, 39.1 cm²/g for 68 keV versus 0.241 cm²/g for 82 keV at 5 mg/mL. This high mass attenuation coefficient results in a complete absorption of any incoming photon within the tumor region surface, reducing the transmission probability and any further increase due to subsequent absorption beyond. This saturation of the absorption is also observed at high gold concentrations for the 82-keV beam, asymptotically reaching the same value as the 68-keV beam (extending the results in Figure 9.9 to high, though unrealistic, concentrations).

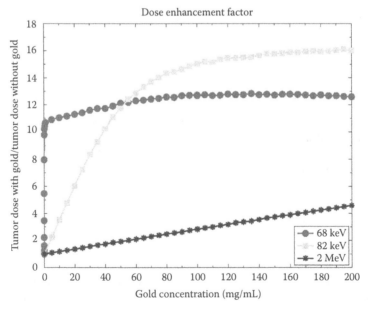

FIGURE 9.9   Dose enhancement factors at three X-ray energy beams, 68 keV, 82 keV, and 2 MeV for different concentration of gold nanoparticles.

9.9 CONCLUSION

The present results based on the RT method could be applicable to in vivo therapy and diagnostics of cancerous tumors using high-Z nanoparticles and monochromatic X-ray. X-ray radiation dose reduction commensurate with resonant enhancements may be realized for cancer theranostics, using high-Z nanoparticles and molecular radiosensitizing agents embedded in malignant tumors. The in situ deposition of X-ray energy, followed by secondary photon and electron emission, should also be localized at the tumor site. We also note the relevance of this work to the development of novel monochromatic or narrow-band X-rays. The underlying mechanism also rests on the biotechnology of high-Z nanoparticles delivered to specific sites, such as cancerous tumors, and then treated with monoenergetic X-rays at resonant atomic and molecular transitions.

Further studies based on the RT method may lead to the use of focused narrow-band or monoenergetic X-ray beams at resonant energies, with a high potential for a much safer treatment of cancerous tumors. However, verification of the absorption enhancement will require collaborative experimental setups with the eventual goal of clinical studies. This multidisciplinary effort involving atomic and molecular physics, chemistry, radiation oncology, nanobiomedicine, and clinical research should lead to advances in the use of narrow-band X-ray using high-Z nanoparticles in biomedical applications.

ACKNOWLEDGMENTS

The work reported herein was partially supported by grants from the Ohio State University and the U.S. National Science Foundation. The computational work was carried out mainly at the Ohio Supercomputer Center in Columbus, Ohio.

REFERENCES

Agostinelli, S. et al. 2003. Physics research section A: Accelerators, spectrometers, detectors and associated equipment. *Nucl Instrum Methods* 506:250–303.

Eissner, W., M. Jones, and H. Nussbaumer. 1974. Techniques for the calculation of atomic structures and radiative data including relativistic corrections. *Comput Phys Commun* 8:270–306.

Hainfeld, J. et al. 2004. The use of gold nanoparticles to enhance radiotherapy in mice. *Phys Med Biol* 49:N309–15.

Larson, D., W. J. Bodell, C. Ling, T. L. Phillips, M. Schell, D. Shrieve, and T. Troxel. 1989. Auger electron contribution to Bromodeoxyuridine cellular radiosensitization. *Int J Radiat Oncol Biol Phys* 16:171–6.

Mass Absorption Coefficients. National Institute of Standards and Technology website: http://physics.nist.gov/PhysRefData/Xcom/Text/download.html.

Monte Carlo Simulation. Palisade Corporation. http://www.palisade.com/risk/monte_carlo_simulation.asp. Accessed July 15, 2011.

Montenegro, M., S. N. Nahar, A. K. Pradhan, Y. Yu, and K. Huang. 2009. Monte Carlo simulations and atomic calculations for Auger processes in biomedical nanotheranostics. *J Phys Chem A* 113:12364–9.

Nahar, S. N. 2006. Atomic data from the iron project - LXI. Radiative E1, E2, E3, and M1 transition probabilities for Fe IV. *Astron Astrophys* 448:779–85.

Nahar, S. N., W. Eissner, G.-X. Chen, and A. K. Pradhan. 2003. Atomic data from the Iron Project - LIII. Relativistic allowed and forbidden transition probabilities for Fe XVII. *Astron Astrophys* 408:789–801.

Nahar, S. N., Y. Luo, I. Le, A. K. Pradhan, E. Chowdhury, and R. Pitzer. 2010. X-ray spectroscopy of bromine compounds and biomedical applications. In WF06, *Abstracts of 65th International Symposium on Molecular Spectroscopy*, Columbus, Ohio, June 21–25, p. 197: http://molspect.chemistry.ohio-state.edu/symposium/abstractbook/absbk10.pdf.

Nahar, S. N., A. K. Pradhan, and C. Sur. 2008. Oscillator strengths and radiative transition rates for Kα lines in gold X-ray spectra: $1s$-$2p$ transitions. *J Quant Spec Rad Transf* 109:1951–9.

Nahar, S. N., A. K. Pradhan, and S. Lim. 2011. Kα transition probabilities for platinum and uranium ions for possible X-ray biomedical applications. *Can J Phys* 89:483–494.

NIST handbook. http://www.nist.gov/physlab/data/handbook/index2.cfm.

Pradhan, A. K., and S. N. Nahar. 2011. *Atomic Astrophysics and Spectroscopy*. Cambridge University Press (Cambridge, UK).

Pradhan, A. K., S. N. Nahar, M. Montenegro, Y. Yu, H. L. Hang, C. Sur, M. Mrozik, and R. M. Pitzer. 2009. Resonant X-ray enhancement of the Auger effect in high-Z atoms, molecules, and nanoparticles: Biomedical applications. *J Phys Chem A* 113:12356–63.

Pradhan, A. K., Y. Yu, S. N. Nahar, E. Silver, and R. Pitzer. 2007. Computational methodology for resonant nano-plasma theranostics for cancer treatment. In *Proceedings of 15th International Conference on the Use of Computers in Radiation Therapy*, Toronto, Ontario, Canada, June 4–7, 2007. http://www.iccr2007.org/.

Seaton, M. J., Y. Yan, D. Mihalas, and A. K. Pradhan. 1994. Opacities for stellar envelopes. *Mon Not R Astron Soc* 266:805–28.

Silver, E., A. K. Pradhan, and Y. Yu. 2008. The X-ray reloaded: Rearming radiography with resonant theranostics. *RT Image* 21:30–4.

Solberg, T. D., K. S. Iwamoto, and A. Norman. 1992. Calculation of radiation dose enhancement factor for dose enhancement therapy of brain tumors. *Phys Med Biol* 37:439–43.

10 Nanomechanical In Vitro Molecular Recognition
Mechanical Resonance–Based Detection

Kilho Eom and Taeyun Kwon

CONTENTS

10.1 INTRODUCTION

To sense and detect the specific biological or chemical species is of high significance [1–4] because such sensing and/or detection are correlated with early diagnosis of diseases such as cancers and/or screening of toxicity in the environment. The past decade has witnessed the emergence of microelectromechanical system (MEMS)/ nanoelectromechanical system (NEMS) devices [2,5], which allow the label-free detection of specific biological or chemical species. Among such devices, micro-/ nanomechanical devices such as nanomechanical resonators [6] have received a lot of attention due to their capability of highly sensitive detection of specific molecules (or atoms) even at single-molecule (atomic) resolutions [7–10]. Further, mechanical resonators are able to quantify the target molecules, for example, their masses [7–10], which implies that nanomechanical resonators are appropriate for not only sensitive label-free detection but also weighing of target molecules [11].

In general, the ability to sense and detect the specific biological molecules even at low concentrations in buffer solution (or blood serum) is critical to early diagnosis. For instance, the blood serum of patients with prostate cancer includes the specific proteins such as "prostate specific antigen (PSA)." It was found that, at the early state of the prostate cancer, blood serum contains PSA in the concentration of ~1 ng/mL, which indicates that detection of small amount of specific proteins is essential for early diagnosis. However, current biosensing tools such as enzyme-linked immunosorbent assay (ELISA) encounter the technological limitations in that they are usually unable to detect the marker proteins at low concentrations (e.g., ~1 ng/mL), which is now called the "diagnostic gray zone" [12]. In recent years, microcantilever sensors as a nanomechanical device have been reported as a device that can overcome such restrictions, that is, the "diagnostic gray zone." Specifically, microcantilevers functionalized with PSA antibodies are able to catch the marker protein PSA even at a concentration of ~1 ng/mL in buffer solution, based on the principle that antigen-antibody binding results in a bending deflection change for microcantilevers [12] (for details, see Section 10.2.1). This indicates that nanomechanical sensors may enable the sensitive detection that implies early diagnosis, which is inaccessible with conventional sensing tools such as ELISA.

The fundamental principle of nanomechanics-based sensing is the direct transduction of molecular binding (or adsorption) on the surface of a nanomechanical sensor into physical property changes [13,14] such as bending deflection changes and/or resonant frequency changes. The bending deflection–based detections have been widely used for studying various biomolecular interactions such as protein antigen–antibody interactions [12,15], DNA hybridization [16–18], DNA–RNA interactions [19], protein–enzyme interactions [20,21], and protein–cell interactions [22]. Despite its wide applications, the bending deflection–based detection has restrictions in that the length of a device should be at least ~100 μm for reliable detection and that it does not enable the quantitative measurement of detected molecules (for details, see Section 10.2). On the other hand, resonance-based detection allows the quantitative measurement of detected molecules because its detection principle is that molecular adsorption increases the effective mass of a device, which results in decreasing the resonant frequency [23,24]. Moreover, the detection principle shows that scaling down of a resonant device is relevant to the increase in the detection sensitivity of a mechanical resonant device even at atomic resolution, since scaling down increases its resonant frequency that is related to its detection sensitivity. For example, Roukes et al. [8,25] first reported that resonant nanobeams are able to sense and detect gas adsorption at atomic resolution, which implies that resonant nanodevices can be used for nanomechanical mass spectrometry [26] that can be realized at lab-on-a-chip level. Nanoresonators and their various sensing applications have recently intrigued researchers.

As nanoresonators have recently been widely used as electromechanical devices for further applications such as sensing or detections, it is necessary to fundamentally understand the dynamic behavior of nanoresonators. Despite its ability to describe the flexural resonance motion of micro-/nanodevices, the continuum elasticity theory in its current form becomes irrelevant to analyze the dynamic behavior of nanoresonators when their spatial scales are reduced to nanometer scales. This is attributed to unanticipated effects such as surface effects. Specifically, as scaled down, nanostructures are well characterized by a large surface-to-volume ratio, which leads to the increase in surface energy [27–29] that is defined as the energetic cost to create the surface due to scaling down. In general, the surface energy depends on the mechanical deformation of the surface, and consequently the mechanical strain of the surface. This results in the surface stress, which is a derivative of surface energy with respect to mechanical strain of the surface, inherent in nanostructures. Moreover, in nanostructures, the surface atoms have fewer bonding neighbors than do the bulk atoms; this leads to nanostructures that are subjected to the surface stress, which results in the reconstruction of the structure [30]. These indicate that the surface stress plays a significant role in the mechanical deformation of nanoscale structures, and consequently the resonance behavior of nanoresonators. This implies that continuum mechanics models in their current form have to be properly modified in order to describe the anomalous dynamic behavior of nanoresonators due to the surface stress effects.

This chapter presents not only the current experimental attempts in the development of micro-/nanoresonators and their sensing applications but also the theoretical and/or computational models that can provide the fundamental insights into the dynamic behavior of nanoresonators and the underlying mechanisms of nanoresonator-based detection. Specifically, we overview the currently suggested continuum mechanics models as well as the multiscale models, which are able to describe the fundamental mechanics of nanoresonators and the basic principles of nanoresonator-based detection. The remainder of this chapter is organized as follows: Section 10.2 reviews the fundamental physics of nanomechanics-based label-free detection based on continuum elasticity theories. Section 10.3 delineates the current experimental attempts in resonator-based chemical or biological detection. Section 10.4 discusses the currently suggested continuum mechanics models that enable the fundamental understanding of the unique dynamic behavior of nanoresonators, which is different from what is anticipated by conventional continuum elasticity theories. Moreover, as a resonator is scaled down to nanometer length scales, the conventional continuum elasticity theory fails to describe the basic principles of nanoresonator-based sensing. This has led us to review the currently suggested novel continuum mechanics model, which includes the unexpected effects such as surface effects, for analysis of the dynamic behavior of nanoresonators and their sensing performances. Section 10.5 describes the currently suggested multiscale models that can explain the underlying mechanisms of nanoresonators and their sensing applications. To the best of our knowledge, the multiscale modeling concept described in Section 10.5 is still premature, so that there will be a lot of room for researchers to contribute to the development of multiscale modeling that allows the fundamental understanding of the underlying mechanisms in the dynamics of nanoresonators and nanoresonator-based detection. In addition, we have provided future guidelines for further development of multiscale models for nanoresonators. Finally, Section 10.6 concludes this chapter with an outlook and future perspectives.

10.2 CONVENTIONAL DETECTION PRINCIPLES

10.2.1 DEFLECTION-BASED DETECTION

Deflection-based sensing was first designed by Stoney to measure the surface stress acting on a thin film. This is usually referred to as "Stoney's formula" [31,32], which relates the surface stress to the surface stress–induced bending curvature of a thin film.

$$\kappa = \frac{6(1-\nu)}{Et^2}\tau \qquad (10.1)$$

where κ is the bending curvature induced by surface stress τ, and E, t, and ν represent the elastic modulus, thickness, and Poisson's ratio, respectively, of a thin film. Stoney's formula has been used to gain insights into the atomic adsorption on the surface (i.e., a thin film) [28,29]. Specifically, atomic adsorption changes the surface state, which results in the change of the surface stress and, consequently, induces the bending curvature of the surface. This principle has been widely employed for cantilever-based sensing of atomic adsorption (for details, see references [28,29]). For instance, Gerber et al. [33] first employed the microcantilever for quantitative understanding of molecular adsorption on the surface. Their model system is the adsorption of alkanethiol onto the gold surface deposited on the microcantilever (Figure 10.1). The surface stress change due to atomic adsorption was measured from the bending deflection motion driven by such adsorption. Moreover, they also studied the effect of conformations of alkanethiol on the surface stress change, which showed that the surface stress induced by molecular adsorption depends on the molecular conformations.

In the last decade, Stoney's formula had been reconsidered for microcantilever-based label-free detection of biological/chemical species in biology or chemistry applications. For label-free detection, the microcantilever must be able to capture the specific biological or chemical species among various species in the sample. For such specific detection, the microcantilever surface should be chemically modified by the introduction of immobilization of specific receptor molecules on the surface. The surface can be chemically modified using (1) a chemical reaction between the thiol group of receptor molecules and gold thin film deposited on the cantilever [34]; (2) a chemical interaction between the amine group of receptor molecules and silicon surface [35]; (3) a self-assembled monolayer that can serve as the cross-linker between the surface and receptor molecules [36]; and (4) single-chain antibody fragments [37]. The details of chemical modification of the surface are well described in references [14,38]. When specific target molecules are bound to the functionalized cantilever surface, such specific molecular binding induces the surface stress changes that lead to a measurable bending deflection change. This detection principle has been widely utilized for cantilever-based biosensing such as protein detection [12,15], DNA detection [16–18,39], RNA detection [19], and protein–drug interactions [40]. For instance, Majumdar et al. [12] have reported the label-free detection of PSA using the measurement of bending deflection change for the microcantilever due to PSA antigen–antibody binding on the cantilever surface. They have suggested that the microcantilever is able to reliably detect PSA even at the

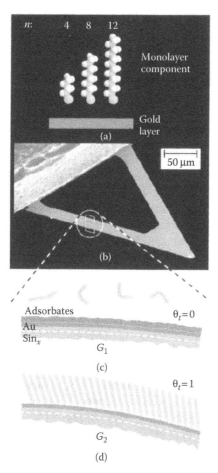

FIGURE 10.1 Schematic illustration on the cantilever-based measurement of surface stress induced by atomic adsorption. (a) Different molecular chains (i.e., alkanethiol) with different lengths are shown, and they can be adsorbed onto a gold layer. (b) Scanning electron microscope (SEM) image of a triangular-shaped microcantilever. (c) Initial configuration of cantilever without any molecular adsorption, and (d) cantilever bending motion induced by the adsorption of molecular chains. Here, the molecular adsorption onto a cantilever leads to the generation of surface stress, which results in the bending deflection of a cantilever. (Adapted from Berger, R., E. Delamarche, H. P. Lang, C. Gerber, J. K. Gimzewski, E. Meyer, and H. J. Guntherodt. 1997. *Science* 276:2021–4. With permission.)

concentration of 1 ng/mL, which is usually referred to as the "diagnostic gray zone", in which conventional toolkit such as ELISA is unable to sense and detect target molecules due to restricted detection sensitivity. This implies that microcantilevers can be an alternative to conventional sensing tools such as ELISA for reliable detection of minute amounts of specific biological species. Gerber et al. [16,39] showed that it is possible to detect specific DNA sequences without any labeling using the microcantilever bending deflection motion. In a similar manner, Majumdar et al. [17] provided that the bending deflection motion of the microcantilever due to DNA hybridization is also dependent on the ionic strength of a solvent, which plays a critical role in the intermolecular interactions between DNA molecules, resulting in a different surface stress state. They also reported the label-free detection of DNA denaturation at the DNA melting temperature using cantilever bending deflection motion. Furthermore, Gerber et al. [19] suggested sensitive label-free detection of specific RNA molecules using cantilever bending motion. In addition, McKendry et al. [40] introduced a nanomechanical drug-screening system based on microcantilevers, on which a specific peptide sequence is functionalized. Specifically, the molecular interactions between drug molecules and peptides (or mutated peptides) were quantitatively detected using cantilever bending deflection motion. These epitomes shed light on the cantilever-based assay for its capability of performing not only label-free detection but also providing fundamental insights into various molecular interactions such as intermolecular interactions, DNA hybridization, conformational changes, and/or drug–biomolecule interactions.

Although the detection principle based on Stoney's formula was widely used for the aforementioned cantilever-based label-free detection, this principle has significant restrictions. First, for reliable detection of biomolecules, the cantilever length should be at least ~100 μm, and if it is much less than 100 μm, then the cantilever bending deflection change due to biomolecular interactions cannot be optically detected due to the resolution of the optical apparatus. This can be elucidated from Stoney's formula, which shows the relationship between deflection change Δz and surface stress τ such as $\Delta z = 6(1 - v)E^{-1}(L/t)^2\tau$, where pure bending motion is assumed. This provides that as the ratio (L/t) decreases, so significantly does the bending deflection change Δz. Second, the bending deflection-based label-free detection does not allow us to quantify the amount of target molecules that are specifically bound to a microcantilever. These two restrictions can be resolved using a cantilever's resonance motion, which is described in Section 10.2.2.

10.2.2 RESONANCE-BASED DETECTION

In the last decade, the label-free detection using a micro-/nanomechanical resonator had been highlighted because of its great potential for sensitive detection even at single-molecule resolution. This is attributed to the principle that scaling down significantly increases the resonant frequency of a device, which is related to its detection sensitivity. Recently, a few researchers at Caltech [8,25] reported label-free detection at atomic resolution using nanomechanical resonators, which implies that nanomechanical resonators can be an alternative to conventional mass spectrometry [26]. In other words, mass spectrometry can be realized at a lab-on-a-chip level. Section 10.3 describes the details of current experimental efforts on resonator-based detection.

The fundamental principle of resonance-based sensing is the direct transduction of molecular binding on the sensing surface into shifts in the resonant frequencies of devices, which arise from the added mass due to molecular binding [23,24]. In the context of continuum mechanics [41–43], the resonant frequency of a micro-/nanoscale beam is described as $\omega_{0,i} = (\lambda_i/L)^2(EI/\rho A)^{1/2}$, where $\omega_{0,i}$ is the ith resonant frequency, λ_i is a constant dependent on the mode index i, and EI, ρ, and A represent the flexural rigidity, mass density, and cross-sectional area of the beam, respectively.

Upon a mass adsorption on a resonant cantilever, the resonant frequency of a cantilever is shifted to $\omega_i^* = (\lambda_i/L)^2\{EI/(\rho A + \Delta m/L\}^{1/2}$, where Δm is the total mass of adsorbed molecules onto a beam. Consequently, the resonant frequency shift $\Delta \omega$ due to molecular adsorption is represented in the following form [23,24]:

$$\frac{\Delta\omega}{\omega_0} = -\frac{1}{2}\frac{\Delta m}{M} \tag{10.2}$$

where ω_0 is the resonant frequency of a bare cantilever, and M is the mass of the cantilever given by $M = \rho AL$. The negative sign in Equation 10.2 indicates the decrease in the resonant frequency when molecules are adsorbed onto a cantilever. Equation 10.2 clearly demonstrates that the more scaling down, the better detection sensitivity a cantilever exhibits. The detection principle depicted in Equation 10.2 is not acceptable in the case of nanoscale resonators due to unanticipated effects such as the surface effect. Specifically, a scaling down of a device leads to the creation of the surface resulting in an increase in the surface energy, that is, the energetic cost to create a surface [27–29]. This surface energy is usually dependent on the mechanical strain, where the surface stress is defined as the derivative of surface energy with respect to strain. In general, this surface stress plays a critical role in the mechanical deformation of nanomaterials [28–30,44], which implies that the dynamic behavior of a nanoresonator is dominated by surface stress effects. As a consequence, the surface stress has an important role in the detection sensitivity of a nanoresonator. This surface stress effect on the detection principle will be described in Section 10.4.2. Moreover, it should be kept in mind that the detection principle in Equation 10.2 assumes that a device undergoes harmonic, flexural motion and that the effect of adsorbed molecules' stiffness on the resonant frequency shift is ignored. Provided the thickness of an adsorbed molecule is comparable to that of a resonant device, the resonant frequency shift cannot be explained by added mass as depicted in Equation 10.2 but depends on the stiffness of adsorbed molecules. These issues on the effect of adsorbate stiffness and surface stress will be considered in Sections 10.4.1 and 10.4.2, respectively.

10.3 EXPERIMENTAL WORKS ON RESONANCE-BASED DETECTION

Resonator-based sensing allows not only the sensitive detection but also the quantification of molecular binding [11,35]. For example, nanomechanical resonators enable the measurement of the molecular weight of specific molecules and/or cells [7,45–47] for further characterization and/or quantification of their biological functions. Despite numerous recent studies on resonator-based detection, this section provides representative epitomes of label-free detection using resonators.

In the early 2000s, resonant microcantilevers were taken into account for sensing and detecting specific cells and/or viruses. For instance, researchers at Cornell [48] reported cell detection using resonant microcantilevers by measuring the resonant frequency shift due to cell adsorption onto a cantilever surface. They estimated the mass of adsorbed cells based on the detection principle described in Equation 10.2.

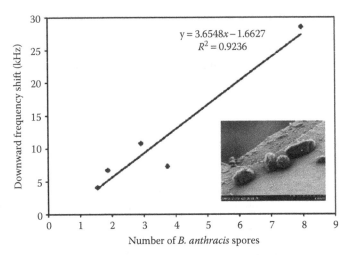

FIGURE 10.2 Resonant frequency shift of a cantilever due to cell adsorption. The inset shows the scanning electron microscope image of a cantilever surface where *Bacillus anthracis* spores are adsorbed. The resonant frequency shift due to the adsorption of spores is linearly proportional to the number of adsorbed spores. (Adapted from Davila, A. P., J. Jang, A. K. Gupta, T. Walter, A. Aronson, and R. Bashir. 2007. *Biosens Bioelectron* 22:3028–35. With permission.)

Similarly, a research group at UIUC [49] provided the label-free detection of cells using cantilevers based on the measurement of resonant frequency shift (Figure 10.2). It should be reminded that these cell detections reported in earlier studies [48,49] are based on the measurement of resonant frequencies of a cantilever operated in dry air rather than an aqueous environment. This indicates that these studies [48,49] are restricted to characterize the cell's function in real time, because a cell performs the function in an aqueous environment. In general, because of the low quality factor (Q-factor) in an aqueous environment [50], resonators are still unfavorable for in situ detection of specific biological species in physiological conditions. Recently, researchers at MIT [7,45,51] developed a novel resonant device, called a *suspended microchannel resonator* (SMR), where a microchannel is embedded in a microcantilever (Figure 10.3). Their SMR is able to exhibit high detection sensitivity due to the detection scheme such that specific biological species are detected inside a microchannel embedded in a cantilever while a cantilever vibrates in dry air with high Q-factor. They showed that the SMR is able to weigh a single cell with high precision [7]. Remarkably, cell growth during a cell cycle is characterized using the SMR. Specifically, based on the measurement of resonant frequency shift, they are able to measure the mass increase rate for cell growth [45].

In recent years, micromechanical or nanomechanical resonators have been widely employed for sensing and detecting biological molecules such as DNA and/or proteins. DNA detection has dated back to the early 2000s during which researchers at Northwestern [52] reported resonant microcantilever-based DNA detection. In their work [52], for sensitive detection, they have used nanoparticles as mass amplification

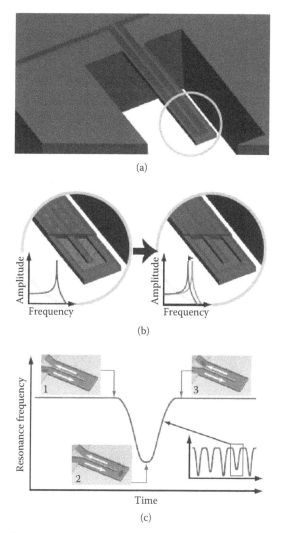

FIGURE 10.3 Schematic of a suspended microchannel resonator (SMR) and its applications in biosensing. (a) A microchannel, in which fluid is filled, is embedded in a cantilever. (b) When there are a few target molecules inside the channel, the resonance curves are shown as the black curves in the left panel. The presence of a lot of target molecules in the buffer solution inside the channel induces an increase in the effective mass of the SMR and, consequently, decreases the resonant frequency as shown in the right panel, where the gray curve indicates the resonance curve for the SMR in the presence of target molecules. (c) Schematic illustration of in situ molecular detections in real time using a SMR. When the buffer solution containing target molecules is injected into the channel, the resonance is decreased due to the molecular weight of target molecules. On the other hand, when pristine buffer solution (which does not contain the target molecules) is injected into the channel, the resonance is increased because of the decrease in overall mass of the SMR due to the reduction in target molecules. (Adapted from Burg, T. P., M. Godin, S. M. Knudsen, W. Shen, G. Carlson, J. S. Foster, K. Babcock, and S. R. Manalis. 2007. *Nature* 446:1066–9. With permission.)

such that when target DNA is chemically conjugated to probe DNA (which is shorter than the target DNA), a secondary complimentary DNA conjugated to a nanoparticle is then injected so as to be conjugated to hybridized DNA between the target DNA and probe DNA. In spite of sensitive detection, nanoparticle conjugation prevents them from measuring the mass of target molecules bound to a functionalized cantilever. A research group at Cornell [11] has recently reported a nanomechanical resonator and its great potential to sense and detect DNA even at single-molecule resolution with measurement of the molecular weight of a single DNA chain with its length of 10^3 base-pairs. Their detection was implemented using a resonator vibrating in a high-vacuum condition. On the other hand, Kwon et al. [34] have recently suggested resonant microcantilever-based in situ, label-free detection of a DNA chain in aqueous environment (Figure 10.4). Specifically, we measured the resonant frequency shift of a microcantilever vibrating in an aqueous environment due to DNA hybridization [34]. Furthermore, the resonant frequency shift due to DNA hybridiza-

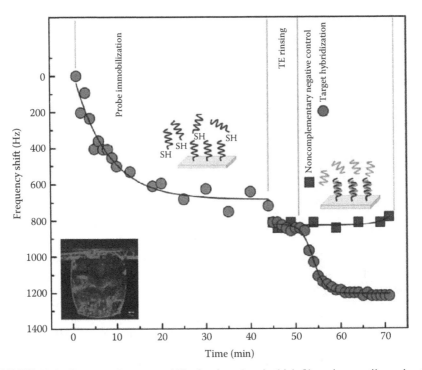

FIGURE 10.4 Resonant frequency shift of a piezoelectric thick-film microcantilever due to single-stranded DNA (ssDNA) adsorption as well as DNA hybridization between probe DNA and target DNA. The inset shows the fluorescence image of the cantilever surface where target DNA molecules are chemically hybridized with probe DNA molecules (which are immobilized on the cantilever surface). The resonant frequency shift due to DNA adsorption and/or DNA hybridization in real time is well depicted by Langmuir kinetics. (Adapted from Kwon, T., K. Eom, J. Park, D. S. Yoon, H. L. Lee, and T. S. Kim. 2008. *Appl Phys Lett* 93:173901. With permission.)

tion is well depicted by the Langmuir kinetic model that allows the extraction of the kinetic rate of DNA hybridization.

Moreover, resonant devices have recently been used for protein detections. For example, a recent study [53] reported a microcantilever vibrating in dry air for sensing specific proteins. The detection principle is the measurement of resonant frequency shift due to loading of proteins onto a cantilever. The study [53] argued that the mass of chemisorbed proteins is insufficient to depict the resonant frequency shift due to protein adsorption and that such resonant frequency shift is attributed to the surface stress originated from protein–protein interactions. This argument will be more delineated in Section 10.4.2. In recent years, Kwon et al. [36] reported the in situ, label-free detection of C-reactive protein in real time using a vibrant microcantilever immersed in buffer solution. The detection of proteins in liquid is ascribed to the relatively high Q-factor for a microcantilever in liquid. Moreover, the resonant frequency shift measured in buffer solution due to protein chemisorption is contributed to by not only the mass of adsorbed proteins but also the hydrophilicity change during protein chemisorptions [36]. Recently, researchers at Cornell [54] provided sensitive, label-free detection using resonant microcantilevers with nanoparticle-based amplifications. In particular, to amplify the resonant frequency shift due to a target a binding onto a cantilever, nanoparticles conjugated to a secondary antibody that was able to capture the target molecules were introduced as an amplifier such that when nanoparticles conjugated to the secondary antibody attached to the target molecules bound to a cantilever, the resonant frequency shift was amplified due to relatively large mass of nanoparticles. Based on this scheme, a detection of proteins at a concentration of ~fg/mL was achieved.

Despite many experimental works on label-free detection of cells, DNA, and proteins, a resonant device has not been widely taken into account to study drug–enzyme–protein interactions, which play a critical role in cellular signal transduction. Specifically, an enzyme takes a dominant role in cleavage of specific proteins and/or peptide sequences so as to produce the substrate (that is cleft protein or peptide), which leads to a cellular signaling cascade that determines the cell function [55,56]. It is important to understand protein–enzyme interactions and/or protein–enzyme–drug interactions for further insight into the cellular signaling mechanism and/or regulation of such signaling. Recently, Kwon et al. [35] have reported the label-free, in situ monitoring of peptide–enzyme interactions using vibrant microcantilevers immersed in buffer solution (Figure 10.5). They have measured the enzyme concentration-dependent proteolysis efficacy based on the measurement of frequency shift due to enzymatic cleavage of peptides on the cantilever surface. Further, they have used the Langmuir kinetic model that is applied to the measured frequency shift due to enzymatic cleavage in real time. It is shown that the Langmuir kinetic model is suitable to describe the enzymatic cleavage mechanism and that the kinetics of enzymatic cleavage was obtained as a function of enzyme concentration. In addition, Kwon, Eom, and colleagues have recently studied the small molecule–mediated inhibition mechanism of proteolysis using resonant microcantilevers operated in buffer solution, which is not published yet but will appear soon. These studies shed light on the great potential of resonant devices for studying the enzymatic activity and/or its inhibition that may provide a key concept of how to regulate cellular signaling.

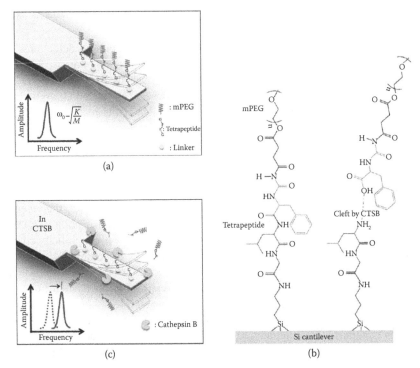

FIGURE 10.5 Schematic illustration of label-free detection of peptide–enzyme interactions using a resonant microcantilever. (a) A cantilever functionalized with specific peptide chains vibrates with resonant frequency of ω_0. (b) Chemical structure of peptide sequence GFLG (GlyPheLysGly). In addition, an enzyme (e.g., Cathepsin B abbreviated as "CTSB") is able to cleave the GFLG sequence such that the chemical bonds between GF and LG are broken by CTSB. (c) When CTSB cleaves the GFLG peptide chains, the overall mass of a functionalized cantilever is reduced, which results in increase in the resonant frequency. (Adapted from Kwon, T., J. Park, J. Yang, D. S. Yoon, S. Na, C.-W. Kim, J. S. Suh, Y. M. Huh, S. Haam, and K. Eom. 2009. *PLoS ONE* 4:e6248. With permission.)

10.4 CONTINUUM MECHANICS MODEL–BASED DETECTION PRINCIPLES

In this section, we consider the continuum mechanics modeling approaches that provide the fundamental insights into the detection principles of resonator-based molecular recognition. Specifically, we account for the cases in which the conventional detection principle described by Equation 10.2 is no longer acceptable. In particular, when a resonator is scaled down to nanoscales, the dynamic response of a resonator to molecular adsorption is governed by not only the mass of adsorbed molecules as described in Equation 10.2 but also the unexpected effects such as the elastic properties of adsorbed molecules and/or the surface stress effects. Here, we discuss the theoretical models that can explain the role of such unexpected effects in the dynamic behavior of nanoresonators.

10.4.1 MASS EFFECT VERSUS STIFFNESS EFFECT

The detection principle demonstrated in Equation 10.2 is suitable to describe the case in which the size of adsorbed molecules and/or cells is much smaller than the transverse dimension (i.e., thickness) of a resonator [36]. For instance, the molecular adsorption onto a thick-film microcantilever does not cause change in the bending rigidity of the cantilever, which implies that the resonant frequency shift of a thick-film cantilever due to molecular adsorption is not affected by the elastic properties of adsorbed molecules, but this is attributed to the mass of added molecules (for details, see reference [36]). On the other hand, when the thickness of a resonator is comparable to the size of adsorbed molecules, the molecular adsorption significantly changes the overall bending rigidity of a resonator [57,58], which implies that the elastic properties of adsorbed molecules or cells play a critical role on the dynamic behavior of a nanoresonator.

Let us consider the continuum mechanics model for a cantilevered resonator with molecular adsorption (see Figure 10.6). The equation of motion for such a case is given by [57]

$$\left[\rho_c T_c + \rho_a T_a(x)\right]\frac{\partial^2 w(x,t)}{\partial t^2} + \frac{\partial^2}{\partial x^2}\left[D(x)\frac{\partial^2 w(x,t)}{\partial x^2}\right] = 0 \qquad (10.3)$$

where ρ_c and T_c represent the density and the thickness of a cantilever, respectively, while ρ_a and $T_a(x)$ indicate the density and the thickness of the adsorbed molecules, respectively, and $D(x)$ is the effective bending rigidity per unit width for the molecule-adsorbed cantilever. Here, when molecules are adsorbed at the location $a < x < b$, $T_a(x)$ is defined as

$$T_a(x) = T_a\left[H(x-a) - H(x-b)\right] \qquad (10.4)$$

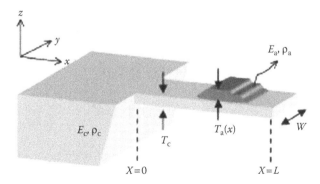

FIGURE 10.6 Schematic illustration of the adsorption of biological species onto a cantilever. (Adapted from Tamayo, J., D. Ramos, J. Mertens, and M. Calleja. 2006. *Appl Phys Lett* 89:224104. With permission.)

where $H(x)$ is the Heaviside unit step function defined as $H(x) = 0$ if $x < 0$; otherwise, $H(x) = 1$. The effective bending rigidity per unit width is given as

$$
D(x) = \frac{D_0}{1 + \left(\dfrac{E_a}{E_c}\right)\left(\dfrac{T_a(x)}{T_c}\right)} \left[1 + \left(\frac{E_a}{E_c}\right)^2 \left(\frac{T_a(x)}{T_c}\right)^4 \right.
$$
$$
\left. + 2\left(\frac{E_a}{E_c}\right)\left(\frac{T_a(x)}{T_c}\right)\left\{ 2 + 3\left(\frac{T_a(x)}{T_c}\right) + 2\left(\frac{T_a(x)}{T_c}\right)^2 \right\} \right]
$$

(10.5)

where $D_0 = E_c T_c^3/12$, and E_c and E_a indicate the elastic moduli of a cantilever and adsorbed molecules, respectively. For the resonance motion, the bending deflection $w(x, t)$ is assumed in the form of $w(x, t) = u(x) \cdot \exp[i\omega t]$, where ω and $u(x)$ represent the natural frequency and its corresponding deflection eigenmode, respectively. Then, the equation of motion depicted as in Equation 10.3 is transformed into the eigenvalue problem as follows:

$$
\Re u = \lambda u \tag{10.6a}
$$

where

$$
\Re(\bullet) \equiv \frac{d^2}{dx^2}\left[D(x)\frac{\partial^2(\bullet)}{\partial x^2} \right] \tag{10.6b}
$$

$$
\lambda \equiv \omega^2 \left[\rho_c T_c + \rho_a T_a(x) \right] \tag{10.6c}
$$

We have used the Ritz method [43] (equivalent to Galerkin's method [59–61]) that assumes the deflection eigenmode $u(x)$ represented in the form of $u(x) = A\psi(x)$, where A and $\psi(x)$ represent a constant and a shape function that satisfies the essential boundary conditions: $\psi(0) = \psi'(0) = \psi''(L) = \psi'''(L) = 0$. By multiplying a shape function by Equation 10.6 followed by integration by parts, the resonant frequency of a resonator with molecular adsorption is given by

$$
\omega^2 = \frac{\displaystyle\int_0^L D(x)\left[\psi''(x)\right]^2 dx}{\displaystyle\int_0^L \left[\rho_c T_c + \rho_a T_a(x)\right]\left[\psi(x)\right]^2 dx} \tag{10.7}
$$

Here, the shape function that satisfies the essential boundary condition can be assumed as the deflection eigenmode for a bare resonator without any adsorption, that is, $\psi(x) = \sin(\lambda x/L) - \sinh(\lambda x/L) + (\sin\lambda + \sinh\lambda)(\cos\lambda + \cosh\lambda)^{-1}[\cosh(\lambda x/L) - \cos(\lambda x/L)]$, where λ is a constant that satisfies the transcendental equation such as $\cos\lambda\cosh\lambda + 1 = 0$.

In the case of homogeneous adsorption, that is, molecular adsorption occurs over the entire length of a cantilever, the resonant frequency of the cantilever with such molecular adsorption is represented in the form [57]

$$\omega = \left(\frac{\lambda}{L}\right)^2 T_c \sqrt{\frac{E_c}{12\rho_c}} \left[1 + \alpha_1 \left(\frac{T_a}{T_c}\right) + \alpha_2 \left(\frac{T_a}{T_c}\right)^2\right] \tag{10.8a}$$

where

$$\alpha_1 = \frac{1}{2}\left(3\frac{E_a}{E_c} - \frac{\rho_a}{\rho_c}\right) \tag{10.8b}$$

$$\alpha_2 = \frac{3}{8}\left[\left(\frac{\rho_a}{\rho_c}\right)^2 + 2\left(\frac{E_a}{E_c}\right)\left(4 - \frac{\rho_a}{\rho_c}\right) - 7\left(\frac{E_a}{E_c}\right)^2\right] \tag{10.8c}$$

As a consequence, the resonant frequency shift $\Delta\omega$ due to molecular adsorption is given by

$$\frac{\Delta\omega}{\omega_0} = \frac{\omega - \omega_0}{\omega_0} = \alpha_1 \left(\frac{T_a}{T_c}\right) + \alpha_2 \left(\frac{T_a}{T_c}\right)^2 \tag{10.9}$$

Here, ω_0 is the resonant frequency of a bare resonator, that is, $\omega_0 = T_c(\lambda/L)^2(E_c/12\rho_c)^{1/2}$. This indicates that the resonant frequency shift due to molecular adsorption is attributed to not only the density (equivalent to mass) of adsorbed molecules but also their elastic properties (elastic modulus). It should be noted that this is consistent with a recent study by Gupta et al. [58,62] who showed that molecular adsorption onto a very thin cantilever increases its resonant frequency presumably due to the significant change in the bending rigidity induced by molecular adsorption (see Figure 10.7).

10.4.2 SURFACE STRESS EFFECTS

In general, the detection principle of resonator-based sensing is the direct transduction of molecular adsorption into the shifts in the resonant frequencies of a resonator due to added mass for adsorbed molecules. This has been well elucidated from Equation 10.2 as described above. However, as a resonator is scaled down, the detection principle depicted by Equation 10.2 is not valid at all because of unexpected effects such as surface effects. In particular, the scaling down of a resonator leads to an increase in the surface energy [27–29], which is defined as the energetic cost to create a surface due to scaling down. This implies that the dynamic behavior of a resonator is critically governed by the surface effect. In other words, the resonance motions are significantly determined by the surface stress [63,64] defined as a derivative of surface energy with respect to a strain. Recent studies [65–68] reported that when a transverse dimension (i.e., diameter) of a nanostructure is less than 50 nm, the surface stress plays a vital role on the bending deformation of the nanostructure.

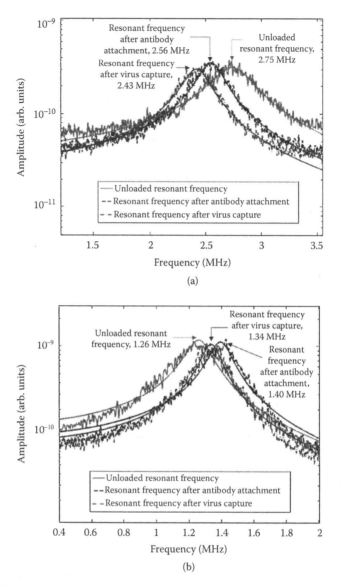

FIGURE 10.7 Resonant frequency shifts for cantilevers due to the adsorption of biological species such as viruses and/or protein antibodies. (a) Resonance curves for a cantilever, whose thickness is relatively larger than the size of the biological species, with the adsorption of biological species are shown. The adsorption of biological species decreases the resonant frequency of the cantilever, implying that the resonant frequency shift is attributed to the mass of the adsorbed biological species. (b) Resonance curves for a cantilever, whose thickness is comparable to the size of the adsorbed biological species, with the adsorption of biological species. The adsorption increases the resonant frequency, which indicates that the elastic stiffness of the adsorbed biological species plays a critical role on the resonant frequency shift due to the adsorption. (Adapted from Gupta, A., D. Akin, and R. Bashir. 2005. In *18th IEEE International Conference on Micro Electro Mechanical System (MEMS)*, 746. IEEE. With permission.)

This suggests that when the diameter of a resonator is less than sub-50 nm, the bending resonance motion would be dominated by the surface stress effect. Moreover, recent studies [28,29,69] provide that molecular adsorption onto the surface of a nanostructure leads to change in the surface state and, consequently, in the surface stress. This indicates that the effect of the surface stress is a key parameter that governs the dynamic response of a resonator to molecular adsorption. In other words, the detection principle described as Equation 10.2 does not hold for gaining insight into the detection principle of nanoresonators, but the surface effect should be taken into account when the continuum mechanics framework is considered to establish the detection principle of nanoresonators. This section is devoted to the currently developed, suggested continuum mechanics framework that takes the surface effect into account to gain fundamental insight into nanoresonator-based detection principles.

Let us denote the surface energy defined as the energetic cost to create a surface due to scaling down as U_S [27–29]. Moreover, we denote the mechanical strain for the nanostructure as ε. Then, the surface stress is defined as $\tau = \partial U_S / \partial \varepsilon = \tau_0 + E_S \varepsilon + O(\varepsilon^2)$ [63,64], where τ_0 and E_S represent the constant surface stress and the strain-dependent surface stress (i.e., surface elastic stiffness), respectively. Here, we first consider the continuum mechanics model that considers the effect of the constant surface stress τ_0 on the resonant frequencies of nanoresonators. We have also reviewed the argument regarding the effect of constant surface stress on the dynamic behavior of nanoresonators. Furthermore, we have briefly overviewed the three-dimensional elastic model that enables the descriptions of the effect of constant surface stress on the resonant frequencies of nanoresonators. In addition, we have also taken into account the continuum mechanics model that shows the vital role of the surface elastic stiffness on the resonant frequencies of nanoresonators.

10.4.2.1 One-Dimensional Beam Model: Gurtin's Argument

An earlier work [70] showed that a conventional continuum mechanics model fails to analyze the resonant frequencies of GaAs thin crystal. They have argued that the surface stress due to the large surface-to-volume ratio for GaAs thin crystal is a key parameter that dominates the resonance motions of GaAs thin film [70]. In a similar manner, recent studies [53,71–74] have reported that the resonant frequencies of micro- or nanoscale resonators in response to molecular adsorptions are critically dominated by the surface stress. In particular, when molecules are adsorbed onto the surface of a resonator, the surface stress induced from molecular adsorption significantly affects the resonant frequencies of a resonator.

To model the effect of surface stress on the resonant frequencies of a nanostructure, the continuum model that accounts for the surface stress effect is provided such that the equation of motion is given by [72–74]

$$EI \frac{\partial^4 w(x,t)}{\partial x^4} + \frac{\partial}{\partial x}\left(N(x)\frac{\partial w(x,t)}{\partial x}\right) + \rho A \frac{\partial^2 w(x,t)}{\partial t^2} = 0 \quad (10.10)$$

where $N(x)$ is given as $N(x) = \tau_0 x$ with τ_0 being a constant surface stress, $w(x,t)$ is the bending deflection of a resonator, and E, I, A, and ρ indicate the elastic modulus, the cross-sectional moment of inertia, the cross-sectional area, and the mass density, respectively, of a resonator. For the oscillatory behavior of bending motion, the bending deflection $w(x, t)$ can be assumed in the form of $w(x,t) = u(x) \cdot \exp[i\omega t]$, where ω and $u(x)$ represent the frequency and its corresponding bending deflection eigenmode, respectively. Consequently, the equation of motion described by Equation 10.10 becomes the eigenvalue problem as follows:

$$\Re u = \lambda u \qquad (10.11a)$$

where

$$\Re(\bullet) \equiv EI \frac{d^4}{dx^4}(\bullet) + \frac{\partial}{\partial x}\left[\tau_0 x \frac{\partial}{\partial x}(\bullet)\right] \qquad (10.11b)$$

and

$$\lambda \equiv \rho A \omega^2 \qquad (10.11c)$$

Here, we consider the cantilevered boundary conditions because micro- or nanoscale cantilevers have been widely used in recent studies for studying the role of surface stress on the resonant frequencies. To solve the eigenvalue problem, we have used the Ritz method [43] (equivalent to Galerkin's method [59–61]) that assumes the bending deflection eigenmode represented in the following form:

$$u(x) = \sum_{n=1}^{N} a_n \psi_n(x) \qquad (10.12)$$

where a_n is the amplitude of the bending deflection eigenmode, and $\psi_n(x)$ is the shape function that satisfies the essential boundary conditions, that is, $\psi_n(0) = \psi_n'(0) = \psi_n''(L) = \psi_n'''(L) = 0$. By multiplying the shape function by Equation 10.11 followed by integration by parts, the eigenvalue problem depicted by Equation 10.11 can be transformed into the form

$$\mathbf{Ka} = \omega^2 \mathbf{Ma} \qquad (10.13)$$

where $\mathbf{K}$ and $\mathbf{M}$ indicate the stiffness matrix and the mass matrix, respectively, and $\mathbf{a}$ is a column vector consisting of the amplitude a_n, that is, $\mathbf{a}^T = [a_1, a_2, \ldots, a_N]$. Here, the stiffness matrix and the mass matrix are defined as

$$K_{ij} = EI \int_0^L \psi_i''(x)\psi_j''(x)\,dx + \tau_0 \int_0^L x\psi_i'(x)\psi_j'(x)\,dx \qquad (10.14a)$$

$$M_{ij} = \rho A \int_0^L \psi_i(x)\psi_j(x)\,dx \qquad (10.14b)$$

Based on the Ritz method, the resonant frequency of a beam that exerts the constant surface stress is given by

$$\omega = \left(\frac{\lambda}{L}\right)^2 \sqrt{\frac{EI}{\rho A}\left[1 + \frac{2}{\pi^2}\frac{\tau_0 L^3}{EI}\right]} \qquad (10.15)$$

Here, λ is an eigenvalue for cantilevered boundary conditions, that is, $\cos\lambda\cosh\lambda + 1 = 0$. Because the resonant frequency of a bare cantilever without any molecular adsorption is given by $\omega_0 = (\lambda/L)^2(EI/\rho A)^{1/2}$, the resonant frequency shift due to a constant surface stress is thus represented in the form [72,73]

$$\frac{\Delta\omega}{\omega_0} = \frac{\omega - \omega_0}{\omega_0} \approx \frac{\tau_0 L^3}{\pi^2 EI} \qquad (10.16)$$

This indicates that the constant surface stress affects the resonant frequency of a resonator such that frequency shift due to a constant surface stress is proportional to the amount of the constant surface stress (Figure 10.8).

However, the continuum mechanics model depicted in Equation 10.10 employed in recent studies [72–74] is invalid because it violates Newton's third law [75–77]. In particular, when the structure exhibits surface stress, there should appear residual stress to satisfy the equilibrium condition. Specifically, the force equilibrium provides the relationship between surface stress and residual stress:

$$\int_{-T_c/2}^{T_c/2} \tau_0 \cdot \delta(y \pm T_c/2)dy + \int_{-T_c/2}^{T_c/2} \sigma_R dy = 0 \quad \text{or} \quad 2\tau_0 + \int_{-T_c/2}^{T_c/2} \sigma_R dy = 0 \qquad (10.17)$$

where T_c is the thickness of a resonator, σ_R is the residual stress, $\delta(x)$ is the Dirac delta function, and y is the coordinate along the thickness. The bending moment exerted on a resonator is

$$M(x,t) = \int_{-T_c/2}^{T_c/2} y\tau_0\delta(y \pm T_c/2)dy + \int_{-T_c/2}^{T_c/2} y\sigma_R dy + \int_{-T_c/2}^{T_c/2} y\sigma_B dy \qquad (10.18)$$

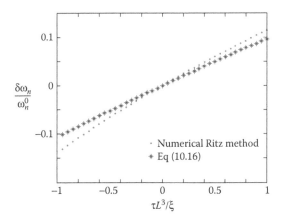

FIGURE 10.8 Resonant frequency shift for a cantilever due to a constant surface stress is estimated from a one-dimensional beam model based on the Ritz method. It is shown that a one-dimensional beam model predicts that the resonant frequency shift due to surface stress is linearly proportional to a constant surface stress. (Adapted from Hwang, K. S., K. Eom, J. H. Lee, D. W. Chun, B. H. Cha, D. S. Yoon, T. S. Kim, and J. H. Park. 2006. *Appl Phys Lett* 89:173905. With permission.)

Here, σ_B is the bending stress given as $\sigma_B = E_c \kappa y$, where E_c is the elastic modulus of a resonator, and κ is the bending curvature represented in the form of $\kappa = \partial^2 w(x,t)/\partial x^2$. Using Equations 10.17 and 10.18, we have the effective bending moment in the form

$$M(x,t) = \int_{-T_c/2}^{T_c/2} y\sigma_B dy = E_c I_c \kappa(x,t) \tag{10.19}$$

where I_c is the cross-sectional moment of inertia. This shows that the effective bending moment is independent of the constant surface stress. In other words, the effective bending rigidity of a resonator is uncorrelated with the constant surface stress, which implies that constant surface stress does not affect the resonant frequency. This argument was first suggested by Gurtin et al. [75] who contradicted the theoretical model of an earlier work [70], which reported that resonance of the GaAs structure depends on a constant surface stress. Similarly, a recent study [69] has developed the continuum model based on surface elasticity, which showed that a constant surface stress does not affect the bending behavior of a beam. In summary, the continuum model to describe the effect of a constant surface stress on the resonance of micro- or nanostructures is unacceptable because of its violation of Newton's third law.

10.4.2.2 Three-Dimensional Elastic Model: Plate Model

As discussed earlier, the one-dimensional beam model [72–74] is insufficient to understand the role of a constant surface stress in the dynamic behavior of a nanoresonator [75]. Specifically, as shown above, the one-dimensional beam model predicts that a constant surface stress does not induce any change in the resonant frequency of a beam. Recently, to gain insights into the role of surface stress in the dynamic behavior of a nanoresonator, Lachut and Sader [76,77] have considered the three-dimensional elastic model, particularly the plate model [78]. Because it is difficult to analytically treat the plate model, Lachut and Sader [76,77] have employed the scaling law as well as finite element numerical simulations to understand the relationship between a constant surface stress and the resonant frequency of a cantilevered plate.

We consider the plate that exerts a constant surface stress but is not constrained (i.e., no-traction boundary conditions). The force equilibrium assuming no traction at all edges provides the uniform in-plane stress in the form

$$\sigma_{xx} = \sigma_{yy} = \frac{\tau_0}{T} \tag{10.20}$$

where τ_0 is a constant surface stress, and T is the thickness of a plate. Here, x is the coordinate along the longitudinal direction, while y is the coordinate along the width of a plate. The displacement field (u_x, u_y), where u_x is the displacement along the coordinate x, whereas u_y is a displacement along the coordinate y, for a plate is given by [41]

$$u_x = -\frac{1-\nu}{E}\left(\frac{\tau_0}{T}\right)x \quad \text{and} \quad u_y = -\frac{1-\nu}{E}\left(\frac{\tau_0}{T}\right)y \tag{10.21}$$

where E and ν indicate the elastic modulus and Poisson's ratio, respectively, of the plate.

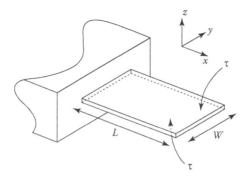

FIGURE 10.9 Schematic description of the plate model for a cantilevered plate under constant surface stress.

Now, let us turn to the original problem, that is, a cantilevered plate that bears the constant surface stress (see Figure 10.9). To simplify the problem, we can decompose the original problem into two subproblems: (1) a plate that bears the constant surface stress but is not constrained by a boundary condition, and (2) a plate that does not exert the constant surface stress but is constrained by a boundary condition such as $u_y(x = 0) = (1 - \nu)\tau_0 y/ET$ and has no traction at all edges except the cantilevered edge. Because the solution to the subproblem (1) shows that the constant surface stress is equilibrated with the residual in-plane stresses given by Equation 10.20 so that the bending deformation of the plate is not dependent on the constant surface stress, the solution to the original problem is identical with that of subproblem (2). Lachut and Sader [76,77] have used finite element numerical simulations in order to solve subproblem (2).

For fundamental insight, Lachut and Sader [76,77] have considered the scaling law for a cantilevered plate that possesses constant surface stress. For such a case, the in-plane stresses are no longer uniform due to essential boundary conditions (i.e., the cantilevered boundary condition). In particular, for $x < O(W)$, where W is the width of the plate, the in-plane stresses exist; otherwise, the in-plane stresses are zero. The scaling law assumes that for $x < O(W/L)$, the in-plane stresses are uniform as given in Equation 10.20. The equation of motion for such a plate is represented in the form [78]

$$D\nabla^4 w - \mathbf{N} \cdot \nabla w = q \tag{10.22}$$

where w is the deflection of the plate, D is the bending rigidity, N is the in-plane stress tensor, and q is the distributed load on the plate. Consequently, from the plate equation given by Equation 10.22, the effective bending rigidity D_{eff} of a plate exerting the uniform in-plane stresses is given as [76]

$$\frac{D_{\mathrm{eff}} - D_0}{D_0} \approx \frac{(1-\nu)\tau_0}{ET}\left(\frac{W}{L}\right)\left(\frac{T}{L}\right)^2 \tag{10.23}$$

Here, D_0 is the bending rigidity of a bare plate, that is, $D_0 = ET^3/12(1 - v^2)$ [78]. Because the resonant frequency of a plate is proportional to the square root of the bending rigidity, the resonant frequency shift due to a constant surface stress of a plate is thus given by [76]

$$\frac{\Delta\omega}{\omega_0} = F(v)\frac{(1-v)\tau_0}{ET}\left(\frac{W}{L}\right)\left(\frac{T}{L}\right)^2 \tag{10.24}$$

where $\Delta\omega$ and ω_0 represent the resonant frequency shift due to the constant surface stress and the resonant frequency of a bare plate, respectively, and $F(v)$ is a constant that depends on Poisson's ratio v. It can be straightforwardly shown that Equation 10.24 does not contradict Gurtin's argument [75] such that for the limiting case of $W/L \rightarrow 0$, the resonant frequency shift induced by the constant surface stress approaches zero. Equation 10.24 shows that the resonant frequency shift driven by the constant surface stress is linearly proportional to the constant surface stress and also the scaling of (W/L) and $(T/L)^2$. Although the three-dimensional elastic model depicted in Equation 10.24 provides fundamental insights into the relationship between the resonant frequency shift and the constant surface stress, this model [76,77] is still insufficient to describe the experimentally observed resonance behaviors of GaAs thin crystal [70] and/or microcantilevers with atomic or molecular adsorption [72–74]. This implies that the resonance behavior of such structures may be attributed to other factors rather than constant surface stress as conjectured from reference [76].

10.4.2.3 Surface Elasticity

As discussed earlier, as a device is scaled down, the surface energy defined as an energetic cost to create the surface due to scaling down increases [28,29], which implies that the surface energy plays a critical role in the mechanical behavior of a nanoscale device. Specifically, the surface stress that is a first derivative of the surface energy with respect to strain is a key parameter that determines the mechanical deformation of a nanostructure. Here, the surface stress is denoted as $\tau = \tau_0 + S\varepsilon$, where τ_0 is a constant surface stress, S is a surface elastic stiffness (i.e., strain-dependent surface stress), and ε is a mechanical strain of a deformed nanostructure. Because a nanoresonator exhibits a dimension such that the transverse dimension (i.e., diameter for cross-sectional area) is much smaller than the longitudinal dimension (i.e., length), a nanoresonator can be modeled as an elastic beam that exerts the surface stress. For a one-dimensional beam, the surface stress can be represented in the form of $\tau = \tau_0 + S\kappa y$, where κ is the bending curvature for the beam and y is the coordinate along the thickness of a resonator. The top surface exerts the surface stress of $\tau_u = \tau_0 + S\kappa T/2$ while the bottom surface exhibits the surface stress of $\tau_l = \tau_0 - S\kappa T/2$, where T is the thickness of the resonator. The equilibrium equations for a beam that bears the surface stress are given by

$$\text{Force: } 2\tau_0 + \int_{-T/2}^{T/2} \sigma_R dy = 0 \tag{10.25a}$$

$$\text{Moment: } M = \int_{-T/2}^{T/2} \tau(y) \cdot y \, dy + \int_{-T/2}^{T/2} \sigma_R y \, dy + \int_{-T/2}^{T/2} \sigma_B y \, dy \qquad (10.25b)$$

where σ_R and σ_B indicate the residual stress (due to surface stress) and the bending stress, respectively, and M is the bending moment of the beam. From Equations 10.25a and 10.25b, the bending moment of the beam is given as [69]

$$M = \left[EI + \frac{SWT^2}{2} \right] \frac{\partial^2 w}{\partial x^2} \qquad (10.26)$$

Here, E, I, W, and w represent the elastic modulus, the cross-sectional moment of inertia, the width, and the bending deflection of a nanoresonator, respectively. Equation 10.26 clearly shows that the bending deformation is governed by the surface elastic stiffness, implying that the dynamic behavior of a nanoresonator is dependent on the surface elastic stiffness. For a doubly clamped nanoresonator, the equation of motion is given by

$$\left(EI + \frac{SWT^2}{2} \right) \frac{\partial^4 w(x,t)}{\partial x^4} - \left[T_0 + \frac{EA}{2L} \int_0^L \left\{ \frac{\partial w(x,t)}{\partial x} \right\}^2 dx \right] \frac{\partial^2 w(x,t)}{\partial x^2}$$
$$+ \rho A \frac{\partial^2 w(x,t)}{\partial t^2} = f(x,t) \qquad (10.27)$$

where T_0 is the mechanical tension applied to the resonator, A is the cross-sectional area of the resonator, ρ is the mass density of a resonator, and $f(x,t)$ is the actuation force per unit length. For harmonic oscillation with small amplitude, the second term can vanish, and it can also be assumed that $f(x,t) \approx 0$. With these assumptions, the resonant frequency of a nanoresonator is obtained as

$$\omega = \left(\frac{\lambda}{L} \right)^2 \sqrt{\frac{EI}{\rho A} \left(1 + \frac{SWT^2}{2EI} \right)} \qquad (10.28)$$

Here, λ is an eigenvalue that depends on the boundary condition. As shown in Equation 10.28, the resonant frequency is dominantly governed by the surface elastic stiffness. This indicates that for effective design of nanoscale resonators as a mass sensor, the surface effect depicted in Equation 10.28 should be taken into account.

10.4.3 Perspectives

As described in Section 10.4.2, the surface stress effect (i.e., both a constant surface stress and a surface elastic stiffness) plays a significant role in the dynamic response of a nanoresonator. As provided in Section 10.4.2.1, the one-dimensional beam model is insufficient to depict the surface stress–driven resonance behavior of a nanostructure, whereas the three-dimensional elastic model (that is, a plate model) described in Section 10.4.2.2 provides fundamental insights into the surface stress–dominated dynamic behavior of a nanoresonator. In particular, it is shown that the bending deformation of the one-dimensional beam model is independent of the constant surface stress as argued by Gurtin et al. [75]. The three-dimensional plate model allows quantitative descriptions of the relationship between the constant surface stress and the resonant frequency of a cantilevered nanoresonator. However, this plate model takes only the constant surface stress into account, which indicates that this model lacks the fundamental physics for the effect of the strain-dependent surface stress (surface elastic stiffness) on the resonance behavior of a nanoresonator. Here, as described in Section 10.4.2.3, it should be noted that the surface elastic stiffness is a key parameter that determines the resonant frequencies of micro- or nanostructures. For instance, recent studies [79–83] reported that the bending deformation and bending resonance of micro- or nanocantilevers are critically governed by the surface elastic stiffness. This indicates that in order to model a nanoresonator, a three-dimensional elastic model such as the plate model should take into account both constant surface stress and surface elastic stiffness effects. That is, for simulation-based design of nanoresonators for specific functions such as actuations (e.g., resonance motions) and/or sensing (e.g., mass detection), it is required to develop a three-dimensional elastic model that includes the surface effects.

For the design of nanoresonators as a mass sensor, the surface effect is presumed to play a critical role in the sensing performance of a resonator. Specifically, as a resonator is scaled down, the surface effect becomes a dominant factor that determines the resonant frequency. In particular, the dynamic behavior of smaller resonators becomes more dominantly governed by the surface effect. This indicates that the sensing performance of a resonator is critically dependent on the size of the resonator due to the surface effect. Moreover, when molecules are adsorbed onto the resonator, the surface state (e.g., surface roughness, surface energy) of the resonator is significantly affected by such adsorption, which implies that mass adsorption can lead to the change in surface stress that is highly correlated with the resonant frequency of the resonator. This implies that unlike earlier studies [8,11,25,50,84] that typically used a conventional continuum beam model (which excludes the surface effect), a realistic model enabling physical insights into the sensing performance of a nanoresonator should be established such that the surface stress effect is considered in the continuum mechanics-based modeling framework.

10.5 MULTISCALE MODELING APPROACHES

Although continuum-modeling approaches are able to provide some important insights into the surface effect in the dynamic behavior of nanoresonators, they still lack physical understanding of the dynamic behavior of nanoresonators in response to atomic adsorption. In particular, a recent study [85] shows that the quality factor of nanoresonators can be significantly tailored by functionalizing the surface of nanoresonators with short molecular chains such as alkanethiol. This indicates that the dynamic response of nanoresonators to molecular adsorption can be affected by not only the mass of added molecules but also other unexpected factors such as molecular interactions and/or binding energy between molecules and the surface of a nanoresonator. Moreover, continuum mechanics models are phenomenological models that are unable to depict the physical origin of the surface stress change due to molecular adsorption. This suggests that we should develop more physically relevant models that enable fundamental insights into the dynamic response of nanoresonators to molecular adsorption. One of the strong candidates for such physically relevant models is the multiscale model that couples the continuum framework to the atomistic model. Specifically, the failure of a conventional continuum mechanics to theoretically predict the role of molecular interactions on the sensing performance of nanoresonators may reside in the inability of a typical continuum mechanics model to capture the physics of atomic interactions that was successfully modeled by atomistic models such as molecular dynamics (MD) simulations. In general, MD simulation is computationally restricted such that the accessible spatial and temporal scales are quite limited so that MD simulation is unsuitable to model nanostructures whose dimension is larger than 100 nm. This implies that in order to develop a more realistic model that allows for descriptions of the dynamic response of nanoresonators to molecular interactions, it is desirable to employ a multiscale modeling concept that combines the features of both continuum mechanics and atomistic mechanics. This section is dedicated to elucidation of the multiscale modeling concept that is able to capture the fundamental physics for resonator-based atomic detection.

10.5.1 CANTILEVER-BASED MOLECULAR DETECTION

As demonstrated earlier, micro- or nanocantilevers have recently been widely employed for studying various molecular interactions such as chemical conjugation (e.g., gold–thiol conjugation) [33], protein–protein interactions [12,15,36], DNA hybridization [16,17,39], protein–DNA interactions [19], and protein–enzyme interactions [35]. Most experimental studies on molecular interactions using cantilevers have considered a simple continuum mechanics model (e.g., Stoney's formula) that enables the measurement of the surface stress change driven by molecular interactions. However, such a continuum mechanics model is a phenomenological model in that it cannot theoretically provide the physical origin of surface stress change due to molecular interactions. There is still a lack of physical models that are able to suggest the direct relationship between molecular interactions and the cantilever's responses such as bending deflection changes and/or frequency shifts. Here, we introduce a currently suggested multiscale model [86] that provides fundamental insights into the effect of molecular interactions on the cantilever's responses.

Now, we consider a cantilever, on whose surface the molecules (e.g., DNA) are adsorbed, that is operated in a solvent. Let us denote the density of adsorbed molecules as $\Xi = N/L$, where N is the total number of adsorbed molecules and L is the length of the cantilever. We assume that molecular adsorption onto the surface of the cantilever affects the cantilever's bending deflection. Specifically, as shown in Figure 10.10, the intermolecular interactions between adsorbed molecules can induce the additional bending deflection of the cantilever. The interspacing distance between adsorbed molecules is given as

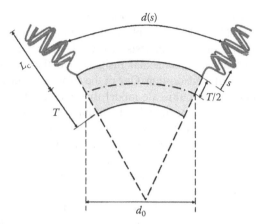

FIGURE 10.10 Schematic illustration of a model for cantilever bending due to DNA adsorption. (Reprinted from Eom, K., T. Y. Kwon, D. S. Yoon, H. L. Lee, and T. S. Kim. 2007. *Phys Rev B* 76:113408. With permission.)

$$d(s) = d_0 \left[1 + \left(\frac{\kappa T}{2} \right) \left(1 + \frac{2s}{T} \right) \right] \qquad (10.29)$$

where d_0 is the interspacing distance in the undeformed configuration given by $d_0 = 1/\Xi$, κ is the bending curvature of the cantilever, T is the thickness of the cantilever, and s is the distance from the surface of the cantilever. We denote $U_{int}(d)$ as the potential energy for the intermolecular interactions between adsorbed molecules. Accordingly, the effective potential energy of a cantilever with molecular adsorption consists of the bending energy, U_B, for the cantilever and the potential energy for intermolecular interactions, $U_{int}(d)$.

$$V = U_B + U_{int} = \frac{1}{2} \int_0^L \frac{EI}{2} \kappa^2 \, dx + \int_0^L \int_0^l \Xi U_{int} \, ds \, dx \qquad (10.30)$$

Here, E and I represent the elastic modulus and the cross-sectional moment of inertia, respectively, of the cantilever, and l is the length of an adsorbed molecule. Now, let us denote $F(\kappa; \Xi)$ as

$$F(\kappa) = \int_0^l \Xi U_{int}(\kappa) \, ds \qquad (10.31)$$

Because the potential energy of intermolecular interactions depends on the interspacing distance, which can be expressed in terms of bending curvature (i.e., Equation 10.29), we re-express the effective potential energy V using a Taylor series expansion such as

$$V = \int_0^L \left[v_0 + \varphi\kappa + (1/2)(EI + \chi)\kappa^2 + O(\kappa^3) \right] dx \qquad (10.32)$$

where $v_0 = F(0)$, $\varphi = \partial F(\kappa)/\partial\kappa|_{\kappa=0}$, and $\chi = \partial^2 F(\kappa)/\partial\kappa^2|_{\kappa=0}$. The bending deflection of a cantilever with molecular adsorption can be found from the minimization of effective potential energy V that consists of both the bending energy of the cantilever at the continuum level and the intermolecular interactions at atomic scale. This was first suggested by Hagan et al. [87] who theoretically predicted the bending deflection change of microcantilevers in response to polymer adsorption. Similarly, several researchers [88,89] have studied the cantilever bending deflection responses to atomic adsorption using a multiscale model based on the effective potential given by Equation 10.32. This indicates that as long as the intermolecular interactions are explicitly known as a function of the interspacing distance, it is straightforward to theoretically predict the bending deflection change due to molecular adsorptions.

To gain an insight into the resonant frequency shift of a cantilever due to molecular adsorption, we also have to consider the kinetic energy of the cantilever with molecular adsorption. The kinetic energy for this case is represented in the form of

$$T = \frac{1}{2}\int_0^L (\rho A + \Xi M_W)\left(\frac{\partial w}{\partial t}\right)^2 dx \tag{10.33}$$

where ρ, A, and w indicate the density, the cross-sectional area, and the bending deflection, respectively, of a cantilever, and M_W is the molecular weight of an adsorbed molecule. For a cantilever that oscillates, we express the bending deflection $w(x,t)$ as $w(x,t) = u(x) \cdot \exp[i\omega t]$, where $u(x)$ is the bending deflection eigenmode, ω is the resonant frequency, and i is the unit of a complex number, that is, $i = \sqrt{-1}$. The mean value of the Hamiltonian per oscillation cycle is given by

$$\langle H \rangle = \langle T \rangle + \langle V \rangle = -\frac{\omega^2}{2}\int_0^L (\rho A + \Xi M_W)u^2\, dx + \int_0^L\left[v_0 + \varphi u'' \frac{EI+\chi}{2}(u'')^2\right]dx \tag{10.34}$$

Here, the angular bracket $\langle\rangle$ indicates the average value per oscillation cycle and prime represents differentiation with respect to the x coordinate. The variational method with $\langle H \rangle$ given by Equation 10.34 provides the weak form of the equation of motion such as

$$\delta\langle H \rangle = \int_0^L\left[-\omega^2(\rho A + \Xi M_W)u + (EI+\chi)\left(\frac{d^4 u}{dx^4}\right)\right]dx\,\delta u$$
$$+ \left[\varphi + (EI+\chi)u''\right]\delta u'\Big|_0^L - (EI+\chi)u'''\delta u\Big|_0^L = 0 \tag{10.35}$$

where the symbol δ indicates the variation, and δu can be regarded as a virtual displacement that satisfies the essential boundary conditions. It should be noted that in Equation 10.36, the first term corresponds to the equation of motion for a resonant cantilever while the remaining terms represent the boundary conditions. That is, the equation of motion can be obtained as

$$(EI+\chi)\frac{d^4 u}{dx^4} - \omega^2(\rho A + \Xi M_W)u = 0 \tag{10.36}$$

Accordingly, the resonant frequency of a cantilever with molecular adsorption can be represented in the form

$$\frac{\omega}{\omega_0} = \sqrt{\frac{1+\chi/EI}{1+\Xi M_W/\rho A}} \tag{10.37}$$

where ω_0 is the resonant frequency of a bare cantilever without any molecular adsorption, that is, $\omega_0 = (\lambda/L)^2(EI/\rho A)^{1/2}$ with λ and L being the eigenvalue and the length of the cantilever, respectively. The resonant frequency shift of a cantilever due to molecular adsorption is, thus, given by

$$\frac{\Delta\omega}{\omega_0} \equiv \frac{\omega - \omega_0}{\omega_0} \approx -\frac{\Xi M_W}{2\rho A} + \frac{\chi}{2EI} \tag{10.38}$$

Equation 10.38 clearly shows that the resonant frequency shift induced by molecular adsorption is attributed to not only the mass of adsorbed molecules but also the intermolecular interactions between adsorbed molecules. In particular, the term χ, the second derivative of intermolecular interactions with respect to bending curvature, is a key parameter that represents a change in the bending rigidity of a cantilever due to molecular adsorption. Duan [89] has also pointed out that the term χ corresponds to the change in the surface elastic stiffness due to molecular adsorption. Eom et al. [86] measured the term χ, that is, the change in bending rigidity, due to the adsorption of double-stranded DNA (dsDNA) molecules onto a cantilever (see Figure 10.11) and predicted the resonant frequency shift for nanocantilevers in response to dsDNA adsorptions. Duan [89] has estimated the term χ for the case in which atoms (e.g., gold atoms) are adsorbed onto a cantilever and has also calculated the resonant frequency shift of a cantilever due to atomic adsorption. It is implied that multiscale modeling is appropriate for theoretical predictions on the mechanical response of a micro- or nanocantilever to molecular adsorptions. Here, it should be noted that the functional form for the potential energy of intermolecular interactions has to be

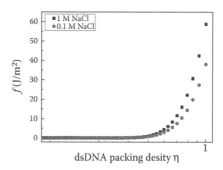

FIGURE 10.11 The change in bending rigidity for a cantilever due to DNA adsorption is computed from a multiscale model. Here, f indicates the normalized bending rigidity change due to DNA adsorption, which is defined as $f = \chi/4IT^2$, while η represents the normalized packing density for adsorbed DNA with definition of $\eta = \Xi/10^9$. (Reprinted from Eom, K., T. Y. Kwon, D. S. Yoon, H. L. Lee, and T. S. Kim. 2007. *Phys Rev B* 76:113408. With permission.)

known a priori for multiscale modeling. Moreover, it should be reminded that the multiscale model described here assumes that intermolecular interaction is a key to the mechanical response of a cantilever to molecular adsorptions. In other words, other possible factors such as binding energy for chemical conjugation between the chemical group of a molecular chain and the surface of a cantilever are not taken into account, albeit these factors may also play a role in the cantilever's response to molecular adsorption.

10.5.2 CARBON NANOTUBE RESONATOR–BASED DETECTION

In recent years, carbon nanotube (CNT) has been highlighted as a nanoscale mechanical building block because CNT is found to exhibit excellent elastic properties such as high stiffness and toughness [90–93]. In particular, experimental studies have shown that CNT can possess the elastic modulus up to ~1 TPa, which is much higher than that of any other man-made materials by several orders in magnitude. Moreover, because of the high elastic modulus and low density of CNT, CNT can bear the ultra-high-frequency (UHF) or very-high-frequency (VHF) dynamic range even up to the ~1 GHz regime [94–98]. This indicates that CNT can be an excellent candidate for a mechanical device that can perform excellent dynamic functions such as high-frequency dynamics. Furthermore, because the sensing performance for resonator-based detection is highly correlated with the resonant frequency of a device as discussed in Section 10.2.2, a few researchers [9,99,100] have theoretically and/or experimentally considered CNT as a resonant mass sensor that enables the label-free detection of gaseous atoms at atomic resolution. This indicates that a CNT resonator can be regarded as a nanomechanical mass spectrometer. In addition, it is known that CNT exhibits high binding affinity for biological molecules such as DNA due to π–π interactions between CNT and biomolecules [101–103], which indicates that a CNT resonator can be considered for nanomechanical biosensing applications.

In this section, we provide a theoretical model [100] for the mechanical response of CNT to DNA adsorption. Here, the theoretical model is constructed based on the multiscale modeling concept such that CNT is regarded as a continuous elastic beam while the interaction between DNA and CNT is depicted by an atomistic model. In the undeformed configuration, the effective potential energy of the CNT-DNA complex is composed of two major contributions: (1) E_{el}, electrostatic repulsion between nucleotide sequences, and (2) E_{bind}, binding energy for the π–π interactions between the CNT surface and DNA. The equilibrium configuration for undeformed CNT bound with DNA can be found from the energy minimization, which was well described in reference [100]. Here, we would like to only focus on the resonant response of CNT to DNA adsorption. We assume that the binding energy is independent of the bending deformation of CNT because the distance between nucleotides and the CNT surface is almost independent of CNT bending deformation. Accordingly, the effective potential energy of DNA-adsorbed CNT that undergoes the bending deformation is composed of the bending energy of CNT and the electrostatic repulsion between nucleotide sequences. The electrostatic repulsion is represented in the form

$$E_{el} = \sum_{n=1}^{N-1} \sum_{m=n+1}^{N} \frac{q^2}{4\pi e_{mn} r_{mn} e_0} \tag{10.39}$$

where r_{mn} is the distance between two nucleotide sequences m and n, e_{mn} is the average permittivity, e_0 is the vacuum permittivity, q is the electric charge, and N is the total number of nucleotide sequences. Under the bending deformation of CNT as shown in Figure 10.12, the distance r_{mn} is given by

$$r_{mn} = R_{mn} \sqrt{1 - P_{mn}\kappa + Q_{mn}\kappa^2} \tag{10.40}$$

where R_{mn} is the distance between nucleotide sequences m and n under the undeformed configuration of CNT, κ is the bending curvature of CNT, and P_{mn} and Q_{mn} are constants that are well described in reference [100]. Consequently, the electrostatic repulsion E_{el} under the bending deformation of CNT is in the form

$$\begin{aligned}
E_{el} &= \sum_{n=1}^{N-1} \sum_{m=n+1}^{N} \frac{q^2}{4\pi e_{mn} R_{mn}} \left(1 - P_{mn}\kappa + Q_{mn}\kappa^2\right)^{-1/2} \\
&\approx \sum_{n=1}^{N} \sum_{m=n+1}^{N} \frac{q^2}{4\pi e_{mn} R_{mn}} \left[1 + \frac{P_{mn}}{2}\kappa + \left(\frac{3}{8}P_{mn}^2 - \frac{Q_{mn}}{2}\right)\kappa^2\right]
\end{aligned} \tag{10.41}$$

The total potential energy V consisting of CNT bending energy and electrostatic interactions is given by

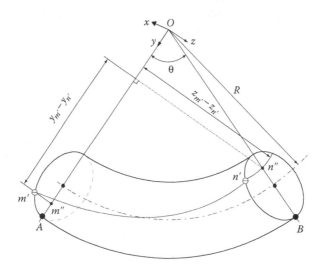

FIGURE 10.12 Schematic illustration of the bending deformation of carbon nanotube (CNT) with DNA adsorption. The dark solid line indicates the DNA chain adsorbed onto the CNT surface, while the dashed line represents the neutral axis. R is the radius of bending curvature. (Adapted from Zheng, M., K. Eom, and C. Ke. 2009. *J Phys D Appl Phys* 42:145408. With permission.)

$$V = \frac{1}{2}\int_0^L D\kappa^2 \, dx + \sum_{n=1}^N \sum_{m=n+1}^N \frac{q^2}{4\pi e_{mn} R_{mn}} \left[1 + \frac{P_{mn}}{2}\kappa + \left(\frac{3}{8}P_{mn}^2 - \frac{Q_{mn}}{2} \right)\kappa^2 \right]$$

$$\equiv \int_0^L \left(v_0 + \alpha\kappa + \frac{\beta + D}{2}\kappa^2 \right) dx$$

(10.42)

where L is the length of CNT, D is the bending rigidity of CNT, and parameters v_0, α, and β are defined as

$$v_0 = \frac{1}{L}\sum_{n=1}^N \sum_{m=n+1}^N \frac{q^2}{4\pi e_{mn} R_{mn}}$$

(10.43a)

$$\alpha = \frac{1}{L}\sum_{n=1}^N \sum_{m=n+1}^N \frac{q^2 P_{mn}}{8\pi e_{mn} R_{mn}}$$

(10.43b)

$$\beta = \frac{1}{L}\sum_{n=1}^N \sum_{m=n+1}^N \frac{q^2}{4\pi e_{mn} R_{mn}} \left(\frac{3}{4}P_{mn}^2 - Q_{mn} \right)$$

(10.43c)

The kinetic energy of CNT with DNA adsorption is given as

$$T = \frac{1}{2}\int_0^L \left(\rho_{CNT} A_{CNT} + \frac{M_{DNA}}{L}\xi(x)\sqrt{1 + \tan^2\theta} \right) dx$$

(10.44)

Here, ρ_{CNT} and A_{CNT} represent the density and the cross-sectional area of CNT, respectively, M_{DNA} is the molecular weight of DNA, $\xi(x)$ indicates the presence of the nucleotide sequences on the CNT surface, that is, $\xi(x) = 1$ for the case in which DNA is adsorbed onto CNT; otherwise, $\xi(x) = 0$, and θ is the wrapping angle for DNA wrapped onto CNT. The resonant frequency of CNT with DNA adsorption can be computed from the Ritz method.

$$\omega^2 = \frac{\int_0^L (D + \beta)[\psi''(x)]^2 \, dx}{\int_0^L \left(\rho_{CNT} A_{CNT} + \frac{M_{DNA}}{L}\xi(x)\sqrt{1 + \tan^2\theta} \right)[\psi(x)]^2 \, dx}$$

(10.45)

where ω is the resonant frequency of CNT with DNA adsorption, and $\psi(x)$ is the admissible shape function that satisfies the essential boundary condition. Figure 10.13 shows the resonant frequency shift of CNT due to DNA adsorptions with respect to the CNT location where DNA is adsorbed.

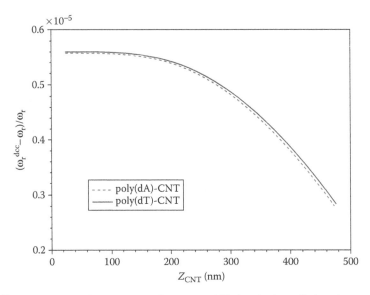

FIGURE 10.13 Normalized resonant frequency shift for single-walled carbon nanotube (SWCNT) due to DNA adsorption [e.g., poly(dA), and/or poly(dT)] with respect to the location of the CNT surface where the DNA chain is adsorbed. It is shown that when the DNA chain is adsorbed onto the clamped end of CNT, the resonant frequency is unaffected by DNA adsorption. On the other hand, when the DNA chain is adsorbed onto the free end of CNT, the resonant frequency shift due to DNA adsorption reaches the maximum. (Adapted from Zheng, M., K. Eom, and C. Ke. 2009. *J Phys D Appl Phys* 42:145408. With permission.)

10.5.3 PERSPECTIVES

The multiscale modeling concept introduced in this section has recently been developed in such a way that a continuum description of the bending deformation of a resonator (e.g., cantilever, CNT) is coupled to the atomistic model that can capture the feature of the intermolecular interactions for adsorbed molecules. Although this multiscale model can depict the role of molecular interactions in the response of a nanoresonator to molecular adsorptions, the model lacks several possible mechanisms, such as binding energy between functional groups of adsorbed molecules and the surface of a resonator, that affect resonant frequencies. Specifically, a recent study [22] reports that the surface stress generated in a cantilever due to adsorption of biological species is dependent on the molecular groups functionalized on the cantilever surface. In particular, the surface stress due to biomolecular adsorption onto the peptide-immobilized cantilever is larger than that onto an antibody-functionalized cantilever [22]. This indicates that the surface stress due to molecular interactions is strongly affected by the binding energy between the cantilever surface and the adsorbed molecule. This implies that the multiscale modeling introduced in this section is insufficient to provide fundamental insights into the role of the binding energy in the mechanical response of a nanocantilever or nanoresonator. One possible way to overcome the restrictions of the multiscale model described here is to develop a more refined multiscale model such that, instead of continuum descriptions of the resonator's bending motion, the resonator has to be described by the multiscale model to implement the atomistic descriptions of the resonator's surface in order to take into account the binding energy between the surface atoms of the resonator and the adsorbed molecule.

10.6 OUTLOOK AND PERSPECTIVES

This chapter discusses the current state of the art in label-free detection using nanoresonators. Specifically, we have reviewed not only the experimental efforts in resonator-based sensing but also the computational or theoretical studies that enable fundamental insights into the detection principles for resonator-based detection. We anticipate that in the next decade, there will be an important attempt to bridge experiments, theories, and computational simulations to gain deep insight into the detection principles of nanoresonator-based detection. Our outlook is based on breakthroughs that may be possible through exploitation of the interplay between experiments, theories, and computational simulations.

One of the interesting issues is the multiscale simulation–based design of nanoresonators for studying the surface effect–incorporated dynamics of nanoresonators, which will be highly correlated with the fundamental physics of nanoresonator-based detection. Recently, Hines and coworkers [85] have experimentally showed that the Q-factors that represent the dissipation mechanism of nanoresonators are strongly dependent on the surface chemistry. In particular, the chemical modification of the surface of nanoresonators enables the improvement of Q-factors, indicating that the dynamic behavior of nanoresonators can be tailored by surface chemistry. A theoretical model by Ru [104] reported the significant effect of surface stress on the Q-factors of nanoresonators. A recent study by Seoanez et al. [105] has used the quantum mechanics approach to study the relationship between the surface roughness and Q-factors of a nanoresonator. Although these experimental [85] and theoretical [104,105] attempts have been made, there is still a gap between experimental and theoretical anticipations on the role of surface effects (e.g., surface stress, surface elastic stiffness) on the dynamic behavior (e.g., Q-factor) of a nanoresonator. Therefore, it still remains elusive to fundamentally understand the surface effect–incorporated dynamic characteristics of nanoresonators. We believe that multiscale modeling-based design may play a vital role in not only the fundamental insights into the experimentally observed surface effect–driven dynamic behavior of nanoresonators but also the construction of novel design concepts of nanoresonators for specific functions such as sensing and/or detection.

Another issue of possible particular interest is the simulation-based design of nanoresonators for specific functions such as actuations. For instance, Craighead et al. [106] have designed MEMS resonators based on finite element simulation in order to achieve high-frequency dynamic performance. In particular, they have used finite element simulations in order to gain insight into the relationship between design parameters such as device geometry and dynamic characteristics such as resonant frequencies. However, finite element simulation is not suitable for designing nanoresonators because of unexpected effects such as surface effects that can significantly affect the dynamic behavior of nanoresonators. This indicates that multiscale modeling–based design techniques, which are based on finite element modeling techniques in order to facilitate usage by scientists and researchers, will receive significant attention for further development of nanoresonators.

Multiscale simulations described in Section 10.5 provide fundamental insights into the detection principles of nanoresonator-based sensing, and thus, multiscale simulations are of particular interest for future applications in nanoresonator-based detection. Continuum mechanics–based simulations, which even take surface effects into account, fail to clearly elucidate the experimental observation that the resonant frequency shift of a micro- or nanocantilever due to biomolecular chemisorption differs from what is expected from continuum theories that even include the surface stress effects. Furthermore, as described above, the chemisorption of molecular chains (e.g., alkanethiol) onto the surface of a nanoresonator significantly influences the dynamic behavior of the nanoresonator. This suggests that multiscale simulation techniques should be developed in order to present an alternative to continuum mechanics–based simulations for theoretical understanding of the detection principles of nanoresonator-based sensing. In addition, the multiscale models discussed in Section 10.5.3 enable the quantitative understanding of the role of molecular interactions in the mechanical response of nanoresonators to biomolecular chemisorption, which implies that multiscale simulation techniques allow the quantification of various molecular interactions, which leads to the implication of nanoresonators for studying various molecular interactions.

Another important issue is to experimentally validate the role of intermolecular interactions in the resonance behavior of a nanoresonator with molecular adsorption, which has been theoretically understood from multiscale simulations. Recently, there has been an effort [107] to bridge the gap between the multiscale simulations and the experiments on the mechanical response of nanocantilevers to molecular adsorptions. Nevertheless, until recently, there have been few attempts to validate the robustness of multiscale models based on the experimental observations. This experimental validation of multiscale models has been made possible due to the recent nanotechnology that facilitates fabrication of nanoscale devices with desired geometry. We expect that it will be much easier to validate the robustness of multiscale models for the case of atomic adsorption onto a nanoresonator rather than biomolecular adsorption because the model system of atomic adsorption can simply be described by a significant factor such as interatomic interactions, whereas the model system for biomolecular adsorption consists of complex, possible factors such as ionic strength [17,108], hydration [18,108,109], and molecular configurations [33,110,111]. Moreover, multiscale modeling–based simulations may also provide a novel concept for experimental design to study the interatomic (or intermolecular) interactions using nanoresonators.

In summary, we expect that there will be a significant effort to bridge the gap between experiments, theories, and computational simulations for understanding the underlying mechanisms of nanoresonator-based detection. This effort will eventually enable us to develop a novel toolkit that can be used for designing the nanoresonators for specific functions such as sensing as well as to broaden our understanding of experimentally observed phenomena in the dynamic behavior of nanoresonators, which is also related to the sensing performance of a nanoresonator.

ACKNOWLEDGMENTS

K. E. gratefully acknowledges the financial support by the National Research Foundation of Korea (NRF) under Grant Nos. NRF-2009-0071246 and NRF-2010-0026223. T. K. acknowledges the support from the NRF under Grant Nos. NRF-2008-313-D00031 and NRF-2010-0009428.

REFERENCES

1. Bashir, R. 2004. BioMEMS: state-of-the-art in detection, opportunities and prospects. *Adv Drug Deliv Rev* 56:1565.
2. Mirkin, C. A., and C. M. Niemeyer. 2007. *Nanobiotechnology II: More Concepts and Applications*. Weinheim, Germany: Wiley-VCH Verlag GmbH & Co.
3. Rosi, N. L., and C. A. Mirkin. 2005. Nanostructures in biodiagnostics. *Chem Rev* 105:1547.
4. Heath, J. R., and M. E. Davis. 2008. Nanotechnology and cancer. *Annu Rev Med* 59:251.
5. Craighead, H. G. 2000. Nanoelectromechanical systems. *Science* 290:1532.
6. Ekinci, K. L. 2005. Electromechanical transducers at the nanoscale: Actuation and sensing of motion in nanoelectromechanical systems (NEMS). *Small* 1:786.
7. Burg, T. P., M. Godin, S. M. Knudsen, W. Shen, G. Carlson, J. S. Foster, K. Babcock, and S. R. Manalis. 2007. Weighing of biomolecules, single cells and single nanoparticles in fluid. *Nature* 446:1066.
8. Yang, Y. T., C. Callegari, X. L. Feng, K. L. Ekinci, and M. L. Roukes. 2006. Zeptogram-scale nanomechanical mass sensing. *Nano Lett* 6:583.
9. Jensen, K., K. Kim, and A. Zettl. 2008. An atomic-resolution nanomechanical mass sensor. *Nat Nanotechnol* 3:533.
10. Gil-Santos, E., D. Ramos, J. Martinez, M. Fernandez-Regulez, R. Garcia, A. San Paulo, M. Calleja, and J. Tamayo. 2010. Nanomechanical mass sensing and stiffness spectrometry based on two-dimensional vibrations of resonant nanowires. *Nat Nanotechnol* 5:641.
11. Ilic, B., Y. Yang, K. Aubin, R. Reichenbach, S. Krylov, and H. G. Craighead. 2005. Enumeration of DNA molecules bound to a nanomechanical oscillator. *Nano Lett* 5:925.
12. Wu, G. H., R. H. Datar, K. M. Hansen, T. Thundat, R. J. Cote, and A. Majumdar. 2001. Bioassay of prostate-specific antigen (PSA) using microcantilevers. *Nat Biotechnol* 19:856.
13. Waggoner, P. S., and H. G. Craighead. Micro- and nanomechanical sensors for environmental, chemical, and biological detection. 2007. *Lab Chip* 7:1238.
14. Goeders, K. M., J. S. Colton, and L. A. Bottomley. 2008. Microcantilevers: Sensing chemical interactions via mechanical motion. *Chem Rev* 108:522.
15. Arntz, Y., J. D. Seelig, H. P. Lang, J. Zhang, P. Hunziker, J. P. Ramseyer, E. Meyer, M. Hegner, and C. Gerber. 2003. Label-free protein assay based on a nanomechanical cantilever array. *Nanotechnology* 14:86.
16. Fritz, J., M. K. Baller, H. P. Lang, H. Rothuizen, P. Vettiger, E. Meyer, H. J. Guntherodt, C. Gerber, and J. K. Gimzewski. 2000. Translating biomolecular recognition into nanomechanics. *Science* 288:316.
17. Wu, G. H., H. F. Ji, K. Hansen, T. Thundat, R. Datar, R. Cote, M. F. Hagan, A. K. Chakraborty, and A. Majumdar. 2001. Origin of nanomechanical cantilever motion generated from biomolecular interactions. *Proc Natl Acad Sci USA* 98:1560.

18. Mertens, J., C. Rogero, M. Calleja, D. Ramos, J. A. Martin-Gago, C. Briones, and J. Tamayo. 2008. Label-free detection of DNA hybridization based on hydration-induced tension in nucleic acid films. *Nat Nanotechnol* 3:301.

19. Zhang, J., H. P. Lang, F. Huber, A. Bietsch, W. Grange, U. Certa, R. McKendry, H. J. Guntherodt, M. Hegner, and C. Gerber. 2006. Rapid and label-free nanomechanical detection of biomarker transcripts in human RNA. *Nat Nanotechnol* 1:214.

20. Pei, J. H., F. Tian, and T. Thundat. 2004. Glucose biosensor based on the microcantilever. *Anal Chem* 76:292.

21. Subramanian, A., P. I. Oden, S. J. Kennel, K. B. Jacobson, R. J. Warmack, T. Thundat, and M. J. Doktycz. 2002. Glucose biosensing using an enzyme-coated microcantilever. *Appl Phys Lett* 81:385.

22. Dhayal, B., W. A. Henne, D. D. Doorneweerd, R. G. Reifenberger, and P. S. Low. 2006. Detection of bacillus subtilis spores using peptide-functionalized cantilever arrays. *J Am Chem Soc* 128:3716.

23. Braun, T., V. Barwich, M. K. Ghatkesar, A. H. Bredekamp, C. Gerber, M. Hegner, and H. P. Lang. 2005. Micromechanical mass sensors for biomolecular detection in a physiological environment. *Phys Rev E* 72:031907.

24. Chun, D. W., K. S. Hwang, K. Eom, J. H. Lee, B. H. Cha, W. Y. Lee, D. S. Yoon, and T. S. Kim. 2007. Detection of the Au thin-layer in the Hz per picogram regime based on the microcantilevers. *Sens Actuat A* 135:857.

25. Li, M., H. X. Tang, and M. L. Roukes. 2007. Ultra-sensitive NEMS-based cantilevers for sensing, scanned probe and very high-frequency applications. *Nat Nanotechnol* 2:114.

26. Greaney, P. A., and J. C. Grossman. 2008. Nanomechanical resonance spectroscopy: A novel route to ultrasensitive label-free detection. *Nano Lett* 8:2648.

27. Freund, L. B., and S. Suresh. 2003. *Thin Film Materials*. Cambridge: Cambridge University Press.

28. Ibach, H. 1997. The role of surface stress in reconstruction, epitaxial growth, and stabilization of mesoscopic structures. *Surf Sci Rep* 29:193.

29. Haiss, W. 2001. Surface stress of clean and adsorbate-covered solids. *Rep Prog Phys* 64:591.

30. Diao, J., K. Gall, and M. L. Dunn. 2003. Surface-stress-induced phase transformation in metal nanowires. *Nat Mater* 2:656.

31. Stoney, G. G. 1909. The tension of metallic films deposited by electrolysis. *Proc R Soc Lond A* 82:172.

32. Zang, J., and F. Liu. 2007. Theory of bending of Si nanocantilevers induced by molecular adsorption: A modified Stoney formula for the calibration of nanomechanochemical sensors. *Nanotechnology* 18:405501.

33. Berger, R., E. Delamarche, H. P. Lang, C. Gerber, J. K. Gimzewski, E. Meyer, and H. J. Guntherodt. 1997. Surface stress in the self-assembly of alkanethiols on gold. *Science* 276:2021.

34. Kwon, T., K. Eom, J. Park, D. S. Yoon, H. L. Lee, and T. S. Kim. 2008. Micromechanical observation of the kinetics of biomolecular interactions. *Appl Phys Lett* 93:173901.

35. Kwon, T., J. Park, J. Yang, D. S. Yoon, S. Na, C.-W. Kim, J. S. Suh, Y. M. Huh, S. Haam, and K. Eom. 2009. Nanomechanical *In Situ* monitoring of proteolysis of peptide by cathepsin B. *PLoS ONE* 4:e6248.

36. Kwon, T. Y., K. Eom, J. H. Park, D. S. Yoon, T. S. Kim, and H. L. Lee. 2007. In situ real-time monitoring of biomolecular interactions based on resonating microcantilevers immersed in a viscous fluid. *Appl Phys Lett* 90:223903.

37. Backmann, N., C. Zahnd, F. Huber, A. Bietsch, A. Pluckthun, H. P. Lang, H. J. Guntherodt, M. Hegner, and C. Gerber. 2005. A label-free immunosensor array using single-chain antibody fragments. *Proc Natl Acad Sci USA* 102:14587.

38. Waggoner, P. S., M. Varshney, and H. G. Craighead. 2009. Detection of prostate specific antigen with nanomechanical resonators. *Lab Chip* 9:3095.
39. McKendry, R., J. Y. Zhang, Y. Arntz, T. Strunz, M. Hegner, H. P. Lang, M. K. Baller, U. Certa, E. Meyer, H. J. Guntherodt, and C. Gerber. 2002. Multiple label-free biodetection and quantitative DNA-binding assays on a nanomechanical cantilever array. *Proc Natl Acad Sci USA* 99:9783.
40. Ndieyira, J. W., M. Watari, A. D. Barrera, D. Zhou, M. Vogtli, M. Batchelor, M. A. Cooper et al. 2008. Nanomechanical detection of antibiotic-mucopeptide binding in a model for superbug drug resistance. *Nat Nanotechnol* 3:691.
41. Landau, L. D., and E. M. Lifshitz. 1986. *Theory of Elasticity*. St. Louis, MO: Butterworth-Heinemann.
42. Timoshenko, S. P., and J. N. Goodier. 1970. *Theory of Elasticity*. New York, NY: McGraw-Hill.
43. Meirovitch, L. 1967. *Analytical Methods in Vibrations*. New York: Macmillan.
44. Miller, R. E., and V. B. Shenoy. 2000. Size-dependent elastic properties of nanosized structural elements. *Nanotechnology* 11:139.
45. Bryan, A. K., A. Goranov, A. Amon, and S. R. Manalis. 2010. Measurement of mass, density, and volume during the cell cycle of yeast. *Proc Natl Acad Sci USA* 107:999.
46. Gfeller, K. Y., N. Nugaeva, and M. Hegner. 2005. Micromechanical oscillators as rapid biosensor for the detection of active growth of Escherichia coli. *Biosens Bioelectron* 21:528.
47. Nugaeva, N., K. Y. Gfeller, N. Backmann, H. P. Lang, M. Duggelin, and M. Hegner. 2005. Micromechanical cantilever array sensors for selective fungal immobilization and fast growth detection. *Biosens Bioelectron* 21:849.
48. Ilic, B., D. Czaplewski, H. G. Craighead, P. Neuzil, C. Campagnolo, and C. Batt. 2000. Mechanical resonant immunospecific biological detector. *Appl Phys Lett* 77:450.
49. Davila, A. P., J. Jang, A. K. Gupta, T. Walter, A. Aronson, and R. Bashir. 2007. Microresonator mass sensors for detection of Bacillus anthracis Sterne spores in air and water. *Biosens Bioelectron* 22:3028.
50. Verbridge, S. S., L. M. Bellan, J. M. Parpia, and H. G. Craighead. 2006. Optically driven resonance of nanoscale flexural oscillators in liquid. *Nano Lett* 6:2109.
51. Burg, T. P., and S. R. Manalis. 2003. Suspended microchannel resonators for biomolecular detection. *Appl Phys Lett* 83:2698.
52. Su, M., S. Li, and V. P. Dravid. 2003. Microcantilever resonance-based DNA detection with nanoparticle probes. *Appl Phys Lett* 82:3562.
53. Lee, J. H., T. S. Kim, and K. H. Yoon. 2004. Effect of mass and stress on resonant frequency shift of functionalized Pb(Zr0.52Ti0.48)O3 thin film microcantilever for the detection of C-reactive protein. *Appl Phys Lett* 84:3187.
54. Varshney, M., P. S. Waggoner, C. P. Tan, K. Aubin, R. A. Montagna, and H. G. Craighead. 2008. Prion protein detection using nanomechanical resonator arrays and secondary mass labeling. *Anal Chem* 80:2141.
55. Kodadek, T. 2008. Biochemistry: Molecular cloaking devices. *Nature* 453:861.
56. Sebolt-Leopold, J. S., and J. M. English. 2006. Mechanisms of drug inhibition of signalling molecules. *Nature* 441:457.
57. Tamayo, J., D. Ramos, J. Mertens, and M. Calleja. 2006. Effect of the adsorbate stiffness on the resonance response of microcantilever sensors. *Appl Phys Lett* 89:224104.
58. Gupta, A. K., P. R. Nair, D. Akin, M. R. Ladisch, S. Broyles, M. A. Alam, and R. Bashir. 2006. Anomalous resonance in a nanomechanical biosensor. *Proc Natl Acad Sci USA* 103:13362.
59. Bathe, K.-J. 1996. *Finite Element Procedures*. Eaglewood Cliffs, New Jersey: Prentice Hall.
60. Oden, J. T., E. B. Becker, and G. F. Carey. 1981. *Finite Elements: An Introduction* (Vol. 1). Eaglewood Cliffs, New Jersey: Prentice Hall.

61. Zienkiewicz, O. C., and R. L. Taylor. 1989. *The Finite Element Method: Basic Formulation and Linear Problems.* London: McGraw-Hill.
62. Gupta, A., D. Akin, and R. Bashir. 2005. Mechanical effects of attaching protein layers on nanoscale-thick cantilever beam for resonant detection of virus particles. In *18th IEEE International Conference on Micro Electro Mechanical System (MEMS),* 746. IEEE.
63. Shuttleworth, R. 1950. The surface tension of solids. *Proc Phys Soc Lond Sect A* 63:444.
64. Rayleigh, L. 1890. On the theory of surface forces. *Phil Mag* 30:285.
65. Park, H. S., W. Cai, H. D. Espinosa, and H. Huang. 2009. Mechanics of crystalline nanowires. *MRS Bull* 34:178.
66. Liang, H., M. Upmanyu, and H. Huang. 2005. Size-dependent elasticity of nanowires: Nonlinear effects. *Phys Rev B* 71:241403.
67. Cuenot, S., C. Fretigny, S. Demoustier-Champagne, and B. Nysten. 2004. Surface tension effect on the mechanical properties of nanomaterials measured by atomic force microscopy. *Phys Rev B* 69:165410.
68. Pirota, K. R., E. L. Silva, D. Zanchet, D. Navas, M. Vazquez, M. Hernandez-Velez, and M. Knobel. 2007. Size effect and surface tension measurements in Ni and Co nanowires. *Phys Rev B* 76:233410.
69. Lu, P., H. P. Lee, C. Lu, and S. J. O'Shea. 2005. Surface stress effects on the resonance properties of cantilever sensors. *Phys Rev B* 72:085405.
70. Lagowski, J., H. C. Gatos, and J. E. S. Sproles. 1975. Surface stress and the normal mode of vibration of thin crystals GaAs. *Appl Phys Lett* 26:493.
71. Cherian, S., and T. Thundat. 2002. Determination of adsorption-induced variation in the spring constant of a microcantilever. *Appl Phys Lett* 80:2219.
72. McFarland, A. W., M. A. Poggi, M. J. Doyle, L. A. Bottomley, and J. S. Colton. 2005. Influence of surface stress on the resonance behavior of microcantilevers. *Appl Phys Lett* 87:053505.
73. Hwang, K. S., K. Eom, J. H. Lee, D. W. Chun, B. H. Cha, D. S. Yoon, T. S. Kim, and J. H. Park. 2006. Dominant surface stress driven by biomolecular interactions in the dynamical response of nanomechanical microcantilevers. *Appl Phys Lett* 89:173905.
74. Dorignac, J., A. Kalinowski, S. Erramilli, and P. Mohanty. 2006. Dynamical response of nanomechanical oscillators in immiscible viscous fluid for in vitro biomolecular recognition. *Phys Rev Lett* 96:186105.
75. Gurtin, M. E., X. Markenscoff, and R. N. Thurston. 1976. Effect of surface stress on the natural frequency of thin crystals. *Appl Phys Lett* 29:529.
76. Lachut, M. J., and J. E. Sader. 2007. Effect of surface stress on the stiffness of cantilever plates. *Phys Rev Lett* 99:206102.
77. Lachut, M. J., and J. E. Sader. 2009. Effect of surface stress on the stiffness of cantilever plates: Influence of cantilever geometry. *Appl Phys Lett* 95:193505.
78. Timoshenko, S. 1940. *Theory of Plates and Shells.* New York: McGraw Hill.
79. Gavan, K. B., H. J. R. Westra, E. W. J. M. van der Drift, W. J. Venstra, and H. S. J. van der Zant. 2009. Size-dependent effective Young's modulus of silicon nitride cantilevers. *Appl Phys Lett* 94:233108.
80. He, J., and C. M. Lilley. 2008. Surface effect on the elastic behavior of static bending nanowires. *Nano Lett* 8:1798.
81. He, J., and C. M. Lilley. 2008. Surface stress effect on bending resonance of nanowires with different boundary conditions. *Appl Phys Lett* 93:263108.
82. Shankar, M. R., and A. H. King. 2007. How surface stresses lead to size-dependent mechanics of tensile deformation in nanowires. *Appl Phys Lett* 90:141907.
83. Wang, G., and X. Li. 2007. Size dependency of the elastic modulus of ZnO nanowires: Surface stress effect. *Appl Phys Lett* 91:231912.

84. Verbridge, S. S., D. F. Shapiro, H. G. Craighead, and J. M. Parpia. 2007. Macroscopic tuning of nanomechanics: Substrate bending for reversible control of frequency and quality factor of nanostring resonators. *Nano Lett* 7:1728.

85. Henry, J. A., Y. Wang, D. Sengupta, and M. A. Hines. 2007. Understanding the effects of surface chemistry on Q: Mechanical energy dissipation in alkyl-terminated (C1-C18) micromechanical silicon resonators. *J Phys Chem B* 111:88.

86. Eom, K., T. Y. Kwon, D. S. Yoon, H. L. Lee, and T. S. Kim. 2007. Dynamical response of nanomechanical resonators to biomolecular interactions. *Phys Rev B* 76:113408.

87. Hagan, M. F., A. Majumdar, and A. K. Chakraborty. 2002. Nanomechanical forces generated by surface grafted DNA. *J Phys Chem B* 106:10163.

88. Dareing, D. W., and T. Thundat. 2005. Simulation of adsorption-induced stress of a microcantilever sensor. *J Appl Phys* 97:043526.

89. Yi, X. and H. L. Duan. 2009. Surface stress induced by interactions of adsorbates and its effect on deformation and frequency of microcantilever sensors. *J Mech Phys Solids* 57:1254.

90. Wong, E. W., P. E. Sheehan, and C. M. Lieber. 1997. Nanobeam mechanics: Elasticity, strength, and toughness of nanorods and nanotubes. *Science* 277:1971.

91. Treacy, M. M. J., T. W. Ebbesen, and J. M. Gibson. 1996. Exceptionally high Young's modulus observed for individual carbon nanotubes. *Nature* 381:678.

92. Lu, J. P. 1997. Elastic properties of carbon nanotubes and nanoropes. *Phys Rev Lett* 79:1297.

93. Schatz, G. C. 2007. Inaugural article: Using theory and computation to model nanoscale properties. *Proc Natl Acad Sci USA* 104:6885.

94. Baughman, R. H., C. Cui, A. A. Zakhidov, Z. Iqbal, J. N. Barisci, G. M. Spinks, G. G. Wallace et al. 1999. Carbon nanotube actuators. *Science* 284:1340.

95. Sazonova, V., Y. Yaish, H. Ustunel, D. Roundy, T. A. Arias, and P. L. McEuen. 2004. A tunable carbon nanotube electromechanical oscillator. *Nature* 431:284.

96. Witkamp, B., M. Poot, and H. S. J. van der Zant. 2006. Bending-mode vibration of a suspended nanotube resonator. *Nano Lett* 6:2904.

97. Sapmaz, S., Y. M. Blanter, L. Gurevich, and H. S. J. van der Zant. 2003. Carbon nanotubes as nanoelectromechanical systems. *Phys Rev B* 67:235414.

98. Ustunel, H., D. Roundy, and T. A. Arias. 2005. Modeling a suspended nanotube oscillator. *Nano Lett* 5:523.

99. Li, C., and T.-W. Chou. 2004. Mass detection using carbon nanotube-based nanomechanical resonators. *Appl Phys Lett* 84:5246.

100. Zheng, M., K. Eom, and C. Ke. 2009. Calculations of the resonant response of carbon nanotubes to binding of DNA. *J Phys D Appl Phys* 42:145408.

101. Zheng, M., A. Jagota, E. D. Semke, B. A. Diner, R. S. McLean, S. R. Lustig, R. E. Richardson, and N. G. Tassi. 2003. DNA-assisted dispersion and separation of carbon nanotubes. *Nat Mater* 2:338.

102. Tu, X., S. Manohar, A. Jagota, and M. Zheng. 2009. DNA sequence motifs for structure-specific recognition and separation of carbon nanotubes. *Nature* 460:250.

103. Nel, A. E., L. Madler, D. Velegol, T. Xia, E. M. V. Hoek, P. Somasundaran, F. Klaessig, V. Castranova, and M. Thompson. 2009. Understanding biophysicochemical interactions at the nano-bio interface. *Nat Mater* 8:543.

104. Ru, C. Q. 2009. Size effect of dissipative surface stress on quality factor of microbeams. *Appl Phys Lett* 94:051905.

105. Seoanez, C., F. Guinea, and A. H. Castro Neto. 2008. Surface dissipation in nanoelectromechanical systems: Unified description with the standard tunneling model and effects of metallic electrodes. *Phys Rev B* 77:125107.

106. Waggoner, P. S., and H. G. Craighead. 2009. The relationship between material properties, device design, and the sensitivity of resonant mechanical sensors. *J Appl Phys* 105:054306.

107. Sushko, M. L., J. H. Harding, A. L. Shluger, R. A. McKendry, and M. Watari. 2008. Physics of nanomechanical biosensing on cantilever arrays. *Adv Mater* 20:3848.

108. Strey, H. H., V. A. Parsegian, and R. Podgornik. 1997. Equation of state for DNA liquid crystals: Fluctuation enhanced electrostatic double layer repulsion. *Phys Rev Lett* 78:895.

109. Kim, S., D. Yi, A. Passian, and T. Thundat. 2010. Observation of an anomalous mass effect in microcantilever-based biosensing caused by adsorbed DNA. *Appl Phys Lett* 96:153703.

110. Mukhopadhyay, R., V. V. Sumbayev, M. Lorentzen, J. Kjems, P. A. Andreasen, and F. Besenbacher. 2005. Cantilever sensor for nanomechanical detection of specific protein conformations. *Nano Lett* 5:2385.

111. Shu, W. M., D. S. Liu, M. Watari, C. K. Riener, T. Strunz, M. E. Welland, S. Balasubramanian, and R. A. McKendry. 2005. DNA molecular motor driven micromechanical cantilever arrays. *J Am Chem Soc* 127:17054.

11 Surface-Enhanced Microcantilever Sensors with Novel Structures

H. L. Duan

CONTENTS

11.1 INTRODUCTION

Microcantilevers have been used as sensors in the processes in physics, chemistry, and biology, and as components in microelectromechanical systems because of their high sensitivity and easy manipulation. The commonly measured output signals of cantilever sensors are their static bending and the resonance frequency shift of vibration. For example, the reaction on the upper or lower surface of a microcantilever beam can bend the cantilever (Figure 11.1), due to the free energy reduction as the result of biomolecular interactions. Therefore, when the free energy on a cantilever surface is reduced, the reduced energy density creates a bending moment that bends the cantilever. Because free energy reduction is the common driving force for all reactions, this mechanical approach has become a common platform for detecting a variety of biomolecular binding such as DNA hybridization, protein–protein interactions, DNA–protein binding, and protein–ligand binding (Kassner et al. 2005). In general, the bending of cantilevers is driven by the eigenstrain and the change of surface stress, while the vibration frequency shift is affected by the mass and surface elasticity. Microcantilever sensors work when contacting with liquid or gas. The adsorbates on them change the surface stress. This change comes from different interacting mechanisms of the adsorbates, such as electrostatic interaction, the van der Waals (vdW) forces, the dipole–dipole interactions, hydrogen bonding, and the changes in the charge distribution of surface atoms, and so on (Yi and Duan 2009). One big challenge in developing microcantilever sensors is to quantify the connections between the properties of adsorbates and the adsorption-induced surface stress. As both static deformation and dynamic frequency are quantities at the continuum level, whereas the interactions exist at the atomic or molecular level, an effective way to address this challenge is to derive the continuum-level descriptions of the surface stress from the molecular-level descriptions of the adsorbate interactions (Yi and Duan 2009).

Developing surface-enhanced microcantilevers with improved sensitivities has long-standing interest in science and technology. It has been demonstrated that nanocantilevers (i.e., the thickness of cantilevers is at the nanoscale) can measure

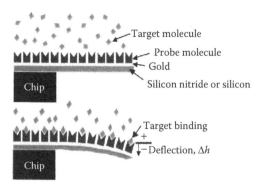

FIGURE 11.1 Microcantilever with specific biomolecular reactions among DNA, proteins, and other chemicals. (Reprinted from Wu, G., H. Ji, K. Hansen, T. Thundat, R. Datar, R. Cote, M. F. Hagan, A. K. Chakraborty, and A. Majumdar. 2001. *Proc Natl Acad Sci USA* 98:1560–4. With permission.)

smaller mass loading than microcantilevers. Furthermore, as the thickness of cantilever sensors reduces to the nanoscale, the effects of the surface stress become more significant because of the high surface-to-volume ratio. Therefore, the sensitivities of cantilevers can be increased with the decrease of their size. Recently, Weissmüller et al. (2003) and Kramer, Viswanath, and Weissmüller (2004) have found that the surface stress can greatly increase the deformation of nanoporous films. This paves the way for designing surface-enhanced microcantilevers made of nano/microporous films. Comparing with the classical solid cantilevers, cantilevers made of nano/microporous films can adsorb a great number of molecules because of their large surface area and thus are expected to have large deformation. Duan (2010) proposed the design of surface-enhanced cantilever sensors using nano/microporous films as surface layers. She has shown that the sensitivities of these novel cantilever sensors for the static deformation and resonance frequencies can be tuned by the porosity, the porous size, and the structure of porous films.

However, the surface roughness can also change the cantilever's sensitivity. Weissmüller and Duan (2008) found that the sensitivity of cantilevers with rough surfaces is significantly dependent on the surface topology. The sensitivities can be decreased all the way to zero or even have the sign reversed for corrugated metal surfaces with large Poisson's ratios. This implies that randomly changing the surface structure could not only tune the magnitude of the cantilever's response but also change its sign (e.g., from + to −). In addition to working in a static mode to measure the deflection, microcantilever sensors can also be used in a dynamic mode to measure the change of resonance frequency. In terms of the dynamic property of microcantilevers, Duan, Xue, and Yi (2009) investigated the influence of surface roughness on the resonance frequency of microcantilevers with consideration of their surface stress, and demonstrated that the surface roughness could increase, decrease, or even eliminate the surface stress, which changes the microcantilevers' resonance frequency. The surface stress is determined by the inclination angle of surface roughness and the Poisson's ratio. It is noted that Wang et al. (2010b) analyzed the mechanics of rough surfaces in a continuum framework by generalizing the approach of Weissmüller and Duan (2008) to solid surfaces with anisotropic elasticity and roughness. It was shown that the coupling of the surface stress at a corrugated surface into a planar substrate depends on the geometry of the corrugation exclusively through the surface orientation distribution function.

Based on the work outlined above (Weissmüller and Duan 2008; Yi and Duan 2009; Duan, Xue, and Yi 2009; Duan 2010; Wang et al. 2010b), this chapter will present a precise characterization of two kinds of surface-enhanced microcantilever sensors with novel structures. First, the mechanics of a cantilever sensor (i.e., the static deformation and resonance frequency shift on the two-layer cantilevers are caused by the simultaneous effects of the eigenstrain, the surface stress, and the mass adsorption) is given. Then, the connection between the surface stress at the continuum level and the adsorbate interactions (adsorbate–adsorbate interactions and adsorbate–surface interactions) at the molecular level for the vdW interaction and the Coulomb interaction is established. In the later part of this chapter, theories and applications of the two kinds of surface-enhanced microcantilever sensors with novel structures are discussed.

11.2 MECHANICS OF MICROCANTILEVER SENSORS

On the basis of the theoretical works of Yi and Duan (2009), we consider a classical cantilever (a thin film/substrate structure) shown in Figure 11.2, which has a wide range of applications as microelectromechanical components (Berger et al. 1997; Wu et al. 2001; Kramer, Viswanath, and Weissmüller 2004). The thickness and Young's modulus of the film (upper layer) are denoted as h_f and E_f, respectively, and those of the substrate (lower layer) are given as h_s and E_s. A perfect bonding is assumed between the two layers. The coordinate system is defined such that the interface of the film and substrate is located at $z = 0$, and the upper and lower surfaces are located at $z = h_f$ and $z = -h_f$, respectively. In the film, there is an eigenstrain $\varepsilon^*(z)$ as a function of coordinate z. The eigenstrain denotes the mismatch strain that may arise from the different lattice constants and/or the hygro-thermal expansion coefficients between the film and the substrate. Surface stresses τ_u and τ_l exist on the upper and lower surfaces of the cantilever, respectively. Here and in the following part, the subscripts u and l denote the quantities related to the upper and lower surfaces of the cantilever, respectively. Depending on the state of the elastic strain, surface stresses τ can be expressed as (e.g., Gurtin and Murdoch 1975)

$$\tau_u = a_u + b_u \varepsilon_u, \ \tau_l = a_l + b_l \varepsilon_l \tag{11.1}$$

where a denotes the constant (strain-independent) surface stress, b is the surface modulus, and ε_u and ε_l are the surface elastic strains. In particular, a constant surface stress $a > 0$ means that the surface tends to contract, and the surface is said to be in tension. On the contrary, $a < 0$ indicates that the surface tends ends to expand and is said to be in compression. We regard the two surfaces and two bulk layers as a whole system; hence, the surface stress τ and the bulk stress in the cross-section are in equilibrium.

11.2.1 STATIC DEFORMATION DUE TO SURFACE STRESS AND EIGENSTRAIN

Assume that the cantilever is so slender ($T, L \gg h_f + h_s$, where T and L are the width and length of the cantilever, respectively) that it is accurate enough to use the Bernoulli–Euler beam theory. For a two-layer cantilever subjected to the surface stress and eigenstrain, the strain in the cantilever can be decomposed into a uniform component and a bending component,

$$\varepsilon = c_0 + (z - h_b)K, \ (-h_s \leq z \leq h_f) \tag{11.2}$$

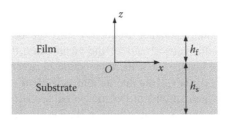

FIGURE 11.2 Schematic diagram of a two-layer cantilever. (Reprinted from Yi, X., and H. L. Duan. 2009. *J Mech Phys Solids* 57:1254–66. With permission.)

where c_0 is the uniform strain component, h_b denotes the position of the bending axis, and K is the curvature of the bending axis. It is noted that the bending axis is the line where the bending strain component is zero, whereas the conventional neutral axis is defined as the line where total strain is zero.

The stresses σ_f and σ_s in the film and substrate can be obtained from Hooke's law,

$$\sigma_f = E_f(\varepsilon_f - \varepsilon^*), \ \sigma_s = E_s\varepsilon_s \tag{11.3}$$

where E is the Young's modulus, and the subscripts f and s identify the quantities related to the film and substrate, respectively. Here, we use the one-dimensional model for simplicity, but the approach holds for the two-dimensional condition. If the two-layer structure is a plate rather than a cantilever, and $\varepsilon^*(z)$ is the in-plane biaxial eigenstrain, then E should be replaced by biaxial modulus $E/(1 - v)$, where v is the Poisson's ratio.

According to the equilibrium equations of the cantilever, that is, the zero resultant forces because of the uniform strain and bending strain components, and zero bending moment, the curvature K can be obtained as

$$K = \frac{6E_f\int_0^{h_f}(z - h_b)\varepsilon^*dz + 6(h_f - h_b)b_u\varepsilon^*(h_f) + 3c_0h_fh_b(E_f + E_s) + 3\Gamma_1}{6(h_s + h_b)^2 b_1 + 6(h_f - h_b)^2 b_u + 2\Gamma_2} \tag{11.4}$$

where

$$\Gamma_1 = (h_s + h_b)(2a_1 + 2c_0b_1 + c_0h_sE_s) - (h_f - h_b)(2a_u + 2c_0b_u + c_0h_fE_f) \tag{11.5a}$$

$$\Gamma_2 = h_f(h_f^2 - 3h_fh_b + 3h_b^2)E_f + h_s(h_s^2 + 3h_sh_b + 3h_b^2)E_s \tag{11.5b}$$

$\varepsilon^*(h_f)$ denotes the eigenstrain at $z = h_f$, and c_0 and h_b are given by

$$c_0 = \frac{E_f\int_0^{h_f}\varepsilon^*dz + b_u\varepsilon^*(h_f) - (a_u + a_1)}{\Gamma_3} \tag{11.6}$$

$$h_b = \frac{E_fh_f^2 - E_sh_s^2 + 2h_fb_u - 2h_sb_1}{2\Gamma_3} \tag{11.7}$$

In Equation 11.4, the assumption in the classical Stoney formula (Stoney 1909) that the film thickness is much less than that of the substrate ($h_f/h_s \ll 1$) is abandoned. Under the condition $h_f/h_s \ll 1$ and a constant eigenstrain (or a constant surface stress), Equation 11.4 reduces to the classical Stoney formula. Equation 11.4 can be regarded as a generalized Stoney formula.

Letting the surface stresses τ_u and τ_1 in Equation 11.4 be zero, the curvature induced by the eigenstrain $\varepsilon^*(z)$ is

$$K = \frac{6E_f\left[2(E_s h_s + E_f h_f)\int_0^{h_f} z\varepsilon^* dz + (E_s h_s^2 - E_f h_f^2)\int_0^{h_f} \varepsilon^* dz\right]}{h_f^4 E_f^2 + h_s^4 E_s^2 + 2E_f E_s h_f h_s(2h_s^2 + 3h_f h_s + 2h_f^2)} \tag{11.8}$$

For a uniform eigenstrain (ε^* = constant), Equation 11.8 becomes

$$K = \frac{6E_f E_s h_f h_s(h_f + h_s)\varepsilon^*}{h_f^4 E_f^2 + h_s^4 E_s^2 + 2E_f E_s h_f h_s(2h_s^2 + 3h_f h_s + 2h_f^2)} \tag{11.9}$$

For $h_f \ll h_s$, Equation 11.9 reduces to the Stoney formula in terms of ε^*, $K = 6E_f h_f \varepsilon^*/E_s h_s^2$. When $\varepsilon^*(z) = 0$, Equation 11.4 reduces to

$$K = \frac{3c_0 h_f h_b(E_f + E_s) + 3\Gamma_1}{6(h_s + h_b)^2 b_1 + 6(h_f - h_b)^2 b_u + 2\Gamma_2} \tag{11.10}$$

where Γ_1 and Γ_2 are given in Equation 11.5, while c_0 and h_b are given by Equations 11.6 and 11.7.

If we only consider the strain-independent surface stresses a_u and a_l, the curvature described by Equation 11.10 becomes

$$K = \frac{6\left[(a_l - a_u)(h_f^2 E_f + h_s^2 E_s) + 2h_f h_s(a_l E_f - a_u E_s)\right]}{h_f^4 E_f^2 + h_s^4 E_s^2 + 2E_f E_s h_f h_s(2h_s^2 + 3h_f h_s + 2h_f^2)} \tag{11.11}$$

With $h_f \ll h_s$ and $b_u = b_l = 0$, the curvature in Equation 11.11 reduces to the classical Stoney formula in terms of surface stress a, that is, $\Delta a = a_u - a_l = -E_s h_s^2 K/6$.

11.2.2 RESONANCE FREQUENCY SHIFT DUE TO MASS ADSORPTION AND SURFACE STRESS

Consider a two-layer cantilever subjected to surface stress τ and mass adsorption, simultaneously. Neglecting the damping effect and shear deformation, the vibration equation of the cantilever loaded by a constant axial end force (F) is (Timoshenko et al. 1974)

$$-\frac{\partial^2 M}{\partial x^2} - F\frac{\partial^2 w}{\partial x^2} + (m + \Delta m)\frac{\partial^2 w}{\partial t^2} = 0 \tag{11.12}$$

where M is the bending moment, and m ($= \rho_f A_f + \rho_s A_s$) and Δm are the effective mass per unit length of the cantilever and mass loading, respectively (the mass of the cantilever is $m_c = mL$, and the total loaded mass is $\Delta m_t = \Delta mL$). In the absence of external force ($F = 0$), Equation 11.12 becomes

$$-\frac{\partial^2 M}{\partial x^2} + (m + \Delta m)\frac{\partial^2 w}{\partial t^2} = 0 \tag{11.13}$$

Using the relation $K = -\partial^2 w/\partial x^2$, the bending moment M on the cross-section of the two-layer cantilever is

$$
\begin{aligned}
\frac{M}{T} &= \int_0^{h_f} \sigma_f(z - h_b)dz + \int_{-h_s}^0 \sigma_s(z - h_b dz) + \tau_u(h_f - h_b) - \tau_1(h_s + h_b) \\
&= -\left[(h_s + h_b)^2 b_1 + (h_f - h_b)^2 b_u + \frac{1}{3}(E_f h_f^3 + E_s h_s^3) \right. \\
&\quad \left. - E_f h_f h_b(h_f - h_b) + E_s h_s h_b(h_s + h_b) \right]\frac{\partial^2 w}{\partial x^2} - \Pi(\varepsilon^*, c_0, h_b)
\end{aligned}
\tag{11.14}
$$

where h_b is given in Equation 11.7 and $\Pi(\varepsilon^*, c_0, h_b)$ is a function of ε^*, c_0, and h_b. By substituting h_b by Equation 11.7, Equation 11.14 can be also expressed as

$$
M = -TD'\frac{\partial^2 w}{\partial x^2} + \text{const}
\tag{11.15}
$$

where D' is the bending stiffness per unit width in the form of

$$
\begin{aligned}
D' &= \frac{E_f h_f + 4b_u}{12\Gamma_3}\left[E_s h_s(3h_f^2 + 3h_f h_s + h_s^2) + E_f h_f^3\right] + \frac{(h_f + h_s)^2 b_u b_1}{\Gamma_3} \\
&\quad + \frac{E_s h_s + 4b_1}{12\Gamma_3}\left[E_f h_f(h_f^2 + 3h_f h_s + 3h_s^2) + E_s h_s^3\right]
\end{aligned}
\tag{11.16}
$$

in which $\Gamma_3 = b_u + b_1 + h_f E_f + h_s E_s$. In the case of $E_f = E_s = E$, Equation 11.16 gives the expression of the bending stiffness of a homogeneous cantilever with the effect of the surface modulus. Furthermore, if the surface moduli of the upper and lower surfaces are equal to each other, that is, $b_u = b_1 = b$, Equation 11.16 reduces to the result ($D' = Eh^3/12 + bh^2/2$) given by Gurtin, Markenscoff, and Thurston (1976). Therefore, the vibration equation of the cantilever in the bending mode (Equations 11.13 and 11.15) is

$$
TD'\frac{\partial^4 w}{\partial x^4} + (m + \Delta m)\frac{\partial^2 w}{\partial t^2} = 0
\tag{11.17}
$$

The solution to Equation 11.17 gives the resonance frequency ($\tilde{f}_i$) of the ith mode of the cantilever clamped on one end with the effects of the changes of the mass and stiffness,

$$
\tilde{f}_i = \frac{1}{2\pi}\left(\frac{\lambda_i}{L}\right)^2 \sqrt{\frac{T(D + \Delta D)}{m + \Delta m}}
\tag{11.18}
$$

where λ_i represents the eigenvalue of the transcendental equation, $\cos\lambda_i\cosh\lambda_i + 1 = 0$, $\Delta D (= D' - D)$ is the change of the bending stiffness, and D is given in Equation 11.16 by letting $b_u = b_1 = 0$. If $\Delta D = \Delta m = 0$, Equation 11.18 is reduced to the resonant frequency f_i without mass loading and the change of bending stiffness.

If $\Delta D \ll D$ and $\Delta m \ll m$, the frequency shift $\left(\Delta f_i = \tilde{f}_i - f_i\right)$ is

$$\Delta f_i \approx \frac{1}{4\pi}\left(\frac{\lambda_i}{L}\right)^2 \sqrt{\frac{TD}{m}}\left(\frac{\Delta D}{D} - \frac{\Delta m}{m}\right) \tag{11.19}$$

Equations 11.17 and 11.19 show that the frequency shift is determined by a combination of the mass loading and the change of the bending stiffness, while the eigenstrain $\varepsilon^*(z)$ and the strain-independent surface stress a do not influence the resonant frequency. Similar phenomena exist in the thermoelastic problem of microscale beam resonators. Based on the dynamic characteristics of the cantilevers with the surface stress and mass loading, the cantilever sensors can be divided into three categories: (1) mass sensors, whose dynamic response is influenced by the adsorbed mass, and whose frequency shift is $\Delta f_i^{(m)} = -(\Delta m/2m)f_i$; (2) surface-stress sensors, whose dynamic response is influenced by the surface modulus, and whose corresponding frequency shift is $\Delta f_i^{(s)} = (\Delta D/2D)f_i$; and (3) mass and surface-stress sensors, whose dynamic response is influenced by the two factors simultaneously, and whose frequency shift is $\Delta f_i = \Delta f_i^{(m)} + \Delta f_i^{(s)}$ depicted by Equations 11.16 and 11.19. For example, Chen et al. (1995) showed by experiments that both the adsorbed mass and surface stress affect the resonant frequency of a gelatin-coated cantilever when it is exposed to water vapor.

For mass sensors, such as those used in the detection of bacteria, nanoparticles, and atom depositions (Illic et al. 2000; Chun et al. 2007), the mass sensitivity is typically defined as the resonant frequency shift per loaded mass

$$\frac{\Delta f_i}{\Delta m_t} = \frac{\Delta f_i}{\Delta mL} \cong -\frac{\sqrt{TDL}}{4\pi m_c^{3/2}}\left(\frac{\lambda_i}{L}\right)^2 \propto \frac{\tilde{E}_{ef}^{1/2}}{\tilde{\rho}_{ef}^{3/2}}\frac{1}{TL^3} \tag{11.20}$$

where m_c is the mass of the cantilever, $\tilde{\rho}_{ef} = (\rho_f h_f + \rho_s h_s)/(h_f + h_s)$ and $\tilde{E}_{ef} = D/(h_f + h_s)^3$ can be regarded as the equivalent mass density and equivalent Young's modulus of the two-layer cantilever, respectively, and ρ_f and ρ_s are the mass densities of the film and the substrate. Equation 11.20 clearly shows that there are three approaches to the increase of the mass sensitivity of cantilever sensors. The first approach is to reduce the size of the cantilever. As indicated in Equation 11.20, if the cantilever shrinks in all three dimensions by a factor α, the mass sensitivity will increase by α^4-fold. Thus, nanocantilevers can measure much smaller mass loading than that of microcantilevers. The other two approaches are to increase the equivalent Young's modulus and to reduce the equivalent mass density of the cantilever.

For surface-stress sensors, the resonance frequency shift induced by molecular adsorption is dominated by the surface elastic modulus, and the contribution of the mass loading is negligible (McFarland et al. 2005). Consequently, if $E_f = E_s = E$, $h = h_s + h_f$, and $b_1 = 0$, Equation 11.19 can be expressed as

$$\frac{\Delta f_i}{f_i} = \frac{3}{2(1 + hE/b_u)} \tag{11.21}$$

It is noted that $\Delta f_i/f_i$ is a function of a nondimensional parameter hE/b_u. As indicated in Equations 11.16 and 11.19, for surface-stress sensors and mass/surface-stress sensors, the strain-independent surface stress a does not change the resonant frequency. This conclusion is different from some results in literatures. In some models, the dynamic analysis of the cantilever is conducted on the bulk layer, while the surface stress a is simply expressed as an equivalent external axial force $(a_u + a_l)T$ or $(a_u + a_l)L$ (Chen et al. 1995; McFarland et al. 2005; Dorignac et al. 2006). However, as made clear by Gurtin, Markenscoff, and Thurston (1976), the distributed loading $q = (a_u + a_l)T\partial^2 w/\partial x^2$ by the surface stress cancels the effect of the equivalent external axial force on the resonant frequency. Hence, the strain-independent surface stress has no effect on frequency shift (Gurtin, Markenscoff, and Thurston 1976), which is shown in Equations 11.16 and 11.19.

11.3 SURFACE STRESS INDUCED BY INTERACTIONS OF ADSORBATES

Surface stress is widely used to characterize the adsorption effects on the mechanical response of nanomaterials and nanodevices. However, quantitative relations between continuum-level descriptions and surface stress and molecular-level descriptions of adsorbate interactions are not well established. Based on the works of Yi and Duan (2009), we first obtain the relations between the adsorption-induced surface stress and the vdW and Coulomb interactions in terms of the physical and chemical interactions between adsorbates and solid surfaces. Then, the adsorption-induced deflection and resonant frequency shift of microcantilevers are numerically analyzed for the vdW and Coulomb interactions. The present theoretical framework quantifies the mechanisms of the adsorption-induced surface stress, and thus provides guidelines to the analysis of the sensitivities, and the identification of the detected substance in the design and application of micro- and nanocantilever sensors.

11.3.1 SURFACE STRESS DUE TO VAN DER WAALS INTERACTION

There exist interatomic/intermolecular forces between the adsorbates on a surface. Because the vdW interaction between the adsorbates is the major driving force for physisorption (Chakarova-Kack et al. 2006; Sony et al. 2007) and the electrostatic interaction is important to the adsorption states of molecules with charging effects (Berger et al. 1997), we investigate these two interactions in terms of the short-range Lennard–Jones (LJ) potential and the long-range Coulomb's law, respectively, and derive the relations between the surface stress at the continuum level and the adsorbate interactions at the atomic/molecular level.

Consider a cantilever with a thickness h and Young's modulus E as shown in Figure 11.3. For simplicity, we assume that the surfaces of the cantilever are chemically homogeneous, so that the adsorbates are distributed statistically uniformly on the upper and lower surfaces, with the mean interspacing distance η between two adsorbates along the length direction as illustrated in Figure 11.3. The origin of the coordinate system is on the geometrical midplane, and the z-axis is perpendicular to the midplane. In this model, we also assume that the first layer of the adsorbates

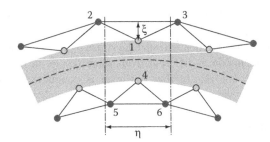

FIGURE 11.3 A cantilever with a uniform distribution of atoms/molecules adsorbed both on the upper and lower surfaces. (Reprinted from Yi, X., and H. L. Duan. 2009. *J Mech Phys Solids* 57:1254–66. With permission.)

(adsorbates 2, 3, 5, and 6 in Figure 11.3) on the cantilever surfaces play a dominant role and the effects of the second and higher layers of the adsorbates are less important (Dareing and Thundat 2005).

For convenience, the strain-independent surface stress $a_{u(1)}$ and the surface elastic modulus $b_{u(1)}$ in Equation 11.1 can be expressed as

$$a_{u(1)} = a_{u(1)}^0 + \Delta a_{u(1)}, \; b_{u(1)} = b_{u(1)}^0 + \Delta b_{u(1)} \tag{11.22}$$

where the quantities with a superscript 0 denote the values before the adsorption, and Δ denotes the changes caused by the adsorption.

We first consider the vdW interactions based on the model used by Dareing and Thundat (2005). The vdW interactions between adsorbates (2, 3, 5, and 6) and the surface atoms (1 and 4), and that between the adsorbates themselves are depicted by the following LJ (6-12-type) potential (cf. Figure 11.3):

$$V(r_{ij}) = -\frac{A}{r_{ij}^6} + \frac{B}{r_{ij}^{12}} \tag{11.23}$$

where r_{ij} is the distance between atoms i and j, and A and B are the LJ constants. The distances between adjacent atoms are

$$r_{23} = \eta\left[1 + \varepsilon(\xi + h/2)\right], \; r_{12} = r_{13} = \sqrt{\left(\frac{r_{23}}{2}\right)^2 + \xi^2}$$

$$r_{56} = \eta\left[1 - \varepsilon(\xi + h/2)\right], \; r_{45} = r_{46} = \sqrt{\left(\frac{r_{56}}{2}\right)^2 + \xi^2} \tag{11.24}$$

where η is taken as the space between two adsorbates, the reciprocal of η is defined as the number density under the undeformed state (number per length), ξ is the distance between the adsorbates and surfaces (cf. Figure 11.3), and ε $(h/2 + \xi)$ and $-\varepsilon$ $(h/2 + \xi)$ are the values of the strains at $z = h/2 + \xi$ and $z = -(h/2 + \xi)$, respectively.

11.3.1.1 Strain-Independent Surface Stress due to van der Waals Interaction

We first establish the connection between the strain-independent surface stress a in Equation 11.1 and the vdW interactions of the distributed adsorbates. Suppose that the adsorbates are only located on one surface of the cantilever (upper surface here) and there is only the surface stress a_u^0 on the upper surface before adsorption. According to Equation 11.23, the potential resulting from vdW interaction over the length L of the cantilever is

$$U_i(K) = \int_L \frac{\vartheta}{\eta} \left[V(r_{12}) + V(r_{13}) + V(r_{23}) \right] dx \qquad (11.25)$$

where ϑ is the number density under the undeformed state (number per width), $V(r_{12}) = V(r_{13}) = -A_1/r_{12}^6 + B_1/r_{12}^{12}$ and $V(r_{23}) = -A_2/r_{23}^6 + B_2/r_{23}^{12}$. K denotes the curvature. Here, A_1 and B_1 are LJ constants for the interaction between the adsorbates and the adjacent cantilever atoms, and A_2 and B_2 are LJ constants for two adjacent adsorbates. The elastic energy (U_{el}) in the bulk and upper surface per unit width over the length L is

$$U_{el}(K) = \frac{1}{2} \int_L \int_{-h/2}^{h/2} \sigma \varepsilon \, dx dz + \int_L \left(a_u^0 \varepsilon_u + \frac{b^0}{2} \varepsilon_u^2 \right) dx \qquad (11.26)$$

where $\sigma = E\varepsilon$, and $\varepsilon = zK$. Here we assume that the extension of the cantilever is very small and makes little contribution to the deformation of the cantilever, which is always true when the thickness h is larger than 2 nm. Hence, we neglect the extension of the cantilever in Equation 11.26. The equilibrium state requires that the total energy U_p $(= U_i + U_{el})$ should be stationary, namely, $\partial U_p/\partial K = 0$. With the modified Stoney formula $a_u = -(Eh^2 + 4hb_u)K/6$ (Yi and Duan 2009), it follows that

$$\frac{a_u^0}{\vartheta} + \frac{3(1+\varsigma K)A_1}{\eta^7 \left[\dfrac{1}{4}(1+\varsigma K)^2 + \left(\dfrac{\xi}{\eta} \right)^2 \right]^4} + \frac{6A_2}{\eta^7(1+\varsigma K)^7} - \frac{6(1+\varsigma K)B_1}{\eta^{13} \left[\dfrac{1}{4}(1+\varsigma K)^2 + \left(\dfrac{\xi}{\eta} \right)^2 \right]^7}$$

$$- \frac{12B_2}{\eta^{13}(1+\varsigma K)^{13}} = \frac{a_u}{\vartheta} \left(\frac{3b_u^0 + hE}{4b_u + hE} \right) \approx \frac{a_u}{\vartheta} \qquad (11.27)$$

where $\varsigma = h/2 + \xi$. It is noted that if the extension of the cantilever is considered, the term $(3b_u^0 + hE)$ on the right side of Equation 11.27 should be $(4b_u^0 + hE)$. Since b_u^0 and b_u are very small compared with hE when $h > 2$ nm, $(3b_u^0 + hE)/(4b_u^0 + hE) \approx 1$, which yields Equation 11.27. Under the condition $\varsigma K \ll 1$, if we ignore ςK in the left-hand side of Equation 11.27, then the relation between adsorption-induced surface stress Δa_u and the interactions of the distributed adsorbates can be obtained as

$$\Delta a_u = \frac{3\vartheta}{\eta^7} \left(2A_2 + \frac{A_1}{r_e^4} \right) - \frac{6\vartheta}{\eta^{13}} \left(2B_2 + \frac{B_1}{r_e^7} \right) \qquad (11.28)$$

where $r_e = 1/4 + (\xi/\eta)^2$. It is observed that Δa_u increases linearly with increasing attractive constants A_1 and A_2, while it decreases linearly with increasing repulsive constants B_1 and B_2. This is consistent with the properties of the surface stress, namely, a positive surface stress tends to contract the surface and a negative one expands the surface. Since A_1 and A_2 (B_1 and B_2) are usually in the same order and $\xi/\eta \approx 1$ when the density of adsorbates is high, from Equation 11.28 we know that the adsorbate–adsorbate interaction (A_2, B_2) plays a more important role than the adsorbate–surface interaction (A_1, B_1) in the magnitude of the adsorption-induced surface stress.

11.3.1.2 Surface Moduli due to van der Waals Interaction

Next, we establish the relation between the vdW interactions of the adsorbates and the surface modulus b in Equation 11.1. Suppose that the same adsorbates are located on both upper and lower surfaces uniformly (cf. Figure 11.3). The potential energy due to the vdW interaction over the length L is

$$U_i(K) = \int_L \frac{\vartheta}{\eta} \left[2V(r_{12}) + V(r_{23}) + 2V(r_{45}) + V(r_{56}) \right] dx = \int_L \frac{\vartheta}{\eta} u_i dx \qquad (11.29)$$

where $V(r_{45}) = -A_1/r_{45}^6 + B_1/r_{45}^{12}$ and $V(r_{56}) = -A_2/r_{56}^6 + B_2/r_{56}^{12}$. The elastic energy ($U_{e2}$) in the bulk and upper and lower surfaces per unit width over L is

$$U_{e2}(K) = \frac{1}{2} \int_L \int_{-h/2}^{h/2} \sigma \varepsilon \, dx dz + \int_L \left[a^0 (\varepsilon_u + \varepsilon_1) + \frac{b^0}{2} (\varepsilon_u^2 + \varepsilon_1^2) \right] dx \qquad (11.30)$$

The kinetic energy of adsorbates and the cantilever per unit width over L is

$$U_k(\dot{w}) = \frac{1}{2} \int_L \left(\rho h + \frac{2 \vartheta m_a}{\eta} \right) \dot{w}^2 dx \qquad (11.31)$$

where ρ is the mass density of the cantilever, w is the deflection of the midplane, the overdot denotes the partial derivative with respect to time t, and m_a is the mass of each adsorbate. Introducing the Lagrangian function $L(w, \dot{w}, t) = U_k(\dot{w}) - U_i(K) - U_{e2}(K)$ into Hamilton's equation $\delta \int_{t_1}^{t_2} L dt = 0$ leads to

$$\frac{\partial^2}{\partial x^2} \frac{\partial}{\partial K} \left(\frac{Eh^3}{24} K^2 + \frac{b^0 h^2}{4} K^2 + \frac{\vartheta}{\eta} u_i \right) - \left(\rho h + \frac{2 \vartheta m_a}{\eta} \right) \frac{\partial^2 w}{\partial t^2} = 0 \qquad (11.32)$$

By ignoring the extension of the cantilever, expanding $\partial u_i/\partial K$ into a Taylor series with respect to K, and keeping the first-order term K, the free vibration equation can be obtained from Equation 11.32,

$$\left(\frac{Eh^3}{12} + \frac{b^0 h^2}{2} + \frac{\vartheta}{\eta} \frac{\partial^2 u_i}{\partial K^2} \bigg|_{K=0} \right) \frac{\partial^4 w}{\partial x^4} + \left(\rho h + \frac{2 \vartheta m_a}{\eta} \right) \frac{\partial^2 w}{\partial t^2} = 0 \qquad (11.33)$$

Comparing Equation 11.33 with the vibration equation in Equation 11.17, one obtains the bending stiffness per unit width, D', as

$$D' = \frac{Eh^3}{12} + \frac{b^0 h^2}{2} + \frac{\vartheta}{\eta} \frac{\partial^2 u_i}{\partial K^2}\bigg|_{K=0} \quad (11.34)$$

where

$$\frac{\partial^2 u_i}{\partial K^2}\bigg|_{K=0} = -\frac{3h^2}{2\eta^6}\left(14A_2 - \frac{A_1}{r_e^4} + \frac{2A_1}{r_e^5}\right) + \frac{3h^2}{12\eta^{12}}\left(52B_2 - \frac{2B_1}{r_e^7} + \frac{7B_1}{r_e^8}\right) \quad (11.35)$$

By letting Equation 11.24 equal $D' = Eh^3/12 + bh^2/2$ (Gurtin, Markenscoff, and Thurston 1976), the adsorption-induced surface modulus Δb is obtained:

$$\Delta b = -\frac{3\vartheta}{\eta^7}\left(14A_2 - \frac{A_1}{r_e^4} + \frac{2A_1}{r_e^5}\right) + \frac{3\vartheta}{\eta^{13}}\left(52B_1 - \frac{2B_1}{r_e^7} + \frac{7B_1}{r_e^8}\right) \quad (11.36)$$

Note that Δb can be either positive or negative, depending on whether the repulsive interaction or the attractive one dominates. Similar to Equation 11.28, the adsorbate–adsorbate interaction terms in Equation 11.36 dominate the adsorption-induced surface modulus when the density of adsorbates is high. Both Equations 11.28 and 11.36 show that the adsorption-induced surface stress Δa and the surface modulus Δb are functions of LJ constants (adsorption mechanism) and the location of adsorbates (surface coverage), and imply that the surface stress provides an insight into the adsorbate interactions as the density of adsorbates varies.

Next, we show some numerical results of the surface stress resulting from vdW interaction. According to Equations 11.28 and 11.36, the contours of the change $\hat{a} = \Delta a \vartheta^{-1} \times 10^9$ of the strain-independent surface stress, and the change $\hat{b} = \Delta b \vartheta^{-1} \times 10^9$ of the surface modulus as functions of the normalized LJ constants A_0 and B_0 are defined by $A_0^1 = A_0^2 \equiv A_0$ and $B_0^1 = B_0^2 \equiv B_0$ in which $A_1 = A_1^0 \times 10^{-77}$ Jm^6, $A_2 = A_0^2 \times 10^{-77}$ Jm^6, $B_1 = B_0^1 \times 10^{-134}$ Jm^{12}, and $B_2 = B_2^0 \times 10^{-134}$ Jm^{12}. The ranges of constants A_0 and B_0 considered in Figure 11.4 are typical for different molecular structures, that is, $A_0 \in [0.02, 10]$ and $B_0 \in [0.02, 4]$. If the number density ϑ is taken as the reciprocal of η ($\vartheta = 1/\eta$), the constant surface stress change (Δa) induced by adsorption varies from about −4.5 to 6 N/m ($\hat{a}$ varies from −1.5 to 2 N), as shown in Figure 11.4a. In the range of $A_0 \in [0.02, 0.15]$ and $B_0 \in [0.02, 0.15]$, which are the representative values of the adsorption of O atoms on a solid surface, the adsorption-induced surface stress is on the order of 0.1 N/m.

As shown in Figure 11.4b, the change of the adsorption-induced surface modulus Δb depicted by the LJ potential is about 1 N/m, which quantitatively agrees with the magnitude (−1 ~ −0.1 N/m) of the surface modulus in the study of Lu et al. (2005). Once Δb is known, the frequency shift induced by the adsorption can be predicted by Equations 11.16 and 11.19. From Figure 11.4a and b, it is clear that when the attractive interaction dominates, a tensile constant surface stress and negative surface modulus are induced. The negative surface modulus will decrease the resonance frequency.

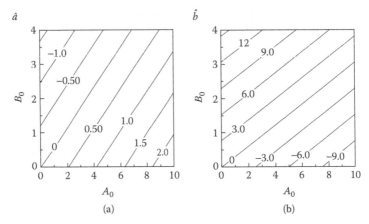

FIGURE 11.4 Contours of (a) $a\vartheta^{-1} \times 10^9$ and (b) $b\vartheta^{-1} \times 10^9$ as functions of A_0 and B_0. (Reprinted from Yi, X., and H. L. Duan. 2009. *J Mech Phys Solids* 57:1254–66. With permission.)

11.3.2 SURFACE STRESS DUE TO COULOMB INTERACTION

Besides the vdW interaction, the Coulomb interaction is another important mechanism for the adsorption of small molecules and gas atoms, especially when there exist charges or charge transfer in the considered system. It has been reported that the sulfur headgroup chemisorbs to the Au surface through the formation of Au–S bonds and the alkyl chains interact through the vdW attraction (Nuzzo, Zegarski, and Dubois 1987). Although the vdW interaction between the akyl chains results in the tilt of the chains, it has little contribution to the adsorption-induced surface stress (Berger et al. 1997; Godin 2004). Moreover, Au–S bonds result in a partial charge transfer of approximately 0.3e between the Au and S atoms (Gronbeck, Curioni, and Andreoni 2000). Due to the Coulomb interaction, the forces exerted between negatively charged sulfur headgroups and positively charged gold atoms contribute to the surface stress. To investigate the relation between the surface stress and the Coulomb interactions of adsorbates, we take the self-assembly of alkanethiols on the Au(111) surface in the vacuum condition as an example. The overlaying structure of alkanethiolates on the Au(111) surface can be described, using Wood's notation, as a $\left(\sqrt{3} \times \sqrt{3}\right)R30°$ overlayer of which a $c(4 \times 2)$ superlattice can also be observed sometimes (Yourdshahyan and Rappe 2002). Since the $\left(\sqrt{3} \times \sqrt{3}\right)R30°$ structure is the primary structure, our investigation focuses on it. For simplicity, we use the one-dimensional parallel configuration (see Figure 11.5) to estimate the contribution of partial charge interactions to the surface stress, and this configuration can be extended to the case of a $c(4 \times 2)$ superlattice. In this model, the partial charge transfer at the Au–S bond is treated as two point charges (Godin 2004). As depicted in Figure 11.5, the adsorbates (2, 3, 6, and 2′) are partially negatively charged sulfur headgroups, whereas the surface atoms (1, 4, 5, and 1′) are positively charged gold atoms.

The electrostatic energy, $W(r_{jk})$, for the Coulomb interaction between two electric charges is given by

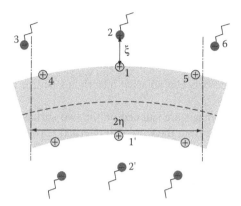

FIGURE 11.5 Schematic illustration of adsorbates situated on the surface of a cantilever, where the alkanethiol molecules are adsorbed on the Au(111) surface. (Reprinted from Yi, X., and H. L. Duan. 2009. *J Mech Phys Solids* 57:1254–66. With permission.)

$$W(r_{jk}) = \frac{1}{4\pi\varepsilon_0} \frac{q_j q_k}{r_{jk}} \tag{11.37}$$

where r_{jk} is the distance between charges j and k, q $(= eZ)$ is the electric charge of the corresponding particle, e $(= 1.602 \times 10^{-19}$ C) is the elementary charge and Z is the ionic valency, and ε_0 $(= 8.8542 \times 10^{-12}$ C^{2}N^{-1}m^{-2}) denotes the permittivity of free space. For the like charges, $W(r_{jk})$ is positive, whereas for the unlike charges, it is negative. The long-range Coulomb energy for a system of N charges is given by $W = \sum_{j=1}^{N} W_j$, where W_j is the electrostatic interaction energy of the charge j with all the other charges, and can be expressed as

$$W_j = \frac{1}{2} \times \frac{1}{4\pi\varepsilon_0} \sum_{k=1,k\neq j}^{N} \frac{q_j q_k}{r_{jk}} \tag{11.38}$$

Similar to Equation 11.25, if the adsorbates are only located on the upper surface, the potential U_i $[= \vartheta (W_1 + W_2)]$ due to the Coulomb interaction over the length η is

$$U_i = \frac{\vartheta q^2}{8\pi\varepsilon_0} \sum_{n=1}^{\infty} \left(\frac{2}{nr_{14}} + \frac{2}{nr_{23}} - \frac{2}{\sqrt{r_{14}^2 n^2 + \xi^2}} - \frac{2}{\sqrt{r_{23}^2 n^2 + \xi^2}} - \frac{2}{\xi} \right) \tag{11.39}$$

where W_1 and W_2 are the electric interaction energies of the Au atom (atom 1) and the sulfur headgroup (adsorbate 2, cf. Equation 11.38), $r_{14} = \eta [1 + \varepsilon (h/2)]$, r_{23} is given in Equation 11.24, and ξ is the distance between the sulfur headgroup and the gold surface.

With the elastic energy U_{el} in Equation 11.26 and the Coulomb interaction in Equation 11.39, the total energy is $U_p = U_i + U_{el}$. From the relation $\partial U_p/\partial K = 0$ and following a similar procedure to that shown above, we obtain the expression of the change Δa of the strain-independent surface stress because of electrostatic interaction,

$$\Delta a = \frac{\vartheta q^2}{2\pi\varepsilon_0 \eta^2} \sum_{n=1}^{\infty} \frac{1}{n} \left[-1 + \left(1 + \frac{\xi^2}{\eta^2 n^2} \right)^{-\frac{3}{2}} \right] \tag{11.40}$$

where η is the distance between adjacent adsorbates (sulfur headgroup) under the undeformed state.

When the same adsorbates are situated on the two surfaces of the cantilever, the potential energy resulting from the Coulomb interaction over the length η is $U_i = \vartheta$ $(W_1 + W_2 + W_1' + W_2')$, where W_j is given in Equation 11.38. Using Hamilton's equation, and the elastic and kinetic energies (cf. Equations 11.30 and 11.31), the surface modulus change Δb can be obtained as

$$\Delta b = \frac{\vartheta q^2}{2\pi\varepsilon_0 \eta^2} \sum_{n=1}^{\infty} \frac{1}{n} \left[2 - 3\left(1 + \frac{\xi^2}{\eta^2 n^2} \right)^{-\frac{3}{2}} + \left(1 + \frac{\xi^2}{\eta^2 n^2} \right)^{-\frac{3}{2}} \right] \tag{11.41}$$

Because of the Coulombic interactions between the charged Au atoms and S headgroups, the adsorption of alkanethiols on the Au(111) surface results in the surface stress. The charge transfer from Au to S is about $q = 0.3e$ for the adsorption of thiols (Gronbeck, Curioni, and Andreoni 2000), the distance between the sulfur headgroups and Au(111) surface is approximately 0.2 nm (Yourdshahyan and Rappe 2002), the distance η between adjacent sulfur headgroups is about $\sqrt{3}a_{Au}$, and the number density ϑ of sulfur headgroups is about $2/(\sqrt{3}a_{Au})$ (the lattice constant of Au is $a_{Au} = 2.884$ A). According to Equation 11.40, the change of the strain-independent surface stress resulting from the adsorption of thiols on the Au(111) surface is $\Delta a = -0.095$ N/m, which is on the order of 0.011–0.019 N/m in the experiment done by Berger et al. (1997). From Equation 11.40, the change of the surface modulus because of the adsorption of thiols on the Au(111) surface is $\Delta b = 0.3522$ N/m.

11.3.3 CURVATURE AND RESONANCE FREQUENCY SHIFT
DUE TO INTERACTIONS OF ADSORBATES

The curvature and normalized frequency shift of a cantilever consisting of a Si substrate and a Au film due to the vdW interaction of adsorbates are plotted in Figure 11.6a and b. The parameters of the cantilever are as follows: $E_s = 180$ GPa, $\rho_s = 2.33$ g/cm^3, $E_f = 90$ GPa, $h_f = 20$ nm, and $\rho_f = 19.3$ g/cm^3. Two cases are considered, namely, mercury (Hg) atoms on Au(100) with $A_0^1 = A_0^2 = 2.8377$, $B_0^1 = B_0^2 = 1.943$, $\xi = 0.45$ nm, and $\eta = 0.4$ nm (Dareing and Thundat 2005); and oxygen (O) atoms on Au(100) surface with $A_0^1 = 0.103$, $B_0^1 = 0.07885$, $A_0^2 = 0.1534$, $B_0^2 = 0.141$, $\xi = 0.282$ nm, and $\eta = 0.324$ nm (A_0^1, A_0^2, B_0^1, and B_0^2 are defined in the above). The mass of a Hg atom is $m_a = 201$ Da and that of an oxygen atom is $m_a = 16$ Da (where 1 Da $= 1.66 \times 10^{-27}$ kg). According to Equations 11.28, 11.36, 11.19, and 11.10, the bending curvature K and the normalized frequency shift $\Delta f_i/f_i$ as functions of the thickness of substrate are shown in Figure 11.6a and b. Here we assume that the surface stresses on upper and lower surfaces are zero before adsorption, taking $\vartheta = \eta^{-1}$ and $\Delta m/m = m_a/[\eta^2(\rho_f h_f + \rho_s h_s)]$.

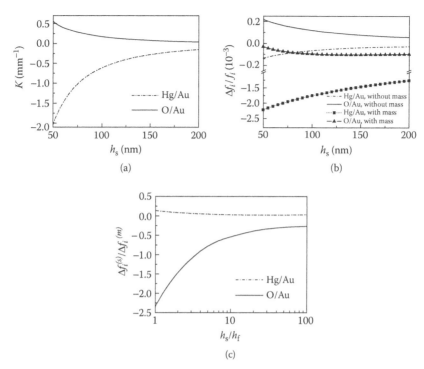

FIGURE 11.6 Variations of (a) the bending curvature and (b) the normalized frequency shift with substrate thickness h_s due to the van der Waals interaction; (c) the ratio of frequency shifts because of surface stress $\left(\Delta f_i^{(s)}\right)$ and mass loading $\left(\Delta f_i^{(m)}\right)$ versus normalized substrate thickness. (Reprinted from Yi, X., and H. L. Duan. 2009. *J Mech Phys Solids* 57:1254–66. With permission.)

As shown in Figure 11.6a and b, for the adsorption of mercury vapor on gold surface, $K < 0$ and $\Delta f_i/f_i < 0$ without the consideration of mass-loading effect, indicating that the adsorption-induced surface stress a is positive, whereas the surface modulus b is negative. For the case of O/Au(100), the adsorption-induced surface stress a is negative but the surface modulus b is positive. It is seen that K increases as the thickness h_s decreases. With consideration of the atom mass, $\Delta f_i/f_i$ changes significantly after the adsorption of Hg or O atoms on gold surface as shown in Figure 11.6b, which indicates that the mass of the adsorbates plays an important role in changing the resonant frequency besides the adsorbate interactions. To elucidate the effects of mass and surface modulus clearly, Figure 11.6c shows the ratio $\Delta f_i^{(s)}/\Delta f_i^{(m)}$, namely, the ratio of frequency shift $\Delta f_i^{(s)}$ resulting from surface modulus to that of $\Delta f_i^{(m)}$ resulting from the mass, as a function of h_s/h_f. It is seen that the ratio $\Delta f_i^{(s)}/\Delta f_i^{(m)}$ depends on the type of the adsorbate and the interacting properties. For example, in the case of O/Au(100), the influence of the adsorption-induced surface modulus on Δf_i is larger than that of the mass on Δf_i under $h_s < 4h_f$. As h_s becomes large, $\Delta f_i^{(s)}/\Delta f_i^{(m)}$ reaches about −0.5, suggesting that the effects of the mass and the adsorbate interactions are comparably important even for large h_s/h_f. For the Hg/Au(100) system, the mass-loading effect dominates the frequency shift and the influence of Hg atom interactions can be neglected.

For the Coulomb interaction, we consider butanethiol molecules adsorbed on the gold surface of the cantilever in Figure 11.2. The variations of ΔK and $\Delta f_i/f_i$ with the space η of adsorbates for $q = 0.3e$ is shown in Figure 11.7a and b. The cantilever parameters are the same as those used for the vdW interaction in Figure 11.6, except $h_s = 200$ nm. The mass of a butanethiol molecule, m_a, is about 89 Da, and $\Delta m/m = \vartheta m_a/[\eta \, (\rho_f h_f + \rho_s h_s)]$. As shown in Figure 11.7a and b, the curvature K and normalized frequency shift $\Delta f_i/f_i$ decrease with the increase of the adsorbate space η, indicating that the number density of adsorbates has a significant effect on the static and dynamic properties of cantilever sensors. As the number density ϑ increases, the bending curvature and resonance frequency shift become large. $K > 0$ and $\Delta f_i/f_i > 0$ without the consideration of mass-loading effect indicate that the adsorption-induced surface stress a is negative and the surface modulus b is positive. In Figure 11.7b, we can see that although the Coulomb interaction has an influence on the resonance frequency shift, the large resonance frequency shift arises from the mass of the adsorbates. Figure 11.7c illustrates the dependence of ratio $\Delta f_i^{(s)}/\Delta f_i^{(m)}$ on h_s/h_f for $\eta = 0.3$ nm and 0.45 nm. It is seen that for the case of low density of adsorbates ($\eta = 0.45$ nm),

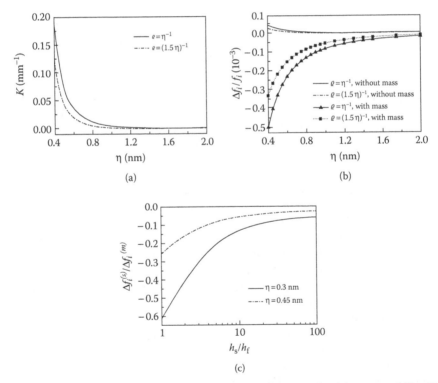

(a)

(b)

(c)

FIGURE 11.7 Variations of (a) the curvature and (b) the normalized frequency shift with adsorbate space η for a two-layer cantilever resulting from the Coulomb interaction between the adsorbates; (c) the ratio of frequency shifts resulting from surface stress $\left(\Delta f_i^{(s)}\right)$ and mass loading $\left(\Delta f_i^{(m)}\right)$ as a function of h_s/h_f. (Data from Yi, X., and H. L. Duan. 2009. *J Mech Phys Solids* 57:1254–66.)

the mass effect dominates the dynamic properties of cantilevers. For the high density of adsorbates ($\eta = 0.3$ nm), the surface modulus resulting from the adsorbate interactions contributes a lot to the frequency shift, especially for a small h_s/h_f.

11.4 SURFACE-ENHANCED CANTILEVER SENSORS WITH POROUS FILMS

Development of surface-enhanced microcantilevers with improved sensitivities is of long-standing interest. Based on the work of Duan (2010), the design of surface-enhanced cantilever sensors using nano/microporous films as surface layers is presented here. The static deformation and resonance frequencies of these surface-enhanced sensors with the simultaneous effects of the eigenstrain, the surface stress, and the adsorbed mass are analyzed.

11.4.1 MECHANICS OF CANTILEVERS WITH POROUS FILMS

Comparing with solid layers, the porous films, which not only have a lower effective density $\tilde{\rho}_{ef}$, but also a greater surface area to accommodate more adsorbates, may increase the sensitivity of cantilevers. Therefore, we can design surface-enhanced cantilever sensors with nano/microporous film layers (Figure 11.8a). The deformation and resonance frequency of these surface-enhanced cantilevers depend on the structure of nanoporous films. Nanoporous films with three different structures are shown in Figure 11.8b through d. Figure 11.8b shows the nanoporous film with randomly distributed spherical pores in a closed form. Comparing with the solid film, the nanoporous film shown in Figure 11.8b not only has the same surface area to accommodate the adsorbates, but also has a lower effective elastic moduli. Therefore, only the nanoporous film with geometry shown in Figure 11.8b is used for comparison. In the surface-enhanced cantilever sensors, we consider two kinds of porous geometry (Figure 11.8c and d). The first is a film with ordered aligned cylindrical

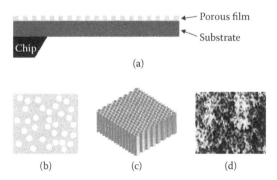

FIGURE 11.8 (a) Schematic diagram of a surface-enhanced cantilever sensor with a nano/microporous film. (b) Film containing randomly distributed spherical pores. (c) Film with aligned cylindrical pores. (d) Disordered nanoporous Au film obtained by dealloying. (Reprinted from Duan, H. L. 2010. *Acta Mech Solida Sinica* 23:1–12. With permission.)

nanopores (Figure 11.8c), and the second is a porous film with a disordered structure (cf. Figure 11.8d) obtained by dealloying. Dealloying, the selective dissolution of the less noble component from a solid solution, is well known to make nanoporous structures (Kramer, Viswanath, and Weissmüller 2004). Figure 11.8d shows scanning electron microscopy images of the nanoporous gold microstructures, which were obtained by dealloying $Ag_{75}Au_{25}$ master alloy sheets (e.g., Kramer, Viswanath, and Weissmüller 2004).

11.4.1.1 Deformation and Resonance Frequency

For microcantilevers with porous films, Equations 11.4, 11.16, and 11.18 that characterize the deformation and resonance frequency still apply. However, the mass density and the Young's modulus E_f in Equations 11.4, 11.16, and 11.18 should be replaced by the corresponding effective properties of porous films, which depend on the porosity and the geometries of porous films. Also, the constant surface stress on the porous surface would cause the porous film to shrink or expand (Weissmüller et al. 2003). Therefore, the effect of the constant surface stress on the cantilever deformation can be equivalent to that of an eigenstrain (or self-strain, cf. Equation 11.4). In the following, we will give a theoretical framework for calculating the self-strain of the nanoporous films, while considering the effect of the surface elasticity of the nanopores (e.g., Duan et al. 2005, 2006; He 2006).

As in other works studying the effect of surface stress on the properties of heterogeneous media containing nano-inhomogeneities (e.g., Sharma, Ganti, and Bhate 2003; Duan et al. 2005), we consider an isotropic surface constitutive equation for the porous surfaces

$$\tau = \tau_0 + 2\mu^s \varepsilon^s + \lambda^s (tr\varepsilon^s)\mathbf{I} \tag{11.42}$$

where τ $(=\tau\mathbf{I})$ is the surface-stress tensor, τ_0 is the constant surface stress, $\mathbf{I}$ is the second-order unit tensor in two-dimensional space, λ^s and μ^s are the surface moduli, and ε^s is the surface-strain tensor. Thus, the effect of the surface elastic moduli (λ^s and μ^s) on the deformation and resonance frequencies of the cantilevers can be considered by replacing the stiffness of the solid surface layer in the above sections with that of the porous one. The framework to predict the effective stiffness of nanoporous materials can be found in the studies of Duan et al. (2005, 2006). The effect of constant surface stress τ_0 on the cantilever deformation can be equivalent to that of an eigenstrain (cf. Equation 11.4). By substituting the self-strain into Equation 11.4, the static deformation of the cantilever with the porous film can be obtained.

Consider an elastic solid containing nanopores with arbitrary shapes and spatial dispersions. If there exists a uniform stress field σ^0 in the porous solid, obviously, the uniform stress must satisfy the boundary condition on the surfaces of pores,

$$\sigma_0 \cdot \mathbf{n} = \nabla_s \cdot \tau \tag{11.43}$$

where $\mathbf{n}$ is the outward unit normal vector to the surfaces of pores Γ, and $\nabla_s \cdot \tau$ denotes the surface divergence of a tensor field τ (e.g., Gurtin, Weissmüller, and Larche 1998).

Next, the porous solid (an elastic solid containing nanopores with arbitrary shapes and spatial dispersions) as a whole can be regarded as a macroscopically homogeneous material. Consider a representative volume element (RVE) of a volume V of this porous solid (including pores), and let V_1 and V_2 denote the volumes of the pore and solid phase, respectively. Denoting the overall self-strain resulting from the constant surface stress (τ_0) and effective elasticity modulus as ε^{s*} and $\bar{\mathbf{C}}$, respectively, the macroscopic constitutive equation of the porous solid is (He 2006)

$$\bar{\sigma} = \bar{\mathbf{C}} : (\bar{\varepsilon} - \varepsilon^{s*}) \tag{11.44}$$

in which $\bar{\sigma}$ and $\bar{\varepsilon}$ are the volume-averaged stress and strain, respectively,

$$\bar{\varepsilon} = \frac{1}{2V} \int_S (\mathbf{N} \otimes \mathbf{u} + \mathbf{u} \otimes \mathbf{N}) \mathrm{d}S, \quad \bar{\sigma} = \frac{1}{V} \int_S (\sigma \cdot \mathbf{N}) \otimes \mathbf{X} \mathrm{d}S \tag{11.45}$$

where S is the outer surface of the RVE, $\mathbf{N}$ is the outward unit normal vector to S, $\mathbf{x}$ is the position vector, and $\mathbf{u}$ and σ are the displacement vector and stress tensor in the bulk, respectively. Under the homogeneous stress field σ^0, the average stress $(\bar{\sigma})$ of the RVE is

$$\bar{\sigma} = (1 - p)\sigma^0 - \frac{p}{V} \int_\Gamma (\sigma^0 \cdot \mathbf{n}) \otimes \mathbf{x} \mathrm{d}\Gamma \tag{11.46}$$

where p denotes the porosity of the film, and Γ is the surfaces of pores.

First, we find a homogeneous stress field σ^0 and the corresponding $\bar{\sigma}$ for a certain τ from Equations 11.42, 11.43, and 11.46 (He 2006). Subsequently, with this homogeneous stress field, and Equations 11.44 and 11.46, the self-strain ε^{s*} can be obtained. The detailed procedure for obtaining ε^{s*} can be found for an example case, that is, the nanoporous material with aligned cylindrical pores.

Equations 11.42 through 11.46 constitute the general framework to calculate the self-strain induced by the constant surface stress τ_0 for the nanoporous solids. However, the analytical solution for self-strain only exists for pores with constant curvatures, for example, spherical and aligned cylindrical shapes. In the following, we will give the self-strain for films with spherical and cylindrical pores (cf. Figure 11.8b and c) by using this framework. For the solids with disordered porous structures (cf. Figure 11.8d), we will use another method presented by Weissmüller and Cahn (1997).

Using the framework mentioned earlier in this section, it is found that the self-strain ε^{s*} of the films with spherical pores is hydrostatic $\left(\varepsilon_{xx}^{s*} = \varepsilon_{yy}^{s*} = \varepsilon_{zz}^{s*} = \varepsilon^{s*}\right)$ and given by

$$\varepsilon^{s*} = -\frac{p(3\kappa_f + 4\mu_f)\tau_0}{R_0 \mu_f \left[(4p\mu_f + 3\kappa_f)\Sigma^* + 6\kappa_f(1 - p)\right]} \tag{11.47}$$

in which $\Sigma^* = 2(\lambda_s + \mu_s)/(R_0\mu_f)$, μ_f and κ_f are the shear modulus and bulk modulus of the matrix of the film, and R_0 is the radius of the spherical pores. The self-strain in Equation 11.47 is the same as that given by He (2006). When $\Sigma^* \to 0$, Equation 11.47 reduces to the result without the effect of the surface moduli.

Equations 11.4, 11.16, and 11.18 characterizing the deformation and resonance frequency of the two-solid-layer cantilevers can be applied to the cantilevers without porous films. However, the eigenstrain (ε^*) and Young's modulus E_f in Equations 11.4, 11.16, and 11.18 should be replaced by $\varepsilon^* + \varepsilon^{s*}$ and the corresponding effective Young's moduli of the porous films with spherical pores. The effective elastic moduli of the porous film with spherical pores at the nanoscale have been given by an equation labeled (36) in the study of Duan et al. (2005).

Because the films with embedded spherical pores have lower effective moduli than those of the solid ones, the cantilevers with these kind of films have a larger bending and frequency shift than those of the solid ones. However, because of the closed form of the spherical pores, no new surface area has been created for the adsorbates, and the adsorbates are situated only on the upper surface. However, the porous films in opened forms can both create new surface area for the adsorbates and reduce the elastic moduli (e.g., Figure 11.8c and d). For these kinds of geometry (e.g., Figure 11.8c and d), both the upper surfaces and the surfaces of pores adsorb molecules, and influence the bending and dynamic response of the cantilever sensors. We introduce a parameter β, the ratio of the area of the (planar) upper surface to that of the pore surfaces, to characterize the contributions of surface stress on the upper surface and that on the surface of the pores to the bending and dynamic response of the cantilever sensors. For films with aligned cylindrical pores, $\beta = (1 - p)r_0/(2ph_f)$. Numerical evaluations of Equations 11.4 and 11.21 show that the surface stress on the solid part of the upper surface (a_u and b_u) can be neglected if $\beta < 0.05$. Otherwise, a_u and b_u in Equations 11.4 and 11.21 should be replaced by $a_u^p = a_u(1 - p_s)$ and $b_u^p = b_u(1 - p_s)$, in which p_s is the porosity of the upper surface. For the aligned cylindrical pores, the porosity of the upper surface is equal to that of the porous film ($p_s = p$). Generally, the ligament size in the nanoporous films obtained by dealloying is less than 20 nm, and $\beta < 0.05$. Therefore, the surface stress on the upper surface (a_u^p and b_u^p) can be neglected when evaluating the bending and dynamic response (Equations 11.4 and 11.21) of the cantilever sensors with the disordered porous films (Figure 11.8d).

11.4.1.2 Film with Aligned Cylindrical Pores

Based on the micromechanical model with surface-stress effects (Duan et al. 2006), the effective transverse Young's modulus $\bar{E}_{Tf}$ of the porous film with aligned cylindrical pores is

$$\bar{E}_{Tf} = \frac{4\bar{\mu}_{Tf}\bar{\kappa}_{Tf}\bar{E}_{Lf}}{\bar{E}_{Lf}(\bar{\kappa}_{Tf} + \bar{\mu}_{Tf}) + 4\bar{v}_{Lf}^2\bar{\mu}_{Tf}\bar{\kappa}_{Tf}} \tag{11.48}$$

where the effective longitudinal Young's modulus $\bar{E}_{Lf}$, the transverse shear modulus $\bar{\mu}_{Tf}$, the longitudinal Poisson ratio $\bar{v}_{Lf}$, and the transverse bulk modulus $\bar{\kappa}_{Tf}$ are given in equation 10, equation 13a and b, and equation 22 in the study of Duan et al. (2006), respectively.

For the film containing the aligned cylindrical nanopores, it is found that the homogeneous stress field σ^0 ($= \sigma^0 \mathbf{I}$) that satisfies the boundary condition (Equation 11.43) on the surfaces of pores is $\sigma^0 = \mathbf{I}[\tau_0 + 2(\lambda^s + \mu^s)\varepsilon^0]/r_0$, where $\varepsilon^0 = \sigma^0/(3\kappa_f)$, r_0 is the radius of the cylindrical pores, and $\mathbf{I}$ is the second-order identity tensor in the three-dimensional space. With these relations, σ^0 and ε^0 can be obtained as

$$\sigma^0 = \frac{3\kappa_f \tau_0}{3r_0 \kappa_f - 2(\lambda^s + \mu^s)}, \quad \varepsilon^0 = \frac{\tau_0}{3r_0 \kappa_f - 2(\lambda^s + \mu^s)} \tag{11.49}$$

According to Equation 11.46, the volume-averaged stress $\bar{\sigma}$ is

$$\bar{\sigma}_{zz} = (1 - p_0)\sigma^0 + 2p\sigma^0, \quad \bar{\sigma}_{xx} = \bar{\sigma}_{yy} = \sigma_0 \tag{11.50}$$

which indicates that if $\bar{\sigma}$ is chosen as Equation 11.50, the boundary condition on the void surface is satisfied, and then uniform stress field σ^0 in the solid is ensured. The overall self-strain ε^{s*} due to the surface stress can be obtained from Equation 11.44. By Equation 11.50 and the expressions for $\bar{E}_{Lf}$, $\bar{\nu}_{Lf}$, $\bar{\kappa}_{Tf}$ in the study of Duan et al. (2006), it is found that the self-strain is hydrostatic ($\varepsilon_{xx}^{s*} = \varepsilon_{yy}^{s*} = \varepsilon_{zz}^{s*} = \varepsilon^{s*}$) and given by

$$\varepsilon^{s*} = \frac{2p\tau_0}{r_0 \mu_f \Xi}\left[(1 - \nu_f)(1 - p) + 2p\mu^* - (1 + p)\nu_f \mu^*\right] \tag{11.51}$$

where

$$\Xi = -2(1 - p)^2(1 + \nu_f + \nu_f \mu^*) + \nu_f\left[8p^2(\mu^*)^2 + \lambda^*(-1 + 6p - 5p^2 + 8p^2\mu^*)\right] \\ -(1 + 2p - 3p^2)(\lambda^* + \mu^*) - 4p\mu^*(1 + p)(\lambda^* + \mu^*) \tag{11.52}$$

$\lambda^* = \lambda^s/(r_0 \mu_f)$, $\mu^* = \mu^s/(r_0 \mu_f)$, and ν_f is the Poisson's ratio of the matrix in the film. If the effect of the surface moduli is neglected, Equation 11.51 reduces to

$$\varepsilon^{s*} = -\frac{2p(1 - \nu_f)\tau_0}{r_0(1 - p)E_f} \tag{11.53}$$

By replacing the Young's modulus E_f of the solid film with the effective Young's modulus $\bar{E}_{Tf}$ of the porous one (Equation 11.48) and substituting ε^* with $\varepsilon^* + \varepsilon^{s*}$ in Equation 11.4, in which the eigenstrain ε^* may arise from the different lattice constants between the film and the substrate, the bending curvature and resonance frequency of the cantilever with porous film containing aligned cylindrical pores can be obtained. For the cantilever with a nanoporous film, the surface moduli (λ^s and μ^s) of the pore surfaces and that of the upper porous surface b_u are of the same order because they are the surface moduli of the same material. The constant surface stress τ_0 of the pore surfaces and that of the upper porous surface a_u are of the same order for the same reason. Moreover, λ^s, μ^s, and τ_0 are different before and after adsorption, and so are a_u and b_u. After adsorption, λ^s, μ^s, and τ_0 can be expressed as $\lambda^s = \lambda_0^s + \Delta\lambda^s$, $\mu^s = \mu_0^s + \Delta\mu^s$, and $\tau_0 = \tau_0^s + \Lambda\tau^s$, where λ_0^s, μ_0^s, and τ_0^s represent surface properties with no adsorption, and $\Delta\lambda^s$, $\Delta\mu^s$, and $\Delta\tau^s$ denote the change of surface property induced by adsorption.

11.4.1.3 Disordered Porous Film

Generally, the elastic deformation of the porous materials depends on the geometrical property (e.g., microstructure) and elastic moduli of the matrix material. It is very difficult to obtain the exact elastic deformation, while considering the constant surface stress and surface elasticity simultaneously for the disordered porous film. To obtain the self-strain for films with disordered pores, we assume that the surface stress is independent of the strain (negligible excess surface elasticity). The governing equation for the surface stress in the disordered porous film is (Weissmüller and Cahn 1997)

$$V_2 \langle tr\sigma \rangle_{V_2} + A_2 \langle tr\tau_0 \rangle_{A_2} = 0 \tag{11.54}$$

where σ denotes the bulk stress tensor, V_2 and A_2 refer to the volume and total surface area of the solid phase, and the brackets $\langle \bullet \rangle_{V_2}$ and $\langle \bullet \rangle_{A_2}$ represent the averages over the volume and all the surfaces of the solid phase, respectively. By analogy with the definition of hydrostatic loading, the mean pressure $\langle P \rangle_{V_2}$ is the porous film can be obtained,

$$\langle P \rangle_{V_2} = -\frac{\langle tr\sigma \rangle_{V_2}}{3} = \frac{2}{3}\alpha\tau_0 \tag{11.55}$$

in which $\alpha\ (= A_2/V_2 = pA_2/[(1-p)V_1])$ is the specific surface area (the area per volume of the solid phase), and V_1 is the volume of the pores. Equation 11.55 can be used to calculate the mean pressure in the solid phase. In particular, for the aligned cylindrical pores, the specific surface area $a = 2p/[(1-p)r_0]$, and $\langle P \rangle_{V_2} = 4p\tau_0/[3(1-p)r_0]$.

For the nanoporous materials with the disordered structure in Figure 11.8c, the overall elastic response is isotropic (Kramer, Viswanath, and Weissmüller 2004). There exist a volume change ΔV_2 of the solid phase and a volume change ΔV_1 of the pores because of the constant surface stress τ_0. The overall self-strain is

$$\varepsilon^{s*} = \frac{\Delta V_1 + \Delta V_2}{3V} = \left(1 + \frac{\Delta V_1}{\Delta V_2}\right)\frac{(1-f)\Delta V_2}{3V_2} \tag{11.56}$$

and ΔV_2 can be deduced by the following equation:

$$\frac{\Delta V_2}{V_2} = -\frac{\langle P \rangle_{V_2}}{\kappa_f} = -\frac{2\alpha\tau_0}{3\kappa_f} \tag{11.57}$$

However, it is difficult to obtain the volume change ΔV_1 of the pores of the disordered structure. For the disordered nanoporous Pt materials ($\kappa_f = 283$ GPa, $\mu_f = 74$ GPa, and $p = 0.887$) used by Weissmüller et al. (2003), $\Delta V_1/\Delta V_2 = 2.867$ has been obtained from the experimental data.

With a negligible excess surface elasticity, the volume change of the nanoporous material with spherical voids of radius R_0 is

$$\frac{\Delta V_0}{V} = -\frac{2p\tau_0}{\kappa_f R_0}, \quad \frac{\Delta V_1}{\Delta V_2} = \frac{3\kappa_f + 4p\mu_f}{4(1-p)\mu_f} \tag{11.58}$$

$\Delta V_1/\Delta V_2$ in Equation 11.58 is a function of the porosity p and the elastic moduli of the solid phase.

One way to predict the effective elastic moduli of the bicontinuous porous media is based on the three-point approximation, which was given by Equations 4.9 and 4.10 in the article of Torquato (1998). It has been proved that the three-point approximation is a good estimate in predicting the effective moduli for a media embedded with identical overlapping spherical voids. Moreover, the effective elastic moduli of the bicontinuous porous media with different configurations can be predicted by adjusting the geometrical parameters (η_2 and ζ_2) in Equations 4.9 and 4.10 in the study of Torquato (1998). Therefore, we use the three-point approximation to predict the effective elastic modulus of the disordered porous film.

Noting that the present theory can apply to the films with pore sizes at different scales (macro-, micro-, and nanoscales), it has been proved that the effects of λ^s and μ^s diminish when the sizes of the pores increase. Thus, the theory with pore sizes at the macro- and microscales can be obtained by letting λ^s and μ^s equal to zero (Duan et al. 2006). For example, the effective elastic moduli of the porous film with the aligned cylindrical pores (with pore sizes at macro- and microscales) can be obtained from Equation 11.48 by letting $\lambda^s = \mu^s = 0$.

11.4.2 DEFORMATION OF CANTILEVERS WITH POROUS FILMS

In the following numerical calculations, the deformation of cantilevers with three types of gold film layers, namely, the solid gold film, the film with aligned cylindrical pores at the microscales, and the film with aligned cylindrical pores at the nanoscales, are calculated and compared. The elastic moduli of the solid gold film are $E_f = 90$ GPa, and $\nu_f = 0.42$, and the Young's modulus of the substrate is $E_s = 10$ GPa and the substrate thickness is $h_s = 1$ μm. The porosity of films with nano- and microscale pores is $p = 0.7$, and the radius of the nanoscale pore is 4 nm. The constant surface stress and surface modulus for the Au(111) surface are $\tau_0 = 1.64205$ N/m, $\lambda^s = -2.70738$ N/m, and $\mu^s = -2.62728$ N/m (Shenoy 2005). Assume $a_u = \tau_0$ and $b_u = \lambda^s + 2\mu^s$. The surface stress and surface modulus of the porous film on the upper planar surface are $a_u^p = (1 - p)a_u = 0.4926$ N/m and $b_u^p = (1 - p)b_u = -2.38858$ N/m. The variations of the curvatures of the cantilevers with three types of gold film layers resulting from the surface stress with respect to $\log_{10} h_f/h_s$ are plotted in Figure 11.9, where K_s, K_m, K_n^a, and K_n^0 denote the curvatures of the cantilevers with a solid film, a microporous film, and a nanoporous film with and without consideration of a_u^p and b_u^p, respectively. For the cantilevers with a microporous film, the effective transverse Young's modulus of the film (obtained from Equation 11.48) and the surface stress a_u^p are both smaller than the corresponding quantities of the solid film. The curvature of the cantilever with the microporous film (K_m) is only slightly different from that of the solid film (cf. Figure 11.9).

However, the surface stress τ makes a remarkable contribution to the static deformation of the cantilever with a nanoporous film. According to Equation 11.51, the surface stress (τ_0) and surface moduli (λ^s and μ^s) lead to an eigenstrain (self-strain, $\varepsilon^{s*} = -0.014131$) in the film layer. Meanwhile, the effective transverse Young's modulus of the nanoporous film (obtained from Equation 11.48), the surface stress a_u^p, and the surface modulus b_u^p make contributions to the static bending. With these influences, the cantilever with the nanoporous film exhibits a much larger deformation

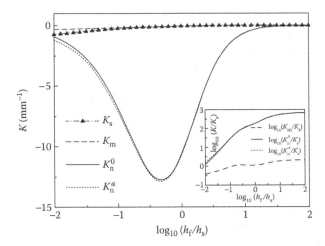

FIGURE 11.9 Variations of curvatures with $\log_{10}(h_f/h_s)$ for silicon cantilevers with gold films (K_s for solid, K_m for microporous film, K_n^a and K_n^0 for nanoporous film). (Reprinted from Duan, H. L. 2010. *Acta Mech Solida Sinica* 23:1–12. With permission.)

than those coated with the solid and the microporous films, especially when h_f/h_s are in the same order of magnitude (cf. Figure 11.9). For the nanoporous film considered here, $\beta = 8.6 \times 10^{-4} \ll 0.05$. It can be seen from Figure 11.9 that there is little difference between K_n^a and K_n^0, which is consistent with the conclusion that the surface stress a_u^p and surface modulus b_u^p on the upper planar surface can be neglected when $\beta < 0.05$, but the surface effect on the surfaces of the pores play a dominant role. The inset of Figure 11.9 shows that as the thickness of the film layer increases, the magnitudes of K_n^a and K_n^0 increase more rapidly than those of K_m and K_s. The magnitude of K_m/K_s nearly keeps constant with the increase of h_f (cf. the inset of Figure 11.9). Therefore, the nanoporous film has the largest static deformation. Additionally, the deformation of the cantilever with the porous film can be tuned by varying the densities and sizes of the pores.

Next, we compare the curvatures of the cantilevers with nanoporous films of three different kinds of geometry, that is, the nanoporous film with randomly distributed spherical pores shown in Figure 11.8b, the nanoporous film with aligned cylindrical pores shown in Figure 11.8c and that with disordered geometry shown in Figure 11.8d. Each of these cantilevers consists of a solid Au substrate cantilever made by dealloying (Kramer, Viswanath, and Weissmüller 2004). The variations of the curvatures (K_n) a as function of α (nm^{-1}) are shown in Figure 11.10 (predicted by Equation 11.4). The elastic constants of gold are $E_f = 90$ GPa and $v_f = 0.42$, and there is no eigenstrain ε^*. The constant surface stresses are taken as $a_u = a_l = \tau_0 = 1.64205$ N/m (Shenoy 2005), and the surface moduli are neglected. The porosity of the three kinds of porous film is $p = 0.6$. Since $\beta < 0.05$, the surface stress a_u^p on the upper planar surface for the films with aligned cylindrical pores and disordered geometry can be neglected. The effective Young's moduli and self-strains of the porous films are needed to predict the bending curvatures. The effective elastic moduli of the nanoporous films with the randomly distributed spherical pores (solid

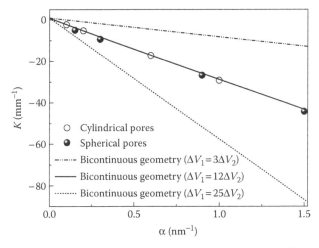

FIGURE 11.10 Variations of curvatures with α for two-layer gold cantilevers with nanoporous films. (Reprinted from Duan, H. L. 2010. *Acta Mech Solida Sinica* 23:1–12. With permission.)

circle) are predicted by Equation 11.36 in the study of Duan et al. (2005) by letting $\lambda_s = \mu_s = 0$ and those of the nanoporous film with the aligned cylindrical pores (open circle) are predicted using Equation 11.48 by letting $\lambda_s = \mu_s = 0$. For the nanoporous film containing aligned cylindrical pores, the effective transverse Young's modulus $\bar{E}_{Tf}$ is 18.11 GPa when $p = 0.6$. The effective elastic moduli of the nanoporous film with a disordered bicontinuous geometry are predicted from Equations 4.9 and 4.10 in the article of Torquato (1998). For $p = 0.6$, $\eta_2 = 0.439$, $\zeta_2 = 0.351$, and $\bar{E}_f = 15$ GPa, the self-strain ε^{s*} for the disordered nanoporous film can be obtained from Equation 11.56 for a given $\Delta V_1/\Delta V_2$. Three cases for $\Delta V_1/\Delta V_2 = 3$, 12, and 25 are plotted in Figure 11.10. As $\Delta V_1/\Delta V_2$ increases, the magnitude of curvature K increases. Figure 11.10 shows that for a specific α, the curvature of the two-layer cantilever with the nanoporous film containing spherical voids is almost the same as that of the cantilever with the nanoporous film containing aligned cylindrical pores, and they are also nearly equal to that of the cantilever with the disordered porous film with $\Delta V_1 = 12\Delta V_2$.

11.4.3 RESONANCE FREQUENCY SHIFT OF CANTILEVERS WITH POROUS FILMS

The changes of dynamic response due to mass loading are analyzed for cantilevers consisting of a substrate with a length of 80 μm, a width of 20 μm, and a thickness of 200 nm coated with a 50-nm-thick gold film. According to Equations 11.20 and 11.48, Figure 11.11 plots the first and second modes of the mass sensitivity for the silicon nitride (SiN$_x$, $E_s = 290$ GPa, $\nu_s = 0.24$, $\rho_s = 3.1$ g/cm^3) substrates with three types of thin gold films ($E_f = 90$ GPa, $\nu_f = 0.42$, $\rho_f = 19.3$ g/cm^3): the solid film, and the film with the aligned micro- and nanoscale cylindrical pores (porosity $p = 0.7$). The radius and surface moduli of the nanoscale cylindrical pores are given as $r_0 = 2.5$ nm, $\lambda^s = -2.70738$ N/m, and $\mu^s = -2.62728$ N/m, respectively (Shenoy 2005).

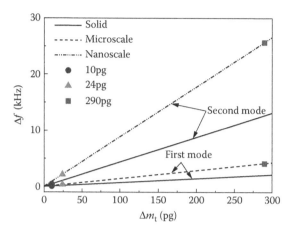

FIGURE 11.11 Variations of first and second modes of resonance frequency shift with loaded mass. (Reprinted from Duan, H. L. 2010. *Acta Mech Solida Sinica* 23:1–12. With permission.)

Since the value of $\tilde{E}_{\text{ef}}^{1/2}/\tilde{\rho}_{\text{ef}}^{3/2}$ of the porous film is larger than that of the solid film, cantilevers with micro- and nanoporous films are more sensitive than that with the solid film under the mass loading Δm_t (cf. Equation 11.20). It can be seen from Figure 11.11 that the mass sensitivities for the cantilevers with the micro- and nanoporous films are numerically indistinguishable from each other and are about twice that of the cantilever with the solid film for both the first and second modes of vibration. If the number (n_a) of adsorbates per unit area is the same for the three types of cantilevers (the mass loading adsorbed on the cantilever is $\Delta m_t = n_a S_a (1 + 2ph_f/r_0)$, where S_a is the area of the adsorbing surface), the cantilever with the nanoporous film adsorbs much more mass loading than those with the microporous and solid films. For example, if the radii of the micro- and nanopores are 50 nm and 2.5 nm, respectively, and the mass loading for solid films is 10 pg, then the mass loadings for the micro- and nanoporous films can be as high as 24 pg and 290 pg. Therefore, the induced frequency shift Δf for the cantilever with the nanoporous film is about 60 times that of the solid film, and Δf for the microporous film is about five times that of the solid film (cf. Figure 11.11).

To study the sensitivity of the surface-stress sensors, the normalized frequency shift $\Delta f_i/f_i$ of a two-layer cantilever (consisting of a gold film ($E_f = 90$ GPa, $h_f = 10$ nm) and a polymethylmethacrylat (PMMA) substrate ($E_s = 3$ GPa)) as a function of the porosity p and substrate thickness h_s is shown in Figure 11.12. Three different kinds of gold films are considered, that is, the nano- and microporous films with aligned cylindrical pores and the solid film. The adsorption-induced surface modulus is taken as $b_u = 0.5$ N/m ($b_l = 0$, $\Delta m = 0$). For the nano- and microporous films, $b_u^p = (1 - p)b_u$. The surface moduli of the nanoscale pores are given by $\lambda_0^s = -2.70738$ N/m and $\mu_0^s = -2.62728$ N/m ($\Delta\lambda^s = \Delta\mu^s = b_u/3$ and $r_0 = 2.5$ nm). It is seen from Figure 11.12 that $\Delta f_i/f_i$ of the cantilever with the nanoporous film is largest, and that of the classical solid cantilever is the lowest. This is because the effective Young's modulus of the nanoporous film is the lowest and that of the solid film is the largest. It is also shown in Figure 11.12 that the cantilevers coated with the nano- and microporous films display

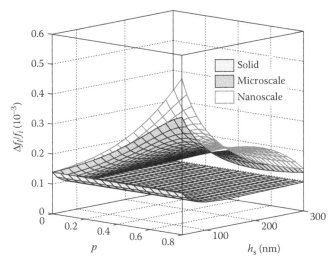

FIGURE 11.12 Variations of normalized frequency shift ($\Delta f_i/f_i$) with porosity p and substrate thickness h_s. (Reprinted from Duan, H. L. 2010. *Acta Mech Solida Sinica* 23:1–12. With permission.)

larger sensitivity enhancements than that of the solid one as the substrate thickness decreases. For instance, for $p = 0.6$, the cantilevers with the nano- and microporous films and 100-nm thick substrates display sensitivity enhancements up to 3.2-fold and 2.8-fold, respectively, compared with the cantilevers with the solid film. However, for a 300-nm thick substrate, the sensitivity enhancements are about 1.9-fold and 1.8-fold for the cantilevers with the nano- and microporous films, respectively.

Design of novel surface-enhanced cantilever sensors consisting of nano- and microporous films is presented, and the sensitivities of these cantilevers are compared with that of the classical solid one. It is shown that the sensitivities, the static deformation, and resonance frequencies of these novel cantilever sensors can be tuned by the density, size, and structure of the porous films. By comparing three kinds of cantilever consisting of the solid film, and the films with aligned cylindrical pores at the microscale and nanoscale, it is shown that if the effect of the surface stress is considered only, the nanoporous one has the highest static and dynamic sensitivities, and the solid one has the lowest. If only the effect of mass is considered, the dynamic sensitivities of the cantilevers with pores at the microscale and nanoscale are larger than that of the solid one.

11.5   SURFACE STRESS ON ROUGH SURFACES

Experimental studies of the surface stress of solids typically work with surfaces that are not perfectly planar. The experiment then probes an effectively averaged surface stress. The evolution of the surface morphology, for instance during film growth or reconstruction, is also affected by the surface stress acting on a corrugated surface. Based on the work of Wang et al. (2010b), we give the mechanics theory of the rough surfaces in a continuum framework.

11.5.1 THEORY OF EFFECTIVE SURFACE STRESS

Consider a solid body of finite-size and denote $\hat{B}$ and $\hat{S}$ as its bulk and surface, respectively (Figure 11.13). As the derivative of a surface free energy density with respect to the strain tangential to the surface, the surface stress $\hat{s}$ is a tangential superficial tensor. The orientation and magnitude of $\hat{s}$ may vary along the surface. The generalized capillary equation for solids (Weissmüller and Cahn 1997) relates $\hat{s}$ to the bulk stress $\hat{T}_b$ at equilibrium through

$$0 = \int_{\hat{B}} \hat{T}_b d\hat{V} + \int_{\hat{S}} \hat{s} d\hat{A} \tag{11.59}$$

We allow for variation of the surface orientation on two distinctly different scales. Corrugation or roughness is present at the microscopic scale, which is much smaller than the scale at which the macroscopic surface orientation varies. We may thus introduce the notion of a "smooth" surface, S, that is identical to the actual surface but lacks the roughness of $\hat{S}$. The smooth surface S coincides with $\hat{S}$ in a coarse-grained picture that does not resolve the roughness. The position of S is chosen so that the body B, which is enclosed by S, has the same volume as the actual body $\hat{B}$.

Here and in the following, the quantities with a hat, for instance, the local outer surface normal $\hat{n}(\hat{x})$ and the projection tensor $\hat{P}(\hat{x})$ and $\hat{N}(\hat{x})$, are defined in the actual body $\hat{B}$ or on its rough surface $\hat{S}$, whereas the quantities without a hat such as $n(x)$, $P(x)$, and $N(x)$ refer to B and to its smooth surface S. Here, $\hat{P}(\hat{x})$ and $P(x)$ are "superficial tensors" (Gurtin and Murdoch 1975; Gurtin, Weissmüller, and Larche 1998). We take the position coordinate, x, to refer to positions on S. Our study is restricted to surfaces S that can be described by a stepwise differentiable height function, $h(x)$, as indicated in Figure 11.13. This construction excludes, in particular, surfaces with reentrant angles. It also implies that the position $\hat{x}$ on $\hat{S}$ may be labeled by the corresponding x on S.

We investigate the effect of the roughness on the effective surface stress, s, which needs to be attributed to S in order to obtain the coarse-grained picture that removes the distinction between $\hat{B}$ and B—the same bulk stress in both bodies. In particular, the stress T_b in B then satisfies

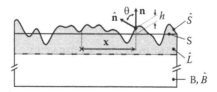

FIGURE 11.13 Sketch of a solid body $\hat{B}$ with a rough surface $\hat{S}$ (normal vector $\hat{n}(x)$, profile of the rough surface $h(x)$); its equivalent body is B with planar surface S (normal $n(x)$). (Reprinted from Wang, Y., J. Weissmüller, and H. L. Duan. 2010b. *J Mech Phys Solids* 58:1552–6. With permission.)

$$0 = \int_B \mathbf{T}_b dV + \int_S \mathbf{s} dA \tag{11.60}$$

where $\mathbf{T}_b = \hat{\mathbf{T}}_b$ at a point that is far enough from the surface when the influence of the microscopic details is negligible.

We denote $\hat{L}$ as a layer near the surface that is sufficiently thick to contain all the relevant differences between $\hat{\mathbf{T}}_b$ and $\mathbf{T}_b$ and at the same time thin enough so that within $\hat{L}$ the value of $\mathbf{T}_b$ can be approximated as independent of the distance from the surface. At any position on S, the limit of $\mathbf{T}_b$ at S is denoted by $\mathbf{T}_S$. In terms of this concept, combining Equations 11.59 and 11.60 suggests that the surface stresses are related by

$$\int_S \mathbf{s} dA = \int_{\hat{S}} \hat{\mathbf{s}} dA + \int_{\hat{L}} (\mathbf{T}_b - \mathbf{T}_S) d\hat{V} \tag{11.61}$$

The computation of the effective stress starts out with exploiting the assumption that the curvature of the macroscopic surface is negligible. We may therefore approximate $\mathbf{T}_S$ as a tangential stress. In terms of $\mathbf{T}_s \cdot \mathbf{n} = div_S \mathbf{s}$ (cf. Gurtin and Murdoch 1975), for a planar surface, we have

$$\mathbf{T}_S \cdot \mathbf{n} = 0 \tag{11.62}$$

Multiplying Equation 11.61 by $\mathbf{n}$ and noting Equation 11.62 and $\mathbf{s} \cdot \mathbf{n} = 0$, we get

$$\int_{\hat{L}} \hat{\mathbf{T}}_b \cdot \mathbf{n} d\hat{V} = -\int_{\hat{S}} \hat{\mathbf{s}} \cdot \mathbf{n} d\hat{A} \tag{11.63}$$

This equation illustrates that the bulk stress in the layer $\hat{L}$ near $\hat{S}$ has a component which is normal to S. This normal component is required to equilibrate the out-of-plane component (i.e., normal to S) of the surface stress $\hat{\mathbf{s}}$ and $\mathbf{s}$.

Next, we shall inspect the stress and strain in the tangent plane of S. Assume the linear elastic response throughout $\hat{L}$ so that, using Hooke's law, Equation 11.61 can be rewritten as

$$\int_S \mathbf{s} dA = \int_{\hat{S}} \hat{\mathbf{s}} d\hat{A} + \int_{\hat{L}} \mathbf{C} : (\hat{\mathbf{E}} - \mathbf{E}_S) dV \tag{11.64}$$

Here, $\mathbf{C}$ denotes the stiffness tensor in the bulk, and $\hat{\mathbf{E}}$ and $\mathbf{E}_S$ are the strains conjugate to $\hat{\mathbf{T}}_b$ and $\mathbf{T}_S$, respectively.

Consistent with our earlier assumption on uniformity of $\mathbf{T}_b$ near the surface, we consider only such situations where the tangential strain in $\hat{L}$ is uniform along the thickness and satisfies

$$\mathbf{P} \cdot \hat{\mathbf{E}} \cdot \mathbf{P} = \mathbf{P} \cdot \mathbf{E}_S \cdot \mathbf{P} \text{ in } \hat{L} \tag{11.65}$$

Equation 11.65 embodies the condition that the layer remains coherent with the underlying bulk. An important consequence of Equation 11.65 is that the strain difference, $\mathbf{E} - \mathbf{E}_S$, which occurs in Equation 11.64, represents a *normal* strain. We emphasize that Equation 11.65 is the most restrictive simplifying assumption here: Lateral relaxation may be significant in surface features with a large aspect ratio, and the present theory will then become inaccurate. By aspect ratio we mean the proportion of height to lateral extension of the surface features.

The tangential part of Equation 11.64 is

$$\int_S s \, dA = \int_{\hat{S}} \mathbf{P} \cdot \hat{\mathbf{s}} \cdot P d\hat{A} + \int_{\hat{L}} \mathbf{P} \cdot \left[\mathbf{C} : (\hat{\mathbf{E}} - \mathbf{E}_S) \right] \cdot P d\hat{V} \tag{11.66}$$

Equations 11.63 and 11.66 embody the central physics of our argument. From Equation 11.66, we know that the effective surface stress is composed of two terms that are both tangential to the local macroscopic surface. The first term on the right-hand side of Equation 11.66 represents the tangential projection—onto the macroscopic surface plane—of the surface stress $\hat{\mathbf{s}}$ acting at the corrugated surface. The second term on the right-hand side of Equation 11.66 represents the transverse elastic response to the stress component which is normal to S—as determined by Equation 11.63—in the bulk region underneath the surface. As a consequence of our assumption of coherency between layer and underlying bulk, we had found that the strain $\hat{\mathbf{E}} - \mathbf{E}_S$ is normal to S (cf. Equation 11.65). By multiplying this normal strain with the stiffness tensor and taking the tangential (relative to S) part of the resulting stress, we extract the tangential stress by which the material near the surface reacts to the normal component of $\hat{\mathbf{s}}$. Equilibrium requires that the normal components of the bulk and surface stresses in the layer cancel each other, see Equation 11.63. For a given magnitude of $\hat{\mathbf{s}}$, it is seen that the first stress component in Equation 11.66 depends only on the geometry of the surface, whereas the second depends also on the elastic properties of the bulk solid.

In some instances, it may be convenient to change the reference frame and to convert to integration over S. To this end, we introduce the parameter F with the property $F = \delta\hat{A}/\delta A$, accounting for the fact that the actual surface area, $\delta\hat{A}$, associated with a segment of the layer with the macroscopic area δA is increased because of its inclination. For any quantity $Y(\mathbf{x})$, one then has

$$\int_{\hat{S}} Y(\hat{\mathbf{x}}) d\hat{A} = \int_S Y(\mathbf{x}) F(\mathbf{x}) dA \tag{11.67}$$

Let us denote $h(\mathbf{x})$ (see Figure 11.13) as the height of the surface profile at position $\mathbf{x}$ on S, and $\theta(\mathbf{x})$ as the angle included by the outer normal, $\mathbf{n}(\mathbf{x})$, of S at $\mathbf{x}$ and the normal, $\hat{\mathbf{n}}(\mathbf{x})$, of the corresponding segment of $\hat{S}$. Then, the normal vectors are related by

$$F(\mathbf{x})\hat{\mathbf{n}}(\mathbf{x}) = \mathbf{n} - \nabla h(\mathbf{x}) \tag{11.68}$$

where

$$F = \sqrt{1 + |\nabla h|^2} = \sec\theta \tag{11.69}$$

and

$$|\nabla h|^2 = \tan\theta \tag{11.70}$$

With Equations 11.67 through 11.69, any integration over the rough surface $\hat{S}$ can be readily converted into one over the planar surface S.

Since the surface stress is a second-rank tensor, its local value is isotropic in the plane of $\hat{S}$ if the local segment of surface has at least threefold crystallographic symmetry. This applies in particular to the (111)- and (100)-oriented surfaces of cubic crystals. If we assume that $\hat{\mathbf{s}}$ is isotropic everywhere on $\hat{S}$, that is

$$\hat{\mathbf{s}} = \hat{f}\hat{\mathbf{P}} \tag{11.71}$$

with $\hat{f}$ being a scalar surface stress, and if we assume isotropic elasticity then—according to Equations 11.66 and 11.70—the apparent local value of $\mathbf{s}$ is

$$\mathbf{s} = \hat{f}F\left[\mathbf{P}\cdot\hat{\mathbf{P}}\cdot\mathbf{P} - \frac{v}{1-v}\mathbf{P}(\mathbf{n}\cdot\hat{\mathbf{P}}\cdot\mathbf{n})\right] \tag{11.72}$$

This expression is readily evaluated in terms of the geometry relations mentioned in this section . Expressed within an orthonormal set of coordinates with the x- and y-axis in the plane, the result is

$$s_{xx} = \frac{\hat{f}}{\sqrt{1+|\nabla h|^2}}\left(1 + h_h^2 - \frac{v}{1-v}|\nabla h|^2\right)$$

$$s_{yy} = \frac{\hat{f}}{\sqrt{1+|\nabla h|^2}}\left(1 + h_x^2 - \frac{v}{1-v}|\nabla h|^2\right) \tag{11.73}$$

$$s_{xy} = \frac{\hat{f}}{\sqrt{1+|\nabla h|^2}}h_x h_y$$

where h_x and h_y refer to the x- and y-component of ∇h.

Because $|\nabla h|^2 = \tan^2\theta$, Equation 11.73 implies that the local value of the effective scalar surface stress, $f = (tr\mathbf{s})/2$, on S depends on the surface morphology only through the angle of inclination. The value of f is given by

$$f = \frac{1}{2}\hat{f}\left(\frac{1+v}{1-v}\cos\theta + \frac{1-3v}{1-v}\sec\theta\right) \tag{11.74}$$

which is the result of Weissmüller and Duan (2008).

11.5.2 Impact of Surface Geometry

In many problems of film growth and surface stress measurement, the experimentalist disposes of detailed information on the surface geometry. For instance, a statistical description of the roughness or the shape of islands, as well as the crystallographic orientation of their facets may be available. By contrast, the interaction of the capillary forces at the corrugated surface with the underlying bulk is typically measured only in terms of a laterally averaged tangent stress. This implies that area-averages over the local projected surface-stress values $\mathbf{s}$ are, here, the relevant parameters for comparison to experiment.

In general, the area-averages can be obtained from our theory by integration over the area, considering the variation in surface orientation and in the magnitude and anisotropy of the surface-stress tensor. However, if we restrict our attention to isotropic surface stress on $\hat{S}$ mentioned in Section 11.5.1, and if we consider S a *planar*

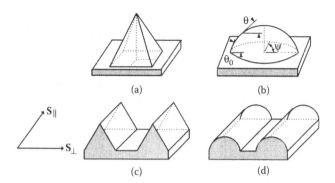

FIGURE 11.14 Examples for morphologies of corrugated surfaces: (a) pyramidal islands; (b) dome-shaped islands; (c) acute ridges; and (d) round ridges. On surfaces with an in-plane isotropic elastic response, (a) and (b) in the top row give isotropic surface stress, whereas (c) and (d) in the bottom row give surface stresses that can be enhanced in the direction along the ridges (surface stress component $s_{\parallel}$) or normal to those features ($s_{\perp}$), depending on the magnitude of the transverse elastic coupling. (Reprinted from Wang, Y., J. Weissmüller, and H. L. Duan. 2010b. *J Mech Phys Solids* 58:1552–6. With permission.)

surface, then the effective surface-stress values depend on the position exclusively through the orientation of the surface normal, $\hat{\mathbf{n}}$. The integration in position space can then be converted into one in orientation space. If the orientation of $\hat{\mathbf{n}}$ is parameterized by the inclination (or colatitude) θ and by the longitude ψ (see Figure 11.14), the effective surface stress will be

$$\langle \mathbf{s} \rangle_S = \frac{1}{A} \int_S \mathbf{s}(\mathbf{x}) dA = \frac{1}{A} \int_0^{\pi/2} \int_0^{2\pi} \mathbf{s}(\theta, \psi) G(\theta, \psi) \cos \theta \sin \theta d\psi d\theta \qquad (11.75)$$

with A being the area of S and with $\mathbf{s}$ given in Equation 11.72. The function $G(\theta, \psi)$ is a surface orientation distribution function defined so that the area of all segments of $\hat{S}$ with orientation between (θ, ψ) and $(\theta + \delta\theta, \psi + \delta\psi)$ is given by $\delta\hat{A} = G(\theta, \psi) \sin \theta \delta\psi\delta\theta$. The cosine in Equation 11.75 accounts for the projection of $\delta\hat{A}$ onto S.

Equation 11.75 takes on a particularly convenient form when one is only interested in the scalar surface stress, f, on a planar surface S or when the corrugation is one-dimensional as mentioned earlier in this section. The relevant orientation variable is then simply the inclination angle θ. The orientation distribution function $g(\theta)$ can be defined in θ-space, $d\hat{A} = g(\theta)d\theta$, and we have

$$\langle f \rangle_S = \frac{1}{A} \int_0^{2\pi} f(\theta) g(\theta) \cos \theta d\theta \qquad (11.76)$$

where f is given by Equation 11.74.

Equations 11.75 and 11.76 embody a nontrivial aspect of our theory: The coupling of the surface stress from a corrugated surface into a planar surface substrate depends on the geometry of the corrugation exclusively through the surface orientation distribution function. Additional details, such as the height of features or the spacing between them, do not affect the effectively averaged interaction of the surface with the underlying bulk once the orientation distribution function is given.

For the purpose of illustration, let us apply the theory to the surface morphologies displayed in Figure 11.14, namely, arrays of pyramidal- and dome-shaped (more precisely, spherical cap) islands, as well as the acute and round (with the cross-section of a circular arc) parallel ridges. In experiments such features typically cover the surface only partly; this is accounted for by the coverage parameter, α, which is the fraction of S covered by the islands or ridges. Computing the effective surface stress requires Equation 11.76 along with the relevant orientation distribution functions. The latter are readily obtained as follows:

- For pyramidal islands,

$$g(\theta) = A\left[(1-\alpha)\delta(\theta) + \alpha \sec \theta_0 \delta(\theta - \theta_0)\right] \tag{11.77}$$

- For dome-shaped islands,

$$g(\theta) = A\left[(1-\alpha)\delta(\theta) + 2\alpha \csc^2 \theta_0 H(\theta - \theta_0) \sin \theta\right] \tag{11.78}$$

- For acute ridges,

$$g(\theta) = A\left[(1-\alpha)\delta(\theta) + \alpha \sec \theta_0 \delta(\theta - \theta_0)\right] \tag{11.79}$$

- For round ridges,

$$g(\theta) = A\left[(1-\alpha)\delta(\theta) + \alpha \csc \theta_0 H(\theta - \theta_0)\right] \tag{11.80}$$

Here, $\delta(\beta)$ represents the Dirac delta function, which has the value zero everywhere except at $\beta = 0$, where its value is infinite. $H(\beta)$ is the Heaviside step function, $H = 0$ for $\beta < 0$ and $H = 1$ for $\beta \geq 0$. For the acute features, θ_0 denotes the inclination angle, whereas for the rounded features, θ_0 refers to the "wetting angle," that is, the value of θ at the junction with the substrate. Using the above results along with Equation 11.76, we have inspected special cases designed to exemplify the role of the key parameters, such as surface inclination, anisotropy of the corrugation, anisotropy of the elastic response, and the strength of transverse coupling.

Figure 11.15 is designed to illustrate the impact of anisotropic elastic response on the effective surface stress. The figure was computed for pyramidal islands, assuming isotropic $\hat{s}$. Here, the coupling is stronger along the [001] in-plane direction than that along [1 $\overline{1}$0]. Figure 11.15 illustrates that the anisotropic elastic response induces an anisotropic effective surface stress, even for the isotropic surface geometry.

Figure 11.16 depicts an aspect of the impact of the surface morphology on the effective surface stress. The graphs compare the magnitude f of **s** for surfaces covered, on the one hand, by facetted features (islands or ridges) with a single value of θ and, on the other hand, for surfaces with rounded features (see Figure 11.14). As in Equations 11.78 and 11.80, the shape of the rounded features is parameterized by the

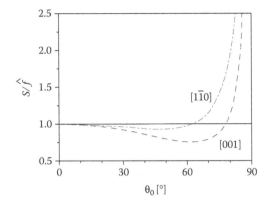

FIGURE 11.15 Principal components of normalized effective surface stress, **s**, for a surface with isotropic local surface stress, $\hat{f}$, isotropic roughness (pyramidal islands, $\alpha = 1$), and anisotropic elastic response. The transverse coupling is stronger along the [001] in-plane direction than along $\left[1\bar{1}0\right]$. Graphs labeled [001] and $\left[1\bar{1}0\right]$ show projected values of **s** in the respective directions. (Reprinted from Wang, Y., J. Weissmüller, and H. L. Duan. 2010b. *J Mech Phys Solids* 58:1552–6. With permission.)

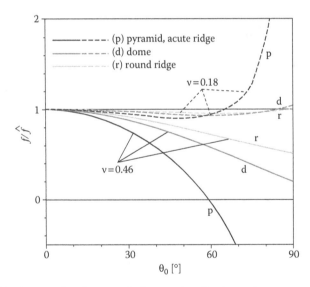

FIGURE 11.16 Impact of surface morphology on the effective surface stress, **s**. Normalized magnitude, $f/\hat{f}$, of **s** for surfaces with isotropic local surface stress, $\hat{f}$, and isotropic elastic response. The surface is entirely covered ($\alpha = 1$) by facetted (pyramidal islands or acute ridges) or round (dome-shaped islands or round ridges) features. Inclination of facets or wetting angle at the base of the features is designated by θ_0. Solid and dashed lines refer to different Poisson's ratios ($v = 0.46$ for solid lines and $v = 0.18$ for dashed lines). Facetted ridges and islands have identical effective surface stress in spite of their different symmetry. Rounding diminishes the effect of corrugation on the effective surface stress. (Reprinted from Wang, Y., J. Weissmüller, and H. L. Duan. 2010b. *J Mech Phys Solids* 58:1552–6. With permission.)

wetting angle at the base of the feature. On average, the inclination is then less than for the facetted features. As a consequence, the effect of rounding will reduce the change in f; this is readily observed by inspection of Figure 11.16.

Although the question of rounded or facetted features has only qualitative consequences for the effective surface stress, the dimensionality of the corrugation has more interesting ramifications. Figure 11.17 shows the principal components of **s** for surfaces with isotropic local surface stress, $\hat{f}$, and isotropic elastic response. Under these conditions, **s** is isotropic for the "island" geometries inspected above. Yet, the effective surface stress is seen to be strongly anisotropic for one-dimensional corrugation, such as the acute ridges that are represented by the data in Figure 11.17. This effect is pronounced even when there is no transverse coupling, as the graphs for $v = 0$ in Figure 11.17 illustrate. In fact, the anisotropy in the example arises entirely from the projection of the surface stress onto S: the component of $\hat{s}$ that is *normal* to the ridges is inclined relative to the macroscopic surface plane S; the projection onto S therefore reduces the surface stress in that direction, and the effect overcompensates the increase in surface area because of the inclination. The net reduction is seen in Figure 11.17a. By contrast, the component of $\hat{s}$ that is parallel to the ridges (namely, $s_{\parallel}$) is also parallel to S. The projection onto S does not reduce the magnitude of this component, and the increase in surface area because of the inclination therefore dominates the θ-dependent of $s_{\perp}$ (Figure 11.17b). The treatment of 1D roughness by Spaepen (2000) neglected $s_{\parallel}$ and therefore overestimated the drop in f with inclination angle.

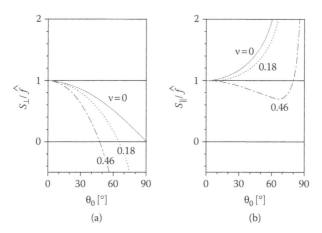

FIGURE 11.17 Principal components of the normalized effective surface stress, s, for a surface with isotropic local surface stress, $\hat{f}$, and isotropic elastic response. The surface is entirely covered ($\alpha = 1$) by acute ridges with inclination angle θ. (a) and (b) show effective surface stress components in two orthogonal directions, $s_{\perp}$ is normal to the ridges and $s_{\parallel}$ is along the ridges. Transverse coupling strengths are parameterized by Poisson's ratio, v, as indicated in the graph. Values refer to no transverse coupling ($v = 0$) and to the surfaces Si(111) ($v = 0.18$) and Au(100) ($v = 0.46$). Note that $s_{\parallel}$ increases with θ_0 while $s_{\perp}$ decreases and can even invert its sign. (Reprinted from Wang, Y., J. Weissmüller, and H. L. Duan. 2010b. *J Mech Phys Solids* 58:1552–6. With permission.)

11.6 SURFACE-ENHANCED CANTILEVER SENSORS WITH ROUGH SURFACES

Based on the works of Weissmüller and Duan (2008), and Duan et al. (2009), we study the effect of surface roughness on the deformation and the resonance frequency of microcantilever sensors. The analysis demonstrates that surface roughness can enhance, decrease, or even annul the effect of surface stress on the curvature and resonance frequency, depending on the surface inclination angle and the Poisson's ratio of the coating film on the cantilever.

11.6.1 CANTILEVER BENDING WITH ROUGH SURFACES

The curvature of the cantilever, K ($K = 1/R$, cf. Figure 11.18a), in response to an effective biaxial stress of magnitude s_{eff}, acting on tangentially in layer, is

$$K = -\frac{6(1-v_s)s_{eff}h_f}{E_s H^2} \tag{11.81}$$

Motivated by experiment, we take $Kh_f \ll 1$ (experimental values can be as small as 10^{-10}), so that the bending strain within the layer can be taken as radially uniform.

For use in Equation 11.81 we have in particular

$$s_{eff} = \frac{E_f}{1-v_f}\varepsilon_f + T \tag{11.82}$$

where ε_f denotes the strain in the film, measured relative to a stress-free reference state, and T is the effective in-plane stress independent of the strain. In general, measurements of substrate bending alone are not sufficient to separate the two contributions to s_{eff} and to isolate T. However, ε_f can be determined separately from diffraction-based lattice parameter data (Spaepen 2000). We focus on processes that change the surface stress only, while maintaining a coherent substrate-layer interface. Substrate bending is then controlled by changes in T alone, and we restrict our attention to this quantity.

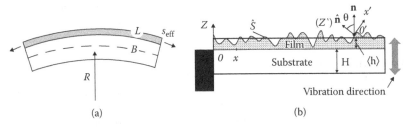

FIGURE 11.18 (a) Illustration of substrate bending with radius $R = 1/\kappa$ in response to stress s_{eff} in L. (Reprinted from Weissmüller, J., and H. L. Duan. 2008. *Phys Rev Lett* 101:146102. With permission.) (b) Vibration of cantilever. (Reprinted from Duan H. L., Y. H. Xue, and X. Yi. 2009. *Acta Mech Solida Sinica* 22:550–4. With permission.)

For isotropic and uniform surface stress, according to Equation 11.74, the sensitivity ratio, T/T_0, of a cantilever subject to changes in its surface stress is given by

$$\frac{T}{T_0} = \frac{1}{2}\left\langle \frac{1+v_f}{1-v_f}\cos\theta + \frac{1-3v_f}{1-\mu_f}\sec\theta \right\rangle \tag{11.83}$$

where T_0 $(= \hat{f}/h_f)$ is the effective stress for a planar surface. The Taylor expansion of Equation 11.82 is $T/T_0 = 1 - A_2<\theta^2> - A_4<\theta^4> + O(<\theta^6>)$, where $A_2 = v_f/(1 - v_f)$ and $A_4 = (7v_f - 3)/24[(1 - v_f)]$. For a sufficiently small θ, one then has

$$\frac{T}{T_0} \approx 1 - \frac{v_f}{1-v_f}\langle\theta^2\rangle \tag{11.84}$$

The "roughness factor," ρ, is the ratio of total surface area of the layer over its projected area, $\rho = \hat{A}/A$. For a small roughness, $\rho \approx 1 + (1/2)<\theta^2>_S$. Standard data analysis software in atomic force microscopy (AFM) evaluates ρ, connecting the mean-square inclination angle in Equation 11.84 to experimentally accessible parameters. As an example, Figure 11.19 shows an AFM image of $h(\mathbf{r})$ for a 50-nm-thick gold film that was used in a wafer-bending study of the surface stress change, during potential cycling in electrolyte. Its analysis yields $\rho = 1.061$. Thus, the root-mean-square inclination angle $\langle\theta^2\rangle_A^{1/2} = 19°$ and, since $-v_f/(1 - v_f)<\theta^2> = -0.10$ (based on $v_f = 0.44$ for gold), Stoney's equation in its conventional form will here underestimate the true mean value of the surface stress by 10%.

It is noted that Ergincan, Palasantzas, and Kooi (2010) have pointed out that the misorientation has a natural link to the statistical description of surface roughness,

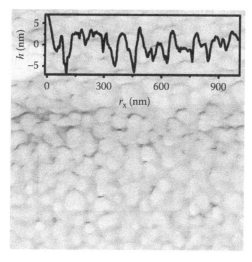

FIGURE 11.19 Atomic force microscope image of a 50-nm-thick gold film on Si(100). Image area is 1 μm × 1 μm. Inset shows film thickness h_f versus position x for a representative horizontal cross-section. (Data from Weissmüller, J., and H. L. Duan. 2008. *Phys Rev Lett* 101:146102.)

which connects to surface stress measurement. They introduces a root-mean-square local slope, ρ_{rms}, which satisfies $\rho_{rms}^2 \approx \langle \theta^2 \rangle$ in the limit of "weak roughness," $\langle \theta^2 \rangle \ll 1$, and can be expressed as a function of w, ξ, and H. Wang et al. (2010a) have shown that Equation 11.84 makes a good approximation to Equation 11.83 in the limit of "weak roughness" and that there may be a significant deviation between the exact and approximate results when $\theta \geq 45°$.

Next, we will show that the cantilever sensitivity is significantly dependent on the topology of the surface when larger inclination angles are admitted. As an example, we consider a surface covered completely by pyramidal hillocks with faces of identical $\hat{f}$ and identical inclination angle θ, as in Figure 11.20. A similar geometry can be deliberately produced with etching. Since $\hat{f}$ and θ are uniform, Equation 11.83 can here be evaluated without averaging. As can be seen in Figure 11.20, the geometry here has a decisive effect on the sensitivity of the cantilever: At small θ, the effective in-plane stress, T, in layers with a rough surface is always diminished as compared with that for a planar surface with the same $\hat{f}$, T_0. However, as θ increases, T can be enhanced if $\nu_f < 1/3$, as for the example of Si. By contrast, when $\nu_f > 1/3$ as for Cu and Au, the value of T is reduced until it vanishes at a finite θ. At this value, the cantilever is entirely insensitive to surface stress. When θ is further increased, the cantilever response will even change sign. Note that for very acute pyramids (i.e., for large aspect ratio), our assumption of a uniform bending strain ceases to approximate the elastic response near the tip. Thus, Equation 11.83 will not accurately predict T for pyramids with θ near 90°. Experiments investigating surface features with more moderate aspect ratio, for which the uniform bending strain approximation is appropriate, find only slight reduction in sensitivity for artificially structured Si(100) surfaces, and a much more pronounced reduction for rough Au. This trend agrees with the predicted dependence on the material as displayed in Figure 11.20.

The roughness has an important effect on the cantilever sensitivity. Although the stress state in a thin film is typically taken to be planar, the leading correction term arises from out-of-plane surface-stress components, which couple into the in-plane

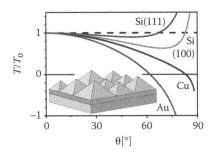

FIGURE 11.20 Cantilever sensitivity, T/T_0, for a rough surface entirely covered by pyramidal hillocks (see illustration in inset) versus inclination angle θ of the pyramid faces. Poisson's ratios for the graphs are $\nu_f = 0.18$ [Si(111)], 0.28 [Si(100)], 0.34 (Cu), and 0.44 (Au). (Data from Weissmüller, J., and H. L. Duan. 2008. *Phys Rev Lett* 101:146102.)

stress through transverse elastic coupling. For nearly planar surfaces with small roughness, the surface corrugation affects the sensitivity through the mean square of the inclination angle, θ. Our example suggests a significant correction to Stoney's equation even when the surface was carefully prepared for planarity. Because $<\theta^2>_S$ is readily measured, the impact of roughness may be assessed and the correction performed routinely in experiments. This would seem all the more desirable because cantilever bending studies sustain the major part of the current experimental data base for surface stress. In other instances, it may be desirable to deliberately intro- duce large corrugation in the cantilever. In Si, this may enhance the sensitivity, a finding that may be beneficial for the emerging applications in biochemical sensing. By contrast, the sensitivity can be decreased all the way to zero or even inverted sign for corrugated surfaces of metals with large Poisson's ratio. Therefore, the impact of surface roughness on the cantilever response requires a correction in the case of nominally planar surfaces, and it may provide a tool for intentionally tuning the cantilever sensitivity.

Consider a cantilever consisting of two layers: the substrate of thickness H and the film (upper layer) of mean thickness $<h> \ll H$, as shown in Figure 11.18b. Here, $<\bullet>$ denotes the spatial average over all positions and directions. It is assumed that the two layers are well bonded. The global coordinate system is defined such that the interface of the film and substrate is the xy-plane, and the z-axis is upward. The origin of the global coordinate system is at the fixed end. The profile (height) of the rough surface $h(\mathbf{r})$ is a continuous and differentiable function of the coordinates x and y. The local coordinate system is defined at each point on the rough surface such that the z'-axis is parallel to, and the $x'y'$-plane is perpendicular to, the outer normal ($\hat{\mathbf{n}}$), as shown in Figure 11.18.

Surface stress τ exists on the rough surface of the film. For isotropic surfaces, τ is expressed by Equation 11.42. Note that $\tau^s (= 2 \mu^s \varepsilon^s + \lambda^s (\mathrm{tr} \varepsilon^s) \mathbf{I})$ is the strain-dependent surface stress. Since $<h> \ll H$, we assume that the midplane is at $z = -H/2$. Compared with the bending caused by the change of the surface stress, the overall stretching of the cantilever can be neglected, namely, it can be assumed that $c_0 = 0$ in Equation 11.2. Then from Equation 11.2, the strain components are given by

$$\varepsilon_{xx} = -\left(z + \frac{H}{2}\right)\frac{\partial^2 w}{\partial x^2}, \quad \varepsilon_{yy} = 0 \qquad (11.85)$$

Based on Hooke's law, Equation 11.85, and the plane stress condition $\sigma_{zz} = 0$, we have $\sigma_{xx} = \hat{E}\varepsilon_{xx}$, where $\hat{E} = E/(1-v^2)$. The surface strain ε^s can be calculated by means of coordinate transformation from the global coordinate system to the local inclined one at $z = h(\mathbf{r})$ (Figure 11.18b). Then the surface stress τ^s can be obtained.

The kinetic energy, U_T, of the cantilever is

$$U_T = \frac{1}{2}\int_V \rho\left[\left(\frac{\partial u}{\partial t}\right)^2 + \left(\frac{\partial w}{\partial t}\right)^2\right]dV \qquad (11.86)$$

where ρ is density, V is the volume of the cantilever, t is time, and u, $v = 0$, and w are the displacements in the x-, y-, and z-directions, respectively. Here the rotational kinetic energy is ignored. The elastic energy, U_E, composed of the bulk energy and surface energy, is given by

$$U_E = \frac{1}{2} \int_V \sigma_{ij} \varepsilon_{ij} dV + \int_{\hat{S}} \tau_0 \varepsilon_{ii}^s d\hat{S} + \frac{1}{2} \int_{\hat{S}} \tau_{ij}^s \varepsilon_{ij}^s d\hat{S} \tag{11.87}$$

where σ_{ij} and ε_{ij} ($i, j = x, y, z$) are the bulk stress and strain components, τ_{ij}^s and ε_{ij}^s indicate the surface stress and strain components on $\hat{S}$, and $\hat{S}$ denotes the rough surface of the film. Introducing the Lagrangian function, $L = U_T - U_E$ into Hamiltonian equation $\delta \int_0^t L dt = 0$ leads to the vibration equation for the cantilever

$$\frac{\hat{E}^* H^3}{12} \frac{\partial^4 w}{\partial x^4} + \rho_s H \frac{\partial^2 w}{\partial t^2} = 0 \tag{11.88}$$

where

$$\hat{E}^* = \left[1 + \frac{3b_u}{\hat{E}_s H (1+g^2)^{3/2}} \left(1 - \frac{\nu_f}{1-\nu_f} g^2 \right)^2 \right] \hat{E}_s \tag{11.89}$$

Because $\eta = h/H \ll 1$, we neglect the term η and its higher order terms in Equation 11.88. Moreover, we assume that the derivates of $(\nabla h)^2$ with respect to x are very small. Therefore, we replace $(\nabla h)^2$ with its mean value g^2 where $g \left(= \sqrt{\langle (\nabla h)^2 \rangle} \right)$ is the average surface slope. As shown in Equation 11.88, the constant surface stress τ_0 does not influence the resonance frequency, but only affects the equilibrium position of vibration.

The solution to Equation 11.88 gives the fundamental resonance frequency $\tilde{f}$ of the microcantilever with the influence of the surface stress and surface roughness. The frequency shift $\Delta f_r = \tilde{f} - f_0$ is

$$\frac{\Delta f_r}{f_0} \approx \frac{3b_u}{2\hat{E}_s H (1+g^2)^{3/2}} \left(1 - \frac{\nu_f}{1-\nu_f} g^2 \right)^2 \tag{11.90}$$

where f_0 denotes the classical fundamental resonance frequency of the microcantilever without the surface stress and surface roughness. Equation 11.90 shows that $\Delta f_r/f_0$ is a function of the thickness and elastic modulus of the substrate, the surface moduli, the Poisson's ratio of the film ν_f, and the average surface slope g. If there is no roughness ($g = 0$), Equation 11.90 reduces to

$$\frac{\Delta f_s}{f_0} \approx \frac{3b_u}{2\hat{E}_s H} \tag{11.91}$$

where Δf_s is the resonance frequency shift only due to effect of surface stress. From Equations 11.90 and 11.91, we obtain

$$\frac{\Delta f_r}{\Delta f_s} \approx (1+g^2)^{-3/2} \left(1 - \frac{\nu_f}{1-\nu_f} g^2 \right)^2 \tag{11.92}$$

In the case of slight roughness ($g \ll 1$), Equation 11.92 shows that the surface roughness weakens the influence of the surface stress on the frequency shift ($\Delta f_r/\Delta f_s < 1$). If Δf_s for a flat surface is larger than Δf_r, then the roughness is always slight, which can be treated as a perturbed state of the flat surface. For strong roughness, $\Delta f_r/\Delta f_s$, depending on v_f and g_f, can be bigger, smaller than 1, or even equal to 0 when $g = \sqrt{(1-v_f)/v_f}$, which means the surface roughness can enhance, decrease, or annul the effect of surface stress on the resonance frequency.

In order to estimate the frequency shift caused by the surface topology, we consider a silicon substrate completely covered by pyramidal hillocks with an identical inclination angle, θ. Figure 11.21 shows the cantilever sensitivity $\Delta f_r/f_0$ as a function of θ for different Poisson's ratios $v_f = 0.44$ (Au), 0.34 (Cu), 0.28 [Si(100)], and 0.18 [Si(111)]. Because θ is uniform, we have $g = \tan\theta$. The parameters of the substrate are $H = 1$ μm, $E_s = 130$ GPa, and $v_s = 0.28$. The surface moduli are chosen as the ones of gold film $\lambda^s = -2.70738$ N/m and $\mu^s = -2.62728$ N/m. As θ increases approximately to $45°$ for the case of $v_f = 0.44$, Δf_r vanishes, which means that changing the surface roughness could completely eliminate the surface stress. When θ becomes large (about $75°$ for the case of $v_f = 0.44$, strong roughness), the surface roughness starts to enhance the effect of surface stress on the resonance frequency shift. For strong roughness, the frequency shift will be enhanced as θ increases.

It is found that the resonance frequency of microcantilevers is affected by the surface roughness, besides the conventional factors such as the surface and bulk elasticity of the material. The effect of the surface roughness on the curvature and the resonance frequency depend on the profile of the surface roughness including the wavelength, amplitude, the inclination angle, and the Poisson's ratio; the surface roughness can enhance, decreases, or even annul the effect of the surface stress on the curvature and the resonance frequency. As real surfaces often have various roughness, this effect may have profound implications for the design of analysis of the sensitivities of microcantilever sensors when they operate in the static and dynamic modes.

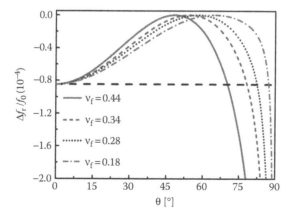

FIGURE 11.21 Cantilever sensitivity $\Delta f_r/f_0$ for a rough surface covered by pyramidal hillocks versus the inclination angle, θ. (Reprinted from Duan H. L., Y. H. Xue, and X. Yi. 2009. *Acta Mech Solida Sinica* 22:550–4. With permission.)

11.7 CONCLUSION

Surface stress is widely used to characterize the effects that the adsorption and surface morphology have on the mechanical response from nanomaterials and nanodevices. We first established the connections between the surface stress at the continuum level and the adsorbate interactions (e.g., vdW interaction and Coulomb interaction) at the molecular level. Then, we proposed two strategies that use porous films and rough surfaces to improve the sensitivity of microcantilever sensors. We also analyzed the bending and resonance frequency shift, which are caused by the eigenstrain, the surface stress, and the adsorbed mass, on these novel microcantilever sensors. Meanwhile, the relationship between the surface stress and the surface corrugation was established. We showed that both rough surfaces with optimized topology and nanoporous films can dramatically enhance the sensitivity of these novel microcantilever sensors.

REFERENCES

Berger, R., E. Delamarche, H. P. Lang, C. Gerber, J. K. Gimzewski, E. Meyer, and H.-J. Guntherodt. 1997. Surface stress in the self-assembly of alkanethiols on gold. *Science* 276:2021–4.

Chakarova-Kack, S. D., E. Schroder, B. I. Lundqvist, D. C. Langreth. 2006. Application of van der Waals density functional to an extended system: Adsorption of benzene and naphthalene on graphite. *Phys Rev Lett* 96:146107.

Chen, G. Y., T. Thundat, E. A. Wachter, and R. J. Warmack. 1995. Adsorption-induced surface stress and its effects on resonance frequency of microcantilevers. *J Appl Phys* 77:3618–22.

Chun, D. W., K. S. Hwang, K. Eom, J. H. Lee, B. H. Cha, W. Y. Lee, D. S. Yoon, and T. S. Kim. 2007. Detection of the Au thin-layer in the Hz per picogram regime based on the microcantilevers. *Sens Actuat A* 135:857–62.

Dareing, D. W., and T. Thundat. 2005. Simulation of adsorption-induced stress of a microcantilever sensor. *J Appl Phys* 97:043526.

Dorignac, J., A. Kalinowski, S. Erramilli, and P. Mohanty. 2006. Dynamical response of nanomechanical oscillators in immiscible viscous fluid for in vitro biomolecular recognition. *Phys Rev Lett* 96:186105.

Duan, H. L., J. Wang, Z. P. Huang, and B. L. Karihaloo. 2005. Size-dependent effective elastic constants of solids containing nano-inhomogenieties with interface stress. *J Mech Phys Solids* 53:1574–96.

Duan, H. L., J. Wang, B. L. Karihaloo, and Z. P. Huang. 2006. Nanoporous materials can be made stiffer than non-porous counterparts by surface modification. *Acta Mater* 54:2983–90.

Duan, H. L., Y. H. Xue, and X. Yi. 2009. Vibration of cantilevers with rough surfaces. *Acta Mech Solida Sinica* 22:550–4.

Duan, H. L. 2010. Surface-enhanced cantilever sensors with nano-porous films. *Acta Mech Solida Sinica* 23:1–12.

Ergincan, O., G. Palasantzas, and B. J. Kooi. 2010. Influence of random roughness on cantilever curvature sensitivity. *Appl Phys Lett* 96:041912.

Godin, M. 2004. Surface stress, kinetics, and structure of alkanethiol self-assembled monolayer. Ph.D. Thesis, McGill University, Canada.

Gronbeck, H., A. Curioni, and W. Andreoni. 2000. Thiols and disulfides on the Au(111) surface: The headgroup-gold interaction. *J Am Chem Soc* 122:3839–42.

Gurtin, M. E., X. Markenscoff, and R. N. Thurston. 1976. Effect of surface stress on the natural frequency of thin crystals. *Appl Phys Lett* 29:529–30.

Gurtin, M. E., and A. I. Murdoch. 1975. A continuum theory of elastic material surfaces. *Arch Ration Mech Anal* 57:291–323.

Gurtin, M. E., J. Weissmüller, and F. Larche. 1998. A general theory of curved deformable interfaces in solid at equilibrium. *Philos Mag A* 78:1093–109.

He, L. H. 2006. Self-strain of solids with spherical nanovoids. *Appl Phys Lett* 88:151909.

Illic, B., D. Czaplewski, H. G. Craighead, P. Neuzil, C. Campagnolo, and C. Batt. 2000. Mechanical resonant immunospecific biological detector. *Appl Phys Lett* 77:450–2.

Kassner, M. E. et al. 2005. New directions in mechanics. *Mech Mater* 37:231–59.

Kramer, D., R. N. Viswanath, and J. Weissmüller. 2004. Surface stress-induced macroscopic bending of nanoporous gold cantilevers. *Nano Lett* 4:793–6.

Lu, P., H. P. Lee, C. Lu, and S. J. O'Shea. 2005. Surface stress effects on the resonance properties of cantilever sensors. *Phys Rev B* 72:085405.

McFarland, A. W., M. A. Poggi, M. J. Doyle, L. A. Bottomley, and J. S. Colton. 2005. Influence of surface stress on the resonance behavior of microcantilevers. *Appl Phys Lett* 87:053505.

Nuzzo, R. G., B. R. Zegarski, and L. H. Dubois. 1987. Fundamental studies of the chemisorption of organosulfur compounds on Au(111). Implications for molecular self-assembly on gold surfaces. *J Am Chem Soc* 109:733–40.

Sharma, P., S. Ganti, and N. Bhate. 2003. Effect of surfaces on the size-dependent elastic state of nano-inhomogenieties. *Appl Phys Lett* 82:535–7.

Shenoy, V. B. 2005. Atomistic calculations of elastic properties of metallic fcc crystal surfaces. *Phys Rev B* 71:094104.

Sony, P., P. Puschnig, D. Nabok, and C. Ambrosch-Draxl. 2007. Importance of van der Waals interaction for organic molecule-metal junctions: Adsorption of thiophene on Cu(110) as a prototype. *Phys Rev Lett* 99:176401.

Spaepen, F. 2000. Interfaces and stresses in thin films. *Acta Mater* 48:31–42.

Stoney, G. G. 1909. The tension of metallic films deposited by electrolysis. *Proc Roy Soc A* 82:172–5.

Timoshenko, S., D. H. Young, and W. Weaver. *Vibration Problems in Engineering*, fourth ed. New York: John Wiley & Sons.

Torquato, S. 1998. Effective stiffness tensor of composite media: II. Applications to isotropic dispersions. *J Mech Phys Solids* 46:1411–40.

Wang, Y., J. Weissmüller, and H. L. Duan. 2010a. Comment on "Influence of random roughness on cantilever curvature sensitivity." *Appl Phys Lett* 96:226101.

Wang, Y., J. Weissmüller, and H. L. Duan. 2010b. Mechanics of corrugated surfaces. *J Mech Phys Solids* 58:1552–6.

Weissmüller, J., and J. W. Cahn. 1997. Mean stresses in microstructures due to interface stresses: A generalization of capillary equation for solids. *Acta Mater* 45:1899–906.

Weissmüller, J., and H. L. Duan. 2008. Cantilever bending with rough surfaces. *Phys Rev Lett* 101:146102.

Weissmüller, J., R. N. Viswanath, D. Kramer, P. Zimmer, R. Wurschum, and H. Gleitler. 2003. Charge-induced reversible strain in a metal. *Science* 300:312–5.

Wu, G., H. Ji, K. Hansen, T. Thundat, R. Datar, R. Cote, M. F. Hagan, A. K. Chakraborty, and A. Majumdar. 2001. Origin of nanomechanical cantilever motion generated from biomolecular interactions. *Proc Natl Acad Sci USA* 98:1560–4.

Yi, X., and H. L. Duan. 2009. Surface stress induced by interactions of adsorbates and its effects on deformation and frequency of microcantilever sensors. *J Mech Phys Solids* 57:1254–66.

Yourdshahyan, Y., and A. M. Rappe. 2002. Structure and energetics of alkanethiol adsorption on the Au(111) surface. *J Chem Phys* 117:825–33.

12 Nanoscale Adhesion Interactions in 1D and 2D Nanostructure-Based Material Systems

Changhong Ke and Meng Zheng

CONTENTS

12.1 INTRODUCTION

Recent advances in one-dimensional (1D) and two-dimensional (2D) nanostructures have fueled the development of nanotechnology in a number of science and engineering disciplines. For many of their applications, 1D and 2D nanostructures are either assembled into micro- or macroscale structures (e.g., thin films and yarns), or integrated with other bulk materials to form heterogeneous material systems and devices (e.g., nanocomposites and solid-state electronics). As a result of their exceptionally large surface-to-volume ratios, the interfaces formed between nanostructures themselves and between nanostructures and other material surfaces play important roles in the functionality and performance of nanostructure-based material systems. Characterization of the interfacial interactions in 1D and 2D nanostructure-based material systems is a critical step to the complete understanding of the nanoscale interface and to the tuning of their design and manufacturing for optimal functionality and performance. For instance, for 1D nanostructure-based bistable nanoswitches [1–4], the adhesion strength between the nanostructure and the substrate is of importance not only in defining the stable states of the device, but also in controlling the switching processes between the stable states. For nanostructure-reinforced polymer composites, the interfacial strength between the nanostructure and the polymer matrix determines the load-transfer efficiency between the nanostructure and the bulk polymer material, and it has a profound influence on the bulk mechanical properties (e.g., Young's modulus and yield strength) of the polymer nanocomposite. Interfacial failure may significantly compromise the reinforcement effect such that the potential of high-strength nanostructures as reinforcing additives may not be fully realized. In this chapter, we will present an overview about the nanoscale interfacial interactions in 1D and 2D nanostructure-based material systems and the recent theoretical and experimental advances in characterizing the nanoscale adhesion interactions in 1D and 2D nanostructures. This chapter is organized as follows: in the first part, we review the structures, properties, and applications of 1D and 2D nanostructures; in the second part, we discuss the nanoscale adhesion interactions in 1D and 2D nanostructures; and in the third part, we review the theoretical and experimental techniques and research advances of characterizing three major types of interfacial interactions, including binding interactions between nanostructures, adhesion interactions between nanostructures and flat surfaces, and interfacial bonding interactions between nanostructures and supporting polymer matrices.

12.2 OVERVIEW OF 1D AND 2D NANOSTRUCTURES

1D and 2D nanostructures are mostly in the forms of nanotubes and nanowires (including nanorods) and nanosheets, respectively, which are macromolecules composed of single or multiple chemical elements and are less than 100 nm in lateral or thickness dimension.

Carbon nanotubes (CNTs) [5] are the most exciting and researched nanotube structures because of their exceptional material properties and application prospects. Boron nitride nanotubes (BNNTs) are another type of promising 1D nanostructure and have recently received increasing attention from the research community [6,7]. From the structural perspective, CNTs and BNNTs are highly similar and are formed by rolling a graphene sheet and hexagonal boron nitride (BN) nanosheet, respectively. The CNT is composed of seamless covalent C–C hexagonal networks, while the BNNT is composed of repeated and partially ionic B–N bonding networks. The rolling direction of the graphene or the BN nanosheet is denoted as the chirality vector (n, m). The indexes n and m of the chirality vector denote the number of unit vectors along two directions in the honeycomb lattice of graphene or the BN nanosheet. The nanotubes are called "zig-zag" if $m = 0$, and called "arm-chair" if $m = n$. Both CNTs and BNNTs can crystallize in single-walled and multiwalled nanophases. The interlayer distances are about 0.35 nm for CNTs and 0.33 nm for BNNTs, respectively. Both single-walled CNTs (SWCNTs) and multiwalled CNTs (MWCNTs) can be synthesized using arc discharge [8,9], laser ablation [10], and chemical vapor deposition (CVD) [11] methods. It is noted that the CVD method is capable of producing directionally aligned CNTs by applying external electrical fields [12]. Several methods have been proposed to synthesize BNNTs, such as catalyst-based CVD methods [13,14], laser heating methods [15–18], and recently reported catalyst-free pressurized vapor/condenser methods [19]. Both CNTs and BNNTs possess many extraordinary material properties and hold great potential for a number of applications. For instance, both CNTs and BNNTs have excellent mechanical properties, thermal conductivity, and chemical stability [20]. The elastic modulus of individual CNTs is found to be ~1 TPa [21], whereas an elastic modulus for BNNTs as high as 1.3 TPa [22] has been reported. The reported thermal conductivity of SWCNTs is nearly 3500 $W \cdot m^{-1} \cdot K^{-1}$ [23], whereas the thermal conductivity of single-walled BNNTs was predicted to be more than 3000 $W \cdot m^{-1} \cdot K^{-1}$ [24]. However, CNTs and BNNTs possess distinct electrical properties. SWCNTs are either metallic or semiconductive depending on the tube chirality vector (n, m). If n-m is an integer of 3, the SWCNT is metallic. Otherwise, the SWCNT is semiconductive. MWCNTs typically show metallic behaviors. In contrast, BNNTs possess a large bandgap (~5–6 eV) [7,25–27] that is independent of their chirality vectors, and are considered as excellent electrical insulators. BNNTs are very resistant to oxidation at high temperature [28] and inert to almost all harsh chemicals [29]. CNTs hold tremendous potential for a variety of applications, such as composites, electronics, sensors, and biomedicines [30–33], whereas BNNTs are promising for a number of applications, such as composites in which BNNTs are mechanical and/or thermal reinforcing additives [34], protective shields/capsules [35], and electrical insulators. Compared with the extensive studies and literature on the material properties and applications of CNTs, the research on BNNTs is still in its very early stage, which is mainly ascribed to difficulties in the synthesis of high-quality BNNTs [20].

Nanowires are 1D nanostructures of solid cross-sections. The materials of nanowires include silicon [36–40], gold [41,42], silver [43–45], platinum [46], germanium [39,47–50], and zinc oxide (ZnO) [51,52], just to name a few. The electrical properties of nanowires can usually be controlled in a predictable manner either during synthesis or by postsynthesis treatments, whereas only very limited success has been achieved for CNTs. In contrast to nanotubes, nanowires usually do not exhibit the same degree of mechanical flexibility, which affects their integration and reliability for many of their applications, such as nanowire-based nanoscale electromechanical systems (NEMS).

The family of 2D nanostructures contains only a few members, such as graphene and BN nanosheets. Graphene is a monolayer of carbon atoms arranged in a honeycomb network and can be considered as an unrolled SWCNT [53–57]. Following its debut in 2004 [58], graphene has received a great deal of attention because of its exceptional mechanical, electrical, and thermal properties. The Young's modulus and strength of single-layer graphene sheets are reported to be ~1 TPa and 130 GPa, respectively [59]. Studies have shown that graphene exhibits thermal conductivity in the range of 3080–5150 $W \cdot m^{-1} \cdot K^{-1}$ [60,61], and possesses an exceptionally high specific surface area of 2630 m^2/g [62]. Due to its extraordinary material properties, graphene holds promise for a number of potential applications, including field-effect transistors [24,63–65], electromechanical resonators [66], solar cells [4,67–70], ultracapacitors [71,72], polymer composites [73–79], and biomedical devices [80–82]. The reported production methods of graphene sheets in the literature include bottom-up synthesis, CVD growth, and graphite exfoliation methods [83–89]. Because many applications of graphene are based on monolayer graphene sheets, production of monolayer graphene has been intensively studied and several synthesis methods have been demonstrated. Exfoliating graphite by micromechanical cleavage or chemical intercalation methods is capable of producing monolayer graphene sheets [53,55,90–100]. It is reported that monolayer graphene can be synthesized by the deposition of ethylene on the nickel surface through controlling temperature and pressure in the CVD method [83,84]. Production of graphene on SiC and ruthenium surfaces was also demonstrated [85,86]. It is noted that monolayer graphene can be obtained from bulk graphite through chemical exfoliation. Bulk graphite can be first intercalated to graphite oxide in acids, and then exfoliated to graphene oxide sheets with the aid of ultrasonication or surfactant [55,93,94,101,102]. The produced graphene oxide can be reduced to graphene with treatments of hydrazine, and graphene nanosheets can be stabilized in the aqueous solution with the aid of surfactants [54,70,103]. To recover graphene conductivity, the chemical groups on the graphene surface can be easily removed by thermal annealing [54,67]. Recently, it was demonstrated that monolayer or few-layer graphene sheets can be produced by unzipping MWCNTs using plasma etching [104] or oxidation [105] methods. It is noted that the mechanical cleavage and chemical exfoliation approaches can also be used to produce BN nanosheets from bulk hexagonal BN materials. BN nanosheets have good mechanical properties and are useful for polymer composite applications by acting as mechanical and thermal reinforcing additives [34]. Compared with graphene, the properties and applications of BN nanosheets are little explored, which is mainly because of the limited availability of this material.

12.3 OVERVIEW OF NANOSCALE ADHESION INTERACTIONS

The strength of nanoscale adhesion interactions can be ascribed to two fundamental types of interface phenomena: (1) the physical adsorption based on nonbonded *van der Waals* interactions and (2) the chemical adsorption based on chemical bonding interactions. For pristine and defect-free CNT and graphene structures that are made of chemically stable sp^2 covalent C–C bonding networks, the van der Waals interaction-based physical adsorption accounts for the adhesion interactions between nanostructures themselves (e.g., nanotube–nanotube and graphene–graphene) and their adhesion interactions with other material surfaces (e.g., flat substrate surfaces or polymer matrices). It is noted that the hybridized sp^3 bond, which is chemically more active than the sp^2 bond, may be formed in the structure of CNTs either by chemical treatments [106] or by introducing significant local mechanical strains [107,108]. The existence of sp^3 bonds leads to the formation of chemical bonds on the interface between CNTs and other material surfaces (e.g., gold), resulting in substantially higher adhesion strength than the van der Waals interaction-based adsorption [109]. For nanotube- and nanosheet-based polymer composites, the interface between nanostructures and polymer matrices is critical to their bulk mechanical properties, and thus their functionalities and performance. Efficient load transfer requires adequate interfacial strength. Therefore, the van der Waals interaction-based interface between nanostructures and polymer matrices is usually far from satisfying the interface strength requirement for the full utilization of the high-strength properties of nanostructures. To enhance the interface strength, the surface of nanotubes or nanosheets needs to be chemically functionalized by introducing additional functional groups. The surface functionalization of nanostructures increases the interface strength by enhancing the nonbonded van der Waals interaction-based adsorption interaction and, more importantly, forms high-strength chemical bonds between nanostructures and polymer matrices.

As a result of their high-aspect-ratio characteristics, the surface-to-volume ratios for nanotubes and nanowires are extremely high. Therefore, the surface effect has a significant impact on their properties and applications. For instance, as-grown CNTs are typically in the form of bundles, rather than individual tubes, which is because of the strong van der Waals interaction among neighboring tubes in the bundle. The elastic modulus and yield strength of bundled CNTs are found to be much lower than those of individual tubes because the van der Waals interaction between neighboring tubes is much weaker than the covalent C–C bond [110–112]. This phenomenon also occurs for BNNTs. It is reported that BNNTs are likely to possess higher surface adhesion energy compared with CNTs as a result of the strong electrostatic interaction from the permanent dipole interactions between boron and nitrogen atoms [113]. Therefore, the interfacial strength may play an even more critical role in the properties and applications of BNNTs, compared with CNTs. It is noted that the mechanical properties of MWCNTs [114] and bundled SWCNTs [115] can be significantly increased when the nanotubes are exposed to moderate electron beam irradiation. This mechanical strength enhancement is attributed to the electron beam–induced cross-links between neighboring graphite layers, which significantly enhances adhesion interactions in the nanotube structures.

The interfacial interaction between individual CNTs or BNNTs and polymer matrices is important to the mechanical properties of the nanotube-reinforced polymer composites. In particular, studies on the polymerization dynamics in the presence of CNTs [116–118] and CNT functionalization [119] suggest that these high-strength nanostructures are likely much more than passive contributors to the mechanical enhancement of polymers. It is very likely that nanotubes are not only just mixing into the polymer, but also initiating/participating in reactions that change completely the polymer in the neighborhood of nanotubes. In addition, studies have shown that adhesion also plays important roles in the catalyst-based nanostructure synthesis [120–122], in which the adhesion between the catalyst particle and the nanostructure needs to be strong enough to support the nucleation-based nanostructure growth.

Because of the important roles of aforementioned interfacial interactions in the synthesis, properties, and applications of 1D and 2D nanostructure-based material systems, these nanoscale interfacial phenomena have been under intensive studies during the past decades by using a variety of theoretical and experimental techniques. The theoretical modeling techniques, either based on classical continuum mechanics (CM) or atomistic-level molecular dynamics (MD) and quantum mechanics, usually predict the interfacial interaction based on nanostructures of either perfect structural configurations (i.e., defect-free) or predefined defects, such as Stone–Wales transformation. Stone–Wales transformation is a crystallographic defect as a result of the chemical bond rotation, and may either preexist or be generated during the nanotube and nanosheet deformation process [123]. Theoretical modeling studies are capable of providing useful insights into the nature of the interfacial adhesions and their roles in the properties and applications of 1D and 2D nanostructures. However, because the interfacial interaction is greatly affected by many factors such as inevitable and hard-to-quantify surface imperfections (e.g., defects and molecule contamination) [64,124–126], experimental characterization of interfacial interactions is of great importance not only to the verification of the accuracy and reliability of theoretical predictions, but also to the development of future advanced prediction models. However, because of their nanoscale dimensions, it is technically difficult to manipulate and perform quantitative characterization of the interfacial interaction for 1D and 2D nanostructures, which demands precise positioning of the nanostructure and concurrent measurements of both the applied load and the mechanical deformation of the nanostructure with adequate resolutions. In Section 12.4, we review the recent advances in both the theoretical and experimental studies of the nanoscale adhesion interactions. Our review focuses on adhesion interactions in CNTs and graphene, whereas adhesion interactions in BNNTs and nanowires are also discussed.

12.4 THEORETICAL AND EXPERIMENTAL STUDIES OF NANOSCALE ADHESION INTERACTIONS

12.4.1 BINDING INTERACTIONS BETWEEN NANOSTRUCTURES

For one pair of parallel SWCNTs, their binding interaction is based on the van der Waals interaction among the carbon atoms on both nanotubes, and can be calculated using the Lennard-Jones (LJ) potential [127]. LJ potential defines the interaction between two atoms as $u(r) = \dfrac{A}{r^{12}} - \dfrac{B}{r^{6}}$, in which r is the atom distance, and A and B are material constants (for carbon–carbon interactions, $A = 15.2$ eV·Å^6 and $B = 24.1$ keV·Å^{12}) [110]. It is noted that the r^{-6} term represents the attractive interaction, whereas the r^{-12} term represents the repulsive interaction. The equilibrium distance between two atoms, r_0, which corresponds to the minimum van der Waals energy, is given by $r_0 = \left(\dfrac{2A}{B} \right)^{\frac{1}{6}}$. By using a continuum model based on the LJ potential, the per-unit-length van der Waals energy between two identical, parallel, and undeformed SWCNTs, whose cross-sections are illustrated by the left drawing in Figure 12.1a, is given by [110]

$$W(r) = \frac{3\pi n_\sigma^2}{8R^3} \left(-A I_A + \frac{21B}{32R^6} I_B \right) \tag{12.1}$$

in which R is the radius of the nanotube, r is the perpendicular distance between tube centers, $n_\sigma = 38/\text{nm}^2$ is the graphene surface density of carbon atoms, and I_A and I_B are two double integrals and are given by

$$I_A = \int_0^{2\pi} \int_0^{2\pi} \left[(\cos\theta_2 - \cos\theta_1)^2 + (\sin\theta_2 - \sin\theta_1 + r/R)^2 \right]^{-5/2} d\theta_1 d\theta_2 \tag{12.2}$$

$$I_B = \int_0^{2\pi} \int_0^{2\pi} \left[(\cos\theta_2 - \cos\theta_1)^2 + (\sin\theta_2 - \sin\theta_1 + r/R)^2 \right]^{-11/2} d\theta_1 d\theta_2 \tag{12.3}$$

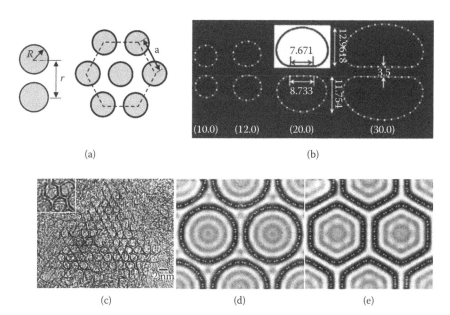

FIGURE 12.1 (a) Schematics of the cross-sections of (left) one pair of contacting single-walled carbon nanotubes and (right) one single-walled carbon nanotube (SWCNT) bundle in a hexagonal assembly configuration. (b) Deformed cross-section conformations of a series of one pair of $(n, 0)$ carbon nanotubes. No flat contacting interface for $n < 12$. (From Tang, T., A. Jagota, and C. Y. Hui. 2005. *J Appl Phys* 97:074304. With permission.). (c) High-resolution transmission electron microscopy image of a lattice of polygonized SWCNTs. (From Lopez, C. M. J., A. Rubio, J. A. Alonso, L. C. Qin, and S. Iijima. 2001. *Phys Rev Lett* 86:3056–9. With permission.) (d) and (e) Simulated transmission electron microscope images of a lattice of (12, 12) SWCNTs with circular and rounded hexagonal cross-sections. (From Lopez, M. J., A. Rubio, J. A. Alonso, L. C. Qin, and S. Iijima. 2001. *Phys Rev Lett* 86:3056–9. With permission.)

Studies have shown that only nanotubes of small radius may preserve their circular cross-sections under the intertube binding interaction. By using a nonlinear elastic model, Tang and coworkers [128] reported that the cross-section of the nanotube remains intact if the nanotube radius $R \leq R_{cr} = \sqrt{\dfrac{Et^3}{12W(1-v^2)}}$, in which t, E, and v are the effective wall thickness, Young's modulus, and Poisson's ratio of the nanotube, respectively. For SWCNTs, E ranges from 3.859 to 5.5 TPa, t ranges from 0.066 to 0.089 nm, and $v \approx 0.19$ [129]. For nanotubes of $R > R_{cr}$, a flat contact interface is formed as a result of the tube–tube adhesion interaction. The lateral deformation of one pair of identical nanotubes was demonstrated by MD simulations based on the LJ potential, as shown in Figure 12.1b. It is shown that both (10, 0) and (12, 0) nanotubes are able to preserve their circular cross-sections when binding to another identical tube, while for (20, 0) and (30, 0) tubes, the binding interfaces are formed by two flat segments on the cross-sections of both tubes.

Studies have shown that directly synthesized SWCNTs tend to self-organize into crystalline bundles of 2D triangular lattices on their cross-sections [9,10], which are illustrated by the right drawing of Figure 12.1a. Assuming that the nanotubes in the bundle remain undeformed, the equilibrium lattice constant is estimated to be $a = 2r + 0.313$ nm based on Equation 12.1, which gives $a = 1.67$ nm for a bundle of (10,10) tubes that have a tube radius of 0.6785 nm [110]. For bundles consisting of tubes of large diameters, the intertube van der Waals force may induce significant tube deformation in the radial direction, resulting in the formation of facets. Experimental results obtained by high-resolution transmission electron microscopy [130], as shown in Figure 12.1c, clearly capture the radial deformation of the tubes in the bundle, which have a diameter of 1.7 nm. Two images in Figure 12.1d and e show two possible cross-section configurations of a bundle of (12, 12) tubes (1.63 nm in diameter) based on MD simulations. It is noted that the lattice of the rounded hexagonal tube cross-sections correlates well with the experimental image shown in Figure 12.1c.

In addition to the radial deformation of the nanotube, the longitude deformation of the nanotube may be greatly affected by the adhesion interaction between the contact surfaces of different nanotubes or different segments of the same nanotube. The MD simulation results shown in Figure 12.2a reveal the formation of the racket-shape structures through the self-binding of different portions of a high aspect-ratio CNT [131]. The intratube association and dissociation can be initiated by tuning the ambient temperature. Figure 12.2b shows a Y junction formed by two partially bound double-walled CNTs (DWCNTs) [132]. The separation of these two tubes at the junction is considered as a result of a balanced equilibrium between the tube–tube binding interaction and the bending deformations of the tubes. Therefore, the binding energy between two joined tubes at the junction point can be estimated based on the bending stiffness of the tube (EI) and the separation angle 2θ. By using a continuum model as illustrated by the lower drawing in Figure 12.2b that considers the nanotubes as Euler beams, the binding energy per unit tube length is given by [132]

$$\gamma = \frac{4EI}{l^2}\left(\theta - 3\frac{\Delta}{l}\right)^2 \tag{12.4}$$

where l is the separation length of the tube and Δ is one-half of the opening width. For the DWCNTs shown in Figure 12.2b that have an estimated bending stiffness $EI = 1.07 \times 10^{-23}$ N·m^2, γ is calculated to be 0.36 nJ/m, which is close to the value obtained from atomistic modeling, 0.32 nJ/m [132].

It is noted that the binding interaction involving structural deformations can effectively associate the binding interaction with the mechanical properties of the nanostructure. Therefore, it is possible to estimate the mechanical properties of the nanostructure, such as Young's modulus, from the deformation configuration and the binding interaction. Figure 12.2c illustrates the adhesion-driven buckling of CNTs, in which one SWCNT bundle buckles on another free-standing SWCNT bundle purely based on the adhesion interaction on their binding interfaces. The adhesion-driven buckling of nanotubes was demonstrated by the transmission electron microscope (TEM) image shown in Figure 12.2d. Imaging analysis shows that the deformation curvature of the buckled nanotube bundle is identical to that of the first buckling mode of a fixed–fixed column, meaning that the deformation curvature

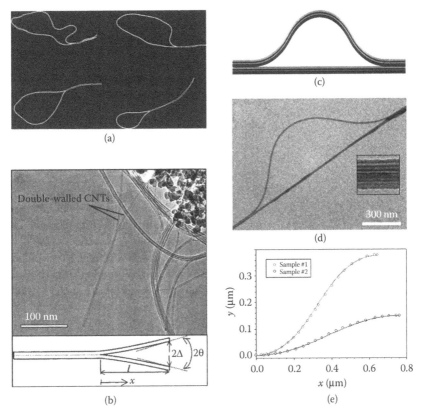

FIGURE 12.2 (a) Self-folding of carbon nanotubes (CNTs) with aspect ratio around 500 into a tennis racket-like shape because of the van der Waals interaction between different portions of the same nanotube. (Reprinted from Buehler, M. J., Y. Kong, H. Gao, and Y. Huang. 2006. *J Eng Mater Technol* 128:3–10. With permission.) (b) (upper) Transmission electron microscope (TEM) image of double-walled CNTs grown by chemical vapor deposition methods; (lower) a double-beam model for two joined CNTs. (Reprinted from Chen, B., M. Gao, J. M. Zuo, S. Qu, B. Liu, and Y. Huang. 2003. *Appl Phys Lett* 83:3570–1. With permission.) (c) Schematic of the adhesion-driven buckling of one single-walled CNT (SWCNT) bundle partially bound to another straight SWCNT bundle; (d) TEM image of a buckled SWCNT bundle that is partially bound on another free-standing straight SWCNT bundle (the inset shows tubes in the buckled bundle); (e) the comparison between the experimental measurements (circle curves) and the theoretical predictions (solid curves) on the deformation curvatures for two different buckled nanotube samples (only one-half of the deformation curvatures are plotted due to symmetry). The gray curves represent the buckled nanotube bundle shown in (d). (Reprinted from Ke, C.-H., M. Zheng, and I.-T. B. G.-W. Zhou. 2010. *J Appl Phys* 107:104305. With permission.)

of the buckled nanotube bundle is completely symmetric with respect to its central line and that the deformation curvature of one-half of the buckled bundle is completely antisymmetric with respect to its inflection point. Based on a nonlinear continuum model that assumes the SWCNT bundle as an inextensible elastica rod, the per-unit-length binding energy between two nanotube bundles at the binding interface is given by $\gamma = \dfrac{\beta^2 I}{2} E$, in which I is the moment of the inertia of the buckled nanotube bundle, and β is a factor that is purely related to the deformation curvature of the buckled nanotube. By assuming an ideal binding interaction between nanotubes on the interfaces between two bundles and the undeformed cross-section of all the tubes in the bundles, the elastic modulus of the buckled nanotube bundle can be estimated. For the buckled nanotube bundle shown in Figure 12.2d, its Young's modulus is estimated to be 140 GPa if the tubes in the bundle are assumed to form a rectangular cross-section, and 212 GPa if the tubes in the bundles are assumed to form a semicircular cross-section. Both estimated values for the Young's modulus are substantially lower than the reported values of individual SWCNTs.

It is noted that, for the experimental studies shown in Figure 12.2b through 12.e, the strength of the binding interaction between nanotubes can be only estimated at one single location (i.e., the separation or delamination point) because of their static equilibrium configuration nature. Recently, our research group proposed a nanomechanical peeling scheme to study the binding interaction between two nanotube bundles at various binding locations, which is illustrated in Figure 12.3a and is exemplified by the scanning electron microcopy (SEM) snapshot shown in Figure 12.4a. The peeling measurements are performed on two partially bound nanotube bundles, including one fixed–fixed, free-standing bundle. The floating end of the other bundle is attached to the tip of a nanomanipulator probe, which is mounted to a 3D piezo-driven nanomanipulator stage. The nanomanipulator stage is incorporated with a closed-loop, feedback-controlled mechanism and is capable of moving in the X–Y–Z directions at a resolution of 1 nm. By controlling the displacement of the probe, the grabbed nanotube bundle can be gradually peeled off from the other fixed–fixed bundle. This peeling process can be considered as a quasistatic process and the delamination of the bundle at the separation interface (i.e., peel front) can be considered because of a balanced competition between the adhesion interaction (between the nanotubes on the binding interface) and the local bending deformation of the delaminated bundle. The peeling

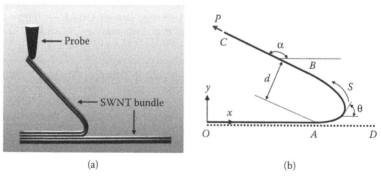

(a)   (b)

FIGURE 12.3   (a) Schematic of peeling off one single-walled carbon nanotube (SWCNT) bundle from another free-standing SWCNT bundle by means of nanomanipulation inside a high-resolution scanning electron microscope. (b) Schematic of the nonlinear elastic model for the nanomechanical peeling measurements illustrated in (a). (Reprinted from Ke, C., M. Zheng, G. Zhou, W. Cui, N. Pugno, and R. N. Miles. 2010. *Small* 6:438–45. With permission.)

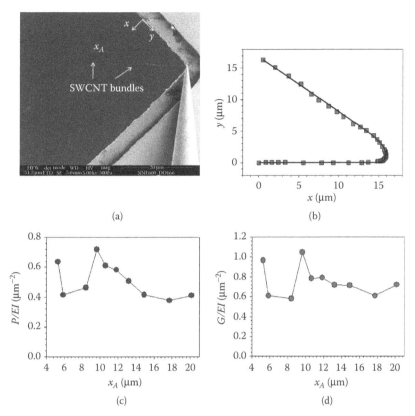

(a)

(b)

(c)

(d)

FIGURE 12.4 (a) One representative scanning electron microcopy snapshot of peeling off a single-walled carbon nanotube (SWCNT) bundle. (b) Comparison between experimental measurements (circle curve) and theoretical predictions (solid curve) of the deformation curve of the delaminated SWCNT bundle shown in (a). (c) The predicted peeling force for peeling measurements performed at a variety of peel front positions. (d) The predicted adhesion energy per unit length between the SWCNT bundles at a variety of peel front positions. Both the peeling force and the adhesion energy are normalized by the bending rigidity of the delaminated bundle *EI*. (From Ke, C., M. Zheng, G. Zhou, W. Cui, N. Pugno, and R. N. Miles. 2010. *Small* 6:438–45. With permission.)

test is interpreted using a continuum model, as illustrated in Figure 12.3b, based on nonlinear elastica theory, which models the bundled nanotubes as an inextensible rod. This simplification is consistent with the earlier experimental observation that the nanotube could be repeatedly bent to large angles and strain with no permanent distortion of the tube topography [133]. As shown in Figure 12.3b, the peeled nanotube bundle can be divided into three segments: *OA*, *AB*, and *BC*. Point *A* denotes the peel front. Segment *AB* is considered under pure bending, whereas segment *BC* is considered under pure stretching. In the (s, θ) coordinate system, the governing equation of the deformed rod segment *AB* is given by

$$EI \frac{d^2\theta}{ds^2} - P\sin(\alpha - \theta) = 0 \qquad (12.5)$$

where EI is the flexural rigidity of the rod and is assumed to be constant along the rod, s is the arc length along the deformed rod measured from the fixed end O, θ is the angle between the tangent of the rod at s and the x-axis, and α is the slope angle of segment BC. In the Cartesian coordinate system, the corresponding governing equation for segment AB is given by [134]

$$x(\theta) = x_A + \frac{1}{\sqrt{P/EI}} \int_0^\theta \frac{\cos\theta}{\sqrt{2(1 - \cos(\alpha - \theta))}} \, d\theta \quad 0 \leq \theta \leq \alpha \qquad (12.6)$$

$$y(\theta) = \frac{1}{\sqrt{P/EI}} \int_0^\theta \frac{\sin\theta}{\sqrt{2(1 - \cos(\alpha - \theta))}} \, d\theta \quad 0 \leq \theta \leq \alpha \qquad (12.7)$$

where x_A is the x coordinate of the peel front. By fitting the measured deformation curvatures from the captured high-resolution SEM images as exemplified by Figure 12.4a, the peeling force P, normalized by the bending stiffness of the delaminated nanotube bundle EI, can be obtained. The comparison between the measured deformation curvature for the nanotube in the peeling test shown in Figure 12.4a, and the theoretical predictions based on Equations 12.6 and 12.7 is shown in Figure 12.4b, which displays a good agreement with the normalized peeling force, P/EI, as the only fitting parameter in the theoretical prediction. The per-unit-length adhesion energy or energy-release rate at the peel front can be calculated as [135,136] $G = P(1 - \cos\alpha)$, in which the slope angle α is measured directly from the SEM images. Based on a series of peeling experiments at various peel front positions for the same nanotube sample, the corresponding peeling forces and the adhesion energies between two nanotube bundles are calculated and presented in Figure 12.4c and d, respectively. The estimated adhesion energy between two SWCNT bundles from the peeling measurements was expressed in a normalized form, $G = EI \cdot (0.75 \pm 0.15 \, \mu m^{-2})$, which indicates a 20% fluctuation from the mean value of the predicted adhesion energy. Such variation in the adhesion energy may be attributed to several possible reasons, such as defects and/or surface contamination on the nanotube structure. We want to highlight that our in-situ mechanical peeling techniques provide a new approach of delaminating bundled nanotubes without compromising their material properties.

12.4.2 ADHESION INTERACTIONS BETWEEN NANOSTRUCTURES AND FLAT SURFACES

Similar to the effect of the tube–tube adhesion interactions, the adhesion interactions between nanostructures and flat substrate surfaces may also induce significant lateral (radial) and/or longitude structural deformation for 1D and 2D nanostructures. Hertel et al. [137] reported a molecular-mechanics study of the elastic deformation of SWCNTs and MWCNTs on a flat silicon substrate. Their results show that the lateral deformation of the nanotube cross-section is dependent on both the outer diameter and the bending stiffness of the nanotube. As shown in Figure 12.5a, for arm-chair SWCNTs, the (5, 5) and (10, 10) tubes can keep their circular cross-section configurations, whereas tubes of larger diameters such as (20, 20) tubes experience noticeable lateral deformations. The radial deformation of the nanotube becomes more prominent

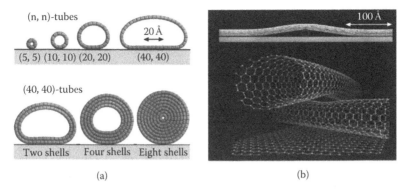

FIGURE 12.5 (a) Molecular-mechanics calculation of the radial deformation of carbon nanotubes on flat silicon substrates. The upper images show the radial compressions of single-walled carbon nanotubes with respect to tube radius. The lower images show the radial deformations of carbon nanotubes with respect to different numbers of shells for (40, 40) tubes. (b) Molecular-mechanics calculation of the axial and radial deformations of two crossover (10, 10) nanotubes. The lower image is a magnified view of the nanotube junction. (From Hertel, T., R. E. Walkup, and P. Avouris. 1998. *Phys Rev B* 58:13870–3. With permission.)

with the increase in tube radius. A flat segment in the nanotube cross-section comes into being as a result of its binding interaction with the substrate. The radial deformation of the nanotube can be diminished or even vanished by adding inner graphite layers, thus forming double-walled or MWCNTs as shown by the three lower images in Figure 12.5a. The increase of the inner graphite layers effectively increases the tube's bending rigidity, while their contribution to the adhesion interaction with the substrate is fairly minimal because the van der Waals interaction decreases rapidly with the increase of the distance. Figure 12.5b illustrates the deformation of two cross-bar (10, 10) SWCNTs on the graphite substrate obtained by MD simulations [137]. The upper nanotube deforms as a result of its adhesion interactions with the substrate and the lower nanotube. The lower image in Figure 12.5b clearly reveals the atomistic-level deformation of the nanotubes at the crossover junction. The MD results show that the force applied on the lower tube by the upper one is about 5.5 nN, and the cross-sections of both tubes are compressed by 20%. These simulation results clearly demonstrate that the adhesion between the nanotube and the substrate plays an important role in the radial deformation of the nanotube. It is noted that the substantial local deformation of the hexagonal C–C network of the nanotube has a considerable influence on the electrical properties of the nanotube by changing its bandgap [108]. Therefore, a variety of novel electronic applications, such as molecular switches [107], can be exploited by tuning the nanotube–substrate adhesion interaction.

The surface morphology of monolayer graphene on a flat substrate that is regulated by the underlying patterned nanowires on the substrate [138], as illustrated in Figure 12.6a, was recently studied based on CM modeling. The topography of the deformed graphene membrane was obtained by minimization of the total free energy of the system that is composed of the adhesion interaction energy and strain energy. The respective adhesion interactions of the graphene–substrate and graphene–nanowire interfaces were calculated using continuum models based on

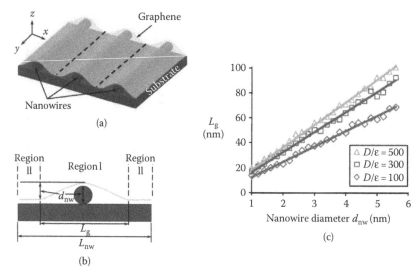

FIGURE 12.6 (a) Schematic of parallel nanowires patterned on a flat substrate surface covered with a blanket graphene and (b) their cross-section configuration. (c) The equilibrium width of the corrugated graphene region I, L_g, as a function of the nanowire diameter d_{nw} for various values of D/ε. The straight fitting lines denote the linear dependence of L_g on d_{nw}. (Reprinted from Zhang, Z., and T. Li. 2010. *J Appl Phys* 107:103519. With permission.)

the LJ potential. The regulated graphene morphology is shown to be governed by several factors including the nanowire diameter d_{nw}, the nanowire spacing L_{nw}, and the interfacial binding energies between graphene and the underlying nanowires and substrate. If the nanowire spacing L_{nw} is much larger than the nanowire diameter d_{nw}, the graphene sheet is expected to corrugate to wrap around the nanowire in a region of width L_g (region I) and remain flat on the substrate in region II, as illustrated in Figure 12.6b. The amplitude of the graphene corrugation is equal to the nanowire diameter d_{nw}. However, for small nanowire spacing, the graphene sheet is expected to corrugate to only partially conform to the envelope of the nanowire surfaces, with a periodic amplitude much smaller than d_{nw}. Figure 12.6c shows that the equilibrium width of the corrugated graphene region I, L_g, has a linear dependence on the nanowire diameter d_{nw} for various values of D/ε, in which D is the graphene bending stiffness and ε is the binding energy between two atoms at the equilibrium distance. Using a similar approach, the general morphology of graphene on the substrate with periodic surface grooves was also investigated [139].

    Several studies have been reported in the experimental characterization of the adhesion interaction between nanostructures and flat substrates [140–144]. The quantification of the adhesion interaction requires the separation of the interface between the nanostructure and the substrate, which may occur either in the normal direction or tangent direction on the nanostructure–substrate binding interface. Many of the proposed testing schemes are based on atomic force microscopy (AFM) [145], which is a versatile technique for high-resolution imaging, manipulation, and force sensing. By using a slender cantilever with a sharp tip, AFM can be used as a powerful imaging tool to capture the topography and the position of nanostructures at an ultrahigh spatial resolution. The sharp AFM probe can be controlled to move the nanostructure around on the flat substrate by applying a lateral force or to peel the nanostructure from the substrate by applying a vertical force. The applied lateral and vertical forces can be quantified by detecting the respective torsional and vertical deflections of the AFM cantilever using optical techniques or piezoresistive film-based self-sensing techniques [143]. AFM has been widely used in the nanoscale adhesion and tribology researches [146–150].

The mechanical peel test is a technique of separating the binding interface along its normal direction and has been widely used to measure macroscopic adhesive strength [151] and microscopic bonding strength in thin films [152]. Strus et al. [141] proposed an AFM-based nanomechanical peeling scheme to measure the binding interaction between an individual CNT and a flat substrate, as illustrated in Figure 12.7a. In this nanomechanical peeling scheme, one CNT, which is attached to a tipless AFM cantilever, is first brought into contact with a flat graphite substrate. Then the nanotube is gradually peeled off from the substrate by lifting the AFM cantilever. The vertical component of the applied peeling force on the nanotube at each peeling stage is quantified through measuring the deflection of the AFM cantilever using a laser reflection scheme. It is noted that the nanomechanical peeling experiments were performed at room temperature inside a chamber backfilled with dry nitrogen to minimize the influence of humidity on the nanotube–substrate adhesion interaction. A nonlinear elastica model was developed to theoretically predict the peeling process and to interpret the experimental measurements. Figure 12.7b shows the theoretically predicted peeling process for both approaching (forming the nanotube–substrate interface) and retracting (separating the nanotube–substrate interface) processes based on representative parameters of the tested nanotubes [142]. The peeling force, the work needed to peel a nanotube off the substrate, and the energy-release rate per unit tube length can be obtained from the recorded force–displacement curve that is acquired the displacement and the deflection of the AFM cantilever. The solid black curve in Figure 12.7b illustrates all of the possible peeling forces as a function of the peeling point displacement, while the dashed and solid arrowed curves represent the respective accessible forces that would be captured experimentally during the processes of the CNT approaching and retracting from the substrate. Their results interestingly reveal that the contact between the nanotube and the substrate may exist in two distinct configurations: line contact (the nanotube is in an S-shape) and point contact (the nanotube is in an arc-shape), as illustrated in Figure 12.7c, suggesting that substantial structural transitions exist for the deformed nanotube during the nanomechanical peeling process. The work added to the system by the external force that peels the CNT during the CNT peeling process are represented by the solid gray and stripped-shaded areas, which are under the respective line-contact and point-contact force–displacement curves during retraction. It is reported that the required peeling force to lift a MWCNT (1.5 μm in length, 10 and 41 nm for inner and outer diameters, respectively) from a highly order pyrolitic graphite substrate is 14 nN. By using this nanomechanical peeling scheme, the adhesion interactions between CNTs and a variety of material surfaces have been investigated. The per-unit-length CNT–substrate interfacial binding energies are reported to be 0.6 pJ/m for polyimide, 1.1 pJ/m for graphite, and 1.7 pJ/m for epoxy [142].

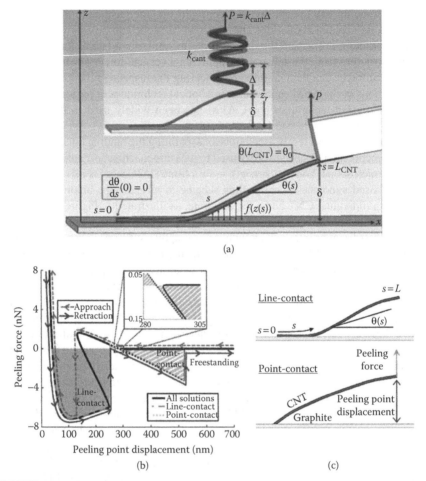

FIGURE 12.7 (a) Schematic of peeling off one carbon nanotube from a flat substrate by atomic force microscopy and the elastica-theory based computational model. (Reprinted from Strus, M. C., L. Zalamea, A. Raman, R. B. Pipes, C. V. Nguyen, and E. A. Stach. 2008. *Nano Lett* 8:544–50. With permission.) (b) Theoretically predicted process of peeling carbon nanotubes from graphite surface based on the representative experimental parameters. (c) Schematics of the line-contact and point-contact configurations when the nanotube is peeled off the substrate. (Reprinted from Strus, M. C., C. I. Cano, R. B. Pipes, C. V. Nguyen, and A. Raman. 2009. *Compos Sci Technol* 69:1580–6. With permission.)

It is noted that the deformation of the nanotube during the nanomechanical peeling process was not directly measured as a result of lack of means of visualizing the deformation of the nanotube in the aforementioned AFM-based peeling scheme. The quantification of the adhesion interaction between the nanotube and the substrate has to rely on the postulated nanotube deformation curve obtained by theoretical modeling. Ishikawa et al. [143] have recently developed a method to perform nanomechanical peeling tests inside a high-resolution SEM. A self-detecting AFM cantilever was used to measure the peeling force at a resolution of 0.1 nN. The mechanical deformation of the nanotube during the peeling process was measured using the high-resolution electron beam. Their results confirm that the mechanical peeling of a nanotube from the substrate involves multistable states of the nanotube between the line and point contacts, as previously suggested by Strus et al. [141,142] By using this in-situ SEM-AFM-based nanomechanical peeling technique, the needed peeling forces to pull off one same MWCNT from the substrates of graphite, mica, and NaCl (001) were investigated and reported to be 2.2, 1.8, and 1.9 nN, respectively, suggesting that the nanotube–substrate interaction strength is strongest for the graphite surface, while comparable for the mica and NaCl (001) surfaces.

The chemical bonding-based adhesion interactions between CNTs and substrates were also studied by AFM-based chemical force microscopy techniques. The chemical bonds on the nanotube–substrate interface are typically formed by the additional functional groups introduced to the surfaces of the nanotube and/or the substrate through chemical functionalization and are expected to possess higher adhesion strength than the van der Waals interaction. Li et al. [153] investigated the adhesion interaction between individual CNTs and a variety of polypeptides by using the testing scheme as illustrated in Figure 12.8a. For the AFM measurement, the AFM tip is first modified with the attachment of polypeptide chains. The adhesion interactions of the polypeptide chain–coated AFM tip to pristine CNTs and carboxylated CNTs were then studied through the force–distance curve of the AFM, which operates in the force modulation mode. The adhesion between the AFM tip and the

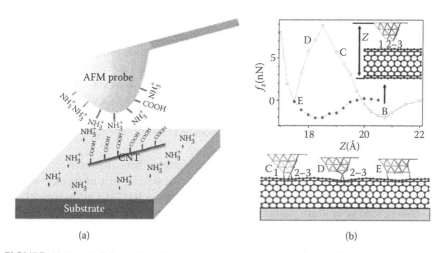

(a) (b)

FIGURE 12.8 (a) Schematic of interactions between polylysine and carboxylated carbon nanotubes measured by atomic force microscopy (AFM). (Reprinted from Li, X. J., W. Chen, Q. W. Zhan, L. M. Dai, L. Sowards, M. Pender, and R. R. Naik. 2006. *J Phys Chem B* 110:12621–5. With permission.) (b) Force–distance curve for the binding interaction when an Au-coated AFM tip is pushed/pulled on a (8, 8) single-walled carbon nanotube lying on a SiO₂ substrate. Black and gray curves represent the pushing and pulling traces, respectively. The tip-nanotube contact geometries for different points B, C, D, and E are shown by the bottom images, which are obtained using density function theory-based simulations. (Reprinted from Gonzalez, C., J. Ortega, F. Flores, D. Martinez-Martin, and J. Gomez-Herrero. 2009. *Phys Rev Lett* 102:106801. With permission.)

nanotube is reflected in the attractive force zone in the tip retraction portion of the captured force–distance curves. It is found that both the pH value of the solution and the degree of the CNT oxidation have strong influence on the adhesion interaction. Their measurements also reveal the existence of the π–π stacking interaction between aromatic moieties in the peptide chains and the nanotube surface, which is found to be independent of the type of CNTs. It is noted that the adhesion interaction is also dependent on the thiol terminal groups on the attachment interface [154]. An AFM-based chemical force microscopy study of the adhesion between thiolated AFM cantilever tips and SWCNT thin-film paper reveals a direct correlation of adhesion force with respect to the thiol terminal groups, whose strength are in the order of $NH_2^- > CH_3^- > OH^-$ [154].

Studies have suggested that the C–C bond of sp^2 hybridization in the pristine CNT can be converted to chemically more active sp^3 hybridization as a result of the local distortion strain induced by external forces [107,108]. The hybridized sp^3 carbon atoms can form strong short-range chemical-bonding interactions with other material surfaces, such as gold. The chemical bonding force between the hybridized sp^3 C–C bond network and the gold surface was recently studied by Gonzalez et al. [109] They used a combined experimental–theoretical approach to investigate the chemical force between individual SWCNTs lying on a flat SiO_2 substrate and Au-coated AFM tips. The adhesion experiments were performed inside an ultra-high vacuum AFM. The Au-coated AFM tips were controlled to first push against and then pull from the nanotube surface. The nanotube–Au adhesion interaction was measured through the recorded force–distance curves. Density function theory (DFT)-based atomistic-level simulation was utilized to predict the bonding strength between the carbon nanotube and the gold surface. The upper image in Figure 12.8b shows the simulation results of the interaction force as a function of the distance between the AFM tip and an (8, 8) SWCNT. The lower images show the simulated tip-nanotube contacting geometry for different points B, C, D, and E in the upper plot. Their experimental measurements indicate that the adhesion force between the gold layer coated on the AFM tip and the nanotube is about 2 nN, which is consistent with their DFT-based theoretical predictions.

Our research group recently investigated the binding interaction between a nanotube in the closed circular ring configuration and a flat gold surface using a combined experimental–theoretical nanomechanical testing approach [144]. Our nanomechanical testing scheme is illustrated in Figure 12.9a, where a vertically placed CNT ring is pushed against (or pulled from) the substrate by an external force P. Figure 12.9b shows a tested CNT rings, 7 µm in diameter, that is partially attached to a probe mounted to a 3D piezo-driven nanomanipulator. The CNT ring was formed by self-folding of a long and thin bundle of SWCNTs, which were originally synthesized by CVD methods, through nanomanipulation inside a high-resolution SEM. By carefully adjusting the orientation of the manipulator probe, the CNT ring was placed horizontal, meaning that the normal direction of the ring was parallel to that of the electron beam. Therefore, its in-plane motion and deformation could be captured from the high-resolution electron beam. The CNT ring was controlled to vertically push into or pull from the flat substrate, which was a piece of Si wafer evaporated with a 30-nm gold film. Our measurements revealed significant compressive deformation

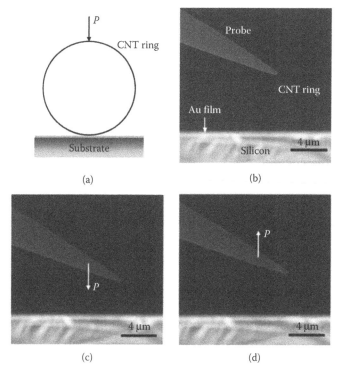

FIGURE 12.9 (a) Schematic of a carbon nanotube (CNT) ring on a flat substrate under a point load. (b) Scanning electron microcopy image of a CNT ring attached to a manipulator probe being positioned close to a vertically placed flat Si-chip coated with a thin Au film on top. (c, d) Representative SEM snapshots of one CNT ring being pushed against and pulled away from the substrate. (Reprinted from Zheng, M., and C. H. Ke. 2010. *Small* 6:1647–55. With permission.)

of the CNT ring when it was pushed onto the substrate, as shown in Figure 12.9c. When the CNT ring was gradually pulled from the substrate, the CNT ring was observed to sustain substantial tensile deformation before it was fully detached from the substrate. Figure 12.9d captures the tensile deformation of the CNT ring for the moment right before the CNT ring was released from the substrate. Our image analysis shows that the deformation of the CNT ring was purely elastic because there was no visible permanent plastic deformation observed after three repeated cycles of pushing and pulling of the same CNT ring. The mechanical deformations of the CNT nanoring in both compression and tension were interpreted using nonlinear elastica model [155]. Figure 12.10a and b show the comparison between experimental measurements (circle curves) and theoretical predictions (solid curves) on the deformation curvatures of the CNT ring in both compression and tension based on the respective SEM snapshots shown in Figure 12.9c and d. The corresponding applied forces on the CNT ring, which were not directly measured in our tests, were predicted based on the experimental measurements on the deformation curvatures of the CNT ring. Our results show that the predicted compressive load on the CNT ring

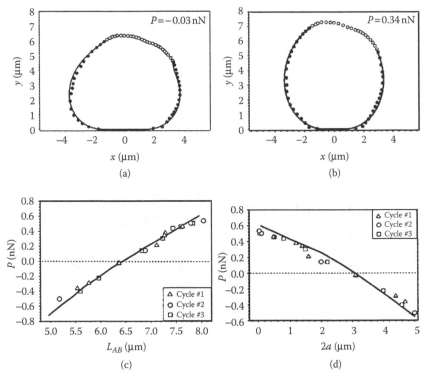

FIGURE 12.10 (a, b) Comparison between experimental measurements (circles) and theoretical predictions (solid curves) for the deformed carbon nanotube (CNT) ring shown in Figure 12.9a and b, respectively. The theoretically predicted applied force P is marked on each figure. (c) Relationship between experimentally measured ring height and theoretically predicted applied load for three repeated pushing/pulling cycles of the same CNT ring. (d) Relationship between theoretically predicted applied load and contact length of the deformed ring for three repeated pushing/pulling cycles of the same CNT ring. (Reprinted from Zheng, M., and C. H. Ke. 2010. *Small* 6:1647–55. With permission.)

shown in Figure 12.9c is 0.03 nN, whereas the predicted tensile load on the CNT ring shown in Figure 12.9d is 0.34 nN.

Our measurements reveal that the van der Waals interaction between the CNT ring and the substrate has a profound effect on the mechanical deformation of the CNT ring. Figure 12.10c shows the relationship between the applied force and the height of the deformed CNT ring, whereas Figure 12.10d shows the relationship between the applied force and the contact length between the CNT ring and the substrate. Our results show that the CNT ring of large ring diameter may experience significant deformation purely because of its adhesion interaction with the substrate. A linear relationship between the applied force and the CNT ring deformation is shown in Figure 12.10c when the CNT ring is under compressive loads, suggesting that CNT nanorings can be pursued for a number of applications, such as ultrasensitive force sensors and flexible and stretchable structural components in nanoscale mechanical and electromechanical systems.

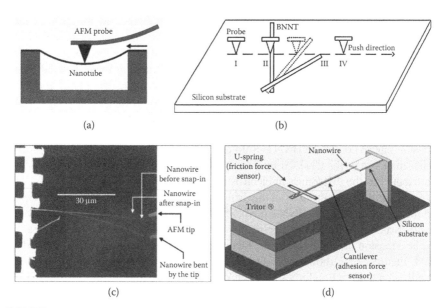

FIGURE 12.11 (a) Schematic of deflecting a free-standing carbon nanotube by an atomic force microscopy (AFM) cantilever until slip occurs on the nanotube–substrate interface. (Reprinted from Whittaker, J. D., E. D. Minot, D. M. Tanenbaum, P. L. McEuen, and R. C. Davis. 2006. *Nano Lett* 6:953–7. With permission.) (b) Schematics of manipulating individual boron nitride nanotubes (BNNTs) on the flat substrate by an AFM probe. In position I to II, the probe moved toward the BNNT in the lateral direction. At position II, the probe collided with the BNNT. In positions II to III, the probe contacted the BNNT and pushed it for a distance. In positions III to IV, the probe moved away from the BNNT. (Reprinted from Hsu, J. H., and S. H. Chang. 2010. *Appl Surf Sci* 256:1769–73. With permission.) (c) Testing the adhesion interaction by pushing the free end of a nanowire staying on top of a transmission electron microscope grid with a silicon AFM cantilever to initiate relative slip. (Reprinted from Desai, A. V., and M. A. Haque. 2007. *Appl Phys Lett* 90:033102. With permission.) (d) Schematic of testing the adhesion interaction of a nanowire using a nanofabricated adhesion–friction force sensor that is mounted on a Tritor piezo-actuator. (Reprinted from Manoharan, M. P., and M. A. Haque. 2009. *J Phys D Appl Phys* 42:095304. With permission.)

The adhesion interactions between nanostructures and flat substrates were also investigated by measuring their friction forces, which are in tangent direction to the contact interface. Figure 12.11a illustrates one experimental scheme of measuring the friction force between one SWCNT and a silicon dioxide substrate [156]. An AFM probe was used to apply a vertical force on the middle point of the suspended nanotube, which was grown directly across trenches, until slip occurred along the tube axis. Based on measurements on four different suspended nanotubes with the contact length ranging from 140 to 228 nm, the tension forces in the deflected nanotube were measured to be 7~8 nN when slip occurred, suggesting that the nanotube–substrate adhesion force is independent of the contact length. The surface adhesion between BNNTs and silicon substrates was also studied by AFM manipulation [113], as illustrated in Figure 12.11b. A silicon AFM probe with a pyramidal-shape tip, operating in the contact scanning mode, was controlled to be in contact with the substrate and move in a straight-line manner to push the BNNT on the substrate by a certain angle. Two different BNNT samples, which have diameters of 14 and 17 nm and lengths of 640 and 850 nm, respectively, were used for the measurements. The shear stress between the nanotube and the substrate resulting from their adhesion interaction and the corresponding sliding energy loss were calculated to be about 0.5 GPa and 0.2 J/m^2, respectively, from the recorded force–distance curves. For comparison, the shear stress between a CNT of 8 nm in diameter and 368 nm in length and a silicon substrate was reported to be 0.01 GPa [157]. Therefore, the measured adhesion interaction of the BNNT–Si interface is more than one order of magnitude higher than that of the CNT–Si interface. The observed surface adhesion difference is attributed to the difference in electrostatic interaction in CNTs and BNNTs. The permanent dipoles resulting from the difference in electron negativity between boron and nitrogen atoms in the BNNT lead to higher electrostatic interaction than that of the all-carbon made CNTs, which results in higher surface adhesion for BNNTs.

The adhesion interactions between nanowires and substrates were also studied by AFM. Figure 12.11c shows an in-situ electron microscopy experiment of measuring the adhesion and friction forces between ZnO nanowires and silicon substrates [158]. The tested ZnO nanowires were synthesized by the vapor–liquid–solid mechanism using gold particles as catalyst on silicon substrates. The directly grown ZnO nanowires were first transferred onto a TEM grid. Some nanowires protruding from the TEM grid were deflected by an AFM cantilever mounted to a nanomanipulator, as shown in Figure 12.11c. The friction force between the silicon AFM cantilever and the nanowire is calculated from the deformation profiles of the AFM cantilever and the nanowire. The van der Waals and friction forces between the ZnO nanowire and the silicon AFM cantilever were reported to be 81.05 pN and 7.7 nN, respectively. It is noted that the exposure to the electron beam in the above measurements may affect the adhesion force measurement. In order to avoid the possible influence of the electron beam on the adhesion and friction measurements, a custom-designed adhesion-friction force sensor was developed to study the shear strength of the nanowire–substrate interface at the ambient environment, and the testing scheme is shown in Figure 12.11d [159]. The nanofabricated adhesion–friction force sensor is mounted on a Tritor piezo-actuator. The adhesion–friction interaction of the nanowires–substrate contact can be obtained from the deflection of the force sensing component. By using this apparatus, the friction force between ZnO nanowires (~100 nm in diameter) and Si surfaces were measured to be a few micro-Newtons, which is ascribed to a molecularly thin surface moisture layer formed at the nanowire–substrate contact interface. It is noted that the adhesion and friction interactions of InAs nanowires on silicon nitride substrates were also studied based on AFM manipulation [160,161].

12.4.3 INTERFACIAL INTERACTION BETWEEN NANOSTRUCTURES AND POLYMER MATRICES

For 1D and 2D nanostructure-based polymer nanocomposites, the interfacial interaction between nanostructures and the supporting polymer matrices plays a key role in the mechanical properties of bulk nanocomposites and the realization of the reinforcing effect of the high-strength nanostructures. This is because the load-transfer efficiency in nanocomposite materials depends on the interfacial shear strength between the nanostructure and the matrix. A high interfacial shear stress transfers the applied load to the nanotube over a short distance. Understanding the interfacial shear stress is an essential step to the optimal design of the nanostructure-reinforced polymer composites. Because pristine nanostructures such as CNTs and graphene have very smooth and chemically stable surfaces, the mechanical strength of the nanostructure–polymer interface based on pure van der Waals interaction is fairly weak, thus greatly downgrading their reinforcing effects. Therefore, surface functionalization is very important for nanostructure-based polymer composites because the added functional groups on the nanostructure surface can significantly strengthen the nanostructure–polymer interactions, thus increasing the interfacial shear strength.

The interfacial interactions between nanostructures and polymers are usually investigated by performing pull-out tests, in which the embedded nanostructure is stretched out of the polymer matrix by external mechanical loadings. Figure 12.12a shows a fractured specimen of CNT–epoxy nanocomposites, indicating the CNTs being pulled out of the polymer matrix [162].

Figure 12.12b shows an AFM-based pull-out experiment to measure the fracture energy for stretching out one CNT from a polymer matrix [163]. One CNT that was premounted to an AFM tip was brought down to the surface of a liquid phase epoxy polymer until a substantial segment of the nanotube was embedded into the polymer. After curing, the polymer surface became solid while the nanotube was kept in the position. The nanotube was pulled out from the polymer by lifting off the AFM cantilever and the force–distance profile of the AFM cantilever during the pull-out process was recorded. The total interfacial fracture energy between the nanotube and the polymer is calculated based on the maximum pull-out force and the embedded nanotube length. Both pristine and carboxyl-modified CNTs were used in the tests. It is found that the strong interface, which corresponds to higher fracture energy, occurs for smaller diameter nanotubes and longer embedded length. In addition, it is reported that the chemical modification of the nanotube surface provided a substantially higher interfacial strength, which was ascribed to the strong chemical bonding between the modified CNT and the polymer matrix.

The interfacial interactions between CNTs and polymers have been also investigated by both continuum and atomistic-level simulation techniques. A cohesive law for interfaces between MWCNTs and polymers based on pure van der Waals interactions was developed using CM based on the LJ potential [164]. The strength of the nanotube–polymer interface σ^{int} is given by

$$\sigma^{int} = 3.076\,\sigma_{max}\left[\left(1+0.682\frac{\sigma_{max}}{\phi_{total}}[u]\right)^{-4} - \left(1+0.682\frac{\sigma_{max}}{\phi_{total}}[u]\right)^{-10}\right] \quad (12.8)$$

where $[u]$ is the normal displacement at the nanotube–polymer interface. Both the cohesive strength σ_{max} and total cohesive energy ϕ_{total} are related to the parameters in the LJ potential, as well as the area density of the CNT and the volume density of the polymer. Based on the interfacial strength model given by Equation 12.8, it is found that the CNTs can improve the mechanical behavior of the composite only at small strain, while the improvement disappears at relatively large strain [165].

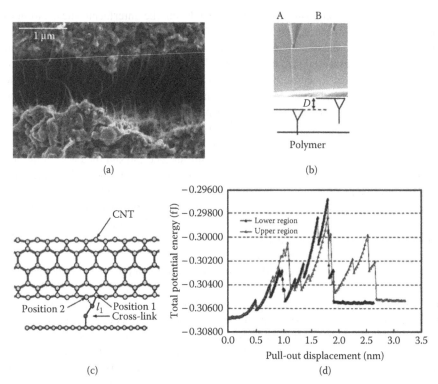

FIGURE 12.12   (a) Scanning electron microcopy (SEM) image of a fractured specimen of carbon nanotube (CNT)–epoxy nanocomposites showing the CNTs being pulled out of the polymer matrix. (Reprinted from Ajayan, P. M., L. S. Schadler, C. Giannaris, and A. Rubio. 2000. *Adv Mater* 12:750–3. With permission.) (b) SEM images of pulling out an individual CNT from the polymer. The CNT was mounted onto an atomic force microscopy (AFM) tip and partially embedded in a polymer matrix. The CNT was pulled out from the polymer by lifting the AFM cantilever. (Reprinted from Barber, A. H., S. R. Cohen, and H. D. Wagner. 2003. *Appl Phys Lett* 82:4140–2. With permission.) (c) Schematic of position switching of cross-links on the CNT–polymer interface. (d) Energy variation during the pull-out of a CNT with four cross-links at the nanotube–polymer interface. (Reprinted from Chowdhury, S. C., and T. Okabe. 2007. *Compos Part A Appl Sci Manuf* 38:747–54. With permission.)

This observation is explained by the fact that the completely debonded nanotubes behave like voids in the matrix, thus resulting in weakening of the composite. At large strain, the increase of the interface adhesion between CNTs and polymer matrices may help to significantly improve the mechanical properties of the composite.

The adhesion and interfacial interactions between CNTs and a variety of polymers, such as polyethylene [166,167], epoxy [168] and polystyrene [169] matrices, have been investigated by using molecular modeling. In particular, the effect of the chemical functionalization of the nanotube surface on the nanotube–polymer interfacial strength has been intensively studied. Figure 12.12c illustrates the polyethylene cross-links on the interface of the nanotube and the bulk polyethylene polymer. The nanotube–polymer interfacial strength was characterized by performing pull-out tests based on MD simulations [166]. It is found that the interfacial strength is affected by not only the presence of the cross-links but also the location of the cross-links. The latter factor is because of the fact that the cross-links on the interface may shift locations during the nanotube pull-out process, thus affecting the potential energy of the whole system. Figure 12.12d shows the variation in total potential energy as a function of the pull-out displacement for interfaces with four polyethylene cross-links located in two regions: upper region and lower region, which are defined based on their distances to the nanotube-exposed end. It was reported that an interface having cross-links closer to the nanotube-exposed end (upper region) transfers energy up to a longer pull-out displacement because the cross-links have a longer traveling distance on the CNT. The fluctuation in the total energy is caused by the switching of the cross-link bonds, while the energy variation for the interface having all cross-links in the upper region is smaller than that for the interface having all cross-links in the lower region. The calculated interfacial shear stress for interfaces having cross-links in the lower region is higher compared with the case when the cross-links are in the upper region. The MD simulations also revealed the effect of pentagon–heptagon defects on the interfacial stress transfer [166]. It is found that pentagon–heptagon defects, which reportedly reduce the tensile strength of the CNT [170], also have negative impact on the nanotube–polymer interfacial shear stress because of the improper connections of cross-links in the region close to the defects.

The effect of the interfacial interaction between graphene and polymer on the graphene-based polymer nanocomposites has also been investigated both experimentally and theoretically. Similar to CNTs, the chemical functionalization of graphene surface may significantly enhance its bonding strength with polymer. In a study reported by Ramanathan et al. [96], the mechanical properties of reinforced poly(methyl methacrylate) (PMMA) polymer by 1 wt% functionalized graphene sheets (FGS) that are partially oxygenated significantly exceed the mechanical performance of pure PMMA, as well as its composites with SWCNTs and expanded graphite (EG) plates, which were produced by heating sulfuric acid-intercalated graphite. The substantial mechanical performance of FGS-based PMMA composites is ascribed to the fact that oxygen functionalities enhance graphene sheets' bonding interactions with polymer matrices. Figure 12.13b shows an SEM image of the EG–PMMA interface. It is shown that the thicker protruding plates from the simple expanded graphite exhibits poor bonding to the polymer matrix. Figure 12.13c shows an SEM image of the FSG–PMMA interface. The size scale (nanosheet thickness) and morphology (wrinkled texture) of the FGS, as well as their surface chemistry, lead to strong interfacial interactions with the supportive polymer matrix, as illustrated by polymer adhesion to the pulled-out FGS.

The van der Waals force-based interfacial binding interaction between graphene and the supporting polymer matrix [171] and the role of the surface functionalization on the interfacial strength [172] were recently investigated using MD simulations. Figure 12.14 shows the results of the interfacial strength between graphene and polyethylene polymer by performing a peeling test to remove the graphene from the polymer surface through applying a force normal to the van der Waals interaction-based graphene–polymer binding interface [171]. Three cases were investigated for different boundary constraints imposed at the beginning of the simulations: *Case A*: a portion of the polymer at the top is fixed in space; *Case B*: a portion of the polymer

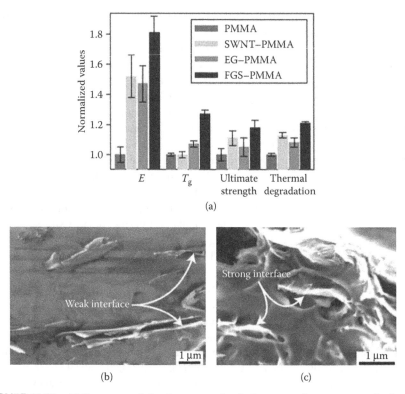

FIGURE 12.13 (a) Summary of the thermo-mechanical property improvements for 1 wt% functionalized graphene sheets–poly(methyl methacrylate) (FGS–PMMA), compared with single-walled carbon nanotube (SWCNT)–PMMA and expanded graphite (EG)–PMMA composites. All property values are normalized to the values of pure PMMA. (b) Scanning electron microcopy (SEM) image of EG–PMMA, in which the thick graphite plates exhibit poor bonding to the polymer matrix. (c) SEM image of FGS–PMMA, in which the graphene nanosheets exhibit strong bonding to the polymer matrix. (Reprinted from Ramanathan, T., A. A. Abdala, S. Stankovich, D. A. Dikin, M. Herrera-Alonso, R. D. Piner, D. H. Adamson et al. 2008. *Nat Nanotechnol* 3:327–31. With permission.)

near the graphene surface is fixed in space; *Case C: all* polymer atoms are fixed. Figure 12.14 shows the comparison of the load–displacement response for three cases *A*, *B*, and *C* in opening mode separation of the graphene sheet from the supportive polymer matrix. The required separation forces for cases *B* and *C* are comparable and significantly higher than that for case *A*. Figure 12.15 shows the effect of the surface functionalization on the graphene–polymer interfacial bonding strength [172]. Figure 12.15a illustrates the functionalized graphene with carboxyl groups randomly chemisorbed to 2.5% of the carbon atoms, while Figure 12.15b shows the interaction energy during the pullout of the graphene sheet from the PMMA matrix as a function of the degree of graphene surface functionalization. The MD simulation results show that the chemical functionalization of graphene surface can dramatically increase the interfacial bonding strength between the graphene and the polymer matrix, which has a positive correlation with the degree of graphene surface functionalization.

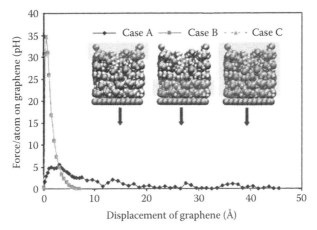

FIGURE 12.14 Comparison of the load–displacement responses for cases *A*, *B*, and *C* in opening mode separation of the monolayer graphene sheet from the supportive polymer matrix. (Reprinted from Awasthi, A. P., D. C. Lagoudas, and D. C. Hammerand. 2009. *Model Simul Mater Sci Eng* 17:015002. With permission.)

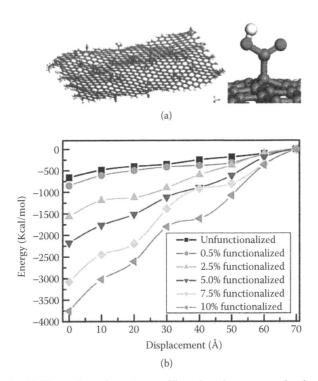

FIGURE 12.15 (a) Illustration of graphene with carboxyl groups randomly chemisorbed to 2.5% of the carbon atoms. (b) Interaction energy plots during the pullout of the graphene sheet from the poly(methyl methacrylate) matrix. (From Lv, C., Q. Z. Xue, D. Xia, M. Ma, J. Xie, and H. J. Chen. 2010. *J Phys Chem C* 114:6588–94. With permission.).

12.5 SUMMARY

In this chapter, we have reviewed the current state-of-the-art in characterization of nanoscale adhesive interactions between nanostructures and substrates, where such a characterization plays a significant role in the design of nanoscale functional devices such as nanoscale switches. In particular, the experimental efforts to characterize the various nanoscale adhesive interactions have been reviewed along with the theoretical framework based on the CM theory. As stated in Section 12.4, there have been current attempts to use computational simulation on nanoscale adhesions using atomistic simulations and/or CM-based simulations. Despite these computational efforts, we believe that there is much room for researchers to make contributions to the development of novel modeling that allows the simulation of nanoscale adhesion complementary to the experimental observations. Specifically, atomistic simulations that enable fundamental insights into atomic-scale frictions (and/or adhesions) are computationally prohibitive for simulating the adhesive interaction between nanostructures and larger-scale structures (e.g., microstructures or bulk structures), which requires simulations at large spatial and temporal scales that are inaccessible with conventional atomistic simulation. Moreover, the classical CM models lack the details of atomic-scale phenomena in frictions and/or adhesions. This suggests that for gaining insight into the adhesive interactions between nanostructures and larger-scale structures, it is required to develop de novo models that can be established from coupling between two characteristic features of atomistic models and CM models in order to simulate the adhesive behaviors at large spatial and temporal scales while maintaining the atomic-scale characteristics.

ACKNOWLEDGMENTS

The authors acknowledge the support from State University of New York at Binghamton, American Chemical Society-Petroleum Research Fund, and the U.S. Air Force Office of Scientific Research.

REFERENCES

1. Rueckes, T., K. Kim, E. Joselevich, G. Y. Tseng, C. L. Cheung, and C. M. Lieber. 2000. Carbon nanotube-based nonvolatile random access memory for molecular computing. *Science* 289:94–7.
2. Ke, C. H., and H. D. Espinosa. 2004. Feedback controlled nanocantilever device. *Appl Phys Lett* 85:681–3.
3. Ke, C. H., and H. D. Espinosa. 2006. In situ electron microscopy electromechanical characterization of a bistable NEMS device. *Small* 2:1484–9.
4. Jang, J. E., S. N. Cha, Y. Choi, G. A. J. Amaratunga, D. J. Kang, D. G. Hasko, J. E. Jung, and J. M. Kim. 2005. Nanoelectromechanical switches with vertically aligned carbon nanotubes. *Appl Phys Lett* 87:163114.
5. S. Iijima. 1991. Helical microtubules of graphitic carbon. *Nature* 354:56–8.
6. Chopra, N. G., R. J. Luyken, K. Cherrey, V. H. Crespi, M. L. Cohen, S. G. Louie, and A. Zettl. 1995. Boron-nitride nanotubes. *Science* 269:966–7.
7. Rubio, A., J. L. Corkill, and M. L. Cohen. 1994. Theory of graphitic boron-nitride nanotubes. *Phys Rev B* 49:5081–4.

8. Ebbesen, T. W., and P. M. Ajayan. 1992. Large-scale synthesis of carbon nanotubes. *Nature* 358:220–2.

9. Journet, C., W. K. Maser, P. Bernier, A. Loiseau, M. L. delaChapelle, S. Lefrant, P. Deniard, R. Lee, and J. E. Fischer. Large-scale production of single-walled carbon nanotubes by the electric-arc technique. *Nature* 388:756–8.

10. Thess, A., R. Lee, P. Nikolaev, H. J. Dai, P. Petit, J. Robert, C. H. Xu, et al. 1996. Crystalline ropes of metallic carbon nanotubes. *Science* 273:483–7.

11. Li, W. Z., S. S. Xie, L. X. Qian, B. H. Chang, B. S. Zou, W. Y. Zhou, R. A. Zhao, and G. Wang. 1996. Large-scale synthesis of aligned carbon nanotubes. *Science* 274:1701–3.

12. Zhang, Y. G., A. L. Chang, J. Cao, Q. Wang, W. Kim, Y. M. Li, N. Morris, E. Yenilmez, J. Kong, and H. J. Dai. 2001. Electric-field-directed growth of aligned single-walled carbon nanotubes. *Appl Phys Lett* 79:3155–7.

13. Su, C. Y., Z. Y. Juang, K. F. Chen, B. M. Cheng, F. R. Chen, K. C. Leou, and C. H. Tsai. Selective growth of boron nitride nanotubes by the plasma-assisted and iron-catalytic CVD methods. *J Phys Chem C* 113:14681–8.

14. Lourie, O. R., C. R. Jones, B. M. Bartlett, P. C. Gibbons, R. S. Ruoff, and W. E. Buhro. 2000. CVD growth of boron nitride nanotubes. *Chem Mater* 12:1808–10.

15. Golberg, D., Y. Bando, M. Eremets, K. Takemura, K. Kurashima, and H. Yusa. 1996. Nanotubes in boron nitride laser heated at high pressure. *Appl Phys Lett* 69:2045–7.

16. Yu, D. P., X. S. Sun, C. S. Lee, I. Bello, S. T. Lee, H. D. Gu, K. M. Leung, G. W. Zhou, Z. F. Dong, and Z. Zhang. 1998. Synthesis of boron nitride nanotubes by means of excimer laser ablation at high temperature. *Appl Phys Lett* 72:1966–8.

17. Laude, T., Y. Matsui, A. Marraud, and B. Jouffrey. 2000. Long ropes of boron nitride nanotubes grown by a continuous laser heating. *Appl Phys Lett* 76:3239–41.

18. Arenal, R., O. Stephan, J. L. Cochon, and A. Loiseau. 2007. Root-growth mechanism for single-walled boron nitride nanotubes in laser vaporization technique. *J Am Chem Soc* 129:16183–9.

19. Smith, M. W., K. C. Jordan, C. Park, J. W. Kim, P. T. Lillehei, R. Crooks, and J. S. Harrison. 2009. Very long single- and few-walled boron nitride nanotubes via the pressurized vapor/condenser method. *Nanotechnology* 20:505604.

20. Golberg, D., Y. Bando, Y. Huang, T. Terao, M. Mitome, C. C. Tang, and C. Y. Zhi. 2010. Boron nitride nanotubes and nanosheets. *Acs Nano* 4:2979–93.

21. Dresselhaus, M. S., G. Dresselaus, and P. Avouris. 2001. *Carbon Nanotubes*. Berlin: Springer.

22. Chopra, N. G., and A. Zettl. 1998. Measurement of the elastic modulus of a multi-wall boron nitride nanotube. *Solid State Commun* 105:297–300.

23. Pop, E., D. Mann, Q. Wang, K. Goodson, and H. J. Dai. 2006. Thermal conductance of an individual single-wall carbon nanotube above room temperature. *Nano Lett* 6:96–100.

24. Xiao, Y., X. H. Yan, J. X. Cao, J. W. Ding, Y. L. Mao, and J. Xiang. 2004. Specific heat and quantized thermal conductance of single-walled boron nitride nanotubes. *Phys Rev B* 69:205415.

25. Blasé, X., A. Rubio, S. G. Louie, and M. L. Cohen. 1994. Stability and band-gap constancy of boron-nitride nanotubes. *Europhys Lett* 28:335–340.

26. Lee, C. H., M. Xie, V. Kayastha, J. S. Wang, and Y. K. Yap. 2010. Patterned growth of boron nitride nanotubes by catalytic chemical vapor deposition. *Chem Mater* 22:1782–7.

27. Lee, C. H., J. S. Wang, V. K. Kayatsha, J. Y. Huang, and Y. K. Yap. 2008. Effective growth of boron nitride nanotubes by thermal chemical vapor deposition. *Nanotechnology* 19:455605.

28. Chen, Y., J. Zou, S. J. Campbell, and G. Le Caer. 2004. Boron nitride nanotubes: Pronounced resistance to oxidation. *Appl Phys Lett* 84:2430–2.

29. Golberg, D., Y. Bando, K. Kurashima, and T. Sato. 2001. Synthesis and characterization of ropes made of BN multiwalled nanotubes. *Scr Mater* 44:1561–5.

30. Craighead, H. G. 2000. Nanoelectromechanical systems. *Science* 290:1532–5.

31. Mahar, B., C. Laslau, R. Yip, and Y. Sun. 2007. Development of carbon nanotube-based sensors — A review. *Ieee Sens J* 7:266–84.

32. Coleman, J. N., U. Khan, and Y. K. Gun'ko. 2006. Mechanical reinforcement of polymers using carbon nanotubes. *Adv Mater* 18:689–706.

33. Liu, Z., S. Tabakman, K. Welsher, and H. J. Dai. 2009. Carbon nanotubes in biology and medicine: In vitro and in vivo detection, imaging and drug delivery. *Nano Res* 2:85–120.

34. Zhi, C. Y., Y. Bando, T. Terao, C. C. Tang, H. Kuwahara, and D. Golberg. 2009. Towards thermoconductive, electrically insulating polymeric composites with boron nitride nanotubes as fillers. *Adv Funct Mater* 19:1857–62.

35. Li, Y. B., P. S. Dorozhkin, Y. Bando, and D. Golberg. 2005. Controllable modification of SiC nanowires encapsulated in BN nanotubes. *Adv Mater* 17:545–9.

36. Badzey, R. L., G. Zolfagharkhani, A. Gaidarzhy, and P. Mohanty. 2004. A controllable nanomechanical memory element. *Appl Phys Lett* 85:3587–9.

37. Hu, J. T., O. Y. Min, P. D. Yang, and C. M. Lieber. 1999. Controlled growth and electrical properties of heterojunctions of carbon nanotubes and silicon nanowires. *Nature* 399:48–51.

38. Hu, J. T., T. W. Odom, and C. M. Lieber. 1999. Chemistry and physics in one dimension: Synthesis and properties of nanowires and nanotubes. *Acc Chem Res* 32:435–45.

39. Morales, A. M., and C. M. Lieber. 1998. A laser ablation method for the synthesis of crystalline semiconductor nanowires. *Science* 279:208–11.

40. Yu, D. P., C. S. Lee, I. Bello, X. S. Sun, Y. H. Tang, G. W. Zhou, Z. G. Bai, Z. Zhang, and S. Q. Feng. 1998. Synthesis of nano-scale silicon wires by excimer laser ablation at high temperature. *Solid State Commun* 105:403–7.

41. Ji, C. X., and P. C. Searson. Fabrication of nanoporous gold nanowires. *Appl Phys Lett* 81:4437–9.

42. Wong, T. C., C. P. Li, R. Q. Zhang, and S. T. Lee. Gold nanowires from silicon nanowire templates. *Appl Phys Lett* 84:407–9.

43. Bhattacharyya, S., S. K. Saha, and D. Chakravorty. 2000. Silver nanowires grown in the pores of a silica gel. *Appl Phys Lett* 77:3770–2.

44. Barbic, M., J. J. Mock, D. R. Smith, and S. Schultz. 2002. Single crystal silver nanowires prepared by the metal amplification method. *J Appl Phys* 91:9341–5.

45. Malandrino, G., S. T. Finocchiaro, and I. L. Fragala. 2004. Silver nanowires by a sonoself-reduction template process. *J Mater Chem* 14:2726–8.

46. Husain, A., J. Hone, H. W. C. Postma, X. M. H. Huang, T. Drake, M. Barbic, A. Scherer, and M. L. Roukes. 2003. Nanowire-based very-high-frequency electromechanical resonator. *Appl Phys Lett* 83:1240–2.

47. Heath, J. R., and F. K. Legoues. 1993. A liquid solution synthesis of single-crystal germanium quantum wires. *Chem Phys Lett* 208:263–8.

48. Greytak, A. B., L. J. Lauhon, M. S. Gudiksen, and C. M. Lieber. 2004. Growth and transport properties of complementary germanium nanowire field-effect transistors. *Appl Phys Lett* 84:4176–8.

49. Ziegler, K. J., D. M. Lyons, J. D. Holmes, D. Erts, B. Polyakov, H. Olin, K. Svensson, and E. Olsson. 2004. Bistable nanoelectromechanical devices. *Appl Phys Lett* 84:4074–6.

50. Wu, Y. Y., and P. D. Yang. 2000. Germanium nanowire growth via simple vapor transport. *Chem Mater* 12:605–7.

51. Banerjee, D., J. Y. Lao, D. Z. Wang, J. Y. Huang, Z. F. Ren, D. Steeves, B. Kimball, and M. Sennett. 2003. Large-quantity free-standing ZnO nanowires. *Appl Phys Lett* 83:2061–3.

52. Dai, Y., Y. Zhang, Y. Q. Bai, and Z. L. Wang. 2003. Bicrystalline zinc oxide nanowires. *Chem Phys Lett* 375:96–101.

53. Novoselov, K. S., D. Jiang, F. Schedin, T. J. Booth, V. V. Khotkevich, S. V. Morozov, and A. K. Geim. 2005. Two-dimensional atomic crystals. *Proc Natl Acad Sci U S A* 102:10451–3.
54. Li, D., M. B. Muller, S. Gilje, R. B. Kaner, and G. G. Wallace. 2008. Processable aqueous dispersions of graphene nanosheets. *Nat Nanotechnol* 3:101–5.
55. Stankovich, S., D. A. Dikin, G. H. B. Dommett, K. M. Kohlhaas, E. J. Zimney, E. A. Stach, R. D. Piner, S. T. Nguyen, and R. S. Ruoff. 2006. Graphene-based composite materials. *Nature* 442:282–6.
56. Geim, A. K., and K. S. Novoselov. 2007. The rise of graphene. *Nat Mater* 6:183–91.
57. Li, D., and R. B. Kaner. 2008. Graphene-based materials. *Science* 320:1170–1.
58. Novoselov, K. S., A. K. Geim, S. V. Morozov, D. Jiang, Y. Zhang, S. V. Dubonos, I. V. Grigorieva, and A. A. Firsov. 2004. Electric field effect in atomically thin carbon films. *Science* 306:666–9.
59. Lee, C., X. D. Wei, J. W. Kysar, and J. Hone. 2008. Measurement of the elastic properties and intrinsic strength of monolayer graphene. *Science* 321:385–8.
60. Teweldebrhan, D., and A. A. Balandin. 2009. Modification of graphene properties due to electron-beam irradiation. *Appl Phys Lett* 94:013101.
61. Ghosh, S., I. Calizo, D. Teweldebrhan, E. P. Pokatilov, D. L. Nika, A. A. Balandin, W. Bao, F. Miao, and C. N. Lau. 2008. Extremely high thermal conductivity of graphene: Prospects for thermal management applications in nanoelectronic circuits. *Appl Phys Lett* 92:151911.
62. Dreyer, D. R., S. Park, C. W. Bielawski, and R. S. Ruoff. 2010. The chemistry of graphene oxide. *Chem Soc Rev* 39:228–40.
63. Li, X. L., X. R. Wang, L. Zhang, S. W. Lee, and H. J. Dai. 2008. Chemically derived, ultrasmooth graphene nanoribbon semiconductors. *Science* 319:1229–32.
64. Lee, G. D., C. Z. Wang, E. Yoon, N. M. Hwang, and K. M. Ho. 2008. The formation of pentagon-heptagon pair defect by the reconstruction of vacancy defects in carbon nanotube. *Appl Phys Lett* 92:043104.
65. Jozsa, C., M. Popinciuc, N. Tombros, H. T. Jonkman, and B. J. van Wees. 2008. Electronic spin drift in graphene field-effect transistors. *Phys Rev Lett* 100:236603.
66. Bunch, J. S., A. M. van der Zande, S. S. Verbridge, I. W. Frank, D. M. Tanenbaum, J. M. Parpia, H. G. Craighead, and P. L. McEuen. 2007. Electromechanical resonators from graphene sheets. *Science* 315:490–3.
67. Wang, X., L. J. Zhi, and K. Mullen. 2008. Transparent, conductive graphene electrodes for dye-sensitized solar cells. *Nano Lett* 8:323–7.
68. Zhou, J., P. Fei, Y. F. Gao, Y. D. Gu, J. Liu, G. Bao, and Z. L. Wang. 2008. Mechanical-electrical triggers and sensors using piezoelectric micowires/nanowires. *Nano Lett* 8:2725–30.
69. Hong, W. J., Y. X. Xu, G. W. Lu, C. Li, and G. Q. Shi. 2008. Transparent graphene/PEDOT-PSS composite films as counter electrodes of dye-sensitized solar cells. *Electrochem Commun* 10:1555–8.
70. Wu, J. B., H. A. Becerril, Z. N. Bao, Z. F. Liu, Y. S. Chen, and P. Peumans. 2008. Organic solar cells with solution-processed graphene transparent electrodes. *Appl Phys Lett* 92:263302.
71. Stoller, M. D., S. J. Park, Y. W. Zhu, J. H. An, and R. S. Ruoff. 2008. Graphene-based ultracapacitors. *Nano Lett* 8:3498–502.
72. Vivekchand, S. R. C., C. S. Rout, K. S. Subrahmanyam, A. Govindaraj, and C. N. R. Rao. 2008. Graphene-based electrochemical supercapacitors. *J Chem Sci* 120:9–13.
73. Wang, S. R., M. Tambraparni, J. J. Qiu, J. Tipton, and D. Dean. 2009. Thermal expansion of graphene composites. *Macromolecules* 42:5251–5.
74. Ganguli, S., A. K. Roy, and D. P. Anderson. 2008. Improved thermal conductivity for chemically functionalized exfoliated graphite/epoxy composites. *Carbon* 46:806–17.

75. Sanchez-Paisal, Y., D. Sanchez-Portal, N. Garmendia, R. Munoz, I. Obieta, J. Arbiol, L. Calvo-Barrio, and A. Ayuela. 2008. Zr-metal adhesion on graphenic nanostructures. *Appl Phys Lett* 93:053101.

76. Wang, S. R., Y. Zhang, N. Abidi, and L. Cabrales. 2009. Wettability and surface free energy of graphene films. *Langmuir* 25:11078–81.

77. Kalaitzidou, K., H. Fukushima, and L. T. Drzal. 2007. Mechanical properties and morphological characterization of exfoliated graphite–polypropylene nanocomposites. *Compos Part A Appl Sci Manuf* 38:1675–82.

78. Cho, D., S. Lee, G. M. Yang, H. Fukushima, and L. T. Drzal. 2005. Dynamic mechanical and thermal properties of phenylethynyl-terminated polyimide composites reinforced with expanded graphite nanoplatelets. *Macromol Mater Eng* 290:179–87.

79. Xie, S. H., Y. Y. Liu, and J. Y. Li. 2008. Comparison of the effective conductivity between composites reinforced by graphene nanosheets and carbon nanotubes. *Appl Phys Lett* 92:243121.

80. Mohanty, N., and V. Berry. 2008. Graphene-based single-bacterium resolution biodevice and DNA transistor: Interfacing graphene derivatives with nanoscale and microscale biocomponents. *Nano Lett* 8:4469–76.

81. Xu, M. S., D. Fujita, and N. Hanagata. 2009. Perspectives and challenges of emerging single-molecule DNA sequencing technologies. *Small* 5:2638–49.

82. Hu, W. B., C. Peng, W. J. Luo, M. Lv, X. M. Li, D. Li, Q. Huang, and C. H. Fan. 2010. Graphene-based antibacterial paper. *Acs Nano* 4:4317–23.

83. Berger, C., Z. M. Song, X. B. Li, X. S. Wu, N. Brown, C. Naud, D. Mayou, T. B. Li, J. Hass, A. N. Marchenkov, E. H. Conrad, P. N. First, and W. A. de Heer. 2006. Electronic confinement and coherence in patterned epitaxial graphene. *Science* 312:1191–6.

84. Hass, J., W. A. de Heer, and E. H. Conrad. 2008. The growth and morphology of epitaxial multilayer graphene. *J Phys-Condens Matter* 20:323202.

85. Emtsev, K. V., F. Speck, T. Seyller, L. Ley, and J. D. Riley. 2008. Interaction, growth, and ordering of epitaxial graphene on SiC{0001} surfaces: A comparative photoelectron spectroscopy study. *Phys Rev B* 77:155303.

86. Sutter, P. W., J. I. Flege, and E. A. Sutter. 2008. Epitaxial graphene on ruthenium. *Nature Mater* 7:406–11.

87. de Parga, A. L. V., F. Calleja, B. Borca, M. C. G. Passeggi, J. J. Hinarejos, F. Guinea, and R. Miranda. 2008. Periodically rippled graphene: Growth and spatially resolved electronic structure. *Phys Rev Lett* 100:056807.

88. Zhou, W., Y. Huang, B. Liu, J. Wu, K. C. Hwang, and B. Q. Wei. 2007. Adhesion between carbon nanotubes and substrate: mimicking the gecko foothair. *Nano* 2:175–9.

89. Zhi, L. J., and K. Mullen. 2008. A bottom-up approach from molecular nanographenes to unconventional carbon materials. *J Mater Chem* 18:1472–84.

90. Stoberl, U., U. Wurstbauer, W. Wegscheider, D. Weiss, and J. Eroms. 2008. Morphology and flexibility of graphene and few-layer graphene on various substrates. *Appl Phys Lett* 93:051906.

91. Shao, Q., G. Liu, D. Teweldebrhan, and A. A. Balandin. 2008. High-temperature quenching of electrical resistance in graphene interconnects. *Appl Phys Lett* 92:202108.

92. Dietzel, D., T. Monninghoff, L. Jansen, H. Fuchs, C. Ritter, U. D. Schwarz, and A. Schirmeisen. 2007. Interfacial friction obtained by lateral manipulation of nanoparticles using atomic force microscopy techniques. *J Appl Phys* 102:084306.

93. Jeong, H. E., and K. Y. Suh. 2009. Nanohairs and nanotubes: Efficient structural elements for gecko-inspired artificial dry adhesives. *Nano Today* 4:335–46.

94. Kashiwase, Y., T. Ikeda, T. Oya, and T. Ogino. 2008. Manipulation and soldering of carbon nanotubes using atomic force microscope. *Appl Surf Sci* 254:7897–900.

95. Wang, G. X., X. L. Gou, J. Horvat, and J. Park. 2008. Facile synthesis and characterization of iron oxide semiconductor nanowires for gas sensing application. *J Phys Chem C* 112:15220–5.

96. Ramanathan, T., A. A. Abdala, S. Stankovich, D. A. Dikin, M. Herrera-Alonso, R. D. Piner, D. H. Adamson, et al. 2008. Functionalized graphene sheets for polymer nanocomposites. *Nature Nanotechnol* 3:327–31.

97. Jang, B. Z., and A. Zhamu. 2008. Processing of nanographene platelets (NGPs) and NGP nanocomposites: A review. *J Mater Sci* 43:5092–101.

98. Schniepp, H. C., K. N. Kudin, J. L. Li, R. K. Prud'homme, R. Car, D. A. Saville, and I. A. Aksay. 2008. Bending properties of single functionalized graphene sheets probed by atomic force microscopy. *Acs Nano* 2:2577–84.

99. Szabo, T., O. Berkesi, P. Forgo, K. Josepovits, Y. Sanakis, D. Petridis, and I. Dekany. 2006. Evolution of surface functional groups in a series of progressively oxidized graphite oxides. *Chem Mater* 18:2740–9.

100. Xu, C., X. D. Wu, J. W. Zhu, and X. Wang. 2008. Synthesis of amphiphilic graphite oxide. *Carbon* 46:386–9.

101. Lotya, M., Y. Hernandez, P. J. King, R. J. Smith, V. Nicolosi, L. S. Karlsson, F. M. Blighe, et al. 2009. Liquid phase production of graphene by exfoliation of graphite in surfactant/water solutions. *J Am Chem Soc* 131:3611–20.

102. Vadukumpully, S., J. Paul, and S. Valiyaveettil. 2009. Cationic surfactant mediated exfoliation of graphite into graphene flakes. *Carbon* 47:3288–94.

103. Dikin, D. A., S. Stankovich, E. J. Zimney, R. D. Piner, G. H. B. Dommett, G. Evmenenko, S. T. Nguyen, and R. S. Ruoff. 2007. Preparation and characterization of graphene oxide paper. *Nature* 448:457–60.

104. Jiao, L. Y., X. J. Xian, Z. Y. Wu, J. Zhang, and Z. F. Liu. 2009. Selective positioning and integration of individual single-walled carbon nanotubes. *Nano Lett* 9:205–9.

105. Kosynkin, D. V., A. L. Higginbotham, A. Sinitskii, J. R. Lomeda, A. Dimiev, B. K. Price, and J. M. Tour. 2009. Longitudinal unzipping of carbon nanotubes to form graphene nanoribbons. *Nature* 458:872–5.

106. Tsai, J.-L., S.-H. Tzeng, and Y.-J. Tzou. 2010. Characterizing the fracture parameters of a graphene sheet using atomistic simulation and continuum mechanics. *Int J Solids Struct* 47:503–9.

107. Tombler, T. W., C. W. Zhou, L. Alexseyev, J. Kong, H. J. Dai, L. Lei, C. S. Jayanthi, M. J. Tang, and S. Y. Wu. 2000. Reversible electromechanical characteristics of carbon nanotubes under local-probe manipulation. *Nature* 405:769–72.

108. Minot, E. D., Y. Yaish, V. Sazonova, J. Y. Park, M. Brink, and P. L. McEuen. 2003. Tuning carbon nanotube band gaps with strain. *Phys Rev Lett* 90:156401.

109. Gonzalez, C., J. Ortega, F. Flores, D. Martinez-Martin, and J. Gomez-Herrero. 2009. Initial stages of the contact between a metallic tip and carbon nanotubes. *Phys Rev Lett* 102:106801.

110. Girifalco, L. A., M. Hodak, and R. S. Lee. 2000. Carbon nanotubes, buckyballs, ropes, and a universal graphitic potential. *Phys Rev B* 62:13104–10.

111. Lu, J. P. 1997. Elastic properties of carbon nanotubes and nanoropes. *Phys Rev Lett* 79:1297–300.

112. Ke, C.-H., M. Zheng, I.-T. Bae, and G.-W. Zhou. 2010. Adhesion-driven buckling of single-walled carbon nanotube bundles. *J Appl Phys* 107:104305.

113. Hsu, J. H., and S. H. Chang. 2010. Surface adhesion between hexagonal boron nitride nanotubes and silicon based on lateral force microscopy. *Appl Surf Sci* 256:1769–73.

114. Peng, B., M. Locascio, P. Zapol, S. Y. Li, S. L. Mielke, G. C. Schatz, and H. D. Espinosa. 2008. Measurements of near-ultimate strength for multiwalled carbon nanotubes and irradiation-induced crosslinking improvements. *Nat Nanotechnol* 3:626–31.

115. Kis, A., G. Csanyi, J. P. Salvetat, T. N. Lee, E. Couteau, A. J. Kulik, W. Benoit, J. Brugger, and L. Forro. 2004. Reinforcement of single-walled carbon nanotube bundles by intertube bridging. *Nat Mater* 3:153–7.

116. Jia, Z. J., Z. Y. Wang, C. L. Xu, J. Liang, B. Q. Wei, D. H. Wu, and S. W. Zhu. 1999. Study on poly(methyl methacrylate)/carbon nanotube composites. *Mater Sci Eng A Struct Mater Prop Microstruct Proc* 271:395–400.

117. Haggenmueller, R., F. M. Du, J. E. Fischer, and K. I. Winey. 2006. Interfacial in situ polymerization of single wall carbon nanotube/nylon 6,6 nanocomposites. *Polymer* 47:2381–8.

118. Leelapornpisit, W., M. T. Ton-That, F. Perrin-Sarazin, K. C. Cole, J. Denault, and B. Simard. 2005. Effect of carbon nanotubes on the crystallization and properties of polypropylene. *J Polym Sci Part B Polym Phys* 43:2445–53.

119. Chen, J., A. M. Rao, S. Lyuksyutov, M. E. Itkis, M. A. Hamon, H. Hu, R. W. Cohn, P. C. Eklund, D. T. Colbert, R. E. Smalley, and R. C. Haddon. 2001. Dissolution of full-length single-walled carbon nanotubes. *J Phys Chem B* 105:2525–8.

120. Ding, F., P. Larsson, J. A. Larsson, R. Ahuja, H. M. Duan, A. Rosen, and K. Bolton. 2008. The importance of strong carbon-metal adhesion for catalytic nucleation of single-walled carbon nanotubes. *Nano Lett* 8:463–8.

121. Ribas, M. A., F. Ding, P. B. Balbuena, and B. I. Yakobson. 2009. Nanotube nucleation versus carbon-catalyst adhesion—Probed by molecular dynamics simulations. *J Chem Phys* 131:224501.

122. Larsson, P., J. A. Larsson, R. Ahuja, F. Ding, B. I. Yakobson, H. M. Duan, A. Rosen, and K. Bolton. 2007. Calculating carbon nanotube-catalyst adhesion strengths. *Phys Rev B* 75:115419.

123. Jiang, H., X. Q. Feng, Y. Huang, K. C. Hwang, and P. D. Wu. 2004. Defect nucleation in carbon nanotubes under tension and torsion: Stone-Wales transformation. *Comput Methods Appl Mech Eng* 193:3419–29.

124. Samsonidze, G. G., G. G. Samsonidze, and B. I. Yakobson. 2002. Energetics of Stone-Wales defects in deformations of monoatomic hexagonal layers. *Comput Mater Sci* 23:62–72.

125. Tserpes, K. I., and P. Papanikos. 2007. The effect of Stone-Wales defect on the tensile behavior and fracture of single-walled carbon nanotubes. *Compos Struct* 79:581–9.

126. Wong, H. S., C. Durkan, and N. Chandrasekhar. 2009. Tailoring the local interaction between graphene layers in graphite at the atomic scale and above using scanning tunneling microscopy. *Acs Nano* 3:3455–62.

127. Lennard-Jones, J. E. 1930. Perturbation problems in quantum mechanics. *Proc R Soc A* 129:598–615.

128. Tang, T., A. Jagota, and C. Y. Hui. 2005. Adhesion between single-walled carbon nanotubes. *J Appl Phys* 97:074304.

129. Pantano, A., D. M. Parks, and M. C. Boyce. 2004. Mechanics of deformation of single- and multi-wall carbon nanotubes. *J Mech Phys Solids* 52:789–821.

130. Lopez, M. J., A. Rubio, J. A. Alonso, L. C. Qin, and S. Iijima. 2001. Novel polygonized single-wall carbon nanotube bundles. *Phys Rev Lett* 86:3056–9.

131. Buehler, M. J., Y. Kong, H. Gao, and Y. Huang. 2006. Self-folding and unfolding of carbon nanotubes. *J Eng Mater Tech* 128:3–10.

132. Chen, B., M. Gao, J. M. Zuo, S. Qu, B. Liu, and Y. Huang. 2003. Binding energy of parallel carbon nanotubes. *Appl Phys Lett* 83:3570–1.

133. Falvo, M. R., G. J. Clary, R. M. Taylor, V. Chi, F. P. Brooks, S. Washburn, and R. Superfine. 1997. Bending and buckling of carbon nanotubes under large strain. *Nature* 389:582–4.

134. Ke, C., M. Zheng, G. Zhou, W. Cui, N. Pugno, and R. N. Miles. 2010. Mechanical peeling of free-standing single-walled carbon nanotube bundles. *Small* 6:438–45.

135. Goussev, O. A., P. Richner, and U. W. Suter. 1999. Local bending moment as a measure of adhesion: The cantilever beam test. *J Adhesion* 69:1–12.
136. Landau, L. D., and E. M. Lifshitz. 1986. *Theory of Elasticity.* 3rd ed. Oxford, UK: Pergamon Press.
137. Hertel, T., R. E. Walkup, and P. Avouris. 1998. Deformation of carbon nanotubes by surface van der Waals forces. *Phys Rev B* 58:13870–3.
138. Zhang, Z., and T. Li. 2010. Graphene morphology regulated by nanowires patterned in parallel on a substrate surface. *J Appl Phys* 107:103519.
139. Li, T., and Z. Zhang. 2010. Substrate-regulated morphology of graphene. *J Phys D Appl Phys* 43:075303.
140. Akita, S., H. Nishijima, T. Kishida, and Y. Nakayama. 2000. Influence of force acting on side face of carbon nanotube in atomic force microscopy. *Jpn J Appl Phys Part 1 Regul Pap Short Notes Rev Pap* 39:3724–7.
141. Strus, M. C., L. Zalamea, A. Raman, R. B. Pipes, C. V. Nguyen, and E. A. Stach. 2008. Peeling force spectroscopy: Exposing the adhesive nanomechanics of one-dimensional nanostructures. *Nano Lett* 8:544–50.
142. Strus, M. C., C. I. Cano, R. B. Pipes, C. V. Nguyen, and A. Raman. 2009. Interfacial energy between carbon nanotubes and polymers measured from nanoscale peel tests in the atomic force microscope. *Compos Sci Technol* 69:1580–6.
143. Ishikawa, M., R. Harada, N. Sasaki, and K. Miura. 2009. Adhesion and peeling forces of carbon nanotubes on a substrate. *Phys Rev B* 80:193406.
144. Zheng, M., and C. H. Ke. 2010. Elastic deformation of carbon-nanotube nanorings. *Small* 6:1647–55.
145. Binnig, G., C. F. Quate, and C. Gerber. 1986. Atomic force microscope. *Phys Rev Lett* 56:930–3.
146. Bhushan, B., X. Ling, A. Jungen, and C. Hierold. 2008. Adhesion and friction of a multiwalled carbon nanotube sliding against single-walled carbon nanotube. *Phys Rev B* 77:165428.
147. Bhushan, B., and X. Ling. 2008. Adhesion and friction between individual carbon nanotubes measured using force-versus-distance curves in atomic force microscopy. *Phys Rev B* 78:045429.
148. Mate, C. M., G. M. Mcclelland, R. Erlandsson, and S. Chiang. 1987. Atomic-scale friction of a tungsten tip on a graphite surface. *Phys Rev Lett* 59:1942–5.
149. Lee, C., X. D. Wei, Q. Y. Li, R. Carpick, J. W. Kysar, and J. Hone. 2009. Elastic and frictional properties of graphene. *Phys Status Solidi B Basic Solid State Phys* 246:2562–7.
150. Enachescu, M., R. J. A. van den Oetelaar, R. W. Carpick, D. F. Ogletree, C. F. J. Flipse, and M. Salmeron. 1998. Atomic force microscopy study of an ideally hard contact: The diamond(111) tungsten carbide interface. *Phys Rev Lett* 81:1877–80.
151. Cook, J. W., S. Edge, and D. E. Packham. 1997. The adhesion of natural rubber to steel and the use of the peel test to study its nature. *Int J Adhes Adhes* 17:333–7.
152. Bagchi, A., and A. G. Evans. 1996. The mechanics and physics of thin film decohesion and its measurement. *Interface Sci* 3:169–93.
153. Li, X. J., W. Chen, Q. W. Zhan, L. M. Dai, L. Sowards, M. Pender, and R. R. Naik. 2006. Direct measurements of interactions between polypeptides and carbon nanotubes. *J Phys Chem B* 110:12621–5.
154. Poggi, M. A., P. T. Lillehei, and L. A. Bottomley. 2005. Chemical force microscopy on single-walled carbon nanotube paper. *Chem Mater* 17:4289–95.
155. Zheng, M., and C. Ke. 2011. Mechanical deformation of carbon nanotube nano-rings on flat substrate. Accepted for publication in *J Appl Phys* 109:074304.
156. Whittaker, J. D., E. D. Minot, D. M. Tanenbaum, P. L. McEuen, and R. C. Davis. 2006. Measurement of the adhesion force between carbon nanotubes and a silicon dioxide substrate. *Nano Lett* 6:953–7.

157. Hertel, T., R. Martel, and P. Avouris. 1998. Manipulation of individual carbon nanotubes and their interaction with surfaces. *J Phys Chem B* 102:910–5.

158. Desai, A. V., and M. A. Haque. 2007. Sliding of zinc oxide nanowires on silicon substrate. *Appl Phys Lett* 90:033102.

159. Manoharan, M. P., and M. A. Haque. 2009. Role of adhesion in shear strength of nanowire-substrate interfaces. *J Phys D Appl Phys* 42:095304.

160. Conache, G., S. M. Gray, A. Ribayrol, L. E. Froberg, L. Samuelson, H. Pettersson, and L. Montelius. 2009. Friction measurements of InAs nanowires on silicon nitride by AFM manipulation. *Small* 5:203–7.

161. Bordag, M., A. Ribayrol, G. Conache, L. E. Froberg, S. Gray, L. Samuelson, L. Montelius, and H. Pettersson. 2007. Shear stress measurements on InAs nanowires by AFM manipulation. *Small* 3:1398–401.

162. Ajayan, P. M., L. S. Schadler, C. Giannaris, and A. Rubio. 2000. Single-walled carbon nanotube-polymer composites: Strength and weakness. *Adv Mater* 12:750–3.

163. Barber, A. H., S. R. Cohen, and H. D. Wagner. 2003. Measurement of carbon nanotube-polymer interfacial strength. *Appl Phys Lett* 82:4140–2.

164. Jiang, L. Y., Y. Huang, H. Jiang, G. Ravichandran, H. Gao, K. C. Hwang, and B. Liu. 2006. A cohesive law for carbon nanotube/polymer interfaces based on the van der Waals force. *J Mech Phys Solids* 54:2436–52.

165. Tan, H., L. Y. Jiang, Y. Huang, B. Liu, and K. C. Hwang. 2007. The effect of van der Waals-based interface cohesive law on carbon nanotube-reinforced composite materials. *Compos Sci Technol* 67:2941–6.

166. Chowdhury, S. C., and T. Okabe. 2007. Computer simulation of carbon nanotube pull-out from polymer by the molecular dynamics method. *Compos Part A Appl Sci Manuf* 38:747–54.

167. Wei, C. Y. 2006. Adhesion and reinforcement in carbon nanotube polymer composite. *Appl Phys Lett* 88:093108.

168. Xu, X. J., M. M. Thwe, C. Shearwood, and K. Liao. 2002. Mechanical properties and interfacial characteristics of carbon-nanotube-reinforced epoxy thin films. *Appl Phys Lett* 81:2833–5.

169. Liao, K., and S. Li. 2001. Interfacial characteristics of a carbon nanotube-polystyrene composite system. *Appl Phys Lett* 79:4225–7.

170. Hashimoto, A., K. Suenaga, A. Gloter, K. Urita, and S. Iijima. 2004. Direct evidence for atomic defects in graphene layers. *Nature* 430:870–3.

171. Awasthi, A. P., D. C. Lagoudas, and D. C. Hammerand. 2009. Modeling of graphene-polymer interfacial mechanical behavior using molecular dynamics. *Model Simul Mater Sci Eng* 17:015002.

172. Lv, C., Q. Z. Xue, D. Xia, M. Ma, J. Xie, and H. J. Chen. 2010. Effect of chemisorption on the interfacial bonding characteristics of graphene-polymer composites. *J Phys Chem C* 114:6588–94.

13 Advances in Nanoresonators

Towards Ultimate Mass, Force, and Molecule Sensing

Changhong Ke and Qing Wei

CONTENTS

13.1 INTRODUCTION

Nanoresonators are one of the essential members in the family of nanoelectro-mechanical systems (NEMS). Made of vibrating components with characteristic dimensions in the sub-100 nm regime, nanoresonators present many unique physical characteristics, which are superior to their micro- and macroscale predecessors. For instance, nanoresonators could have their resonance frequencies in the gigahertz range, quality factors (Q-factors) of more than 10,000, and active masses on the order of femtograms (fg, 10^{-15} g) and even attograms (ag, 10^{-18} g). These extraordinary properties and features of nanoresonators enable a number of applications, such as mass and force sensors [1–6], molecule and chemical detectors [4,7–15], and signal processing [16–18] with unprecedented resolutions and performance. For instance, the detection limit of mass sensing using nanoresonators has been achieved even up to the yocotogram (yg, 10^{-24} g), which implies that nanoresonators can be regarded as mass spectroscopy with single atom resolution [19]. The development of nanoresonators sprouted in the middle of the 1990s and these are based on tiny single-crystal silicon (Si) beams manufactured from bulk materials using top–down micro- and nanofabrication approaches [20]. Such single-crystal material-based nanoresonators have been under continuous development during the past decades and have been popularly used, up to date, due to their ability to perform specific functions such as actuation and detection. Rapid advances in the synthesis, manipulation, and characterization of one-dimensional (1D) nanostructures such as carbon nanotubes [21] and a variety of nanowires fueled the development of nanoresonators during the past decade (2000s). Scientific breakthroughs in two-dimensional (2D) nanostructures, largely ascribed to the unveiling of single-layer and few-layer graphene in 2004 [22], allow nanoresonators to be built with atomically thin membranes [23]. Both 1D and 2D nanostructures are ideal building blocks for nanoresonators and other NEMS devices because of their high-elastic moduli, excellent flexibilities, low densities, and tiny dimensions. Critical device characteristics of nanoresonators based on carbon nanotubes and nanowires, such as resonant frequency, Q-factor, energy dissipation, device size, and integration level, are comparable, or even superior in some aspects, to their counterparts made of single-crystal materials, leading to a similar conclusion in the comparison of their application performance. In particular, advances in the direct integration of 1D and 2D nanostructures [24–26] are promising to enable large-scale, low-cost, bottom–up manufacturing of nanoresonators with high yields and uniform performance [27,28], thus avoiding the time-consuming and expensive top–down manufacturing of nanosize single-crystal beams. By using multiplexing actuation and detection schemes, the motion of each resonator in the device array can be individually controlled, leading to a dramatic increase in efficiency for many of their sensing and signal-processing applications.

Nonetheless, the extraordinary device behaviors and applications of nanoresonators pose many scientific and technical challenges, which are mainly ascribed to their intrinsic nature of tiny dimension and motion ranges, as well as high-resonant frequency. Unlike their macroscale counterparts that generally operate at unresonant frequencies, micro- and nanoscale resonators usually prefer to work in resonance states. Therefore, it is very challenging to detect the nanometer or even subnanometer motion of the vibrating component at a frequency of ~100 MHz or more. The motion detection becomes an even more demanding task when nanoresonators are required to operate in liquid environments for applications such as biomolecule detection because the existing motion-detection schemes are mostly for dry and vacuum environments. Fortunately, many fantastic breakthroughs in various aspects of nanoresonator research have taken place during the past few years, which keep pushing the development of nanoresonators to new stages and toward their ultimate application limits. For instance, newly emerged motion-detection schemes enable the detection of resonance motion of individual nanotubes in liquid [29], paving the way for in vitro resonance detection of chemical species and biomolecules. New configuration designs and actuation schemes of nanoresonators enable novel characteristics, such as sustainable self-oscillation [30,31] and symmetric tunable oscillation with enhanced stability [32]. Significant advances have also been witnessed for many applications of nanoresonators. For instance, mass sensitivity of nanoresonators has improved from femogram [33] and attogram [2] levels to the very recent zeptogram (zg, 10^{-21} g) [34,35] and yoctogram [19] levels.

The purpose of this chapter is to highlight some of the recent key advances in nanoresonator research, focusing on the new developments in resonator design and actuation, motion detection, and sensing applications. This chapter is organized into two parts: the first part focuses on advances in nanoresonators based on nanotubes and nanowires. The second part focuses on advances in graphene-based nanoresonators.

13.2   ADVANCES IN NANOTUBE- AND NANOWIRE-BASED NANORESONATORS

Carbon nanotubes are a cylindrical type of 1D nanostructure with extraordinary mechanical, electrical, and chemical properties [36–43]. The elastic modulus and strength of single-walled carbon nanotubes (SWCNTs), which are typically 1~3 nm in diameter, are ~1.2 TPa and ~100 GPa, respectively [42]. Considering their low density (1350 kg/m^3), carbon nanotube-based mechanical and electromechanical resonators possess high natural frequency characteristics that can be depicted by harmonic oscillator model that provides the natural frequency (f_0) in the form,

$$f_0 = \frac{1}{2\pi} \sqrt{\frac{k}{m_{\text{eff}}}} \tag{13.1}$$

where k is the stiffness of the vibrating beam, and m_{eff} is the effective mass of the beam and is dependent on its vibration mode. A natural frequency of more than 1.3 GHz has been reported for carbon nanotube resonators operating in air at room temperature [44]. Based on the fact that k is linearly proportional to the elastic modulus of the beam, the resonance characteristics of nanotubes and nanowires have been used to measure their elastic moduli, where the resonance of individual nanostructures was excited either thermally by controlling the temperature [45], electrostatically by applying alternating voltages [46], or mechanically by using piezo actuators [29]. The Q-factor of carbon nanotubes, a key parameter in assessing the energy dissipation, is reported to be in the range of 50~200 at room temperature [3,35,47] and 800~2000 at low temperatures (~20 K) [35,48]. Recently, Q-factors of more than 10,000 were reported for carbon nanotube resonators operating at milli-Kelvin temperatures [49]. It is noted that the microscopic energy dissipation mechanism in carbon nanotubes that determines their Q-factors is currently still not understood. Ultrahigh-frequencies and Q-factors were also observed in nanowire-based resonators. For instance, Feng et al. reported resonant frequencies of ~200 MHz and Q-factors of ~13,000 for bottom–up synthesized single-crystal Si nanowire resonators [50].

13.2.1 CAPACITIVE MOTION DETECTION

The mass and force sensing mechanisms based on nanoresonators can be easily recognized from Equation 13.1 by correlating the resonant frequency shift with the change of the stiffness and/or the effective mass of the vibrating nanostructure. Precise detection of the resonant frequency shift is critical to achieve the maximum mass and force resolutions. Capacitive sensing with a down-mixing scheme is one of the recently developed motion-detection techniques. This technique was first demonstrated in the study of a carbon nanotube-based tunable resonator [3].

The tunable resonator, as shown in Figure 13.1a, comprises a free-standing single- or few-walled carbon nanotube grown directly by chemical vapor deposition methods and suspended above a 1.2–1.5 μm wide and 500-nm-deep trench between the source (S) and drain (D) electrodes [3]. The nanotube motion is induced and detected using the electrostatic interaction with the gate electrode (G) underneath the nanotube. The driving voltage applied to the gate electrode contains both DC and AC components. The active electrostatic force produced by the DC component of the driving signal V_g^{DC} is used to pull down the nanotube beam, thus controlling its stiffness and resonant frequency. The resonance of the nanotube beam is excited by the AC component of the driving signal δV_g. The vibrational motion of the nanotube was detected by utilizing the transistor properties of semiconducting [51] and small-bandgap semiconducting [52,53], which show that the change of the nanotube's conductance is proportional to the change in the induced electrical charge on the nanotube. It is noted that the induced charge is dependent on the gap distance between the nanotube and the gate electrode. Therefore, the motion of the nanotube can be quantified by measuring the current passing through the nanotube. The nanotube acts as a mixer to down-mix the high-frequency (HF) current signal to the detectable level, thus bypassing the low-pass filtering effect because of the parasitic capacitance in the circuit [54]. This down-mixing scheme was also used in the motion detection of HF nanoresonators using piezoresistive sensing techniques [55]. The experimental setup of detecting the motion of the nanotube is illustrated in Figure 13.1b. A local oscillator is applied to the source electrode at a frequency offsetting the HF signal applied to the gate electrode by an intermediate frequency $\delta\omega$ of 10 KHz. The current going through the nanotube is detected using a lock-in amplifier by using $\delta\omega$ as the reference frequency. The magnitude of the current through the nanotube is given by [3]

$$\delta I = \frac{1}{2\sqrt{2}} \frac{dG}{dV_g}\left(\delta V_g + V_g^{DC}\frac{\delta C_g}{C_g}\right)\delta V_{sd} \tag{13.2}$$

where δV_{sd} is the AC voltage applied on the source electrode, G is the nanotube conductance, C_g is the capacitance between the nanotube and the gate electrode, and δC_g represents the variational capacitance caused by the motion of the nanotube. This capacitive motion-detection scheme requires a closed circuit path for current measurements and can only be used for resonators made of fixed–fixed beams. In other words, this technique may not be applied for motion detection in cantilevered beam-based resonators.

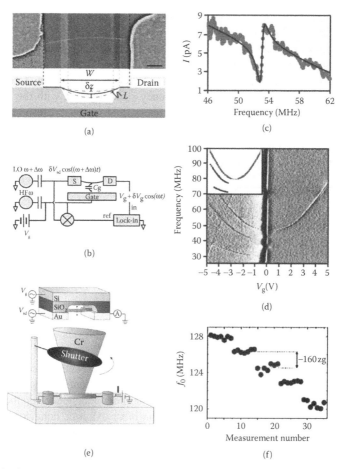

FIGURE 13.1 (a) (Top) Scanning electron microscopy image of a suspended carbon nanotube-based tunable oscillator and (bottom) a schematic drawing of the device configuration. The scale bar is 300 nm. (From Sazonova, V., Y. Yaish, H. Ustunel, D. Roundy, T. A. Arias, and P. L. McEuen. 2004. *Nature* 431:284–7. With permission.) (b) A diagram of the experimental setup for sensing the nanotube motion. (From Sazonova, V., Y. Yaish, H. Ustunel, D. Roundy, T. A. Arias, and P. L. McEuen. 2004. *Nature* 431:284–7. With permission.) (c) The plot of the measured current as a function of driving frequency. The solid black line represents a Lorentzian fit to the experimental data. (From Sazonova, V., Y. Yaish, H. Ustunel, D. Roundy, T. A. Arias, and P. L. McEuen. 2004. *Nature* 431:284–7. With permission.) (d) Detected current as a function of driving frequency and gate voltage. The current is plotted as a derivative in color scale. The data shown in (c) and (d) are for the same nanotube device. (From Sazonova, V., Y. Yaish, H. Ustunel, D. Roundy, T. A. Arias, and P. L. McEuen. 2004. *Nature* 431:284–7. With permission.) (e) Schematic of a carbon nanotube resonator for mass sensing. Chromium (Cr) atoms are deposited onto the nanotube resonator in a Joule evaporator. (From Lassagne, B., D. Garcia-Sanchcz, A. Aguasca, and A. Bachtold. 2008. *Nano Lett* 8:3735–8. With permission.) (f) Stepwise frequency shift-down upon the evaporation of ~160 zg of Cr on a nanotube resonator. The time interval between two points is 40 seconds except the duration of the Cr evaporation, which is 5 minutes. (From Lassagne, B., D. Garcia-Sanchez, A. Aguasca, and A. Bachtold. 2008. *Nano Lett* 8:3735–8. With permission.)

Figure 13.1c shows the measured current in the nanotube as a function of driving frequency, which displays a distinctive feature in the current on top of a slowly changing background. This feature results from the resonant motion of the nanotube, which modulates the capacitance, while the background is modulating gate voltage. Figure 13.1d shows the measured response as a function of the driving frequency and the static gate voltage. The resonant frequency shifts upward with the increase of the DC gate voltage. Several distinct resonances are observed, corresponding to different eigenmodes of the nanotube. The Lorentzian fit to the measured data reveals that the resonant frequency and Q-factor of the tested nanotube are 55 MHz and 80, respectively. It is worth mentioning that the measurements were performed in a vacuum less than 10^{-4} torr at room temperature. For the resonance measurement shown in Figure 13.1c, the estimated electrostatic force and the corresponding effective spring constant are about 60 fN and 4×10^{-4} N·m^{-1}, respectively. The smallest detectable motion of the nanotube was found to be ~0.5 nm on a resonant driving voltage of ~1 mV in the bandwidth of 10Hz. The corresponding force sensitivity was estimated to be ~1 fN Hz$^{-1/2}$. A force sensitivity less than 5 aN at low temperatures (~1 K) was achieved as a result of the minimized thermal-mechanical noises in the nanotube vibration [3].

It is noted that the resonant behavior of the aforementioned nanotube resonator can be significantly affected by the geometry of the driving electrode. Cho et al. [56] have recently demonstrated a tunable and broadband doubly clamped nanotube resonator using a localized electrostatic actuation scheme. The motion of the suspending nanotube is actuated using a needle-shaped electrode placed close to the middle point of the nanotube. The highly localized electrostatic force acting mostly on the central portion of the nanotube has a purely cubic order dependence on the nanotube displacement (middle point), resulting in a nonlinear dynamic behavior that has no favored resonance frequency. The resonance response of this nonlinear resonator is broadband in nature, and its resonant frequency can be tuned through the instantaneous energy of the system [57].

The capacitance-sensing scheme was also used in carbon nanotube-based resonators for ultrasensitive mass sensing [35], as shown in Figure 13.1e. By ignoring the effect of the adsorbate on the stiffness of the vibrating beam (k), the detected mass δm can be obtained from Equation 13.1, and its relationship with the resonant frequency shift δf_0 is given by [35]

$$\delta m = \frac{2m_{\text{eff}}}{f_0} \delta f_0 = \frac{\delta f_0}{\Re} \quad (13.3)$$

in which $\Re = \dfrac{f_0}{2m_{\text{eff}}}$ is defined as the mass responsivity and is a key parameter to evaluate the mass-sensing performance of nanoresonators. The nanofabrication of the resonator shown in Figure 13.1e and its motion actuation and detection schemes are similar to the tunable resonator shown in Figure 13.1a. Chromium (Cr) atoms were used as the adsorbate because of their good adhesion with carbon nanotubes. Cr atoms were deposited onto the surface of the nanotube by joule heating of a Cr bar in a metal evaporator with the evaporation rate set to the lowest detectable level (~10 pm·s^{-1}). Figure 13.1f shows the stepwise shift down of the resonant frequency for a nanotube

(1.2 nm in diameter and 900 nm in length) upon the depositing of ~160 zg of Cr, or ~1860 Cr atoms. The corresponding mass responsivity was reported to be $\Re =$ 11 Hz·yg^{-1}. The standard deviation of the measured resonant frequency of the nanotube at room temperature in their experimental setup is about 280 KHz. The mass resolution of the nanotube resonator was calculated to be 25 zg based on Equation 13.3. It is noted that the mass resolution increases with the measurement resolution of the resonant frequency, which can be increased by lowering the temperature. A mass resolution of 1.4 zg or 15 Cr atoms at temperature of ~5 K was observed [35].

13.2.2 PROBING VIBRATION MOTION BY ATOMIC FORCE MICROSCOPY

For the capacitive motion-sensing techniques discussed in Section 13.2.1, the motion of the vibrating structure is quantified in an average way based on the measured current passing through the structure. It is recognized that it is often a challenging task to assign the measured resonance peaks to their eigenmodes. To overcome this limitation, Garcia-Sanchez et al. [58] proposed a novel mechanical motion-detection scheme based on high-resolution atomic force microscopy (AFM), which is capable of mapping the eigenmode shape of a nanotube resonator in the gigahertz frequency range. The motion of the nanotube resonator, as shown Figure 13.2a, was electrostatically actuated through a side electrode. The scheme of tracking the oscillation motion of the nanotube as a function of time using an AFM probe is illustrated in Figure 13.2b, while the experimental setup for inducing and detecting the oscillation motion of the nanotube resonator is shown in Figure 13.2c. Because the scanning rate of the AFM is much lower compared with the resonant frequency of carbon nanotubes, the vibration profile of the nanotube cannot be measured directly by AFM. A vibration amplitude modulation scheme is used based on multiplying the driving signal $V_{ac} \cos(2\pi f_{RF}t)$ with a term of $1 - \cos(2\pi f_{mod}t)$. The driving frequency f_{RF} is tuned to the resonance frequency of the nanotube to maximize its vibration amplitude. The modulation frequency f_{mod} is used to modulate the vibration amplitude by periodically turning on and off the nanotube oscillation. By using this modulation scheme, the envelope of the vibration amplitude can be imaged using an AFM operating in the tapping mode. The vibration motion detection is optimized by setting the modulation frequency to the resonant frequency of the first eigenmode of the employed AFM cantilever, which has a vibration amplitude that is proportional to that of the nanotube and can be measured using a lock-in amplifier with f_{mod} as the reference frequency. The second eigenmode of the AFM cantilever is used for topography imaging. The four images in Figure 13.2d show the topography and the nanotube vibration images obtained at different eigenmode frequencies for a multiwalled carbon nanotube. The shapes of the measured vibration profiles reveal the respective bending modes based on the number of nodes. The three AFM images shown in (b through d) correspond to the first three eigenmodes of the nanotube and their respective eigenmode frequencies are 0.154, 0.475, and 1.078 GHz. It is noted that this AFM-based mechanical motion-detection scheme was also successfully extended to detect the vibration motion of graphene [23].

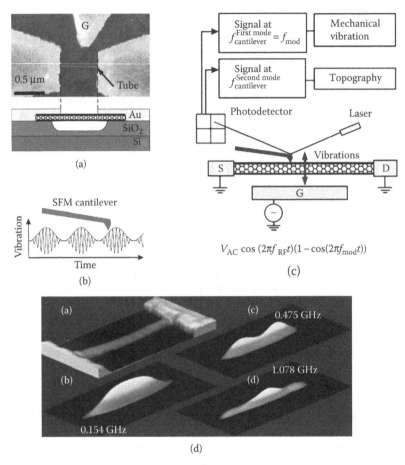

FIGURE 13.2 (a) (Top) Scanning electron microscopy image and (bottom) schematic of a doubly clamped carbon nanotube-based electromechanical resonator. The motion of the nanotube is actuated with a side electrode. (b) Schematic of tracking the oscillation motion of the nanotube as a function of time using an atomic force microscopy (AFM) probe. (c) Schematic of the experimental setup for inducing and detecting the oscillation motion of the nanotube resonator. A high-frequency (HF) term f_{RF} is used to match the resonance frequency of the nanotube, and a low-frequency term f_{mod} is used to modulate the HF oscillation. (d) AFM images of the topography and eigenmode vibrations for a 770-nm-long multi-walled carbon nanotube resonator. (From Garcia-Sanchez, D., A. S. Paulo, M. J. Esplandiu, F. Perez-Murano, L. Forro, A. Aguasca, and A. Bachtold. 2007. *Phys Rev Lett* 99:085501. With permission.)

13.2.3 CARBON NANOTUBE RADIO

The mechanical resonance of carbon nanotubes was utilized in the design of a novel carbon nanotube radio [59]. Unlike other proposed carbon nanotube radios [60,61] that are fully based on the electrical properties of carbon nanotubes, the nanotube radio proposed by Jensen et al. [59] relies on the mechanical resonance of the nanotube to fulfill the essential components of a radio, which includes antenna,

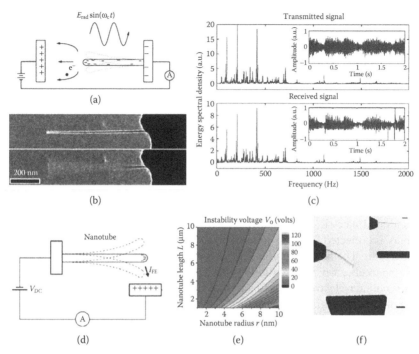

FIGURE 13.3 (a) Schematic of a nanotube radio. The resonant vibration of the nanotube is activated by the transmitted radio signal tuned to the nanotube's resonance frequency. (From Jensen, K., J. Weldon, H. Garcia, and A. Zettl. 2007. *Nano Lett* 7:3508–11. With permission.) (b) Transmission electron microscopy (TEM) images of a nanotube radio staying off and on resonance during a radio transmission. (From Jensen, K., J. Weldon, H. Garcia, and A. Zettl. 2007. *Nano Lett* 7:3508–11. With permission.) (c) Transmitted and received audio waveforms (inset) and frequency spectra for 2 seconds of the song "Good Vibrations" by the Beach Boys. (From Jensen, K., J. Weldon, H. Garcia, and A. Zettl. 2007. *Nano Lett* 7:3508–11.) (d) Schematic of the experimental setup used to actuate sustainable self-oscillations in carbon nanotubes. A DC bias voltage is applied between the nanotube and the counter electrode. A current meter is incorporated into the device circuit for measuring the field-emission current. (From Weldon, J. A., B. Aleman, A. Sussman, W. Gannett, and A. K. Zettl. 2010. *Nano Lett* 10:1728–33. With permission.) (e) Contour plot of the self-oscillation onset voltage for nanotubes with length (*L*) ranging from 1 to 10 µm and radius (*r*) ranging from 1 to 10 nm. (From Weldon, J. A., B. Aleman, A. Sussman, W. Gannett, and A. K. Zettl. 2010. *Nano Lett* 10:1728–33. With permission.) (f) TEM snapshot of the sustaining self-oscillation of a top–down fabricated carbon nanotube nanoresonator at a bias voltage of 40 V; the inset shows the nanotube at 0 V. The scale bar is 1 µm. (From Weldon, J. A., B. Aleman, A. Sussman, W. Gannett, and A. K. Zettl. 2010. *Nano Lett* 10:1728–33. With permission.)

tunable band-pass filter, amplifier, and demodulator. The nanotube radio, as shown in Figure 13.3a, is made of a cantilevered nanotube clamped to one electrode and placed close to another counter-electrode. The device circuit is incorporated with a DC power supply and a current meter. The applied DC voltage negatively charges the tip of the nanotube, which significantly increases its sensitivity to the external electromagnetic field. The nanotube is essentially working as an antenna through its vibration response to the transmitted radio signals. The oscillation of the nanotube becomes significant only when the frequency of the transmitted signal approaches its resonant frequency. The top and bottom transmission electron microscopy (TEM) images in Figure 13.3b show the off and on resonance states of a carbon nanotube radio respectively, during a radio transmission. The charging of the nanotube tip is evidenced by the bright intensity toward the tip portion of the standing-still nanotube in the captured TEM image.

By tuning the resonant frequency of the nanotube using methods such as current-induced length shortening [62], electrostatic force-induced nanotube stiffness [63], or effective mass [64] control, the nanotube radio can be tuned to receive only a preselected band of the electromagnetic spectrum. The amplification and demodulation functions rely on the field-emission properties of carbon nanotubes, which are largely due to their needle-shaped tip geometry that results in highly concentrated electric field. The field-emission current, produced by the applied DC bias, is collected from the counter-electrode and measured by the current meter in the circuit. The field-emission current is modulated by the mechanical vibration of the nanotube, which essentially varies the distance between the nanotube tip and the counter-electrode as a function of time. From the highly nonlinear dependence of the field-emission current on the gap distance, the motion of the nanotube can be faithfully reflected in the measured current signal. Using the mechanical resonance as the intermediate process, the transmitted signal can be captured in the form of electrical current signal, which can be sent to an audio speaker after signal amplification. The function of the nanotube radio was demonstrated by the results shown in Figure 13.3c that the nanotube radio faithfully reproduced the transmitted signal. This tiny nanotube radio, with a size that is four to five orders of magnitude smaller than those built with the existing semiconductor technologies, can be used for a number of communication applications in the microwave spectrum range, such as mobile phones, wireless local area network, and global positioning system [64]. It is noted that the correlation between the measured field-emission current and the vibration motion of the nanotube provides a new approach for detecting the vibration motion of nanotube resonators. This motion-detection scheme was used in the demonstration of a carbon nanotube-based active resonator with sustainable self-oscillation [30].

13.2.4 SUSTAINABLE SELF-OSCILLATING RESONATOR

The nanoresonators discussed so far can be categorized as "passive" resonators because alternating driving signals are required in order to drive and maintain their oscillations. By contrast, resonators operating with only DC power supplies are categorized as "active" resonators. Active resonators achieve the sustainable self-oscillation by incorporating an active feedback mechanism in their circuitry [30,31]. For the active nanotube resonator [30] shown in Figure 13.3d, the initial motion of the nanotube is directed toward the electrode due to a balanced interplay between the electrostatic force produced by the counter-electrode underneath the nanotube and the elastic force from the mechanical deflection of the nanotube. If the applied DC bias exceeds a threshold value, a sudden discharge occurs, resulting in significant decrease of charges on the nanotube, thus producing the electrostatic force. Because of the time delay in charging the nanotube as a result of the parasitic capacitance and resistance in the device circuit, the nanotube is pulled away from the counter-electrode by the dominating elastic force. The stepwise electrostatic force acting on the nanotube excites its oscillation, while the rapid repeat of the discharging and the pull-away processes leads to sustainable oscillation of the nanotube. The angle between the nanotube's longitudinal axis and the counter-electrode is identified to be a critical parameter in achieving sustainable self-oscillations. It is reported that a nanotube oriented parallel to the counter-electrode surface, as shown in Figure 13.3d, can self-oscillate, whereas a perpendicularly oriented nanotube as shown in Figure 13.3a with respect to the counter-electrode surface cannot perform the self-oscillation. The onset voltage of self-oscillations can be determined directly from the geometric parameters of the resonator, including the radius (r) and length (L) of the nanotube, its initial gap distance to the counter-electrode (d_0), and its initial orientation angle with respect to the normal direction of the counter-electrode surface (θ). The contour plot shown in Figure 13.3e depicts the dependence of the threshold voltage on parameters r and L by considering $d_0 = L$. Figure 13.3f shows the TEM snapshot of a carbon nanotube resonator in the sustainable self-oscillation stage with a DC bias of 40 V. Feng et al. [31] demonstrated an active NEMS resonator based on single-crystal SiC beams and an active feedback circuitry, which operates with sustaining self-oscillations in the ultrahigh-frequency range (≥ 300 MHz).

13.2.5 DETECTING RESONANCE MOTION IN LIQUID

The majority of the ultrasensitive detection measurements were performed in ultrahigh vacuum environments, in some cases at ultralow temperature to minimize thermal-mechanical noises. For applications of detecting biological molecules, the resonator is usually required to operate in the liquid environment, which poses several challenges: (1) The Q-factor of the resonator will be greatly reduced because of the damping by surrounding molecules in the liquid medium; (2) The detection of the resonator motion in liquid in response to the adsorption of the biomolecules is difficult because most of the existing motion-detection schemes are useful only in the dry and vacuum environments; and (3) Both biomolecules and liquid medium pose a fairly stringent limit on the operation temperature of resonators.

Akita and coworkers recently [29] investigated the motion detection of an cantilevered carbon nanotube-based resonator in water using an optical-detection scheme based on an in-house built inverted optical microscope, as illustrated in Figure 13.4a. The nanotube is mounted on top of a piezoelectric actuator, which drives its oscillation. Once a cantilevered nanotube is immersed into the water, its vibration is detected using optical illumination using a 633-nm laser. The nanotube vibration can be detected with high sensitivity by inducing the laser into the optical microscope to illuminate only the nanotube tip. Figure 13.4b shows the scanning electron microscopy (SEM) image of a representative cantilevered nanotube resonator used in their studies. The nanotube vibration motion is measured through detecting the change of the scattered light intensity in comparison with that of the nonvibrating nanotube. The frequency responses of the nanotube vibration at various frequencies are

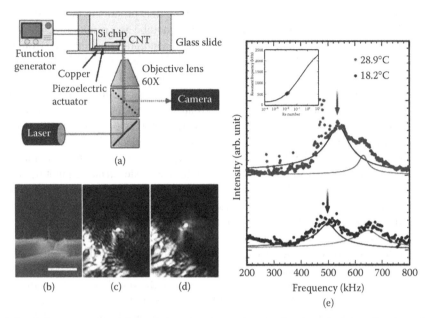

FIGURE 13.4 (a) Schematic of the experimental setup for detecting the vibration of a cantilevered carbon nanotube in liquids. (b) Scanning electron microscopy image and (c, d) optical images of a cantilevered carbon nanotube. The nanotube in (c) is nonvibrating in water, while vibrating in (d). (e) Frequency responses of the nanotube vibration in water at two different temperatures (28.9°C and 18.2°C). (Reprinted from Sawano, S., T. Arie, and S. Akita. 2010. *Nano Lett* 10:3395–8. With permission.)

recorded by capturing the scattered lights from the nanotube using charge-coupled device (CCD) cameras. Figure 13.4c and d shows the captured optical images of a nanotube in water in the nonvibrating and vibrating states, respectively. The vibration motion of the nanotube was obtained through imaging analysis based on the intensity of the captured images. Their results reveal that the nanotube lost its fundamental oscillation mode when immersed in water, which is ascribed to the fact that lower modes of oscillation are more strongly affected by the viscosity of fluids than higher modes as a result of their larger projected frontal areas and higher effective aspect ratios. To better understand the influence of the viscous resistance to the nanotube vibration, the frequency response of the nanotube in water was measured at different temperatures. Figure 13.4e shows the frequency spectra of the second harmonic oscillation of the nanotube in water at two different temperatures. At the temperature of 28.9°C, the spectra could be fitted to two theoretical curves with the resonant frequencies and Q-factors of 540 kHz and 7.6, and 630 kHz and 21.7, respectively. When the water temperature was reduced to 18.2°C, the resonant frequency of the nanotube shifted down to 500 kHz with Q-factor of 7.6. The observed decrease in resonant frequency was explained by the classical theory that the viscosity of water increases with the decrease of the water temperature.

13.2.6 COMBINED MASS AND STIFFNESS SENSING

Most of the existing mass-sensing schemes simply ignore the change of the stiffness of the resonating beam induced by surface stress from the deposition of the adsorbate on the body of the beam. This simplification is largely valid when both the thickness and stiffness of the adsorbate are much smaller compared with those of the resonator beam. For nanoresonators, the thickness of the adsorbate may become comparable to that of the vibrating nanostructure, thus resulting in significant deviations in the quantification of the adsorbate mass. Tamayo and coworkers [65] investigated the effect of the thickness and stiffness of the adsorbate on the frequency behavior of a thin cantilevered beam in response to molecular adsorption. Their studies show that the normalized resonant frequency shift caused by the adsorbate can be approximated by [65]

$$\frac{f_n - f_{0n}}{f_{0n}} \cong \alpha_1 \left(\frac{T_a}{T_c} \right) + \alpha_2 \left(\frac{T_a}{T_c} \right)^2 \tag{13.4}$$

$$\alpha_1 = \frac{1}{2}\left(3\frac{E_a}{E_c} - \frac{\rho_a}{\rho_c} \right) \quad \text{and} \quad \alpha_2 = \frac{3}{8}\left[\left(\frac{\rho_a}{\rho_c} \right)^2 + 2\frac{E_a}{E_c}\left(4 - \frac{\rho_a}{\rho_c} \right) - 7\left(\frac{E_a}{E_c} \right)^2 \right]$$

where T, E, ρ represent thickness, Young's modulus, and density, respectively. Subscripts a and c refer to the adsorbate and the cantilever, respectively. f_{0n} and f_n represent the resonant frequencies of the cantilever before and after the adsorbate deposition, respectively. Equation 13.4 shows that the relative frequency shift can be characterized by the constant α_1 if the adsorbate is much thinner than the cantilever. It is noted that α_1 is composed of two linear components that have an opposite influence on the frequency shift. The ratio of the elastic modulus of the adsorbate and the cantilever produces a positive frequency shift, whereas the ratio of the density (or mass) produces a negative effect. This effect arising from the coupling between the mass and the stiffness of the adsorbate becomes prominent when the thickness of the adsorbate layer is comparable to that of the cantilever. They demonstrated this effect by considering a cantilever made of Si ($\rho_c = 2330$ kg/m^3 and $E_c = 169$ GPa) and the adsorption of alkanethiol self-assembled monolayer ($\rho_a = 675$ kg/m^3 and $E_a = 12.9$ GPa). Results based on Equation 13.4 reveal three distinct behaviors depending on the thickness ratio of the adsorbate and the cantilever. It is found that the mass of the adsorbate dominates for $T_a/T_c < 0.04$, whereas the stiffness of the adsorbate dominates for $T_a/T_c > 0.1$. In between, the effects of the stiffness and mass offset each other, and the resonance frequency shift becomes insensitive to deposited molecules.

Equations 13.1 and 13.3 imply that the mass sensitivity is dependent on the position of the adsorbate on the vibrating structure. Positioning the adsorbate to the location of maximum amplitude, such as the free end of cantilever beams or the middle point of fixed–fixed beams, maximizes the change of the effective mass, and thus the resonant frequency shift. Dohn et al. [66] demonstrated that both the mass and location of the adsorbate can be uniquely determined based on resonant frequency shifts and several eigenmode vibration shapes. Using the energy approach and Rayleigh-Ritz method, the ratio of the added mass Δm to the original cantilever mass m_0 is given by [66]

$$\frac{\Delta m}{m_0} = \frac{1}{U_n^2(z_{\Delta m})}\left(\frac{f_{0n}^2}{f_n^2} - 1\right) \tag{13.5}$$

where U_n is the time-independent mode shape of the beam and $z_{\Delta m}$ is the position of the adsorbate. Once the adsorbate is deposited on the beam surface, both its mass Δm and location $z_{\Delta m}$ are deterministic and unique. Therefore, both the left and right sides of Equation 13.5 are constant, regardless of the bending mode of the beam. Because the measured value of $\frac{f_{0n}}{f_n}$ is dependent on the bending mode, the values of Δm and $z_{\Delta m}$ can be determined as the intersection point among the curves of $\frac{\Delta m}{m_0}$ for all the possible values of $\frac{Z_{\Delta m}}{L}$ for the first few bending modes.

Recently, Gil-Santos et al. [67] proposed a new nanomechanical-sensing scheme for the quantitative detection of the mass, stiffness, and position of the adsorbate by utilizing the imperfection-induced resonant frequency-splitting phenomenon in nanotubes and nanowires. For perfectly axial symmetrical 1D nanostructures, theoretical studies have revealed that their oscillations are composed of two orthogonal vibration components with the same resonant frequency and vibration amplitude [68]. However, any structural imperfection on the nanostructure, such as defects and deposited molecules, breaks the symmetry and leads to two orthogonal vibrations of diverse frequencies. Both the axis orientations and the resonant frequencies of these two orthogonal oscillations are dependent on the degree of the asymmetry of the nanostructure. This unique imperfection-induced resonant frequency-splitting characteristics of 1D nanostructures were experimentally demonstrated using a Si nanowire-based nanoresonator, as shown in Figure 13.5a. The employed Si nanowires with its length of 5–10 μm and diameter of 100–300 nm were grown horizontally from the prepatterned microtrenches on a Si substrate. The thermally induced oscillation of a Si nanowire resonator at room temperature was detected using optical interferometry techniques with a laser beam shining on the nanowire vertically from the top. The reflected lights from the nanowire and the substrate were collected by a photovoltaic Si pin diode. For the tested nanowires, the resonant frequency difference between the two detected vibration components, which is defined as the asymmetric factor (Ω), was measured to be ~0.7% from the captured displacement-frequency spectra.

Amorphous carbon molecules were used as adsorbates in the demonstration of the proposed nanomechanical-sensing scheme. The adsorbate was deposited to the surface of the nanowire using electron beam–induced deposition of hydrocarbon molecules available in the electron microscope chamber [69,70]. Figure 13.5b shows the evolution of the frequency spectra (top to bottom) of the nanowire in response to depositions of amorphous carbon (with its mass of 0.6 fg) near the nanowire clamped end. The dark-gray arrow in Figure 13.5b indicates the deposition angle, which was ~45° with respect to the vertical optical axis. As shown in Figure 13.5a, the measurements show that each deposition rotated the fast vibration plane by 7°–12° toward the deposition direction. Figure 13.5c and d show a Si nanowire of 100 nm in diameter

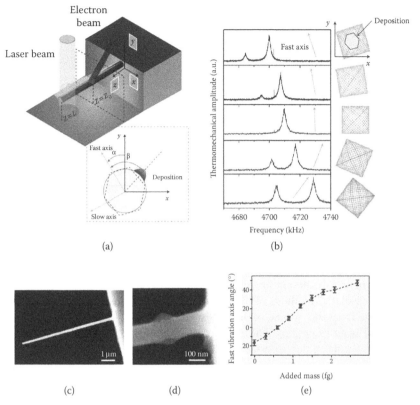

FIGURE 13.5 (a) Schematics of electron beam–induced deposition of carbon atoms on a nanowire and its low-frequency and high-frequency vibration axes (i.e., slow and fast axes). (b) Evolution of the frequency spectra (top to bottom) of the nanowire in response to depositions of atoms (with their mass of 0.6 fg) near the nanowire clamped end. Dark-gray arrow indicates the deposition angle. The other arrows indicate the rotation direction of the fast vibration plane upon deposition. (c, d) Scanning electron microscopy images of a 100-nm-thick nanowire before and after carbon deposition near the clamped end. (e) Angle between the fast vibration axis and the optical axis as a function of mass added to the clamped end of the nanowire. (From Gil-Santos, E., D. Ramos, J. Martinez, M. Fernandez-Regulez, R. Garcia, A. San Paulo, M. Calleja, and J. Tamayo. 2010. *Nat Nanotechnol* 5:641–5. With permission.)

before and after the amorphous carbon deposition near the clamped end, respectively. Figure 13.5e shows the angle between the fast vibration axis and the optical axis (α in Figure 13.5a) as a function of mass added to the clamped end of the nanowire.

To understand the dependences of the resonant frequency shifts and the rotation of the vibration planes on the mass, stiffness, and position of the adsorbate, the following relationships were obtained based on the Ritz formalism [67]:

$$\frac{\Delta f_S}{f_S} + \frac{\Delta f_F}{f_F} \cong \left\{ -\psi(z_0)^2 \frac{\rho_D}{\rho_{NW}} + \phi(z_0)^2 \frac{E_D}{E_{NW}} \right\} \frac{V_D}{V_{NW}} \qquad (13.6a)$$

$$\frac{\Delta f_F}{f_F} - \frac{\Delta f_S}{f_S} \cong \phi(z_0)^2 \frac{E_D}{E_{NW}} \frac{V_D}{V_{NW}} \cos(2\beta) \qquad (13.6b)$$

$$\Delta\alpha \cong \frac{\phi(z_0)^2}{2\Omega_0} \frac{E_D}{E_{NW}} \frac{V_D}{V_{NW}} \sin(2\beta) \qquad (13.6c)$$

where ψ and ϕ represent the nondimensional eigenmode amplitude and curvature, respectively, E is the Young's modulus, ρ is the mass density, V is the volume, Ω_0 is the initial asymmetry factor, and β is the angle between the deposition axis and fast vibration axis. Subscripts F and S refer to the slow and fast vibrations, respectively. Subscripts D and NW refer to the deposited material and the nanowire, respectively. Equation 13.6b and c clearly demonstrate that both the difference in the resonant frequency shift and the rotation angle are only affected by the stiffness of the adsorbate, while Equation 13.6a shows that the mass and stiffness of the adsorbate have opposite effects on the sum of the frequency shift. The influences of the added mass and the stiffness are dependent on the position of the adsorbate on the cantilevered nanowire. The maximum adsorbate stiffness effect occurs at the fixed-end of the nanowire, whereas the maximum adsorbate mass effect occurs at the free-end of the nanowire. The elastic modulus of the adsorbate can be estimated based on Equation 13.6b provided that the volume of the deposited material can be properly quantified. It is implied from Equation 13.6c that the rotation of the vibration axes reaches a maximum value when the deposition angle reaches 45° with respective to the vibration plane, and becomes zero when depositions occur along one of the vibration planes. The rotation degree sensitivity of the resonator reaches maximum when the deposition does not rotate the vibration axes, and was quantified to be ~50 ag per rotation degree. A mass resolution in the order of ~100 zg was envisioned if the measurement uncertainty of the rotation angle is controlled to be less than 0.01°. It was reported that this nanowire resonator-based sensing technique is capable of measuring the Young's modulus of the adsorbate at a resolution of ~0.1 kPa per femotogram of sample, which enables the quantification of the Young's modulus of single dried proteins that fall into the range of 0.1–10 GPa [67].

13.2.7 LARGE-SCALE MANUFACTURING OF NANOWIRE RESONATORS

Up to now, most of the studies on nanotube- and nanowires-based resonators focus on manufacture and characterization of individual prototyped devices. It is strongly desirable if nanoresonators can be manufactured in large scale with high reproducibility and uniform performance. Li et al. [28] investigated the large-scale manufacturing of nanoresonators based on Si and rhodium (Rh) nanowires using a combined top–down microfabrication and bottom–up nanowire self-assembly approach. The fabrication method of the nanowire resonator array is shown in Figure 13.6a. The employed Si and Rh nanowires were presynthesized using the gold catalyst-assisted vapor–liquid–solid method 71 and the anodic aluminum oxide nanopore-based electro-deposition method 72, respectively. The nanowire resonator array was

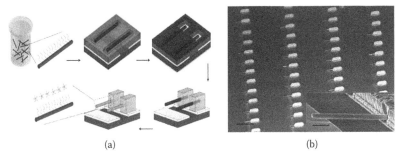

(a)   (b)

FIGURE 13.6 (a) Schematic of the bottom–up integration method to fabricate nanowire resonator arrays. (b) Fabricated nanowire resonator arrays. The scale bars in the resonator array and the inset images are 20 μm and 1 μm, respectively. (Reprinted from Li, M. W., R. B. Bhiladvala, T. J. Morrow, J. A. Sioss, K. K. Lew, J. M. Redwing, C. D. Keating, and T. S. Mayer. 2008. *Nat Nanotechnol* 3:88–92. With permission.)

formed by self-assembly of individual nanowires in solution to the predefined patterns on a Si substrate. An array of metal electrodes, which are used to guide the self-assembly of nanowires and to actuate the nanowire resonators, were first patterned on a Si wafer surface. A thin sacrificial layer of photoresist was then deposited and patterned to form an array of wells covering the underlying electrode patterns. An alternating electrical signal was applied to the electrodes to generate a local electric field to polarize and attract the nanowires in suspension to the patterned photoresist wells. The alignment of the nanowires along the well direction was guided and retained by three types of physical forces: (1) the long-range dielectrophoresis force that is ascribed to the nonlinearity of the electric field; (2) the short-range capacitive force between the nanowires and the guiding electrodes; and (3) the capillary force produced during the evaporation of the suspension liquid within each well. A second photoresist layer was deposited and patterned to define open windows, through which a thick metal layer was electrodeposited to clamp one end of the aligned nanowire in each well. The photoresist layers were subsequently dissolved to suspend the clamped nanowires and to remove those unclamped nanowires resulting from misalignment in the wells. The SEM images in Figure 13.6b show the fabricated nanowire resonator array and a magnified view of one nanowire resonator, clearly indicating that the cantilevered nanowire is straight and parallel to the bottom electrode surface.

The yield of the fabricated nanowire resonator array was reported to be ~80%, with more than 2000 self-assembled well-aligned nanowires. The fabricated nanowire resonator arrays can be easily actuated through applications of AC voltages to the driving electrode. The metal clamp provides the direct electrical connection between the nanowire and the chip, which can be used to drive each resonator using a multiplexing scheme. The reported bottom–up nanowire self-assembly scheme can be extended to the manufacturing of NEMS device arrays based on nanotubes and other types of nanowires with multiplexing capabilities.

13.2.8 DUAL-SIDED ACTUATED NANORESONATORS

Regarding the excitation scheme, resonators can be electrostatically actuated in either a single- or double-side manner. Studies on nanotube- and nanowire-based nanoresonators reported in the literature have focused on single-side-driven devices. Recently, our research group investigated the resonant characteristics of a double-side-driven nanoresonator [32]. The nanoresonator, as shown in Figure 13.7a and b, comprises a conductive carbon nanotube (or nanowire) cantilever with length L and outer diameter R, which is actuated by two parallel-plate electrodes (I and II). Both electrodes have an equal separation with the nanotube, H. The driving signals applied on the two electrodes have the same DC bias V_{dc}, AC amplitude V_{ac}, and period T, while having a 180° difference in phase. The nanotube cantilever is actuated in a "pull and push" manner, and its oscillation is symmetric with respect to its original unactuated position, meaning zero static deflection of the nanotube beam.

The unique characteristic of the dual-side actuated resonator compared with its single-sided counterpart is its symmetric tunable oscillation with larger stable motion range. It is recognized that resonant frequency tuning is a key aspect in the functioning and performance of nanoresonators. This is because challenges associated with nanodevice fabrication and unique nanoscale phenomena, such as van der Waals interactions, quantum effects, and molecular contamination, have a much more significant impact on the nanodevice's dynamic behavior, compared with its micro or larger scale counterparts. The approach of tuning the resonant frequency in such devices is naturally coupled with their actuation schemes. For single-sided tunable resonators, such as the device shown in Figure 13.1a, the nanostructure is actuated in a "pull only" manner by attractive electrostatic and van der Waals forces. In this single-side actuation design, the maximum stable oscillation range of the nanostructure is adversely decreased as a result of the resonant frequency tuning. This is because the constant force component in the electrostatic loading induces a static deflection, pulling down the nanostructure and, thus, reducing its gap distance to the electrode. The maximum stable oscillation amplitude of the nanostructure decreases because the unstable dynamic pull-in phenomenon [32,73–76] occurs at a smaller oscillation amplitude. Consequently, the smaller stable oscillation range of the nanostructure makes sensing of its nanoscale motion an even more challenging task. These limitations do not arise in resonators using the dual-side actuation scheme.

The resonant motion of the dual-side-driven resonator was investigated using the energy method [32]. From a system point of view, the added energy to the resonator is the electrostatic energy provided by the parallel-plate electrodes or the power supply, while the dissipated energy can be approximated by the energy loss resulting from the mechanical damping. For each oscillation period T, if the added energy is equal to the dissipated energy, the oscillation of the nanotube reaches its equilibrium or steady-state. If the added energy cannot be balanced by the dissipated energy, dynamic pull-in [74], or resonant pull-in when the nanotube is in resonance, takes place and the cantilever beam subsequently

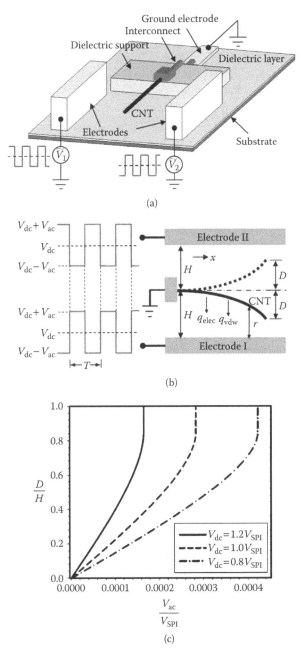

(a)

(b)

(c)

FIGURE 13.7 (a) Three-dimensional blueprint of a double-side-actuated cantilevered nanotube or nanowire-based tunable electromechanical resonator. (b) Schematic illustrations of the double-side-actuated resonator (right) and driving signals (left). (c) The steady-state resonant characteristics, including resonant pull-in, of the double-side-driven resonator for three tuned resonant frequencies by V_{dc}. (Reprinted from Ke, C.-H. 2009. *J Appl Phys* 15:024301. With permission.)

snaps onto one of the electrodes. The tuned resonant frequency of the nanotube, f_{tun}, is determined by the bias voltage V_{dc} and can be approximated by [32]

$$\frac{f_{tun}}{f_n} \approx \sqrt{1 - 0.364\left(\frac{V_{dc}}{V_{SPI}}\right)^2 \left(\frac{2}{\ln(2H/R+2)}+1\right)} \qquad (13.7)$$

where $V_{SPI} = 0.85\dfrac{H+R}{L^2}\ln\left(\dfrac{2(H+R)}{R}\right)\sqrt{\dfrac{EI}{\varepsilon_0}}$ is the quasistatic pull-in voltage of cantilevered nanotube devices with the single-side electrostatic actuation [77]. Here EI is the flexural stiffness of carbon nanotubes, and ε_0 is the permittivity of vacuum. The steady-state resonance amplitude of the nanotube cantilever D can be obtained from the following relationship [32]:

$$\frac{S(D)}{[D/(H+R)]^2} = \frac{0.549\,V_{SPI}^2 L}{QV_{dc}V_{ac}\left[\ln(2H/R+2)\right]^2}\sqrt{1 - 0.364\left(\frac{V_{dc}}{V_{SPI}}\right)^2\left(\frac{2}{\ln(2H/R+2)}+1\right)}:$$

$$(13.8)$$

It is noted that $2\pi\varepsilon_0 S(D)$ represents the capacitance difference between the two nanotube–electrode combinations. Equations 13.7 and 13.8 clearly suggests that tuning the resonant frequency and controlling the steady-state resonance amplitude of the nanotube resonator are decoupled and can be controlled separately by V_{dc} and V_{ac}, respectively. Figure 13.7c shows the steady-state oscillation amplitude of a dual-side-driven nanotube as a function of the AC voltage V_{ac} for three different resonant frequencies tuned by the DC bias V_{dc}. Our results show the maximum stable steady-state oscillation range of the dual-side-driven nanotube resonator is independent of the electrostatic force-based resonant frequency tuning and can reach up to 90% of the gap between the driving electrodes [32].

13.2.9 RESONANCE-BASED DETECTION OF SINGLE DNA MOLECULES

Detection of biological molecules, such as DNA and proteins, is one of the important applications of nanoresonators. DNA, as the genetic information carriers for all living species, has a very unique structure stabilized by a variety of physical interactions. Recent studies show that single-stranded DNA can bind to the side surface of SWCNTs in a helical manner [78,79]. The nanotube-DNA binding interaction is largely due to the π-π stacking interaction between nucleobases and nanotube surfaces. The nanotube-DNA hybrids are promising for a number of important biological and nano science and engineering applications, such as DNA transportation and thermal ablation treatment [80], sorting [78] and patterned placement [81] of nanostructures and many other novel sensing and biomedical applications [82–85].

Detecting single DNA molecules using a carbon nanotube nanoresonator is shown in Figure 13.8a. The attachment of DNA introduces three factors that affect the resonance of the nanotube: (1) increase of the dynamical mass, (2) surface stress resulting from the adhesion interactions between DNA nucleobases and nanotube surfaces, and (3) electrostatic repulsion between negative charges on DNA

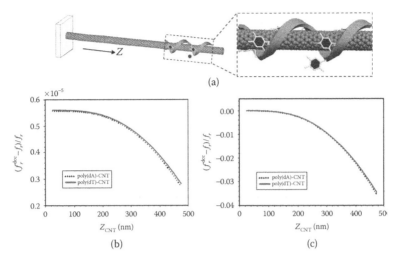

(a)

(b)   (c)

FIGURE 13.8 (a) Schematic of a DNA segment helically bound to a cantilevered carbon nanotube. (b, c) The normalized resonance frequency of a carbon nanotube upon binding of short poly(dT) and poly(dA) segments. The hydrodynamic loading effect is *considered* in the results shown in plot (b), while it is *omitted* in the results shown in plot (c). (Reprinted from Zheng, M., K. Eom, and C.-H. Ke. 2009. *J Phys D Appl Phys* 42:145408. With permission.)

backbones. The biological nature of DNA requires that the detection operation take place in buffer. Therefore, the hydraulic effect may also play an important role in DNA detection using nanoresonators. Our research group investigated the role of the helically wrapped DNA on the dynamical response of nanotubes in an aqueous environment based on a continuum model of nanotube–DNA hybrid structures [13].

The equilibrium helical wrapping configuration of DNA on a straight nanotube segment can be determined by minimizing the potential energy per unit nanotube length with respect to the wrapping angle [86]. We investigated the resonant response of the nanotube–DNA complex based on the variational method with a Hamiltonian, which considers the bending energy and nanotube–DNA binding interactions. If the DNA chain is much shorter than the nanotube, the resonant frequency of the nanotube–DNA complex f_r^{dcc} can be approximated by the following closed-form analytical solutions:

$$\frac{f_r^{\text{dcc}}}{f_r} \approx \sqrt{\frac{1+\eta/EI}{1+\left[v(z_{\text{DNA}})\right]^2 L_{\text{DNA}}\mu_{\text{DNA}}/(\mu_{\text{CNT}}+\mu_{\text{hydro}})}} \tag{13.9}$$

where μ_{CNT}, μ_{hydro}, and μ_{DNA} represent the mass per unit length for nanotube, hydrodynamic loading, and DNA molecule, respectively. $v(z_{\text{DNA}})$ is an admission function of the position of the DNA segment on the nanotube, z_{DNA}. $f_r = f_n(1+\mu_{\text{hydro}}/\mu_{\text{CNT}})^{-1/2}$ is the resonant frequency of the nanotube in water. L_{DNA} is the contour length of the DNA segment. η is a positive quantity related to the electrostatic interaction among charges on the DNA backbone, implying that the electrostatic interaction induces a positive resonant frequency shift. EI is the flexural stiffness of the nanotube. It is noted that the hydrodynamic loading effect is considered to have a more profound influence than the damping effect on the resonance behavior of nanotubes in aqueous environments [87–90]. It is noted that Equation 13.9 is also valid for the nanotube in the fixed–fixed beam configuration provided that an admission function $v(z)$ satisfying the fixed–fixed boundary conditions is used.

We examined the resonant frequency of a nanotube with its diameter of 2 nm and length of 500 nm to the binding of two types of single-stranded DNA homopolymers, such as polydeoxyadenylate (poly(dA)), and deoxythymidylate (poly(dT)), respectively. The DNA segments are assumed to be 45-base in length, that is, $L_{DNA} \approx 31.5$ nm. The molecule weights of dT and dA are 305.901 and 314.911 Da, respectively. The effective adhesion energy for dT-nanotube is 3.3 kcal/mol, and 2.32 kcal/mol for dA-nanotube [91]. The lateral gap distance of the nanotube surface to the DNA is assumed to be 1 nm [91] in the calculation. Figure 13.8b and c show the resonant frequency shifts of the nanotube in response to DNA binding with and without considering the hydrodynamic loading effect, respectively. Our results show that the influence of the DNA-binding interaction on the resonant response of the nanotube is dependent on the binding location of the DNA chain and becomes more prominent when the DNA stays closer to the free end of the nanotube. Our results also reveal that the DNA dynamic mass plays a much more prominent role in the resonant frequency of the nanotube than electrostatic interactions. In addition, the role of the DNA dynamic mass in the resonance behaviors of the nanotube–DNA complex is strongly impacted by the hydrodynamic loading effect. Our studies suggest that nanotube resonator-based DNA-detection measurements should be conducted in a medium with low hydrodynamic loading factors.

13.3 ADVANCES IN GRAPHENE-BASED NANORESONATORS

Single-layer graphene is a 2D nanostructure with the thickness of only one atom. Similar to carbon nanotubes, graphene possesses exceptional mechanical, electrical, and thermal properties, such as high Young's modulus (~1 TPa) and strength (~130GPa) [92] and semimetallic electrical properties [22], which make graphene an ultimate building block for nanoresonators [27,93].

13.3.1 FABRICATION AND ACTUATION OF GRAPHENE RESONATORS

The first graphene-based electromechanical resonator was reported by Bunch et al. in 2007 [93]. The configuration of their graphene resonators is illustrated in Figure 13.9a. A graphene sheet is doubly clamped to a SiO_2 trench by van der Waals force. The graphene resonator is actuated using either an electrical or optical modulation scheme. In the case of electrical modulation, the driving signal contains a time-varying component at frequency f and a constant bias voltage. For optical actuation, the intensity of a diode laser focused on the graphene sheet is modulated at frequency f, causing a periodic contraction/expansion of the graphene membrane that leads to oscillation. In both cases, the motion is detected by monitoring the reflected light intensity from a second laser with a fast photodiode. The employed graphene sheets were obtained using mechanical exfoliation methods [22]. The number of the graphene layers was determined by AFM and Raman spectroscopy. Figure 13.9b

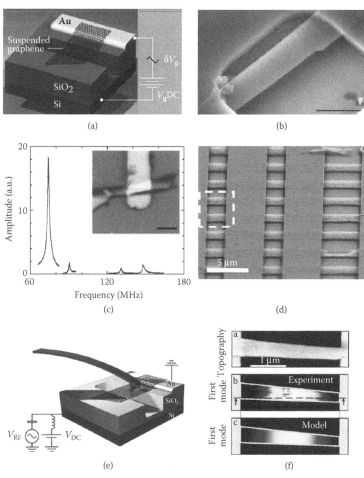

FIGURE 13.9 (a) Schematic and (b) scanning electron microscopy (SEM) image of a suspended graphene resonator. (From Bunch, J. S., A. M. van der Zande, S. S. Verbridge, I. W. Frank, D. M. Tanenbaum, J. M. Parpia, H. G. Craighead, and P. L. McEuen. 2007. *Science* 315:490–3. With permission.) (c) Plot of oscillation amplitude versus frequency for a 15-nm-thick graphene resonator. The inset shows an optical image of the resonator. (From Bunch, J. S., A. M. van der Zande, S. S. Verbridge, I. W. Frank, D. M. Tanenbaum, J. M. Parpia, H. G. Craighead, and P. L. McEuen. 2007. *Science* 315:490–3. With permission.) (d) Scanning electron microscopy image of an array of suspended graphene sheets. (From van der Zande, A. M., R. A. Barton, J. S. Alden, C. S. Ruiz-Vargas, W. S. Whitney, P. H. Q. Pham, J. Park, J. M. Parpia, H. G. Craighead, and P. L. McEuen. 2010. *Nano Lett* 10:4869–73. With permission.) (e) Schematic of measuring the vibration of a graphene resonator using atomic force microscopy. (From Garcia-Sanchez, D., A. M. van der Zande, A. S. Paulo, B. Lassagne, P. L. McEuen, and A. Bachtold. 2008. *Nano Lett* 8:1399–403. With permission.) (f) (Top) The measured topography of a graphene resonator with no buckling. (Middle) The measured fundamental mode shape of the graphene sheet at 31 MHz. (Bottom) The predicted fundamental mode shape of the graphene sheet at 31 MHz obtained using FEM simulations without any stress. (From Garcia-Sanchez, D., A. M. van der Zande, A. S. Paulo, B. Lassagne, P. L. McEuen, and A. Bachtold. 2008. *Nano Lett* 8:1399–403. With permission.)

shows a fabricated graphene resonator. Figure 13.9c shows the measured vibration amplitude-frequency spectrum for a 15-nm-thick graphene sheet suspended over a 5-μm trench. The inset optical image in Figure 13.9c shows the top view of the tested device. Multiple resonances were observed on the amplitude-frequency spectrum. The most prominent peak at the lowest frequency is associated with the fundamental vibrational mode because its detected intensity is the largest when the motion is in-phase across the entire suspended section. The measured amplitude-frequency spectra can be well fitted to Lorentzian curves, from which the resonant frequency and Q-factor of the graphene sheet can be obtained. The resonant frequencies and Q-factors of the tested graphene resonators were reported in the range of 1–170 MHz, and of 20~850, respectively. It is noted that the reported Q-factor of graphene is close to that of SWCNTs [3,35,47]. Their measurements show that the Q-factor of graphene has no discernible dependence on graphene thickness. Similar to carbon nanotubes, the temperature has a substantial influence on the Q-factor of the graphene. High Q-factor operation of graphene resonators is realized at low temperature, which was demonstrated in measurements reported in 2008 [27]. The applications of graphene resonators in force and charge sensing were also examined. A force sensitivity of 0.9 fN/Hz$^{1/2}$ and charge sensitivity of 8×10^{-4} e/Hz$^{1/2}$ were reported for a graphene resonator with its thickness of 5 nm, length of 2.7 μm, and width of 630 nm, operating at room temperature [93]. It was envisioned that graphene resonators are promising to rival single electron transistor electrometers as charge sensors at low temperature.

Large-scale integration of single-layer graphene to form graphene resonator arrays was recently reported [27]. Single-layer graphene sheets were first synthesized on copper foil using chemical vapor–deposition methods [94], and then transferred to substrate using a variety of graphene-transferring techniques [94–96]. Figure 13.9d shows an array of graphene sheets suspended over trenches in Si oxide. The graphene sheets were patterned using photolithography and oxygen plasma techniques [94]. The yield of the fabricated graphene devices was reported to be >80% for graphene sheets with length less than 3 μm and width less than 5 μm. It is noted that complicated conformational structures, such as small-scale ripples and large-scale buckling of the membrane along the length and width directions were observed and reported [27]. Fully clamping all sides of the suspended graphene sheet was reported to improve the reproducibility of the graphene resonator. Quality factors up to 9000 were achieved for the graphene resonators in the fabricated device array at low temperature (10 K) [27], which is desirable for ultrasensitive mass, force, and electrical charge sensing applications.

Detection of the vibration motion of graphene is one of the critical issues in studying the fundamental properties and applications of graphene. Figure 13.9e shows an AFM-based motion-detection scheme [23], which is capable of directly measuring the eigenmode shapes of vibrating graphene sheets. Similar to the scheme used in detecting the vibration of carbon nanotubes [58], the HF driving signal is modulated by a low-frequency signal so that the oscillation profile of the graphene can be measured by tapping mode AFM. The top image in Figure 13.9f shows the topography of a graphene resonator obtained using an AFM. The high-resolution AFM image clearly reveals irregularities on the graphene surface, and contaminations or bulk graphite residues were speculated as the possible sources for the observed structural imperfections. The middle image in Figure 13.9f shows the measured fundamental mode shape of the graphene sheet shown in the top image at 31 MHz, which is in a good agreement with the theoretically predicted mode shape of the graphene (bottom image) using finite element method (FEM) simulations. By using this motion-detection scheme, two distinct types of eigenmodes in suspended graphene sheets were identified and named as "beam modes" and "edge modes" based on the observed vibration profiles [23]. Beam modes refer to vibrations that are uniform across the width of the sheet, whereas edge modes refer to vibrations that have the largest oscillation amplitude along one of the free edges. These eigenmode shapes would be impossible to identify using other measurement techniques, such as optical or capacitive detection, which depend on the average position of the resonator.

13.3.2 MODELING OF GRAPHENE RESONATORS

For single and few-layer graphene resonators as shown in Figure 13.9a and f, the vibration of the graphene sheet is largely in the nonlinear regime as a result of the dominant role of the membrane-stretching effect compared with the membrane-bending effect. Atalaya et al. [97] studied the mechanical behavior of suspended graphene sheets using several continuum elastic models. Their models calculate the stretching and bending energies in graphene sheets based on the atomistic potential interaction between carbon atoms in graphene. A full nonlinear elastic model, which is used to determine the equilibrium shape of the graphene, was developed based on the dynamic behavior of a 2D membrane. The motion of a graphene sheet under an external body force $\bar{F}_0$ is given by [97]

$$\ddot{\bar{u}}(x,y) + c\dot{\bar{u}}(x,y) = \rho_0^{-1}\Re\hat{P}\left[\bar{u}(x,y)\right] + m_c^{-1}\bar{F}_0(x,y,t) \tag{13.10}$$

where $\bar{u}(x,y)$ represents the 3D deformation of the membrane in the Cartesian coordinate system, c is the damping factor, ρ_0 is the graphene mass density, m_c is the carbon mass, $\hat{P}$ is the Piola stress tensor [98], and $\Re$ is a linear differential operator acting on $\hat{P}$.

The general elasticity Equation 13.10 was simplified to the von Karman equations through neglecting the second-order in-plane displacements. By totally eliminating the in-plane displacements, the out-of-plane equation for the out-of-plane displacement $w(x,y,t)$ is obtained as [97]

$$\ddot{w}(x,y,t) + c\dot{w}(x,y,t) - \rho_0^{-1}\sum_{\chi=x,y}\partial_\chi(w_\chi T_\chi) = \frac{F_{0z}}{m_c} \tag{13.11}$$

where T_x and T_y represent the respective tensions in the x and y directions because of the stretching of the graphene sheet, and F_{0z} is the vertical component of the applied external body force. By assuming that the fundamental mode dominates the out-of-plane motion, the out-of-plane equation can be further simplified into a 1D Duffing equation [97]

$$\ddot{w}(t) + c\dot{w}(t) + \omega_0^2 w(t) + \frac{5\pi^4(\lambda + 2\mu)}{128a^4\rho_0} w^3(t) = \frac{F(t)}{m_c} \tag{13.12}$$

where $\lambda = 15.55$ J/m^2 and $\mu = 103.89$ J/m^2 are the Lamé constants for graphene, a is the half length of a square graphene sheet, $F(t)$ is the overlap of the driving force with the fundamental mode shape of a graphene sheet, and ω_0 is the fundamental frequency of graphene.

The mechanical response of a prestretched square graphene of 1 μm in edge length and 0.5% initial strain in both x and y directions was investigated using the above models in both the static and dynamic cases [97]. The upper and lower plots in Figure 13.10a show the stress distribution in the graphene sheet under small and large forces. For the smaller force $P_{dc} = 1$ fN per carbon atom, the calculated deflection of the graphene at equilibrium is small and the tension in the x direction P_{xx} was found to be almost uniform and unchanged from its initial value. For the larger force $P_{dc} = 1$ pN, the nonuniform tension P_{xx} as well as other stress components were obtained and clearly exhibited in the lower plots in Figure 13.10a. The magnitude of the applied load also significantly influences the dynamical behavior of the graphene sheet. Figure 13.10b shows the amplitude-frequency spectra for different driving forces, which were calculated using the aforementioned model depicted in Eq. (13.12). It can be seen that the graphene sheet behaves like a linear harmonic resonator under small driving forces, while exhibiting nonlinear Duffing-type response and instability under large driving forces. Figure 13.10b also reveals a good agreement between the results obtained based on the general elasticity theory depicted in Equation 13.10, and the von Karman approximation given as Equation 13.11.

13.4 SUMMARY

This chapter demonstrates the current state-of-the-art for recently developed nanoscale resonators and their applications such as mass sensing, stiffness measurement, and/or force detection. It has been shown that nanoscale resonators exhibit great potential for various applications such as HF devices and/or ultrasensitive sensors. For effective design of nanoscale resonators for their specific purposes such as sensing and actuation, it is essential to understand the underlying mechanisms in the resonance behavior. It has been shown that the dynamic behavior of nanoscale resonators has been well described from continuum mechanics models such as the Euler–Bernoulli beam model and/or von Karman plate theory. However, it has to be noted that the classical continuum mechanics model lacks the fundamental mechanisms of nanoscale dynamic behavior of nanoresonators, for instance, energy-dissipation mechanisms. This sheds light on the necessity of an atomistic model that can capture such underlying, unique nanoscale dynamics, albeit the atomistic model encounters computational restrictions for simulating structures with spatial scales albeit the atomistic model encounters computational restrictions for simulating structures with spatial scales >O (10 nm) at longer timescales >O (1 μs). This implies that it is required to develop novel computational modeling that enables not

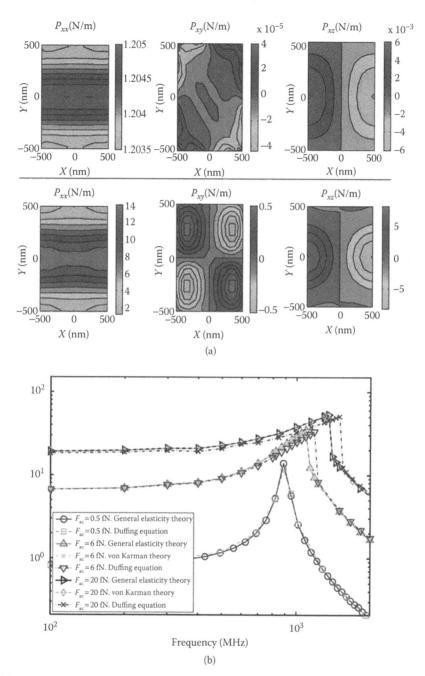

FIGURE 13.10 (a) Tension distributions in the graphene sheet for small deflections with $F_{dc} = 1$ fN per carbon atom (upper plots) and large deflections with $F_{dc} = 1$ pN (lower plots). (b) Amplitude of the fundamental mode of the graphene resonator versus driving frequency for different driving forces. (Reprinted from Atalaya, J., A. Isacsson, and J. M. Kinaret. 2008. *Nano Lett* 8:4196–200. With permission.)

only depictions of atomic-scale characteristics also simulations at large spatial and temporal scales to gain insights into the underlying mechanisms of nanoresonators.

ACKNOWLEDGMENTS

The authors acknowledge the support from State University of New York at Binghamton, American Chemical Society-Petroleum Research Fund, and U.S. Air Force Office of Scientific Research. We also thank M. Zheng and X.-M. Chen for assistance with the preparation of the manuscript.

REFERENCES

1. Kovacs, G. T. A. 1998. *Micromachined Transducers Sourcebook*. New York: McGraw-Hill.
2. Ilic, B., H. G. Craighead, S. Krylov, W. Senaratne, C. Ober, and P. Neuzil. 2004. Attogram detection using nanoelectromechanical oscillators. *J Appl Phys* 95:3694–703.
3. Sazonova, V., Y. Yaish, H. Ustunel, D. Roundy, T. A. Arias, and P. L. McEuen. 2004. A tunable carbon nanotube electromechanical oscillator. *Nature* 431:284–7.
4. DeMartini, B. E., J. F. Rhoads, S. W. Shaw, and K. L. Turner. 2007. Single input-single output mass sensor based on a coupled array of microresonators. *Sens Actuators A Phys* 137:147–56.
5. Dai, M. D., K. Eom, and C. W. Kim. 2009. Nanomechanical mass detection using non-linear oscillations. *Appl Phys Lett* 95:203104.
6. Gil-Santos, E., D. Ramos, A. Jana, M. Calleja, A. Raman, and J. Tamayo. 2009. Mass sensing based on deterministic and stochastic responses of elastically coupled nanocantilevers. *Nano Lett* 9:4122–7.
7. Martin, S. J., M. A. Butler, J. J. Spates, M. A. Mitchell, and W. K. Schubert. 1998. Flexural plate wave resonator excited with Lorentz forces. *J Appl Phys* 83:4589–601.
8. Zalalutdinov, M., B. Ilic, D. Czaplewski, A. Zehnder, H. G. Craighead, and J. M. Parpia. 2000. Frequency-tunable micromechanical oscillator. *Appl Phys Lett* 77:3287–9.
9. Joonhyung Lee, J. J., D. Akin, C. A. Savran, and R. Bashir. 2008. Real-time detection of airborne viruses on a mass-sensitive device. *Appl Phys Lett* 93:013901.
10. N. G. Chopra and A. Zettl. 1998. Measurement of the elastic modulus of a multi-wall boron nitride nanotube. *Solid State Commun* 105:297–300.
11. Aihara, K., J. Xiang, S. Chopra, A. Pham, and A. M. Rao. 2003. GHz carbon nanotube resoantor bio-sensors. In *Proceedings at IEEE Nanotechnologies Conferences*, p. 12–14, August 2003. San Francisco, CA.
12. Vardanega, D., F. Picaud, and C. Girardet. 2007. Towards selective detection of chiral molecules using SWNT sensors. *Surf Sci* 601:3818–22.
13. Zheng, M., K. Eom, and C.-H. Ke. 2009. Calculations of the resonant response of carbon nanotubes to binding of DNA. *J Phys D Appl Phys* 42:145408.
14. Ilic, B., D. Czaplewski, H. G. Craighead, P. Neuzil, C. Campagnolo, and C. Batt. 2000. Mechanical resonant immunospecific biological detector. *Appl Phys Lett* 77:450–2.
15. Naik, A. K., M. S. Hanay, W. K. Hiebert, X. L. Feng, and M. L. Roukes. 2009. Towards single-molecule nanomechanical mass spectrometry. *Nat Nanotechnol* 4:445–50.
16. Nguyen, C. T. C., L. P. B. Katehi, and G. M. Rebeiz. 1998. Micromachined devices for wireless communications. *Proc Ieee* 86:1756–68.
17. Yao, Z., C. L. Kane, and C. Dekker. 2000. High-field electrical transport in single-wall carbon nanotubes. *Phys Rev Lett* 84:2941–4.

18. Rhoads, J. F., S. W. Shaw, K. L. Turner, and R. Baskaran. 2005. Tunable microelectrome-chanical filters that exploit parametric resonance. *J Vib Acoust Trans Asme* 127:423–30.

19. Jensen, K., K. Kim, and A. Zettl. 2008. An atomic-resolution nanomechanical mass sensor. *Nat Nanotechnol* 3:533–7.

20. Cleland, A. N., and M. L. Roukes. 1996. Fabrication of high frequency nanometer scale mechanical resonators from bulk Si crystals. *Appl Phys Lett* 69:2653–5.

21. Iijima, S. 1991. Helical microtubules of graphitic carbon. *Nature* 354:56–8.

22. Novoselov, K. S., A. K. Geim, S. V. Morozov, D. Jiang, Y. Zhang, S. V. Dubonos, I. V. Grigorieva, and A. A. Firsov. 2004. Electric field effect in atomically thin carbon films. *Science* 306:666–9.

23. Garcia-Sanchez, D., A. M. van der Zande, A. S. Paulo, B. Lassagne, P. L. McEuen, and A. Bachtold. 2008. Imaging mechanical vibrations in suspended graphene sheets. *Nano Lett* 8:1399–403.

24. Freer, E. M., O. Grachev, X. F. Duan, S. Martin, and D. P. Stumbo. 2010. High-yield self-limiting single-nanowire assembly with dielectrophoresis. *Nature Nanotechnol* 5:525–30.

25. Burg, B. R., F. Lutolf, J. Schneider, N. C. Schirmer, T. Schwamb, and D. Poulikakos. 2009. High-yield dielectrophoretic assembly of two-dimensional graphene nanostructures. *Appl Phys Lett* 94:053110.

26. Zhang, Y. G., A. L. Chang, J. Cao, Q. Wang, W. Kim, Y. M. Li, N. Morris, E. Yenilmez, J. Kong, and H. J. Dai. 2001. Electric-field-directed growth of aligned single-walled carbon nanotubes. *Appl Phys Lett* 79:3155–7.

27. van der Zande, A. M., R. A. Barton, J. S. Alden, C. S. Ruiz-Vargas, W. S. Whitney, P. H. Q. Pham, J. Park, J. M. Parpia, H. G. Craighead, and P. L. McEuen. 2010. Large-scale arrays of single-layer graphene resonators. *Nano Lett* 10:4869–73.

28. Li, M. W., R. B. Bhiladvala, T. J. Morrow, J. A. Sioss, K. K. Lew, J. M. Redwing, C. D. Keating, and T. S. Mayer. 2008. Bottom-up assembly of large-area nanowire resonator arrays. *Nat Nanotechnol* 3:88–92.

29. Sawano, S., T. Arie, and S. Akita. 2010. Carbon nanotube resonator in liquid. *Nano Lett* 10:3395–8.

30. Weldon, J. A., B. Aleman, A. Sussman, W. Gannett, and A. K. Zettl. 2010. Sustained mechanical self-oscillations in carbon nanotubes. *Nano Lett* 10:1728–33.

31. Feng, X. L., C. J. White, A. Hajimiri, and M. L. Roukes. 2008. A self-sustaining ultrahigh-frequency nanoelectromechanical oscillator. *Nat Nanotechnol* 3:342–6.

32. Ke, C.-H. 2009. Resonant pull-in of a double-sided driven nanotube-based electromechanical resonator. *J Appl Phys* 15:024301.

33. Lavrik, N. V., and P. G. Datskos. 2003. Femtogram mass detection using photothermally actuated nanomechanical resonators. *Appl Phys Lett* 82:2697–9.

34. Yang, Y. T., C. Callegari, X. L. Feng, K. L. Ekinci, and M. L. Roukes. 2006. Zeptogram-scale nanomechanical mass sensing. *Nano Lett* 6:583–6.

35. Lassagne, B., D. Garcia-Sanchez, A. Aguasca, and A. Bachtold. 2008. Ultrasensitive mass sensing with a nanotube electromechanical resonator. *Nano Lett* 8:3735–8.

36. Qian, D., G. J. Wagner, W. K. Liu, M. F. Yu, and R. S. Ruoff. 2002. Mechanics of carbon nanotubes. *Appl Mech Rev* 55:495–533.

37. McEuen, P. L., M. S. Fuhrer, and H. K. Park. 2002. Single-walled carbon nanotube electronics. *Ieee Trans Nanotechnol* 1:78–85.

38. Mintmire, J. W., B. I. Dunlap, and C. T. White. 1992. Are fullerene tubules metallic? *Phys Rev Lett* 68:631–4.

39. Tombler, T. W., C. W. Zhou, L. Alexseyev, J. Kong, H. J. Dai, L. Lei, C. S. Jayanthi, M. J. Tang, and S. Y. Wu. 2000. Reversible electromechanical characteristics of carbon nanotubes under local-probe manipulation. *Nature* 405:769–72.

40. Kuzumaki, T., and Y. Mitsuda. 2004. Dynamic measurement of electrical conductivity of carbon nanotubes during mechanical deformation by nanoprobe manipulation in transmission electron microscopy. *Appl Phys Lett* 85:1250–2.

41. Liu, B., H. Jiang, H. T. Johnson, and Y. Huang. 2004. The influence of mechanical deformation on the electrical properties of single wall carbon nanotubes. *J Mech Phys Solids* 52:1–26.

42. Dresselhaus, M. S., G. Dresselaus, and P. Avouris. 2001. *Carbon Nanotubes*. Berlin: Springer.

43. Huang, J. Y., S. Chen, Z. Q. Wang, K. Kempa, Y. M. Wang, S. H. Jo, G. Chen, M. S. Dresselhaus, and Z. F. Ren. 2006. Superplastic carbon nanotubes—Conditions have been discovered that allow extensive deformation of rigid single-walled nanotubes. *Nature* 439:281.

44. Peng, H. B., C. W. Chang, S. Aloni, T. D. Yuzvinsky, and A. Zettl. 2006. Ultrahigh frequency nanotube resonators. *Phys Rev Lett* 97:087203.

45. Poncharal, P., Z. L. Wang, D. Ugarte, and W. A. de Heer. 1999. Electrostatic deflections and electromechanical resonances of carbon nanotubes. *Science* 283:1513–6.

46. Treacy, M. M. J., T. W. Ebbesen, and J. M. Gibson. 1996. Exceptionally high Young's modulus observed for individual carbon nanotubes. *Nature* 381:678–80.

47. Witkamp, B., M. Poot, and H. S. J. van der Zant. 2006. Bending-mode vibration of a suspended nanotube resonator. *Nano Lett* 6:2904–8.

48. Chiu, H. Y., P. Hung, H. W. C. Postma, and M. Bockrath. 2008. Atomic-scale mass sensing using carbon nanotube resonators. *Nano Lett* 8:4342–6.

49. Hüttel, A. K., G. A. Steele, B. Witkamp, M. Poot, L. P. Kouwenhoven, and H. S. J. van der Zant. 2009. Carbon nanotube as ultrahigh quality factor mechanical resonators. *Nano Lett* 9:2547–52.

50. Feng, X. L., R. R. He, P. D. Yang, and M. L. Roukes. 2007. Very high frequency silicon nanowire electromechanical resonators. *Nano Lett* 7:1953–9.

51. Tans, S. J., A. R. M. Verschueren, and C. Dekker. 1998. Room-temperature transistor based on a single carbon nanotube. *Nature* 393:49–52.

52. Zhou, C. W., J. Kong, and H. J. Dai. 2000. Intrinsic electrical properties of individual single-walled carbon nanotubes with small band gaps. *Phys Rev Lett* 84:5604–7.

53. Minot, E. D., Y. Yaish, V. Sazonova, and P. L. McEuen. 2004. Determination of electron orbital magnetic moments in carbon nanotubes. *Nature* 428:536–9.

54. Knobel, R. G., and A. N. Cleland. 2003. Nanometre-scale displacement sensing using a single electron transistor. *Nature* 424:291–3.

55. Bargatin, I., E. B. Myers, J. Arlett, B. Gudlewski, and M. L. Roukes. 2005. Sensitive detection of nanomechanical motion using piezoresistive signal downmixing. *Appl Phys Lett* 86:133109.

56. Cho, H. N., M. F. Yu, A. F. Vakakis, L. A. Bergman, and D. M. McFarland. 2010. Tunable, broadband nonlinear nanomechanical resonator. *Nano Lett* 10:1793–8.

57. Vakakis, A. F., O. Gendelman, L. A. Bergman, D. M. McFarland, G. Kerschen, and Y. S. Lee. 2008. *Passive Nonlinear Targeted Energy Transfer in Mechanical and Structural Systems*. New York: Springer Verlag.

58. Garcia-Sanchez, D., A. S. Paulo, M. J. Esplandiu, F. Perez-Murano, L. Forro, A. Aguasca, and A. Bachtold. 2007. Mechanical detection of carbon nanotube resonator vibrations. *Phys Rev Lett* 99:085501.

59. Jensen, K., J. Weldon, H. Garcia, and A. Zettl. 2007. Nanotube radio. *Nano Lett* 7:3508–11.

60. Rutherglen, C., and P. Burke. 2007. Carbon nanotube radio. *Nano Lett* 7:3296–99.

61. Kocabas, C., H. S. Kim, T. Banks, J. A. Rogers, A. A. Pesetski, J. E. Baumgardner, S. V. Krishnaswamy, and H. Zhang. 2008. Radio frequency analog electronics based on carbon nanotube transistors. *Proc Natl Acad Sci USA* 105:1405–9.

62. Bonard, J. M., C. Klinke, K. A. Dean, and B. F. Coll. 2003. Degradation and failure of carbon nanotube field emitters. *Phys Rev B* 67:115406.

63. Purcell, S. T., P. Vincent, C. Journet, and V. T. Binh. 2002. Tuning of nanotube mechanical resonances by electric field pulling. *Phys Rev Lett* 89:276103.

64. Dragoman, D., and M. Dragoman. 2008. Tunneling nanotube radio. *J Appl Phys* 104:074314.

65. Tamayo, J., D. Ramos, J. Mertens, and M. Calleja. 2006. Effect of the adsorbate stiffness on the resonance response of microcantilever sensors. *Appl Phys Lett* 89:224104.

66. Dohn, S., W. Svendsen, A. Boisen, and O. Hansen. 2007. Mass and position determination of attached particles on cantilever based mass sensors. *Rev Sci Instrum* 78:103303.

67. Gil-Santos, E., D. Ramos, J. Martinez, M. Fernandez-Regulez, R. Garcia, A. San Paulo, M. Calleja, and J. Tamayo. 2010. Nanomechanical mass sensing and stiffness spectrometry based on two-dimensional vibrations of resonant nanowires. *Nat Nanotechnol* 5:641–5.

68. Conley, W. G., A. Raman, C. M. Krousgrill, and S. Mohammadi. 2008. Nonlinear and nonplanar dynamics of suspended nanotube and nanowire resonators. *Nano Lett* 8:1590–5.

69. Ke, C. H., and H. D. Espinosa. 2006. In situ electron microscopy electromechanical characterization of a bistable NEMS device. *Small* 2:1484–9.

70. Yu, M. F., O. Lourie, M. J. Dyer, K. Moloni, T. F. Kelly, and R. S. Ruoff. 2000. Strength and breaking mechanism of multiwalled carbon nanotubes under tensile load. *Science* 287:637–40.

71. Wang, Y. F., K. K. Lew, T. T. Ho, L. Pan, S. W. Novak, E. C. Dickey, J. M. Redwing, and T. S. Mayer. 2005. Use of phosphine as an n-type dopant source for vapor-liquid-solid growth of silicon nanowires. *Nano Lett* 5:2139–43.

72. Tian, M. L., J. U. Wang, J. Kurtz, T. E. Mallouk, and M. H. W. Chan. 2003. Electrochemical growth of single-crystal metal nanowires via a two-dimensional nucleation and growth mechanism. *Nano Lett* 3:919–23.

73. Fargas-Marques, A., J. Casals-Terte, and A. M. Shkel. 2007. Resonant pull-in condition in parallel-plate electrostatic actuators. *J Microelectromech Syst* 16:1044–53.

74. Nayfeh, A. H., M. I. Younis, and E. M. Abdel-Rahman. 2007. Dynamic pull-in phenomenon in MEMS resonators. *Nonlinear Dyn* 48:153–63.

75. Krylov, S., and R. Maimon. 2004. Pull-in dynamics of an elastic beam actuated by continuously distributed electrostatic force. *J Vib Acoust Trans Asme* 126:332–42.

76. Nayfeh, A. H., and M. I. Younis. 2005. Dynamics of MEMS resonators under superharmonic and subharmonic excitations. *J Micromech Microeng* 15:1840–7.

77. Ke, C. H., N. Pugno, B. Peng, and H. D. Espinosa. 2005. Experiments and modeling of carbon nanotube-based NEMS devices. *J Mech Phys Solids* 53:1314–33.

78. Zheng, M., A. Jagota, M. S. Strano, A. P. Santos, P. Barone, S. G. Chou, B. A. Diner et al. 2003. Structure-based carbon nanotube sorting by sequence-dependent DNA assembly. *Science* 302:1545–8.

79. Zheng, M., A. Jagota, E. D. Semke, B. A. Diner, R. S. Mclean, S. R. Lustig, R. E. Richardson, and N. G. Tassi. 2003. DNA-assisted dispersion and separation of carbon nanotubes. *Nat Mater* 2:338–42.

80. Kam, N. W. S., M. O'Connell, J. A. Wisdom, and H. J. Dai. 2005. Carbon nanotubes as multifunctional biological transporters and near-infrared agents for selective cancer cell destruction. *Proc Natl Acad Sci USA* 102:11600–5.

81. McLean, R. S., X. Y. Huang, C. Khripin, A. Jagota, and M. Zheng. 2006. Controlled two-dimensional pattern of spontaneously aligned carbon nanotubes. *Nano Lett* 6:55–60.

82. Karachevtsev, V. A., A. Y. Glamazda, V. S. Leontiev, O. S. Lytvyn, and U. Dettlaff-Weglikowska. 2007. Glucose sensing based on NIR fluorescence of DNA-wrapped single-walled carbon nanotubes. *Chem Phys Lett* 435:104–8.

83. Xu, Y., P. E. Pehrsson, L. W. Chen, R. Zhang, and W. Zhao. 2007. Double-stranded DNA single-walled carbon nanotube hybrids for optical hydrogen peroxide and glucose sensing. *J Phys Chem C* 111:8638–43.

84. Han, X. G., Y. L. Li, and Z. X. Deng. 2007. DNA-wrapped single-walled carbon nanotubes as rigid templates for assembling linear gold nanoparticle arrays. *Adv Mater* 19:1518–22.

85. Shin, S. R., C. K. Lee, I. So, J. H. Jeon, T. M. Kang, C. Kee, S. I. Kim, G. M. Spinks, G. G. Wallace, and S. J. Kim. 2008. DNA-wrapped single-walled carbon nanotube hybrid fibers for supercapacitors and artificial muscles. *Adv Mater* 20:466–70.

86. Manohar, S., T. Tang, and A. Jagota. 2007. Structure of homopolymer DNA-CNT hybrids. *J Phys Chem C* 111:17835–45.

87. Kirstein, S., M. Mertesdorf, and M. Schonhoff. 1998. The influence of a viscous fluid on the vibration dynamics of scanning near-field optical microscopy fiber probes and atomic force microscopy cantilevers. *J Appl Phys* 84:1782–90.

88. Eom, K., T. Y. Kwon, D. S. Yoon, H. L. Lee, and T. S. Kim. 2007. Dynamical response of nanomechanical resonators to biomolecular interactions. *Phys Rev B* 76:113408.

89. Dareing, D. W., F. Tian, and T. Thundat. 2006. Effective mass and flow patterns of fluids surrounding microcantilevers. *Ultramicroscopy* 106:789–94.

90. Braun, T., V. Barwich, M. K. Ghatkesar, A. H. Bredekamp, C. Gerber, M. Hegner, and H. P. Lang. 2005. Micromechanical mass sensors for biomolecular detection in a physiological environment. *Phys Rev E* 72:031907.

91. Manoharan, M. P., and M. A. Haque. 2009. Role of adhesion in shear strength of nanowire-substrate interfaces. *J Phys D Appl Phys* 42:095304.

92. Lee, C., X. D. Wei, J. W. Kysar, and J. Hone. 2008. Measurement of the elastic properties and intrinsic strength of monolayer graphene. *Science* 321:385–8.

93. Bunch, J. S., A. M. van der Zande, S. S. Verbridge, I. W. Frank, D. M. Tanenbaum, J. M. Parpia, H. G. Craighead, and P. L. McEuen. 2007. Electromechanical resonators from graphene sheets. *Science* 315:490–3.

94. Li, X. S., W. W. Cai, J. H. An, S. Kim, J. Nah, D. X. Yang, R. Piner et al. 2009. Large-area synthesis of high-quality and uniform graphene films on copper foils. *Science* 324:1312–4.

95. Kim, K. S., Y. Zhao, H. Jang, S. Y. Lee, J. M. Kim, K. S. Kim, J. H. Ahn, P. Kim, J. Y. Choi, and B. H. Hong. 2009. Large-scale pattern growth of graphene films for stretchable transparent electrodes. *Nature* 457:706–10.

96. Reina, A., X. T. Jia, J. Ho, D. Nezich, H. B. Son, V. Bulovic, M. S. Dresselhaus, and J. Kong. 2009. Large area, few-layer graphene films on arbitrary substrates by chemical vapor deposition. *Nano Lett* 9:30–5.

97. Atalaya, J., A. Isacsson, and J. M. Kinaret. 2008. Continuum elastic modeling of graphene resonators. *Nano Lett* 8:4196–200.

98. Drozdov, A. D. 1996. *Finite Elasticity and Viscoelasticity*. Singapore: World Scientific.

14 Mechanical Behavior of Monolayer Graphene by Continuum and Atomistic Modeling

Qiang Lu and Rui Huang

CONTENTS

14.1 NONLINEAR CONTINUUM MECHANICS OF MONOLAYER GRAPHENE

A single layer of carbon (C) atoms tightly packed into a two-dimensional hexagonal lattice makes up a graphene monolayer, which is the basic building block for bulk graphite and carbon nanotubes (CNTs). Since isolation and observation of monolayer graphene were first reported in 2005 [Novoselov et al. 2005], extensive studies have been devoted to exploit the unique two-dimensional (2D) lattice structures of graphene for a large variety of applications. Among others, the mechanical properties of graphene are critical for practical applications that often require deformation and mechanical interactions. This chapter focuses on modeling and simulations of mechanical behavior of graphene. Both continuum and atomistic approaches are presented. The two complement each other so that the effective mechanical properties are properly defined in a continuum theory while the quantitative values are determined intrinsically from the discrete models at the atomistic scale.

This chapter is organized as follows. A theoretical framework of nonlinear continuum mechanics is presented in Section 14.1. Section 14.2 discusses common approaches for atomistic simulations. Sections 14.3 to 14.5 consider infinite graphene sheets subjected to uniaxial stretch, uniaxial tension, and cylindrical bending conditions. In Section 14.6, the edge properties of finite graphene ribbons are theoretically examined, including edge energy, edge force, and edge buckling. The nonlinear elastic behavior and fracture of graphene nanoribbons under uniaxial tension are discussed in Section 14.7. The chapter concludes with a brief summary in Section 14.8.

Most mechanical properties, such as Young's modulus, Poisson's ratio, flexural rigidity, and fracture strength, are concepts of continuum mechanics. Specifically for monolayer graphene, these properties originate from its unique two-dimensional (2D) lattice structure. To establish the connection between the atomistic structure of graphene and its intrinsic mechanical properties, a theoretical framework of nonlinear continuum mechanics is presented here to properly define the effective mechanical properties for graphene. For this purpose, the monolayer graphene is treated as a 2D continuum sheet, with no physical thickness. In practice, the thickness of graphene may be defined based on its interactions with the surrounding environment (Aitken and Huang 2010).

14.1.1 KINEMATICS: STRETCH AND CURVATURE

Take a planar graphene sheet as the reference state. The deformation of the sheet is described by a deformation gradient tensor $\mathbf{F}$ that maps an infinitesimal segment $d\mathbf{X}$ at the reference state to the corresponding segment $d\mathbf{x}$ at the deformed state (see Figure 14.1), that is,

$$d\mathbf{x} = \mathbf{F}d\mathbf{X} \text{ and } F_{iJ} = \frac{\partial x_i}{\partial X_J} \tag{14.1}$$

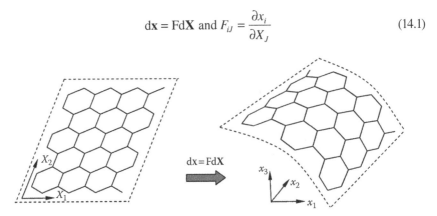

FIGURE 14.1 Schematic illustration of a two-dimensional graphene sheet before and after deformation.

For convenience, we set up the coordinates such that $X_3 = 0$ for the graphene sheet at the reference state and thus the vector d**X** has only two in-plane components ($J = 1, 2$). However, d**x** in general has three components ($i = 1, 2, 3$).

As a measure of deformation, a 2D Green–Lagrange strain tensor is defined as

$$E_{JK} = \frac{1}{2}(F_{iJ}F_{iK} - \delta_{JK}) \tag{14.2}$$

where δ_{JK} is the Kronecker delta, and the standard summation convention over the repeated indices is adopted. With the 2D strain tensor, the linear stretch of an infinitesimal segment d**X** is

$$\lambda = \frac{|dx|}{|dX|} = \sqrt{1 + 2E_{JK}N_J N_K} \tag{14.3}$$

where $N_J = dX_J/|dX|$ is the unit vector in the direction of the line segment d**X**.

A 2D sheet remains planar under a homogeneous deformation with a constant deformation gradient tensor **F**. However an inhomogeneous deformation may induce bending and twisting of the sheet into a corrugated surface in the three-dimensional (3D) space. The curvature of the deformed sheet can be obtained from the first and second fundamental forms of the surface in 3D. Following standard procedures of differential geometry (do Carmo 1976), a curvature tensor is defined as

$$K_{IJ} = -n_i \frac{\partial F_{iI}}{\partial X_J} = -n_i \frac{\partial^2 x_i}{\partial X_I \partial X_J} \tag{14.4}$$

where the unit normal vector of the deformed surface can be written as

$$n_i = \frac{e_{ijk}F_{j1}F_{k2}}{\sqrt{(1 + 2E_{11})(1 + 2E_{22}) - 4E_{12}^2}} \tag{14.5}$$

in which e_{ijk} is the permutation tensor.

With the curvature tensor, the normal curvature at the deformed state for an arbitrary line segment d**X** at the reference state is

$$\kappa_n = \frac{K_{IJ}N_I N_J}{\lambda^2} \tag{14.6}$$

where $N_J = dX_J/|dX|$ and the stretch λ is defined in Equation 14.3. By solving a generalized eigenvalue problem (Arroyo and Belytschko 2004a), two principal curvatures can be obtained at each point, from which the mean curvature and Gaussian curvature can be determined. It is noted that in Equation 14.4, the nonzero curvature occurs only under inhomogeneous deformation with nonzero gradients of the deformation gradient tensor analogous to the strain gradient associated with the bending curvature in the classical theory of plates (Timoshenko and Woinowsky-Krieger 1987).

For the case of infinitesimal deformation, by neglecting the nonlinear terms in Equation 14.2, the 2D Green–Lagrange strain reduces to the linear strain:

$$E_{JK} \approx \varepsilon_{JK} = \frac{1}{2}\left(\frac{\partial u_J}{\partial X_K} + \frac{\partial u_K}{\partial X_J}\right) \tag{14.7}$$

where $u = x - X$ is the displacement. For infinitesimal bending, the unit normal **n** is approximately (0, 0, 1) and the curvature tensor in Equation 14.4 reduces to the familiar form

$$K_{IJ} \approx \frac{\partial^2 u_3}{\partial X_I \partial X_J} \tag{14.8}$$

where u_3 is the lateral displacement normal to the reference plane.

14.1.2 STRESS AND MOMENT TENSORS

Within the theoretical framework of nonlinear elasticity, the mechanical property can be derived from a strain energy density function that depends on the deformation gradient tensor under the condition of homogeneous deformation. For the 2D graphene sheet, it is assumed that the strain energy density is a function of the 2D Green–Lagrange strain and the curvature, namely

$$\Phi = \Phi(E, K) \tag{14.9}$$

Note that the energy density for the 2D sheet has a unit of joule per meter squared, different from that for a 3D solid.

Define the 2D stress (force per unit length) and moment intensity (moment per unit length) as the work conjugates of the 2D Green–Lagrange strain and the curvature, respectively:

$$S_{IJ} = \frac{\partial \Phi}{\partial E_{IJ}} \quad \text{and} \quad M_{IJ} = \frac{\partial \Phi}{\partial K_{IJ}} \tag{14.10}$$

The stress tensor S_{IJ} is analogous to the second Piola–Kirchhoff stress tensor in 3D, whereas the moment tensor M_{IJ} represents a higher order quantity associated with the curvature.

Alternatively, the 2D nominal stress can be defined as the force at the deformed state per unit length of a line segment at the reference state, which can be obtained as the work conjugate of the deformation gradient **F**, namely

$$P_{iJ} = \frac{\partial \Phi}{\partial F_{iJ}} \tag{14.11}$$

The 2D nominal stress is analogous to the first Piola–Kirchhoff stress in 3D. Under the condition of homogeneous deformation, the two stress tensors can be related to each other as

$$P_{IJ} = F_{iI}S_{iJ} \tag{14.12}$$

The 2D true stress may also be defined as the force per unit length of a line segment at the deformed state. However, a general relationship between the true stress and the nominal stress for the 2D sheet has not been established mathematically. The relationship for a special case is given in Section 14.3.

14.1.3 Tangent Modulus

The generally nonlinear elastic properties of the 2D graphene sheet can be described in terms of tangent modulus for both stretching and bending. The material tangent modulus for in-plane stretching deformation is defined as

$$C_{IJKL} = \frac{\partial S_{IJ}}{\partial E_{KL}} = \frac{\partial^2 \Phi}{\partial E_{IJ} \partial E_{KL}} \tag{14.13}$$

Similarly, the tangent-bending modulus of the sheet is

$$D_{IJKL} = \frac{\partial M_{IJ}}{\partial K_{KL}} = \frac{\partial^2 \Phi}{\partial K_{IJ} \partial K_{KL}} \tag{14.14}$$

In addition, there may exist a coupling modulus between stretching and bending, namely

$$\Lambda_{IJKL} = \frac{\partial S_{IJ}}{\partial K_{KL}} = \frac{\partial M_{KL}}{\partial E_{IJ}} = \frac{\partial^2 \Phi}{\partial E_{IJ} \partial K_{KL}} \tag{14.15}$$

Noting the symmetry in the 2D stress, stain, moment, and curvature tensors, the modulus tensors can be written in an abbreviated matrix form, with which an incremental relationship is obtained as follows:

$$\begin{pmatrix} dS_{11} \\ dS_{22} \\ dS_{12} \end{pmatrix} = \begin{bmatrix} C_{11} & C_{12} & C_{13} \\ C_{21} & C_{22} & C_{23} \\ C_{31} & C_{32} & C_{33} \end{bmatrix} \begin{pmatrix} dE_{11} \\ dE_{22} \\ 2dE_{12} \end{pmatrix} + \begin{bmatrix} \Lambda_{11} & \Lambda_{12} & \Lambda_{13} \\ \Lambda_{21} & \Lambda_{22} & \Lambda_{23} \\ \Lambda_{31} & \Lambda_{32} & \Lambda_{33} \end{bmatrix} \begin{pmatrix} dK_{11} \\ dK_{22} \\ 2dK_{12} \end{pmatrix} \tag{14.16}$$

$$\begin{pmatrix} dM_{11} \\ dM_{22} \\ dM_{12} \end{pmatrix} = \begin{bmatrix} D_{11} & D_{12} & D_{13} \\ D_{21} & D_{22} & D_{23} \\ D_{31} & D_{32} & D_{33} \end{bmatrix} \begin{pmatrix} dK_{11} \\ dK_{22} \\ 2dK_{12} \end{pmatrix} + \begin{bmatrix} \Lambda_{11} & \Lambda_{21} & \Lambda_{31} \\ \Lambda_{12} & \Lambda_{22} & \Lambda_{32} \\ \Lambda_{13} & \Lambda_{23} & \Lambda_{33} \end{bmatrix} \begin{pmatrix} dE_{11} \\ dE_{22} \\ 2dE_{12} \end{pmatrix} \tag{14.17}$$

Equations 14.16 and 14.17 present an incremental form of the generally nonlinear and anisotropic elastic behavior for a 2D sheet. Note that the coupling modulus Λ_{IJKL} does not possess the major symmetry and thus the Λ-matrix is not symmetric, whereas both the C- and D-matrices are symmetric. In general, all the tangent moduli in the C-, D-, and Λ-matrices depend on the current state of deformation (stretch and curvature). Under the assumption of infinitesimal deformation, however,

Equations 14.16 and 14.17 reduce to the linear elastic relations, with constant moduli. For an isotropic, linear elastic sheet, the relationship becomes

$$\begin{pmatrix} S_{11} \\ S_{22} \\ S_{12} \end{pmatrix} = \begin{bmatrix} C_{11} & C_{12} & 0 \\ C_{21} & C_{22} & 0 \\ 0 & 0 & C_{33} \end{bmatrix} \begin{pmatrix} E_{11} \\ E_{22} \\ 2E_{12} \end{pmatrix} \tag{14.18}$$

$$\begin{pmatrix} M_{11} \\ M_{22} \\ M_{12} \end{pmatrix} = \begin{bmatrix} D_{11} & D_{12} & 0 \\ D_{21} & D_{22} & 0 \\ 0 & 0 & D_{33} \end{bmatrix} \begin{pmatrix} K_{11} \\ K_{22} \\ 2K_{12} \end{pmatrix} \tag{14.19}$$

where $C_{11} = C_{22}$ and $D_{11} = D_{22}$. The elastic isotropy dictates that $C_{33} = \frac{1}{2}(C_{11} - C_{12})$ and $D_{33} = \frac{1}{2}(D_{11} - D_{22})$ (Huang, Wu, and Hwang 2006). The 2D Young's modulus and Poisson's ratio under infinitesimal deformation can then be obtained as $Y = C_{11} - C_{21}^2/C_{11}$ and $\nu = C_{21}/C_{11}$, respectively.

14.1.4 EDGE EFFECTS

For a finite graphene sheet, in addition to the bulk strain energy density function, an edge energy density function is defined as the excess energy per unit length along the edges (Lu and Huang 2010a), which in general depends on both the stretch and curvature of the edge, namely

$$\gamma = \gamma(\lambda_e, \kappa_e) \tag{14.20}$$

where λ_e and κ_e are defined in Equations 14.3 and 14.6 with the unit vector N_J in the tangential direction of the edge.

In the vicinity of the reference state ($\lambda_e = 1$ and $\kappa_e = 0$), the edge energy may be approximated by expanding in terms of the nominal edge strain, $\varepsilon_e = \lambda_e - 1$, and the edge curvature:

$$\gamma \approx \gamma_0 + f\varepsilon_e + m\kappa_e + \frac{1}{2}\left(\bar{C}\varepsilon_e^2 + \bar{D}\kappa_e^2 + 2\bar{\Lambda}\varepsilon_e\kappa_e\right) \tag{14.21}$$

where γ_0 is the edge energy density at the reference state, $f = \dfrac{\partial\gamma}{\partial\varepsilon_e}$ is the edge force (or edge stress), $m = \dfrac{\partial\gamma}{\partial\kappa_e}$ is the edge moment, $\bar{C} = \dfrac{\partial^2\gamma}{\partial\varepsilon_e^2}$, $\bar{D} = \dfrac{\partial^2\gamma}{\partial\kappa_e^2}$, and $\bar{\Lambda} = = \dfrac{\partial^2\gamma}{\partial\varepsilon_e\partial\kappa_e}$ are edge moduli, with all partial derivatives taken at the reference state. For graphene, all the edge properties depend on the atomistic chirality and bond structures of the edge, which will be discussed in Section 14.6.

14.1.5 A GENERAL FRAMEWORK

As a general continuum model, the constitutive behavior of the 2D graphene sheet is described by the strain energy density function, $\Phi(E,K)$, and the edge energy density function, $\gamma(\lambda_e, \kappa_e)$. The specific forms of the two functions may be determined either phenomenologically by experimental measurements of the mechanical properties or by atomistic modeling. First-principle calculations based on quantum mechanics can also be used to determine these functions (e.g., Wei et al. 2009; Gan and Srolovitz 2010). Once these functions are known, the mechanical behavior of graphene can be described in a general framework of thermodynamics.

Consider a graphene sheet of area A at the reference state, bounded by its edge contour C. The internal potential energy is

$$U = \int_A \Phi(E,K)dA + \oint_C \gamma(\lambda_e, \kappa_e)ds \tag{14.22}$$

Assume that the graphene is subject to forces (t_i) and moments (m_i) along its edge as well as a pressure (p_i) over the area. The work done over a small variation is

$$\delta W = \int_A p_i \delta u_i dA + \oint_C (t_i \delta u_i + m_i \delta \theta_i)ds \tag{14.23}$$

where u_i is the displacement and θ_i is the angle of rotation.

At equilibrium, from the principle of virtual work, the variation of the internal energy is equal to the work done by the external forces, namely,

$$\delta U = \delta W \tag{14.24}$$

Both the equilibrium equations and the boundary conditions can then be deduced from Equation 14.24, similar to the variational analysis for a 2D plate or shell (Timoshenko and Woinowsky-Krieger 1987). Because of the nonlinear kinematics, however, the general form of the equilibrium equations and boundary conditions has not been established. A simple form for a special case is presented in Section 14.7.

14.2 ATOMISTIC SIMULATIONS

Atomistic simulations based on empirical potentials have been used to predict mechanical properties of many materials including carbon nanotubes (CNTs) and graphene. In this chapter, a molecular mechanics (MM) approach is adopted to simulate mechanical behavior of graphene monolayers. The MM simulations are used to determine the static equilibrium state of graphene by minimizing the total potential energy with respect to the atomic positions, subjected to prescribed boundary conditions. Although the standard MM simulations are relatively straightforward, some cautions must be taken in interpreting the simulation results, as discussed in this section.

14.2.1 EMPIRICAL POTENTIAL

Several empirical potential functions have been developed for solid-state C–C interatomic interactions (Tersoff 1988; Brenner 1990; Brenner et al. 2002; Stuart, Tutein, and Harrison 2000), which have enabled both large-scale atomistic simulations and analytical predictions of the elastic properties of graphene. However, the accuracy of each potential is typically limited to a certain set of physical properties. For example, the second-generation reactive empirical bond-order (REBO) potential has been shown to be a reliable potential function for simultaneously predicting bond energy, bond length, surface energy, and bulk elastic properties of diamond (Brenner et al. 2002). However, it considerably underestimates the Young's modulus of CNTs and graphene (Arroyo and Belytschko 2004b; Lu and Huang 2009). By including additional terms for nonbonded intermolecular interaction and torsional interaction, the adaptive intermolecular REBO (AIREBO) potential was shown to predict Young's modulus of graphene in close agreement with first-principle calculations (Stuart, Tutein, and Harrison 2000). However, the predicted equilibrium bond length for graphene is shorter (1.396 vs. 1.420 Å), and it is theoretically unclear how the new terms improved the prediction of Young's modulus. A couple of recent efforts have been reported to improve the Tersoff–Brenner potential for phonon dispersion in graphene and CNTs (Tewary and Yang 2009; Lindsay and Broido 2010). A comparative study of different empirical potentials for graphene is currently in progress, which would guide the selection of potential functions for specific problems and the development of new potential functions with improved overall accuracy.

In this section, the second-generation REBO potential is used for MM simulations of graphene. For completeness, the potential is described briefly as follows. First, the chemical binding energy between two carbon atoms is written in the following form:

$$V_{ij} = V(r_{ij}) = V_R(r_{ij}) - \bar{b}V_A(r_{ij}) \tag{14.25}$$

where r_{ij} is the interatomic distance, V_R and V_A are the repulsive and attractive terms, respectively, as given by

$$V_R(r) = f_c(r)\left(1 + \frac{Q}{r}\right)Ae^{-\alpha r} \tag{14.26}$$

$$V_A(r) = f_c(r)\sum_{n=1}^{3} B_n e^{-\beta_n r} \tag{14.27}$$

and f_c is a smooth cutoff function that limits the range of the covalent interactions within the nearest neighbors, namely

$$f_c(r) = \begin{cases} 1 & r < D_1 \\ \dfrac{1}{2}\left(1 + \cos\left[\dfrac{(r - D_1)\pi}{D_2 - D_1}\right]\right) & D_1 < r < D_2 \\ 0 & r > D_2 \end{cases} \tag{14.28}$$

In addition to the pair potential terms, $\bar{b}$ is an empirical bond-order function, which is a sum of three terms:

$$\bar{b} = \frac{1}{2}\left(b_{ij}^{\sigma-\pi} + b_{ji}^{\sigma-\pi}\right) + b_{ij}^{\pi} \tag{14.29}$$

where the first two terms depend on the local coordination and bond angles, and the third term represents the influence of radical energetics and π-bond conjugation, as well as the dihedral angle for C–C double bonds. Together, the function $\bar{b}$ characterizes the local bonding environment so that the potential can, to some extent, describe multiple-bonding states. As given by Brenner et al. (2002), the analytical forms of these functions are complicated and thus omitted here for brevity.

The parameters for the C–C pair potential terms are $Q = 0.031346$ nm, $A = 10953.5$ eV, $\alpha = 47.465$ nm^{-1}, $B_1 = 12388.8$ eV, $B_2 = 17.5674$ eV, $B_3 = 30.7149$ eV, $\beta_1 = 47.2045$ nm^{-1}, $\beta_2 = 14.332$ nm^{-1}, $\beta_3 = 13.827$ nm^{-1}, $D_1 = 0.17$ nm, and $D_2 = 0.20$ nm. For a planar graphene sheet, these parameters lead to an equilibrium bond length, $r_0 = 0.142$ nm, in close agreement with first-principle calculations (Kudin, Scuseria, and Yakobson 2001) and experimental values (Greenwood and Earnshaw 1984).

The cutoff function in Equation 14.28 is smooth only up to the first derivative. As noted in several studies (Shenderova et al. 2000; Belytschko et al. 2002; Zhang et al. 2004; Zhao, Min, and Aluru 2009), such a cutoff function typically generates spurious bond forces near the cutoff distances, an unphysical result because of discontinuity in the second derivative of the cutoff function. This artifact shall be avoided in the study of nonlinear mechanical properties of graphene under relatively large strains. As suggested by the developers of the original REBO potential (Shenderova et al. 2000), using a larger cutoff distance could remove the unphysical responses. However, to keep the pair interactions within the nearest neighbors, the cutoff distance must not be too large. In the present study, the cutoff function is taken to be 1 within a cutoff distance ($D_1 = 1.9$ Å) and zero otherwise. It is found that the numerical results on the fracture of graphene are essentially unaffected by the choice of the cutoff distance within the range between 1.9 and 2.2 Å.

14.2.2 STRESS CALCULATION

In a discrete atomistic model, forces, rather than stresses, are used for the measure of mechanical interactions. Evaluation of stresses may be carried out subsequently based on the potential energy or directly from the forces. As defined in the continuum theory in Section 14.1, the 2D stress S_{IJ} and moment M_{IJ} can be determined by differentiations of the strain energy density function with respect to the corresponding strain and curvature components, respectively. However, the energy method requires variation of the specific strain and curvature components independently. In the case of uniaxial stretch, for example, only one strain component (E_{11}) is varied, and thus only one stress component (S_{11}) can be determined by the energy method. To evaluate other stress components, we calculate the virial stresses based on a generalization of the virial theorem of Clausius (1870). In particular, the nominal stresses in a graphene sheet are calculated as the average membrane force over a reference area A:

$$P_{iJ} = \frac{1}{2A} \sum_{m \neq n} \left(X_J^{(n)} - X_J^{(m)} \right) F_i^{(mn)} \tag{14.30}$$

where $F_i^{(mn)}$ is the interatomic force between atom m and n at the deformed state, and $X_J^{(m)}$ is the coordinate of the atom m at the reference state. The kinetic part of the virial stresses has been neglected in the static MM simulations.

14.2.3 MOLECULAR MECHANICS VERSUS MOLECULAR DYNAMICS

Atomistic simulations based on molecular dynamics (MD) are commonly used in the study of mechanical behavior of nanomaterials including CNTs and graphene. As opposed to the MM approach, which gives deterministic results based on energy minimization, the MD approach typically yields stochastic results that must be carefully interpreted based on the principles of statistical mechanics. Apparently, the MM approach is limited to simulations of static equilibrium behavior at zero temperature ($T = 0$ K) and cannot account for the effects of temperature, in general. However, the common practice of temperature control in the MD simulations remains problematic as the physical process of heat transfer is often ignored. In addition, in modeling mechanical behaviors, MD simulations often suffer from unrealistically high-loading rates resulting from numerical constraints. Furthermore, care must be taken in stress calculations based on MD simulations (Admal and Tadmor 2010). In this chapter, most results presented are based on MM simulations, whereas MD simulations are used occasionally for comparison.

14.3 UNIAXIAL STRETCH OF GRAPHENE

To study the mechanical behavior of monolayer graphene, we begin by considering an infinite graphene monolayer under the condition of uniaxial stretch, while the direction of stretch varies with respect to the lattice structure of graphene. First, for a continuum description, we set up a coordinate such that the uniaxial stretch is in the X_1-direction. The macroscopically homogeneous deformation of the 2D sheet is represented by the deformation gradient with $F_{11} = \lambda$ and $F_{22} = 1$, whereas the other components of the deformation gradient are all zero. By the definition in Equation 14.2, the 2D Green–Lagrange strain components are $E_{11} = \frac{1}{2}(\lambda^2 - 1)$ and $E_{22} = E_{12} = 0$; the curvature components are all zero in this case, assuming that the sheet remains flat.

By Equation 14.16, the increments of the 2D stresses are as follows:

$$dS_{11} = C_{11}dE_{11} = \lambda C_{11}d\lambda,\ dS_{22} = C_{21}dE_{11} = \lambda C_{21}d\lambda,$$
$$\text{and} \quad dS_{12} = C_{31}dE_{11} = \lambda C_{31}d\lambda \tag{14.31}$$

By the relation in Equation 14.12, the 2D nominal stresses are as follows:

$$P_{11} = \lambda S_{11},\ P_{22} = S_{22},\ P_{12} = \lambda S_{12},\ P_{21} = S_{12},\ P_{13} = P_{31} = 0 \tag{14.32}$$

Noting the symmetry, $S_{12} = S_{21}$, the nominal shear stresses must satisfy the relation, $P_{12} = \lambda P_{21}$, which is required by balance of the angular momentum. The nominal shear stresses (P_{12} and P_{21}) are nonzero when the sheet is anisotropic with a nonzero coupling modulus, C_{31}, which leads to the coupling between stretch and shear.

The nominal strain in the X_1-direction is $\varepsilon = \lambda - 1$. Therefore, the tangent modulus for the nominal stress–strain is

$$\overline{C}_{11} = \frac{dP_{11}}{d\varepsilon} = (1+\varepsilon)^2 C_{11} + \frac{P_{11}}{1+\varepsilon} \qquad (14.33)$$

Only under an infinitesimal nominal strain ($\varepsilon \to 0$), we have $\overline{C}_{11} \approx C_{11}$.

By definition, the 2D true stresses for uniaxial stretch are simply

$$\sigma_{11} = P_{11}, \ \sigma_{22} = \frac{P_{22}}{\lambda}, \ \sigma_{12} = \sigma_{21} = S_{12} \qquad (14.34)$$

In the MM simulations, the strain energy density is calculated as a function of the stretch λ or strain E_{11}, with which the 2D stress S_{11} can be calculated by energy differentiation, $S_{11} = d\Phi/dE_{11}$. To determine the other stress components, the nominal stress components are calculated by the virial approach following Equation 14.30. With the relations in Equations 14.31 and 14.32, the tangent moduli, C_{11}, C_{21}, and C_{31}, can be determined as functions of the stretch λ or strain E_{11}.

A rectangular computational cell of the graphene lattice with periodic boundary conditions is used in the MM simulations. The computational cell is obtained by multiple replications of the smallest rectangular unit cell with one side parallel to the direction of stretch. As shown in Figure 14.2, to simulate uniaxial stretch of graphene in the $(2n, n)$ direction, each unit cell contains 28 atoms in the shaded area, while the computational cell contains an integer number of the unit cells. The direction of stretch is designated by the chiral angle α measured counterclockwise from the zigzag direction of the graphene lattice.

Before each simulation, the graphene lattice is fully relaxed to acquire the equilibrium at the ground state with zero strain. A uniaxial stretch is then applied in two steps. First, all the atoms in the computational cell are displaced according to a homogeneous deformation with the prescribed strain, and the boundaries of the computational cell are stretched accordingly. Second, with the boundaries fixed, the atomic positions are relaxed by internal lattice relaxation to minimize the total

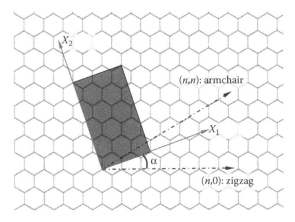

FIGURE 14.2 Illustration of a rectangular unit cell of graphene lattice with a particular chiral direction.

potential energy. The internal relaxation is necessary for the noncentrosymmetric lattice of graphene under a macroscopically homogeneous deformation (Zhou and Huang 2008). A standard quasi-Newton algorithm is used for energy minimization.

Figure 14.3 shows the potential energy per atom as a function of the nominal strain ε for a graphene sheet under uniaxial stretch in the zigzag direction ($\alpha = 0$), along with the snapshots of the atomic structures for $\varepsilon = 0, 0.16, 0.282$, and 0.284. The computational cell in this case contains 160 atoms. The nominal strain is applied with an increment of 0.002. Clearly, the potential energy increases as the strain increases until it reaches a critical point where a sudden drop of the energy occurs. From the snapshots, we see that the atomic lattice is stretched uniformly (with internal relaxation) up to $\varepsilon = 0.282$, while the lattice is fractured spontaneously at the next strain increment. The critical strain for the bond breaking in the zigzag direction is thus estimated to be $\varepsilon_f = 0.283$.

The strain energy density of the graphene monolayer can be calculated as $\Phi = (V - V_0)/A_0$, where V is the energy per atom at the deformed state, V_0 is the energy per atom at the ground state, and A_0 is the area per atom at the ground state. The nominal stress P_{11} can thus be obtained from the potential energy: $P_{11} = \dfrac{\partial \Phi}{\partial \varepsilon}$. Alternatively, the nominal stress as well as the other components ($P_{22}, P_{12},$ and P_{21}) can be calculated by the virial stresses in Equation 14.30. Figure 14.4a shows the nominal stress–strain curves for graphene under uniaxial stretch in the zigzag direction. The nominal stress P_{11} obtained from the potential energy agrees closely with the corresponding virial calculation. In the perpendicular direction, the nominal stress P_{22} is positive as a result of the Poisson's effect. Both the shear components of the nominal stress are zero in this case, indicating no coupling between shear and stretch in the zigzag direction.

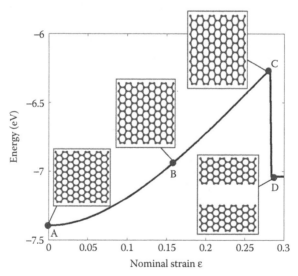

FIGURE 14.3 Energy per atom versus nominal strain of a graphene monolayer under uniaxial stretch in the zigzag direction ($\alpha = 0$), with snapshots of the equilibrium atomic structures at $\varepsilon = 0$ (A), 0.16 (B), 0.282 (C), and 0.284 (D).

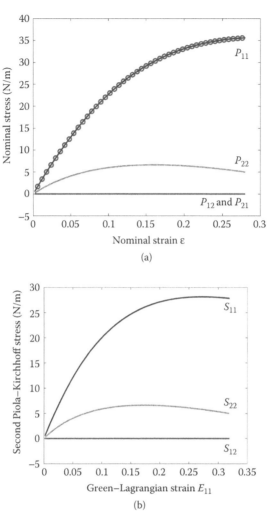

FIGURE 14.4 (a) Nominal stresses versus nominal strain. (b) Second Piola-Kirchhoff stresses versus Green–Lagrange strain for a graphene monolayer under uniaxial stretch in the zigzag direction. The nominal stress P_{11} is calculated from both the energy method (open circles) and the virial method (solid line).

With the nominal stress components, the 2D second Piola–Kirchhoff stress components are obtained by the relationship in Equation 14.32. Figure 14.4b plots the corresponding stresses as a function of the Green–Lagrange strain ($E_{11} = \varepsilon + \varepsilon^2/2$), called S–E curves hereafter, for graphene under uniaxial stretch in the zigzag direction. The tangent moduli C_{11} and C_{21} are then determined from the slopes of the S–E curves, as plotted in Figure 14.5. Both the tangent moduli decrease as the strain increases, demonstrating the nonlinear elastic behavior of graphene under the uniaxial stretch. The other modulus, C_{31}, is zero for $\alpha = 0$.

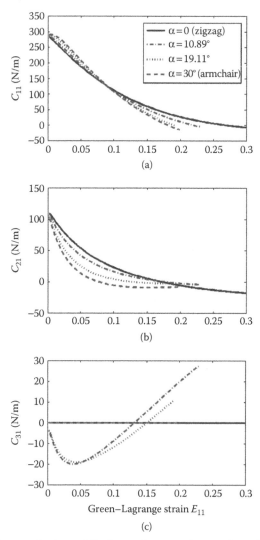

FIGURE 14.5   Tangent elastic moduli of monolayer graphene under uniaxial stretch along different chiral directions. (a) C_{11}; (b) C_{21}; and (c) C_{31}.

The hexagonal symmetry of the graphene lattice dictates that the in-plane elastic property of graphene is isotropic under infinitesimal deformation. However, the non-linear finite deformation of the graphene sheet breaks the hexagonal symmetry of the undeformed lattice, leading to an anisotropic mechanical behavior in the nonlinear regime. As shown in Figure 14.6 for graphene sheets under uniaxial stretch in four different directions, both the tangent modulus and the fracture stress/strain vary with the direction of stretch. The three tangent moduli for the four directions are plotted in Figure 14.5 as functions of the Green–Lagrange strain. Only at infinitesimal strain ($E_{11} \rightarrow 0$) the graphene is isotropic, with $C_{11} = 288.8$ N/m, $C_{21} = 114.9$ N/m, and $C_{31} = 0$ in all directions. These values agree closely with analytical results from

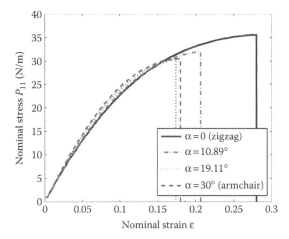

FIGURE 14.6 Nominal stress versus nominal strain for single-atomic-layer graphene sheets under uniaxial stretch along different chiral directions.

earlier studies using the same empirical potential (Arroyo and Belytschko 2004b; Huang, Wu, and Hwang 2006).

Figure 14.5c shows that the tangent modulus C_{31} becomes nonzero under finite stretch in a direction other than zigzag ($\alpha = 0$) or armchair ($\alpha = 30°$). Consequently, a shear stress has to be applied to the graphene sheet in order to maintain uniaxial stretch in the chiral direction ($0 < \alpha < 30°$), an effect resulting from stretch–shear coupling in graphene. In other words, if no shear stress is applied, the shear strain (E_{12}) would become nonzero. The stretch–shear coupling of the planar graphene sheet may have contributed to the reported tension–torsion coupling of single-walled CNTs (Gartstein, Zakhidov, and Baughman 2003; Liang and Upmanyu 2006). Indeed, only CNTs with chirality other than zigzag or armchair were found to exhibit the coupling between tension and torsion.

The nominal stress–strain curves in Figure 14.6 show that the tangent modulus ($\bar{C}_{11}$) as defined in Equation 14.33 decreases as the nominal strain increases. Eventually, $\bar{C}_{11}$ becomes zero at a critical strain, indicating the onset of lattice instability because of loss of ellipticity (Liu, Ming, and Li 2007). This coincides with spontaneous fracture of graphene in the MM simulations, after which the nominal stress becomes zero. As such, the nominal fracture strain and fracture stress of graphene under uniaxial stretch are predicted and plotted in Figure 14.7 as a function of the chiral angle. It is noted that, while the nominal fracture strain varies significantly between 0.178 and 0.283, the nominal fracture stress varies slightly between 30.5 and 35.6 N/m. The MM simulations show that a perfect graphene lattice has the maximum fracture strain and fracture stress in the zigzag direction ($\alpha = 0$). A minimum fracture strength seems to exist in a direction between $\alpha = 19.11°$ and $\alpha = 30°$ (armchair). The strength of suspended monolayer graphene sheets as determined by Lee et al. (2008) based on indentation experiments and a nonlinear membrane model was 42 N/m, with an isotropic fracture strain at 0.25. Although the measured fracture stress is noticeably higher, the fracture strain is well within the range of the MM results.

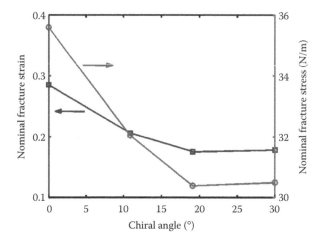

FIGURE 14.7 Nominal fracture strain and stress of single-atomic-layer graphene sheets under uniaxial stretch along different chiral directions.

14.4 INFINITE GRAPHENE UNDER UNIAXIAL TENSION

As discussed in Section 14.3, under the condition of uniaxial stretch, the strain is uniaxial, but the stress is not. However, as a common condition for mechanical testing of materials, uniaxial tensile stress is applied, while the deformation (strain) is not necessarily uniaxial. Consider a 2D sheet under a uniaxial stress in the X_1-direction, meaning that the nominal stress is uniaxial, that is, $P_{iJ} = 0$ except for P_{11}. Assuming that the sheet remains flat, the in-plane components of the deformation gradient are generally nonzero while $F_{31} = F_{32} = 0$. By the relationship in Equation 14.12, we have

$$P_{11} = F_{11}S_{11} + F_{12}S_{21} \tag{14.35}$$

$$P_{22} = F_{22}S_{22} + F_{21}S_{12} = 0 \tag{14.36}$$

$$P_{12} = F_{11}S_{12} + F_{12}S_{22} = 0 \tag{14.37}$$

$$P_{21} = F_{22}S_{21} + F_{21}S_{11} = 0 \tag{14.38}$$

For Equations 14.36 through 14.38 to hold, it requires that

$$F_{21}(F_{11}F_{22} - F_{21}F_{12}) = 0 \tag{14.39}$$

For $P_{11} \neq 0$, we must have $F_{21} = 0$, whereas the other shear component (F_{12}) may not be zero in general. It then follows that $S_{22} = S_{12} = 0$ and $S_{11} = P_{11}/F_{11}$. Thus, the 2D second Piola–Kirchhoff stress is uniaxial.

By Equation 14.16, we have the incremental stress–strain relation for uniaxial stress

$$\begin{pmatrix} dS_{11} \\ 0 \\ 0 \end{pmatrix} = \begin{bmatrix} C_{11} & C_{12} & C_{13} \\ C_{21} & C_{22} & C_{23} \\ C_{31} & C_{32} & C_{33} \end{bmatrix} \begin{pmatrix} dE_{11} \\ dE_{22} \\ 2dE_{12} \end{pmatrix} \tag{14.40}$$

from which we obtain that

$$\begin{pmatrix} dE_{22} \\ 2dE_{12} \end{pmatrix} = -\begin{bmatrix} C_{22} & C_{23} \\ C_{32} & C_{33} \end{bmatrix}^{-1} \begin{pmatrix} C_{21} \\ C_{31} \end{pmatrix} dE_{11} \tag{14.41}$$

and

$$dS_{11} = \left(C_{11} - \begin{pmatrix} C_{12} & C_{13} \end{pmatrix} \begin{bmatrix} C_{22} & C_{23} \\ C_{32} & C_{33} \end{bmatrix}^{-1} \begin{pmatrix} C_{21} \\ C_{31} \end{pmatrix} \right) dE_{11} = \tilde{C}_{11} dE_{11} \tag{14.42}$$

In terms of the nominal stress and nominal strain ($\varepsilon_1 = F_{11} - 1$), we have

$$dP_{11} = (1 + \varepsilon_1) dS_{11} + S_{11} d\varepsilon_1 = \left[(1 + \varepsilon_1)^2 \tilde{C}_{11} + \frac{P_{11}}{1 + \varepsilon_1} \right] d\varepsilon_1 \tag{14.43}$$

Therefore, the tangent Young's modulus is

$$Y = \frac{dP_{11}}{d\varepsilon_1} = (1 + \varepsilon_1)^2 \tilde{C}_{11} + \frac{P_{11}}{1 + \varepsilon_1} \tag{14.44}$$

It is instructive to compare Equation 14.44 with Equation 14.33 to show the difference between uniaxial stress and uniaxial stretch.

In addition to the axial strain (E_{11} or ε_1) in the direction of the uniaxial tension, the graphene sheet deforms with a lateral strain (E_{22} or $\varepsilon_2 = F_{22} - 1$) due to Poisson's effect and with a shear strain (E_{12} or $\gamma = F_{12}$) because of coupling between tension and shear. Define the tangent Poisson's ratio as

$$v = -\frac{d\varepsilon_2}{d\varepsilon_1} \tag{14.45}$$

and define a tangent coupling coefficient as

$$\chi = \frac{d\gamma}{d\varepsilon_1} \tag{14.46}$$

Both the strain increments, $d\varepsilon_2$ and $d\gamma$, can be obtained from Equation 14.41 with the following relationship:

$$dE_{11} = (1 + \varepsilon_1) d\varepsilon_1 \tag{14.47}$$

$$dE_{22} = (1 + \varepsilon_2) d\varepsilon_2 + \gamma d\gamma \tag{14.48}$$

$$dE_{12} = (1 + \varepsilon_1) d\gamma + \gamma d\varepsilon_1 \tag{14.49}$$

For simplicity, consider the case when $C_{13} = C_{23} = 0$ so that there is no coupling between tension and shear, thus the shear strain $\gamma = 0$. The nominal strain in the lateral direction is then

$$d\varepsilon_2 = \frac{dE_{22}}{1+\varepsilon_2} = -\frac{C_{21}}{C_{22}}\frac{1+\varepsilon_1}{1+\varepsilon_2}d\varepsilon_1 \qquad (14.50)$$

In this case, the tangent Poisson's ratio is

$$\nu = \frac{(1+\varepsilon_1)C_{21}}{(1+\varepsilon_2)C_{22}} \qquad (14.51)$$

Only under infinitesimal strain ($\varepsilon_1 \ll 1$), the Young's modulus in Equation 14.44 reduces to $Y_0 = C_{11} - C_{21}^2/C_{22}$, and Poisson's ratio in Equation 14.51 reduces to $\nu_0 = C_{21}/C_{22}$.

Similar to the case of uniaxial stretch in Section 14.3, rectangular computational cells of the graphene lattice are used in the MM simulations of graphene under uniaxial tension. Here, the uniaxial stress condition is achieved by relaxing the constraint on the lateral and shear deformation while maintaining the periodic boundary conditions at all edges. The 2D axial stress can be calculated as a function of the nominal strain by either the energy method or the virial method. Figure 14.8a shows the nominal stress–strain curves for graphene under uniaxial tension in the zigzag and armchair directions, in comparison with first-principle calculations by Wei et al. (2009). Apparently, the MM simulations with the REBO potential underestimate the stiffness of graphene in both directions. The tangent Young's modulus is shown in Figure 14.8b as a function of the nominal strain. The initial Young's modulus (Y_0) predicted by the REBO potential is 243 N/m, whereas the first-principle calculations predict that $Y_0 = 345$ N/m. This discrepancy is the major shortcoming of the REBO potential in modeling mechanical behavior of graphene and CNTs.

As shown in Figure 14.8b, the tangent Young's modulus decreases with increasing strain and eventually becomes zero, at which point fracture occurs as a result of lattice instability (Liu, Ming, and Li 2007). The ideal tensile strength of graphene is thus predicted, which varies with the loading direction, similar to the case of uniaxial stretch in Figure 14.7. It is noted that both the MM simulations and the first-principle calculations predict higher ideal tensile strength in the zigzag direction than in the armchair direction. However, the REBO potential underestimates the ideal strength (fracture stress) in both directions. This discrepancy may be related to the discrepancy in the predictions of the initial Young's modulus. However, the REBO potential overestimates the fracture strain in the zigzag direction, while the predicted fracture strain in the armchair direction compares closely with the first-principle calculation.

As defined in Equation 14.45, the Poisson's ratio of graphene can be readily determined from the MM simulations; in general, the Poisson's ratio is a function of the axial strain (ε_1) and the direction of uniaxial tension. At infinitesimal strain, the initial Poisson's ratio of graphene is isotropic with $\nu_0 = C_{21}/C_{22} = 0.398$ by the REBO

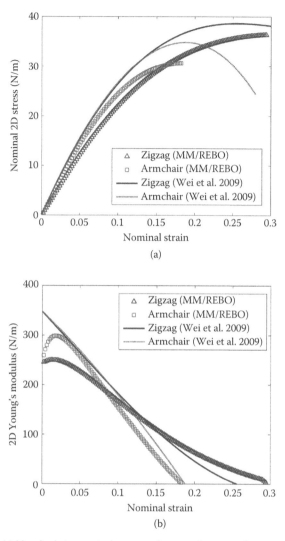

FIGURE 14.8 (a) Nominal stress–strain curves for monolayer graphene under uniaxial tension in the zigzag and armchair directions. (b) Tangent Young's modulus as a function of the nominal strain.

potential. The initial Poisson's ratio of graphene is 0.149 by the first-principle calculations (Kudin, Scuseria, and Yakobson 2001; Liu, Ming, and Li 2007; Wei et al. 2009).

When the direction of uniaxial tension is not in the zigzag or armchair direction of the graphene lattice, coupling between tension and shear leads to a shear strain ($\gamma = F_{12}$) in the nonlinear regime. As defined in Equation 14.46, the coupling coefficient can be determined from MM simulations. However, such simulations have not been reported in literature, possibly because of the challenge to fully relax the shear constraint in the MM simulations. A similar phenomenon was reported for

CNTs under uniaxial tension (Gartstein, Zakhidov, and Baughman 2003; Liang and Upmanyu 2006), where the coupling led to torsional deformation.

14.5   CYLINDRICAL BENDING OF GRAPHENE

As another example, consider rolling of a 2D graphene sheet into a cylindrical tube with the following mapping $(X \to x)$:

$$x_1 = \lambda_1 X_1, \; x_2 = R \sin\left(2\pi \frac{X_2}{L}\right), \quad \text{and} \quad x_3 = R - R \cos\left(2\pi \frac{X_2}{L}\right) \quad (14.52)$$

where λ_1 is the stretch in the axial direction of the tube, R is the tube radius, and L is the width of the undeformed sheet. The deformation gradient in this case is

$$F = \begin{pmatrix} \lambda_1 & 0 \\ 0 & \dfrac{2\pi R}{L} \cos\left(2\pi \dfrac{X_2}{L}\right) \\ 0 & \dfrac{2\pi R}{L} \sin\left(2\pi \dfrac{X_2}{L}\right) \end{pmatrix} \quad (14.53)$$

The corresponding 2D Green–Lagrange strain components are

$$E_{11} = \frac{1}{2}(\lambda_1^2 - 1), \; E_{22} = \frac{1}{2}\left[\left(\frac{2\pi R}{L}\right)^2 - 1\right], \quad \text{and} \quad E_{12} = 0 \quad (14.54)$$

Therefore, the stretch in the circumferential direction of the tube is $\lambda_2 = \dfrac{2\pi R}{L}$. In addition, as defined in Equation 14.4, the curvature tensor has the components

$$K_{22} = \frac{4\pi^2 R}{L^2} \quad (14.55)$$

and $K_{11} = K_{12} = 0$. The normal curvature for a line segment in the circumferential direction is then

$$\kappa_n = \frac{K_{22}}{\lambda_2^2} = \frac{1}{R} \quad (14.56)$$

Thus, a variation in the tube radius simultaneously changes the circumferential stretch and the curvature: $dE_{22} = \left(\dfrac{2\pi}{L}\right)^2 R dR$ and $dK_{22} = \left(\dfrac{2\pi}{L}\right)^2 dR$.

Now consider a special case when $\lambda_1 = \lambda_2 = 1$ so that the graphene monolayer is under pure bending deformation with the radius of curvature R. We then simultaneously change L and R to vary the curvature while maintaining the pure bending condition ($\lambda_1 = \lambda_2 = 1$). As a result, the strain energy density of the graphene is obtained as a function of the curvature only, and the bending moment can be calculated as $M_{22} = d\Phi/dK_{22}$. The bending modulus of the graphene monolayer can then be determined as $D = dM_{22}/dK_{22}$.

The bending modulus of monolayer graphene has been predicted based on empirical potentials (Arroyo and Belytschko 2004b; Huang, Wu, and Hwang 2006; Lu, Arroyo, and Huang 2009) and by first-principle calculations (Kudin, Scuseria, and Yakobson 2001). The fact that the atomically thin graphene monolayer has a finite bending modulus is in contrast with classical theories for plates and shells. For example, the bending modulus of an elastic thin plate scales with the cube of its thickness, namely, $D \sim Eh^3$, where h is the plate thickness and E the Young's modulus of the material. The linear relationship between the bending modulus and the Young's modulus is a result of the classical Kirchhoff hypothesis (Timoshenko and Woinowsky-Krieger 1987), which assumes linear variation of the strain and stress along the thickness of a thin plate. For a graphene monolayer, however, its physical thickness cannot be defined unambiguously in the continuum sense, and the Kirchhoff hypothesis simply does not apply. Therefore, different physical origins must be sought after for the bending moment and bending modulus in monolayer graphene.

Based on the first-generation Brenner potential (Brenner 1990), a simple analytical form was derived for the bending modulus of monolayer graphene under infinitesimal bending curvature (Arroyo and Belytschko 2004b; Huang, Wu, and Hwang 2006),

$$D_0 = \frac{\sqrt{3}}{2} \frac{\partial V_{ij}}{\partial \cos \theta_{ijk}} = \frac{V_A(r_0)}{2} \left(b_0^{\sigma-\pi} \right)' \tag{14.57}$$

where V_{ij} is the interatomic potential as given in Equation 14.25, θ_{ijk} is the angle between two atomic bonds i-j and i-k ($k \neq i, j$), and $(b_0^{\sigma-\pi})'$ denotes the derivative of the bond-order function $b_{ij}^{\sigma-\pi}$ with respect to either one of the two bond angles at the ground state. Equation 14.57 reveals that the physical origin of the bending modulus comes from multibody interactions of the carbon atoms through the bond angle effect in the interatomic potential.

Using the second set of the parameters for the Brenner potential (Brenner 1990), the bending modulus predicted by Equation 14.57 is $D_0 = 0.133$ nN·nm, or equivalently, 0.83 eV. This prediction however is considerably lower than that from first-principle calculations (Kudin, Scuseria, and Yakobson 2001), which gave $D_0 = 3.9$ eV·Å²/atom, or equivalently, 0.238 nN·nm (1.5 eV). Applying the same equation for the second-generation REBO potential (Brenner et al. 2002) leads to an even lower bending modulus: $D_0 = 0.110$ nN·nm or 0.69 eV. The discrepancy between the analytical prediction of Equation 14.57 and the first-principle calculation suggests that the bond angle effect in the empirical potentials does not fully account for the bending stiffness of the monolayer graphene.

In addition to the bond angle effect, the second-generation REBO potential for carbon includes the third nearest neighbors through a bond-order term associated with the dihedral angles (Brenner et al. 2002). The last term of the bond-order function in Equation 14.29 is

$$b_{ij}^{\pi} = b_{ij}^{DH} + \Pi_{ij}^{RC} \tag{14.58}$$

where b_{ij}^{DH} is a function of the dihedral angles, and Π_{ij}^{RC} represents the influence of radical energetics and π-bond conjugation. For a perfect graphene lattice with no vacancy, $\Pi_{ij}^{RC} = 0$, and the dihedral function takes the form

$$b_{ij}^{\mathrm{DH}} = \frac{T_0}{2} \sum_{k,l(\neq i,j)} \left[\left(1 - \cos^2 \Theta_{ijkl}\right) f_c(r_{ik}) f_c(r_{jl}) \right] \tag{14.59}$$

where $T_0 = -0.00809675$ and $f_c(r)$ is the cutoff function that restricts the dihedral function to the four nearest neighbors (k and l) of the atoms i and j, as shown in Figure 14.9. Each C–C bond in the graphene lattice is associated with four dihedral angles, and each dihedral angle accounts for an interaction between one atom (e.g., atom k) and one of its third nearest neighbors (e.g., atom l). For a planar graphene monolayer, the dihedral angles are either 0 or π, and thus $b_{ij}^{\mathrm{DH}} = 0$. However, the dihedral term becomes nonzero upon bending of the graphene monolayer.

By including the dihedral angle effect, a new analytical form for the bending modulus of monolayer graphene was obtained by Lu, Arroyo, and Huang (2009):

$$D_0 = \frac{V_A(r_0)}{2} \left(\left(b_0^{\sigma-\pi}\right)' - \frac{14 T_0}{\sqrt{3}} \right) \tag{14.60}$$

While the first term on the right-hand side of Equation 14.60 is identical to Equation 14.57, the second term results from the effect of dihedral angles in the second-generation REBO potential. With the additional term, Equation 14.60 predicts that $D_0 = 0.225$ nN·nm (1.4 eV), in close agreement with the prediction from the first-principle calculations (Kudin, Scuseria, and Yakobson 2001). This formula establishes a clear connection between the functional form of the empirical potential and the intrinsic bending modulus of monolayer graphene. In particular, the physical origin of the bending modulus is identified as the many-body interactions up to the third nearest neighbors in the graphene lattice, with a significant contribution from the dihedral angle effect.

Atomistic simulations of graphene monolayers rolled into cylindrical tubes of various diameters are performed using the static MM approach. To achieve pure bending (with zero in-plane strain) of the graphene monolayers, the total potential energy is minimized under the constraint that the tube radius and length do not

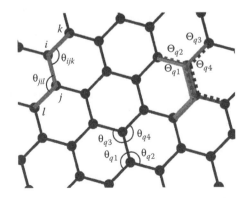

FIGURE 14.9 Illustration of bond angles and dihedral angles in graphene lattice. Each dihedral angle is represented using three bonds connecting four carbon atoms (e.g., i, j, k, l), and each carbon–carbon bond is associated with four bond angles and four dihedral angles.

change during the simulations. The constraint on the tube length is easily applied by the periodic boundary condition along the axial direction of the tube. To enforce the constraint on the tube radius, the graphene is first rolled up by mapping a 2D sheet of width L onto a cylindrical tube of radius $R = L/2\pi$, as prescribed in Equation 14.52. Next, the potential energy is minimized by internal relaxation between the two sublattices of graphene, with one sublattice held in place and the other allowed to relax. In this way, the overall tube radius does not change during the energy-minimization step. Note that the resulting tubes from these simulations are not fully relaxed; in other words, external reaction forces are required to keep the tube dimensions from relaxation, including forces in both the axial and radial directions.

Figure 14.10 plots the strain energy per atom as a function of the curvature for graphene monolayers rolled along the armchair and zigzag directions. For comparison, the results from both the first- and second-generation Brenner potentials (B1 and B2) are shown. To illustrate the effect of dihedral angles, also shown are the results from simulations ignoring the dihedral term in the second-generation Brenner potential (B2*). Clearly, the strain energy for potential B2 is systematically higher than the other two. The dihedral term contributes significantly to the bending energy, especially for large bending curvatures (small tube radius). The corresponding bending moments are obtained by numerically differentiating the strain energy with respect to the bending curvature, $M = dW/d\kappa$, as plotted in Figure 14.11. In all three cases, the bending moment increases almost linearly with the curvature up to 2 nm^{-1}, with slight nonlinearity at large curvatures. By further differentiating the bending moment

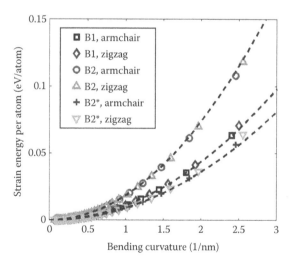

FIGURE 14.10 Strain energy per atom in graphene tubes as a function of bending curvature obtained from different empirical potentials: B1 for the first-generation Brenner potential, B2 for the second-generation Brenner potential, and B2* for the second-generation Brenner potential without considering the dihedral term. Results from molecular mechanics simulations are shown for pure bending of monolayer graphene along armchair and zigzag directions, while the quadratic function, $W = D_0\kappa^2/2$, is plotted as the dashed line using the analytical bending modulus for each potential.

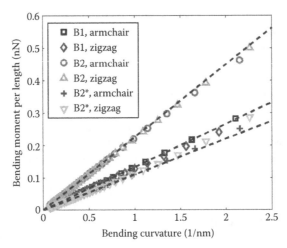

FIGURE 14.11 Bending moment in graphene tubes versus bending curvature along armchair and zigzag directions, obtained from different empirical potentials. The linear elastic bending moment–curvature relation, $M = D_0\kappa$, is plotted as the dashed line using the analytical bending modulus for each potential.

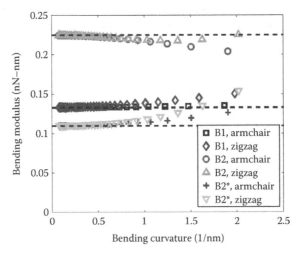

FIGURE 14.12 Tangent bending modulus of monolayer graphene as a function of bending curvature along armchair and zigzag directions, obtained from different empirical potentials. The analytical prediction for the linear elastic bending modulus (independent of curvature) is plotted as the dashed line for each potential.

with respect to the curvature, we obtain the tangent-bending modulus, $D = dM/d\kappa$, as plotted in Figure 14.12. The tangent modulus at small curvatures agrees closely with the analytical prediction by Equation 14.60 as indicted by the dashed lines for the three potentials. At large curvatures (CNTs of small radii), the tangent-bending modulus deviates slightly as a result of nonlinearity. Because of the effect of dihedral angles, the tangent modulus obtained from the potential B2 is considerably higher than those from B1 and B2*.

Figure 14.12 shows that the tangent-bending moduli along the zigzag and armchair directions are essentially identical at small curvatures but become increasingly different as the curvature increases. As expected, the monolayer graphene at the ground state is elastically isotropic because of the hexagonal symmetry of the graphene lattice, and the bending modulus at the linear elastic regime as predicted by Equation 14.60 is independent of the bending direction. However, the lattice symmetry is distorted by the cylindrical bending deformation, and the monolayer graphene becomes slightly anisotropic at the nonlinear regime.

It is noted that, because of the constraint on the tube radius and length, the strain energy of pure bending (Figure 14.10) is slightly higher than the corresponding strain energy in fully relaxed CNTs (Arroyo and Belytschko 2004b). To show the effect of the constraint, Figure 14.13 plots the strain energy as a function of the tube radius for a $(10, 0)$ carbon nanotube. The tube radius R is gradually increased in the MM simulations, whereas the tube length remains fixed. Only the second-generation REBO potential with the dihedral term is used here. As shown in Figure 14.13, the strain energy decreases until it reaches a minimum at $R/R_0 \sim 1.013$, where $R_0 = 0.397$ nm is the tube radius before relaxation. The minimum energy is a few percent lower than the pure bending energy, and it compares closely with the corresponding strain energy by MM calculations without imposing any constraint on the tube radius. Therefore, relaxation of the radial constraint alone leads to increase of the tube radius by about 1.3% or an in-plane strain $\varepsilon = 0.013$ in the circumferential direction of the tube.

To understand the energy reduction and radius increase in the relaxed tube, one may consider the total strain energy as the sum of the bending energy and the in-plane membrane strain energy. As the tube radius increases, the bending energy

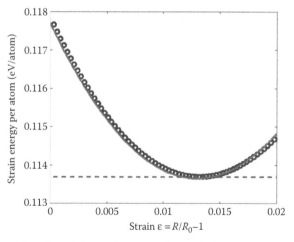

FIGURE 14.13 Relaxation of the strain energy for a $(10, 0)$ carbon nanotube as its radius increases. The atomistic calculations are shown as open circles and the prediction from Equation 14.61 is plotted as the solid curve. The dashed line indicates the strain energy from an atomistic simulation without imposing any constraints on the tube radius.

decreases and the membrane strain energy increases. The competition between the two energy terms leads to a minimum strain energy at the relaxed state. We expand the strain energy with respect to the bending curvature and membrane strain to the leading orders, namely

$$W(\kappa, \varepsilon) \approx \frac{1}{2} D_0 \kappa_0^2 + M_0 (\kappa - \kappa_0) + \sigma_0 \varepsilon + \frac{1}{2} D_0 (\kappa - \kappa_0)^2 + \frac{1}{2} C_0 \varepsilon^2 \qquad (14.61)$$

where $\kappa_0 = 1/R_0$ is the curvature before relaxation, $M_0 = (\partial W / \partial \kappa)_{\varepsilon=0}$ is the bending moment, and $\sigma_0 = (\partial W / \partial \varepsilon)_{\varepsilon=0}$ is the in-plane stress in the circumferential direction for the unrelaxed tube. Noting that at the curvature $\kappa = 1/R = \kappa_0/(1+\varepsilon)$, the strain energy has a minimum at $\varepsilon = (D_0 \kappa_0^2 - \sigma)/(C_0 + 3 D_0 \kappa_0^2)$, where $D_0 = 0.225$ nN $\cdot$ nm and $C_0 = 289$ N/m are the elastic modulus for bending and in-plane stretch, respectively. For the (10, 0) nanotube, it is found that, with $\sigma_0 = -2.38$ N/m, Equation 14.61 agrees closely with the MM calculations in Figure 14.13 for the in-plane strain up to a few percent. The compressive in-plane stress before relaxation may be qualitatively understood as a result of shortening of the bond lengths in the constrained tube, relative to the bond length at the ground state of graphene. The imposed constraint over the tube radius effectively applies an external pressure onto the tube, balancing the circumferential stress. The applied external pressure may be estimated from the Laplace–Young equation, namely $p = \sigma_0/R_0 = 6.0$ GPa, which compares closely with first-principle calculations of CNTs under hydrostatic pressure (Reich, Thomsen, and Ordejon 2002). We note that the presence of the circumferential stress before relaxation is in clear contrast with the classical plate theory that predicts zero membrane force under the pure bending condition. This suggests an intrinsic coupling between bending and in-plane strain because of the discrete nature of the graphene lattice.

14.6 EDGE ENERGY, EDGE FORCE, AND EDGE BUCKLING

As discussed in Section 14.1.4, an edge energy density function can be defined as the excess energy per unit length along the edges of a finite graphene sheet. The physical origin of the edge energy has to do with the change of the bonding environment at the edges. First-principle calculations based on density functional theory (DFT) have suggested in-plane reconstructions as a way to relieve the excess edge energy and thus stabilize the planar edge structure of graphene (Koskinen, Malola, and Hakkinen 2008). However, atomistic simulations using empirical potentials have shown rippling, warping, and twisting, with out-of-plane deformation along free edges of graphene monolayers (Shenoy et al. 2008; Thompson-Flagg, Moura, and Marder 2009; Bets and Yakobson 2009; Lu and Huang 2010).

Based on the REBO potential (Brenner et al. 2002), the chemical binding energy between two carbon atoms is given in the form of Equation 14.25. Take an infinite, planar graphene monolayer as the reference state. The potential energy per atom is

$$U_0 = \frac{3}{2} V(r_0) \qquad (14.62)$$

where $r_0 = 0.142$ nm is the equilibrium bond length and the corresponding bond-order function $\bar{b} = \bar{b}_0 = 0.9510$. The bond energy at the reference state is $V = V(r_0) = -4.930$ eV, and thus the energy per atom is $U_0 = -7.395$ eV.

Now consider an infinitely long ribbon cut out from the graphene monolayer with two parallel edges (zigzag or armchair). Assuming no deformation of the graphene lattice for the moment, the change in the bonding environment along the free edges leads to a change in the bond-order function and thus a change of the bond energy. In particular, for a zigzag edge (Figure 14.14a), the bond-order function for the edge bond becomes $\bar{b}_{Z1} = 0.9478$, while the effect on the other bonds is negligibly small. With the same bond length ($r = r_0$), the bond energy along the zigzag edge is increased from $V_0 = -4.930$ eV to $V_{Z1} = -4.858$ eV. Consequently, the energy per atom in the first row of the zigzag edge increases from $U_0 = -7.395$ eV to $U_{Z1} = -4.858$ eV. Note that each edge atom is associated with two edge bonds instead of three. In addition, the energy per atom in the second row of the zigzag edge also increases because of association with the edge bonds $U_{Z2} = (2V_{Z1} + V_0)/2 = -7.323$ eV. Together, relative to the reference state, the excess energy per unit length of the zigzag edge is

$$\gamma_Z = \frac{1}{\sqrt{3}r_0}(U_{Z1} + U_{Z2} - 2U_0) \tag{14.63}$$

which gives 10.61 eV/nm or 1.70 nN by the REBO potential.

Similarly, for an armchair edge (Figure 14.14b), the bond energy changes in the first and second rows: $V_{A1} = -5.268$ eV and $V_{A2} = -4.858$ eV. The energy per atom is then $U_{A1} = (V_{A1} + V_{A2})/2 = -5.063$ eV in the first row and $U_{A2} = (V_{A2} + 2V_0)/2 = -7.359$ eV in the second row, both greater than the reference value ($U_0 = -7.395$ eV). The excess energy per unit length of the armchair edge is thus

$$\gamma_A = \frac{2}{3r_0}(U_{A1} + U_{A2} - 2U_0) \tag{14.64}$$

which is 11.12 eV/nm or 1.78 nN, slightly higher than that of the zigzag edge.

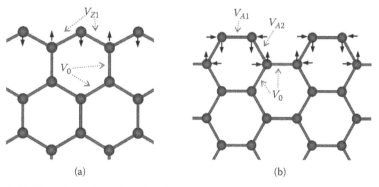

(a)   (b)

FIGURE 14.14 Illustration of the bond structures, bond energies, and unbalanced interatomic forces at (a) the unrelaxed zigzag and (b) armchair edges. Dark arrows indicate the directions of unbalanced forces acting on the edge atoms. Dotted line arrows indicate different bond energies near the edges.

The above analysis of the excess edge energy has assumed no deformation or bond reconstruction along the edges, termed as *unrelaxed* edges. However, the change of the bonding environment also leads to changes in the equilibrium bond length and bond angles along the edges. As a result, the interatomic forces acting on each atom are not balanced along the unrelaxed edges. As illustrated in Figure 14.14, the forces acting on the atoms in the first two rows of the unrelaxed zigzag edge are unbalanced in the direction perpendicular to the edge, while the unbalanced forces are in both perpendicular and parallel directions for atoms along the unrelaxed armchair edge. However, the interatomic forces acting on each atom are balanced in the parallel direction along the zigzag edge because of symmetry. Consequently, the edge atoms tend to displace in the direction of the unbalanced forces, leading to spontaneous deformation of the graphene lattice near the free edges and relaxation of the excess edge energy. Using the REBO potential, we simulate the edge relaxation by a standard MM method, minimizing the total potential energy in a graphene ribbon with two parallel free edges.

For comparison, we first calculate the average potential energy per atom with unrelaxed edges, for which the lattice structure of each ribbon is taken directly from the reference state of a fully relaxed, planar graphene monolayer. As shown in Figure 14.15, the potential energy per atom increases linearly with the inverse of the ribbon width, which can be understood as a result of the excess edge energy, namely,

$$\bar{U}(W) = U_0 + \frac{2\gamma}{N} = U_0 + \frac{S_0}{W}\gamma \tag{14.65}$$

where $\bar{U}$ is the average energy per atom of the ribbon, W is the ribbon width, γ is the excess edge energy per length, $S_0 = \frac{1}{2}\sqrt{3}r_0^2$ is the area of the unit cell of graphene (containing two carbon atoms), and $N = 2W/S_0$ is the number of carbon atoms per unit length of the ribbon. Equation 14.65 reveals the dependence of the average energy on the ribbon width, which agrees closely with the atomistic calculations for the unrelaxed edges.

Next, in the MM simulations, we allow the atoms to move in the direction that reduces the total potential energy. Only in-plane displacements of the atoms are allowed for the moment. Periodic boundary conditions are assumed at both ends of the graphene ribbons, with the end-to-end distance fixed. Upon such relaxation, the ribbon width shrinks slightly, while the ribbon length does not change, thus termed as *1D relaxation*. Figure 14.15 shows that the average energy per atom in each ribbon decreases slightly after the 1D relaxation. By Equation 14.65, the excess edge energy after the 1D relaxation is calculated from the average energy and plotted in Figure 14.16. Before relaxation, the excess edge energies agree closely with those predicted by Equations 14.63 and 14.64, for the zigzag and armchair edges, respectively. After 1D relaxation, both the edge energies are reduced by roughly 2%, that is, $\gamma_Z = 10.41$ eV/nm and $\gamma_A = 10.91$ eV/nm. The excess edge energy is independent of the ribbon width before and after 1D relaxation for the range of the ribbon width shown in Figure 14.16.

Upon 1D relaxation, the interatomic forces acting on each atom are balanced in all directions. However, the mismatch between the equilibrium bond length at the edges

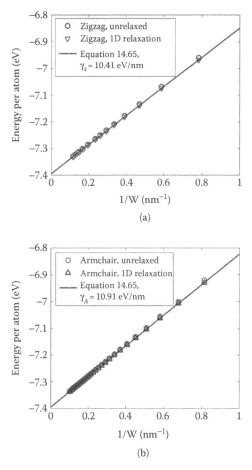

FIGURE 14.15 Average energy per atom of graphene monolayer with (a) zigzag edges and (b) armchair edges, as a function of the ribbon width (W), for unrelaxed edges and after one-dimensional edge relaxation.

and that at the interior of the graphene ribbon leads to a compressive internal force along the free edges, which was called *edge stress* previously (Shenoy et al. 2008; Jun 2008; Huang et al. 2009). The internal edge forces are self-balanced in an infinitely long ribbon, as illustrated in Figure 14.17. To evaluate the magnitude of the edge force, we calculate the total internal force acting on a cross section of the graphene ribbon (with two parallel edges) after the 1D relaxation, which equals twice the corresponding edge force. This calculation gives the following edge forces: $f_Z = -16.22$ eV/nm or -2.60 nN for the zigzag edge and $f_A = -8.53$ eV/nm or -1.36 nN for the armchair edge, both compressive as indicated by the negative sign and independent of the ribbon width. Alternatively, the edge forces can be determined from variation of the excess edge energies in strained graphene ribbons (Shenoy et al. 2008; Jun 2008); both methods predict essentially the same edge forces (Lu and Huang 2010).

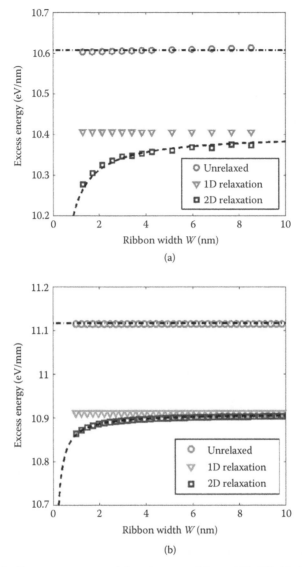

FIGURE 14.16 Excess energy per unit length versus ribbon width (W): (a) zigzag edge and (b) armchair edge.

It is noted that, while the excess edge energies for the zigzag and armchair edges differ slightly (<5%), the edge force of the zigzag edge is nearly twice that of the armchair edge. Table 14.1 compares the edge energies and edge forces predicted by the REBO potential (after 1D relaxation) with other calculations. Several DFT calculations have predicted similar excess edge energies (Koskinen, Malola, and Hakkinen 2008; Jun 2008; Huang et al. 2009). Notably, the excess energy for the zigzag edge (without reconstruction) from the DFT calculations is higher

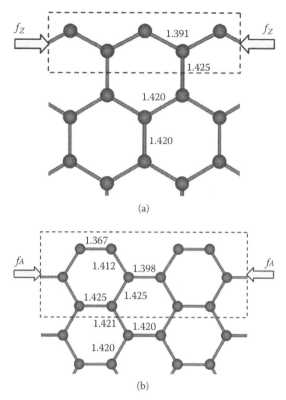

FIGURE 14.17 The bond structures (with bond lengths in angstrom) at (a) zigzag and (b) armchair edges after one-dimensional relaxation. The dashed rectangular boxes represent effective edge layers that are subjected to compressive edge forces (f_Z and f_A).

TABLE 14.1

Comparison of Predicted Excess Edge Energy and Edge Forces (Both in eV/nm) of Monolayer Graphene

	Edge Energy (γ)		Edge Force (F)		r_0 (nm)
	Armchair	Zigzag	Armchair	Zigzag	
DFT (Koskinen, Malola, and Hakkinen 2008)	9.8	13.2	–	–	0.142
DFT (Huang et al. 2009)	10	12	−14.5	−5	0.142
DFT (Jun 2008)	12.43	15.33	−26.40	−22.48	0.142
AIREBO (Shenoy et al. 2008)	–	–	−10.5	−20.5	0.140
MD (Bets and Yakobson 2009)	–	–	−20.4	−16.4	0.146
REBO (Lu and Huang, 2010a)	10.91	10.41	−8.53	−16.22	0.142

DFT = density functional theory; AIREBO = adaptive intermolecular REBO; MD = molecular dynamics; MM = molecular mechanics; REBO = reactive empirical bond-order.

than that for the armchair edge, opposite to the predictions by the REBO potential. As the edge energies are related to the thermodynamically stable shapes of finite graphene sheets or flakes (Gan and Srolovitz 2010), qualitatively different shapes would be predicted based on the DFT edge energies and atomistic simulations using empirical potentials. Although reconstruction of the zigzag edge was predicted by the DFT calculations to have a lower excess energy than that of the armchair edge (Koskinen, Malola, and Hakkinen 2008), no edge reconstruction is observed in the MM simulations.

For the edge forces, the DFT calculations have predicted quite different values (Jun 2008; Huang et al. 2009), possibly because of the uses of different approximations and methods. However, the edge forces predicted by the REBO potential are similar to those predicted by the AIREBO potential (Shenoy et al. 2008), while MD simulations using the Tersoff potential predicted considerably larger edge forces (Bets and Yakobson 2009). The differences may result from different equilibrium bond lengths (r_0) of graphene predicted by the different empirical potentials, as listed in Table 14.1. The REBO potential used in the present study predicts an equilibrium bond length in close agreement with the DFT calculations. Despite the discrepancies, all calculations have predicted compressive edge forces for both the zigzag and armchair edges.

Because of the compressive edge forces, the total potential energy in a graphene ribbon can be partially relaxed either by elongation in the longitudinal direction of the ribbon (namely, 2D relaxation) or by out-of-plane displacement of the atoms (edge buckling). To simulate the 2D relaxation, only in-plane displacements of the atoms are allowed, while the end-to-end distance of the graphene ribbon is varied gradually to impose a longitudinal strain (ε) until the total potential energy reaches a minimum. Figure 14.18 plots the longitudinal strain corresponding to the minimum energy in graphene ribbons with different ribbon widths (W), and the corresponding

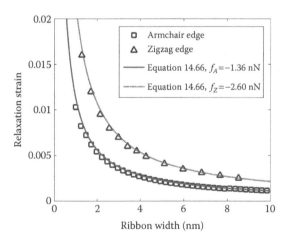

FIGURE 14.18 Longitudinal strain of graphene nanoribbon after 2D relaxation: comparing molecular mechanics calculations (open symbols) and the predictions from Equation 14.66.

excess energy is plotted in Figure 14.16. The longitudinal strain after 2D relaxation is inversely proportional to the ribbon width (Lu and Huang 2010a):

$$\varepsilon = -\frac{2f}{YW} \tag{14.66}$$

where f is the edge force, Y is the Young's modulus of graphene, and W is the ribbon width.

An alternative mode of edge relaxation can be achieved by allowing 3D deformation of the graphene ribbon. The compressive edge force motivates out-of-plane buckling along the free edges, but opposed by the bending stiffness of graphene. The competition leads to an intrinsic wavelength for the edge buckling. Figure 14.19 shows two examples of graphene nanoribbons (GNRs) with edge buckling by MM simulations. Clearly, the buckle amplitude maximizes along the free edges and decays away from the edges. Similar edge buckling was predicted using different empirical potentials (Shenoy et al. 2008; Bets and Yakobson 2009).

For each graphene ribbon, the end-to-end distance is fixed during the MM simulation, with periodic boundary conditions at both ends, and out-of-plane perturbations are introduced to trigger the buckling deformation. The excess energy is calculated as the total energy increase per unit length of the free edges relative to the ground state of graphene, thus including both the edge energy and the interior bending energy of the ribbon due to deformation. It is found that the excess energy depends on both the end-to-end distance (L) and the buckling wave number (n). For each L, different buckling modes are obtained from the MM simulations, indicating that more than one local energy minimum exists. Figure 14.20 plots the excess energy as a function of the buckle wavelength, $\lambda = L/n$, in which the excess energy minimizes at a particular wavelength for each edge configuration.

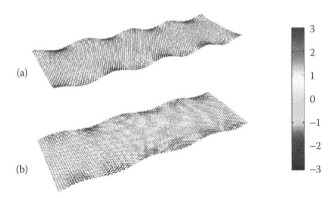

FIGURE 14.19 Edge buckling of graphene nanoribbons with (a) zigzag edges ($W = 7.7$ nm, $L = 23.6$ nm) and (b) armchair edges ($W = 7.9$ nm, $L = 23.4$ nm). W and L are the width and length of the ribbon, respectively.

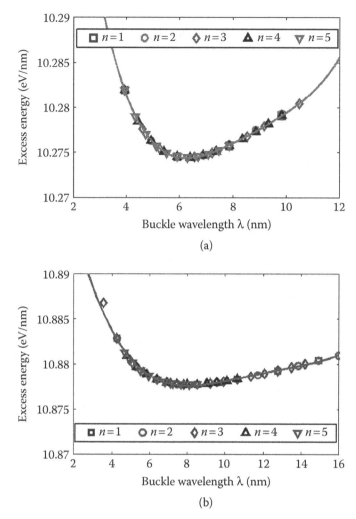

FIGURE 14.20 Excess energy of graphene nanoribbons with edge buckling, against the buckle wavelength. (a) Zigzag edge ($W = 7.7$ nm) and (b) armchair edge ($W = 7.9$ nm). The molecular mechanics results are fitted with fourth-order polynomial functions (plotted as solid lines).

Therefore, an intrinsic wavelength for edge buckling is predicted to be 6.2 nm for the zigzag edge and 8.0 nm for the armchair edge.

It is found that the wavelength of edge buckling is independent of the ribbon width for long graphene ribbons with the width $W > 6$ nm. However, for narrower ribbons, the buckle wavelength changes slightly with the ribbon width, most likely because of the proximity of the two edges that interact with each other. It is also noted that the edge–edge interaction in narrow graphene ribbons ($W < 6$ nm) can lead to antiphase correlation of the buckling waves along the two parallel edges (Figure 14.21), in which case the nanoribbon appears to be twisted.

FIGURE 14.21 Antiphase correlation of edge buckling in a narrow graphene ribbon ($W = 3.4$ nm, $L = 14.8$ nm).

14.7 GRAPHENE NANORIBBONS UNDER UNIAXIAL TENSION

To harvest the unique physical properties of monolayer graphene for potential applications in nanoelectronics and electromechanical systems, graphene ribbons with nanoscale widths ($W < 20$ nm) have been recently produced either by lithographic patterning or by chemically derived self-assembly processes. The edges of the GNRs could be zigzag, armchair, or a mixture of both (Nakada et al. 1996). It has been theoretically predicted that the special characteristics of the edge states leads to a size effect in the electronic state of graphene and controls whether the GNR is metallic, insulating, or semiconducting. The effects of the edge structures on deformation and mechanical properties of GNRs have also been studied to some extent (e.g., Reddy et al. 2009; Faccio et al. 2009; Topsakal and Ciraci 2010; Lu and Huang 2010b). On the one hand, the elastic deformation of GNRs has been suggested as a viable method to tune the electronic structure and transport characteristics in graphene-based devices. On the other hand, plastic deformation and fracture of graphene may pose a fundamental limit for reliability of integrated graphene structures.

Ideally, the mechanical properties of GNRs may be characterized experimentally by uniaxial tension tests. To date, however, no such experiment has been reported, although similar tests were performed for CNTs (Yu et al. 2000). Theoretically, the nonlinear mechanical behavior of GNRs can be studied by combining atomistic simulations and the continuum theory as presented in Section 14.1. Of particular interest is the effect of the edge structures on the mechanical properties of GNRs.

In Section 14.4, atomistic simulations of infinite graphene lattice under uniaxial tension are performed, which predict the nonlinear elastic behavior of monolayer graphene with initial Young's modulus, Poisson's ratio, and the ideal tensile strength in both zigzag and armchair directions. Similarly, to simulate GNRs under uniaxial tension, rectangular ribbons are cut out from a fully relaxed graphene lattice with different edge chirality (zigzag or armchair) and ribbon widths. In each simulation, the tensile strain is applied incrementally in the longitudinal direction of the GNR, until fracture occurs. At each strain level, the statically equilibrium lattice structure of the GNR is calculated to minimize the total potential energy by a quasi-Newton algorithm. Periodic boundary conditions are applied at both ends of the GNR, whereas the two parallel edges of the GNR are free of external constraint. With such boundary conditions, the bond structures are fully relaxed at the edges and the GNR is loaded with an uniaxial stress.

Figure 14.22 shows the results from atomistic simulations for GNRs with unpassivated edges, where the ribbon width (W) is varied between 1 and 10 nm. For each GNR, the average potential energy per carbon atom increases as the nominal strain increases until it fractures at a critical strain. To understand the numerical results, we adopt a simple thermodynamics model for the uniaxially stressed GNRs. For a GNR

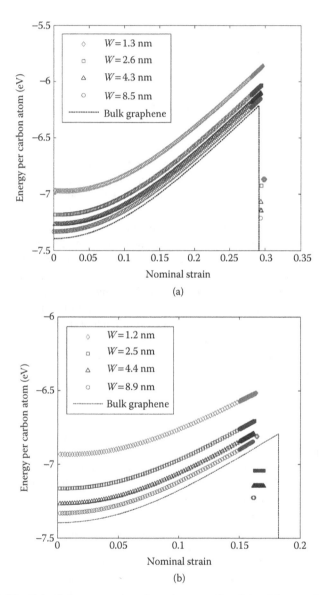

FIGURE 14.22 Potential energy per carbon atom as a function of the nominal strain for graphene nanoribbons under uniaxial tension, with (a) zigzag and (b) armchair edges, both unpassivated. The dashed lines show the results for bulk graphene under uniaxial tension in the zigzag and armchair directions.

of width W and length L, the total potential energy as a function of the nominal strain consists of contributions from deformation of the interior lattice (bulk strain energy) and from the edges (edge energy), namely

$$U(\varepsilon) = U_0 WL + \Phi(\varepsilon)WL + 2\gamma(\varepsilon)L \tag{14.67}$$

where ε is the nominal strain (relative to the bulk graphene lattice at the ground state), U_0 is the potential energy density (per unit area) of graphene at the ground state, $\Phi(\varepsilon)$ is the bulk strain energy density (per unit area), and $\gamma(\varepsilon)$ is the edge energy density (per unit length of the edges). Equation 14.67 is in fact a specialized form of the general Equation 14.22 for the simple case under consideration here. Although the bulk strain energy density as a function of the nominal strain can be obtained directly from the atomistic calculations for the infinite graphene lattice (dashed lines in Figure 14.22), the edge energy density function is determined by subtracting the bulk energy from the total potential energy of the GNRs based on Equation 14.67. Thus, both the energy functions are atomistically determined, and can then be fitted with nonlinear polynomial functions for theoretical purposes.

The GNR under uniaxial tension is subjected to a net force (F) in the longitudinal direction. At each strain increment, the mechanical work done by the longitudinal force equals to the increase of the total potential energy, which can be written in a variational form, that is,

$$\delta U = FL\delta\varepsilon \tag{14.68}$$

Indeed Equation 14.68 is a simple form of the general variational statement in Equation 14.24. Consequently, the axial force (F) can be obtained from the derivative of the potential energy function in Equation 14.67. A 2D nominal stress can then be defined without ambiguity as the force per unit width of the GNR, namely

$$\sigma(\varepsilon) = \frac{F}{W} = \frac{d\Phi}{d\varepsilon} + \frac{2}{W}\frac{d\gamma}{d\varepsilon} \tag{14.69}$$

Alternatively, the stress may be calculated directly from the interatomic forces or the virial method.

Figure 14.23 shows the 2D nominal stress–strain curves of the GNRs obtained by taking the derivative of the potential energy curves in Figure 14.22. Similar stress–strain curves were obtained by MD simulations (Zhao, Min, and Aluru 2009), where the critical strain to fracture is typically lower than that predicted from the static MM simulations because of the effects of temperature and loading rate. Apparently, the stress–strain relation of a GNR is generally nonlinear, for which the tangent modulus as a function of the nominal strain is defined as

$$E(\varepsilon) = \frac{d\sigma}{d\varepsilon} = \frac{d^2\Phi}{d\varepsilon^2} + \frac{2}{W}\frac{d^2\gamma}{d\varepsilon^2} \tag{14.70}$$

The first term on the right-hand side of Equation 14.70 is the tangent Young's modulus of the bulk graphene, which is equivalent to Equation 14.44, and the second

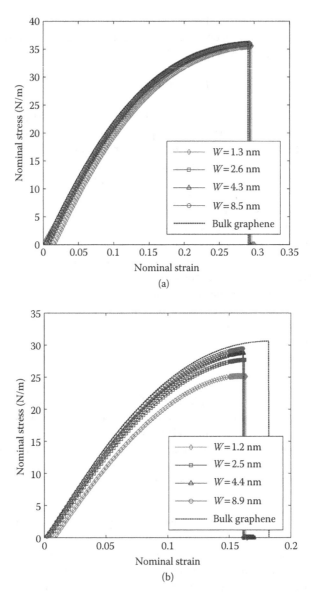

FIGURE 14.23 Nominal stress–strain curves for graphene nanoribbons under uniaxial stretch, with (a) zigzag and (b) armchair edges, both unpassivated. The dashed lines show the results for bulk graphene under uniaxial tension in the zigzag and armchair directions.

term presents the contribution from the edge effect (i.e., edge modulus). Therefore, the elastic modulus of the GNR, in general, depends on the ribbon width (W), as well as the edge chirality. In other words, the nonlinear dependence of the excess edge energy on strain leads to a anisotropic, width-dependent Young's modulus for GNRs. In addition, the Poisson's ratio of the GNRs may also be determined from the same atomistic simulations by calculating the ribbon width as a function of the longitudinal strain, which in general varies nonlinearly with the strain and depends on the edge structure as well.

It is noted from Figure 14.23 that the nominal strain in a GNR is not equal to zero when the nominal stress is zero. As predicted by Equation 14.66, this offset strain is inversely proportional to the ribbon width because of the effect of compressive edge forces for both the zigzag and armchair edges. Recall that the nominal strain is measured relative to the bulk graphene at the ground state. Furthermore, as shown in Section 14.6, a fully relaxed GNR would have periodically buckled edges, which in turn would affect the initial stress–strain behavior of the GNR. However, it is found that the edge buckling of the GNRs essentially disappears under uniaxial tension with the applied nominal strain beyond a fraction of 1%.

The nominal stress–strain curves in Figure 14.23 show approximately linear elastic behavior of all GNRs at relatively small strains (e.g., $\varepsilon < 5\%$). Following Equation 14.70, the initial Young's modulus of the GNRs in the linear regime can be written as

$$E_0 = E_0^b + \frac{2}{W} E_0^e \qquad (14.71)$$

where E_0^b is the initial Young's modulus of the bulk graphene, and E_0^e is the initial edge modulus. Although the bulk graphene is isotropic in the regime of linear elasticity, the edge modulus depends on the edge chirality with different values for the zigzag and armchair edges. As a result, the initial Young's modulus of the GNR depends on the edge chirality and the ribbon width, as shown in Figure 14.24. The REBO potential used in the present study predicts a bulk Young's modulus,

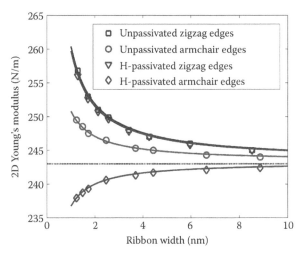

FIGURE 14.24 Initial Young's modulus versus ribbon width for graphene nanoribbons with unpassivated and hydrogen-passivated edges. The horizontal dot-dashed line indicates the initial Young's modulus of bulk graphene predicted from the reactive empirical bond-order potential.

$E_0^b = 243$ N/m, and the predicted edge modulus is $E_0^e = 8.33$ nN (~52 eV/nm) for the zigzag edge and $E_0^e = 3.65$ nN (~22.8 eV/nm) for the armchair edge. With positive moduli for both edges, the Young's modulus of GNRs increases as the ribbon width decreases.

The edges of GNRs are often passivated with hydrogen (H) atoms (Nakada et al. 1996). For a GNR with H-passivated edges, the potential energy in Equation 14.1 is modified to account for the hydrogen adsorption, namely

$$U(\varepsilon) = U_0 WL + \Phi(\varepsilon)WL + 2\gamma(\varepsilon)L - 2\gamma_H(\varepsilon)L \qquad (14.72)$$

where $\gamma_H(\varepsilon)$ is the adsorption energy of hydrogen per unit length along the edges of the GNR, and the negative sign indicates typically reduced edge energy because of H-passivation (Koskinen, Malola, and Hakkinen 2008; Gan and Srolovitz 2010). By comparing the potential energies for the GNRs with and without H-passivation, the adsorption energy can be determined as a function of the nominal strain for both armchair and zigzag edges. At zero strain ($\varepsilon = 0$), our MM calculations predict the hydrogen adsorption energy to be 20.5 and 22.6 eV/nm for the zigzag and armchair edges, respectively, which is closely comparable to that predicted from first-principle calculations (Koskinen, Malola, and Hakkinen 2008). Under uniaxial tension, the adsorption energy varies with the nominal strain. Similar to Equation 14.69, the nominal stress for the H-passivated GNR is obtained as

$$\sigma(\varepsilon) = \frac{d\Phi}{d\varepsilon} + \frac{2}{W}\left(\frac{d\gamma}{d\varepsilon} - \frac{d\gamma_H}{d\varepsilon}\right) \qquad (14.73)$$

Correspondingly, the tangent modulus is

$$E(\varepsilon) = \frac{d\sigma}{d\varepsilon} = \frac{d^2\Phi}{d\varepsilon^2} + \frac{2}{W}\left(\frac{d^2\gamma}{d\varepsilon^2} - \frac{d^2\gamma_H}{d\varepsilon^2}\right) \qquad (14.74)$$

The effect of H-passivation on the initial Young's modulus of GNRs is shown in Figure 14.24. Interestingly, while H-passivation has negligible effect on the initial Young's modulus of GNRs with zigzag edges, the effect is dramatic for GNRs with armchair edges. In the latter case, a negative edge modulus ($E_0^e = -20.5$ eV/nm) is obtained, and thus the initial Young's modulus decreases with decreasing ribbon width, opposite to the unpassivated GNRs.

As discussed in Section 14.4, without any defect, the bulk graphene fractures when the tangent modulus becomes zero (i.e., $d^2\Phi/d\varepsilon^2 = 0$), dictated by the intrinsic lattice instability under tension (Zhang et al. 2004; Liu, Ming, and Li 2007). At a finite temperature, however, fracture may occur much earlier because of thermally activated processes (Zhao, Min, and Aluru 2009). As shown in Sections 14.3 and 14.4, the critical strain to fracture bulk graphene varies with the loading direction. Both first-principle calculations (Liu, Ming, and Li 2007; Wei et al. 2009) and empirical potential models (Lu and Huang 2009) have predicted that the intrinsic critical strain is higher for graphene under uniaxial tension in the zigzag direction than in the armchair direction, suggesting that the hexagonal lattice of graphene preferably fractures along the zigzag directions by cleavage of the C–C bonds. As shown in Figure 14.22a, the GNRs with zigzag edges fracture at a critical strain close to that of bulk graphene loaded in the same direction. In contrast, Figure 14.22b shows that the GNRs with armchair edges fracture at a critical strain considerably lower than bulk graphene. In both cases, the fracture strain slightly depends on the ribbon width, as shown in Figure 14.25. H-passivation of the edges leads to slightly lower fracture strains for zigzag GNRs, but slightly higher fracture strains for armchair GNRs. The apparently different edge effects on the fracture strain imply different fracture mechanisms for the zigzag and armchair GNRs as discussed below.

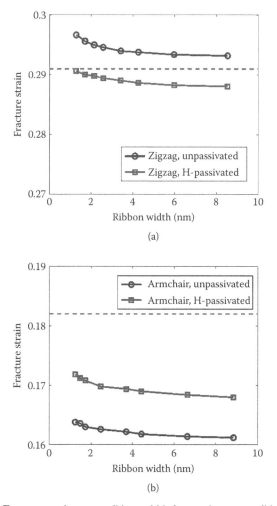

FIGURE 14.25 Fracture strain versus ribbon width for graphene nanoribbons under uniaxial tension, with (a) zigzag and (b) armchair edges. The horizontal dashed line in each figure indicates the fracture strain of bulk graphene under uniaxial tension in the same direction.

The processes of fracture nucleation in GNRs are studied using MD simulations at different temperatures. It should be noted that MD simulations are typically sensitive to the temperature control and the loading rate. In the present study, the MD simulations provide qualitative understanding of the fracture mechanisms, which are consistent with the static MM calculations. The quantitative nature of the MD simulation is thus not essential for this purpose. The MD simulations reveal two distinct mechanisms for fracture nucleation in GNRs of different edge structures at relatively low temperatures ($T < 300$ K). Figure 14.26 shows two fractured GNRs at 50 K. For the GNR with zigzag edges (Figure 14.26a), fracture nucleation occurs stochastically at the interior lattice of the zigzag GNRs. As a result, the fracture strain is very close to that of bulk graphene strained in the same direction, consistent with the MM calculations (Figure 14.22a). However, for the GNR with armchair edges (Figure 14.26b), fracture nucleation occurs exclusively near the edges. Thus, the armchair edge serves as the preferred location for fracture nucleation, leading to a lower fracture strain compared with bulk graphene, as seen also from the MM calculations (Figure 14.22b). Therefore, two distinct fracture nucleation mechanisms are identified as interior homogeneous nucleation for the zigzag GNRs and edge-controlled heterogeneous nucleation for the armchair GNRs. The same mechanisms hold for GNRs with H-passivated edges.

It is evident from Figure 14.26 that cracks preferably grow along the zigzag directions of the graphene lattice in both cases. Although some first-principle calculations have predicted lower edge energy for armchair edges, MM and MD calculations using empirical potentials have predicted the opposite trend, as listed in Table 14.1. However, other first-principle calculations (Liu, Ming, and Li 2007; Wei et al. 2009) have predicted lower fracture strain for bulk graphene under uniaxial tension in the armchair direction, qualitatively consistent with the MM/MD calculations. A simple geometric consideration would also suggest preferred cleavage planes in a hexagonal lattice to be in the zigzag direction rather than the armchair direction, as clearly observed in Figure 14.26. Moreover, formation of suspended atomic chains is observable from the MD simulations for both types of GNRs as shown in

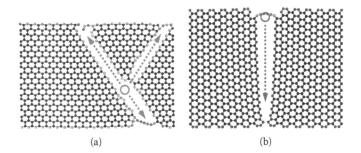

(a)   (b)

FIGURE 14.26 Fracture of graphene nanoribbons (GNRs) under uniaxial tension. (a) Homogeneous nucleation for a zigzag GNR. (b) Edge-controlled heterogeneous nucleation for an armchair GNR. The circles indicate the nucleation sites, and the arrows indicate the directions of crack growth.

Figure 14.26. Similar chain formation was observed in experiments (Jin et al. 2009) and a first-principle study (Topsakal and Ciraci 2010).

In addition to the fracture strain, the nominal fracture stress (uniaxial tensile strength) of the GNRs can be determined from the stress–strain curves in Figure 14.23. As shown in Figure 14.27, as a ribbon width decreases, so does the fracture stress for GNRs with unpassivated edges. H-passivation of the edges slightly increases the fracture stress. The edge effect is relatively small for zigzag GNRs, with all the fracture stresses around 36 N/m, very close to that of bulk graphene. For the armchair GNRs, the fracture stress can be considerably lower (e.g., 27.5 N/m for an unpassivated GNR with $W = 2.5$ nm) comparing with 30.6 N/m for bulk graphene under uniaxial tension in the armchair direction. Again, the lower fracture stress for the armchair GNRs can be attributed to the edge-controlled heterogeneous nucleation mechanism shown in Figure 14.26b.

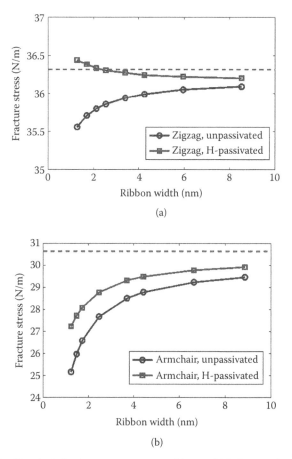

(a)

(b)

FIGURE 14.27 Nominal fracture stress versus ribbon width for graphene nanoribbons under uniaxial stretch, with (a) zigzag and (b) armchair edges. The horizontal dashed line in each figure indicates the fracture stress of bulk graphene under uniaxial tension in the same direction.

14.8 SUMMARY

Despite the discrete nature of the graphene lattice, a 2D continuum theory can be used to describe its mechanical behavior sufficiently well so long as the intrinsic mechanical properties of graphene are properly defined. The unique two-dimensional structure of graphene does require some special treatments in the continuum theory. This chapter has delineated an atomistic approach to determine the in-plane and bending properties of monolayer graphene. The generally anisotropic, nonlinear mechanical behavior is emphasized, while the hexagonal symmetry of the graphene lattice dictates isotropic behavior at the vicinity of the ground state only. Furthermore, the mechanical properties of GNRs, in general, depend on the edge structures. It is noted that the accuracy of the mechanical properties predicted by the atomistic simulations are limited by the empirical potential. The REBO potential used in this chapter does a reasonable job in predicting the bending modulus of graphene, but needs improvements in the predictions of the Young's modulus and edge properties.

ACKNOWLEDGMENTS

The authors gratefully acknowledge funding of this work by the National Science Foundation through Grant No. 0926851. RH is grateful for the support from the Institute for Computational Engineering and Science (ICES), University of Texas at Austin, through the Moncrief Grand Challenge Faculty Awards Program.

REFERENCES

Admal, N. C., and E. B. Tadmor. 2010. A unified interpretation of stress in molecular systems. *J Elast* 100:63–143.

Aitken, Z. H., and R. Huang. 2010. Effects of mismatch strain and substrate surface corrugation on morphology of supported monolayer graphene. *J Appl Phys* 107:123531.

Arroyo, M., and T. Belytschko. 2004a. Finite element methods for the non-linear mechanics of crystalline sheets and nanotubes. *Int J Numer Method Eng* 59:419–56.

Arroyo, M., and T. Belytschko. 2004b. Finite crystal elasticity of carbon nanotubes based on the exponential Cauchy-Born rule. *Phys Rev B* 69:115415.

Belytschko, T., S. P. Xiao, G. C. Schatz, and R. S. Ruoff. 2002. Atomistic simulations of nanotube fracture. *Phys Rev B* 65:235430.

Bets, K. V., and B. I. Yakobson. 2009. Spontaneous twist and intrinsic instabilities of pristine graphene nanoribbons. *Nano Res* 2:161–6.

Brenner, D. W. 1990. Empirical potential for hydrocarbons for use in simulating the chemical vapor deposition of diamond films. *Phys Rev B* 42:9458–71.

Brenner, D. W., O. A. Shenderova, J. A. Harrison, S. J. Stuart, B. Ni, and S. B. Sinnott. 2002. A second-generation reactive empirical bond order (REBO) potential energy expression for hydrocarbons. *J Phys Condens Matter* 14:783–802.

Clausius, R. 1870. On a mechanical theory applicable to heat. *Philos Mag* 40:122–7.

do Carmo, M. P. 1976. *Differential Geometry of Curves snd Surfaces*. Englewood Cliffs, NJ: Prentice-Hall.

Faccio, R., P. A. Denis, H. Pardo, C. Goyenola, and A. W. Mombru. 2009. Mechanical properties of graphene nanoribbons. *J Phys Condens Matter* 21:285304.

Gan, C. K., and D. J. Srolovitz. 2010. First-principles study of graphene edge properties and flake shapes. *Phys Rev B* 81:125445.

Gartstein, Y. N., A. A. Zakhidov, and R. H. Baughman. 2003. Mechanical and electromechanical coupling in carbon nanotube distortions. *Phys Rev B* 68:115415.

Greenwood, N. N., and A. Earnshaw. 1984. *Chemistry of the Elements*. New York: Pergamon Press.

Huang, B., M. Liu, N. Su, J. Wu, W. Duan, B. Gu, and F. Liu. 2009. Quantum manifestations of graphene edge stress and edge instability: A first-principles study. *Phys Rev Lett* 102:166404.

Huang, Y., J. Wu, and K. C. Hwang. 2006. Thickness of graphene and single-wall carbon nanotubes. *Phys Rev B* 74:245413.

Jin, C., H. Lan, L. Peng, K. Suenaga, and S. Iijima. 2009. Deriving carbon atomic chains from graphene. *Phys Rev Lett* 102:205501.

Jun, S. 2008. Density functional study of edge stress in graphene. *Phys Rev B* 78:073405.

Koskinen, P., S. Malola, and H. Hakkinen. 2008. Self-passivating edge reconstructions of graphene. *Phys Rev Lett* 101:115502.

Kudin, K. N., G. E. Scuseria, and B. I. Yakobson. 2001. C_2F, BN, and C nanoshell elasticity from ab initio computations. *Phys Rev B* 64:235406.

Lee, C., X. D. Wei, J. W. Kysar, and J. Hone. 2008. Measurement of the elastic properties and intrinsic strength of monolayer graphene. *Science* 321:385–8.

Liang, H.Y., and M. Upmanyu. 2006. Axial-strain-induced torsion in single-walled carbon nanotubes. *Phys Rev Lett* 96:165501.

Lindsay, L., and D. A. Broido. 2010. Optimized Tersoff and Brenner empirical potential parameters for lattice dynamics and phonon thermal transport in carbon nanotubes and graphene. *Phys Rev B* 81:205441.

Liu, F., P. M. Ming, and J. Li. 2007. Ab initio calculation of ideal strength and phonon instability of graphene under tension. *Phys Rev B* 76:064120.

Lu, Q., M. Arroyo, and R. Huang. 2009. Elastic bending modulus of monolayer graphene. *J Phys D Appl Phys* 42:102002.

Lu, Q., and R. Huang. 2009. Nonlinear mechanics of single-atomic-layer graphene sheets. *Int J Appl Mech* 1:443–67.

Lu, Q., and R. Huang. 2010a. Excess energy and deformation along free edges of graphene nanoribbons. *Phys Rev B* 81:155410.

Lu, Q., W. Gao, and R. Huang. 2010b. Effect of edge structures on elastic modulus and fracture of graphene nanoribbons under uniaxial tension. Posted online at arXiv:1007.3298.

Nakada, K., M. Fujita, G. Dresselhaus, and M. S. Dresselhaus. 1996. Edge state in graphene ribbons: Nanometer size effect and edge shape dependence. *Phys Rev B* 54:17954–61.

Novoselov, K. S., D. Jiang, F. Schedin, T. J. Booth, V. V. Khotkevich, S. V. Morozov, and A. K. Geim. 2005. Two-dimensional atomic crystals. *Proc Natl Acad Sci U S A* 102:10451–3.

Reddy, C. D., A. Ramasubramaniam, V. B. Shenoy, and Y.-W. Zhang. 2009. Edge elastic properties of defect-free single-layer graphene sheets. *Appl Phys Lett* 94:101904.

Reich, S., C. Thomsen, and P. Ordejon. 2002. Elastic properties of carbon nanotubes under hydrostatic pressure. *Phys Rev B* 65:153407.

Shenderova, O. A., D. W. Brenner, A. Omeltchenko, X. Su, and L. H. Yang. 2000. Atomistic modeling of the fracture of polycrystalline diamond. *Phys Rev B* 61:3877–88.

Shenoy, V. B., C. D. Reddy, A. Ramasubramaniam, and Y. W. Zhang. 2008. Edge-stress-induced warping of graphene sheets and nanoribbons. *Phys Rev Lett* 101:245501.

Stuart, S. J., A. B. Tutein, and J. A. Harrison. 2000. A reactive potential for hydrocarbons with intermolecular interactions. *J Chem Phys* 12:6472–86.

Tersoff, J. 1988. Empirical interatomic potential for carbon, with applications to amorphous carbon. *Phys Rev Lett* 61:2879–82.

Tewary, V. K., and B. Yang. 2009. Parametric interatomic potential for graphene. *Phys Rev B* 79:075442.

Thompson-Flagg, R. C., M. J. B. Moura, and M. Marder. 2009. Rippling of graphene. *Europhys Lett* 85:46002.

Timoshenko, S., and S. Woinowsky-Krieger. 1987. *Theory of Plates and Shells*. 2nd ed. New York: McGraw-Hill.

Topsakal, M., and S. Ciraci. 2010. Elastic and plastic deformation of graphene, silicene, and boron nitride honeycomb nanoribbons under uniaxial tension: A first-principles density-functional theory study. *Phys Rev B* 81:024107.

Wei, X., B. Fragneaud, C. A. Marianetti, and J. W. Kysar. 2009. Nonlinear elastic behavior of graphene: Ab initio calculations to continuum description. *Phys Rev B* 80:205407.

Yu, M. F., O. Lourie, M. J. Dyer, K. Moloni, T. F. Kelly, and R. S. Ruoff. 2000. Strength and breaking mechanism of multiwalled carbon nanotubes under tensile load. *Science* 287:637–40.

Zhang, P., H. Jiang, Y. Huang, P. H. Geubelle, and K. C. Hwang. 2004. An atomistic-based continuum theory for carbon nanotubes: Analysis of fracture nucleation. *J Mech Phys Solids* 52:977–98.

Zhao, H., K. Min, and N. R. Aluru. 2009. Size and chirality dependent elastic properties of graphene nanoribbons under uniaxial tension. *Nano Lett* 9:3012–5.

Zhou, J., and R. Huang. 2008. Internal lattice relaxation of single-layer graphene under in-plane deformation. *J Mech Phys Solids* 56:1609–23.

Index